CAMBRIDGE STUDIES IN ADVANCED MATHEMATICS 36

Representation Theory of Artin Algebras

T0328848

Already published

Representation Theory of Artin Algebras

Maurice Auslander

Professor of Mathematics, Brandeis University

Idun Reiten and Sverre O. Smalø

Professors of Mathematics, University of Trondheim

CAMBRIDGE
UNIVERSITY PRESS

CAMBRIDGE UNIVERSITY PRESS
Cambridge, New York, Melbourne, Madrid, Cape Town, Singapore, São Paulo

Cambridge University Press
The Edinburgh Building, Cambridge CB2 2RU, UK

Published in the United States of America by Cambridge University Press, New York

www.cambridge.org
Information on this title: www.cambridge.org/9780521411349

First published 1995
First paperback edition, with corrections, 1997

A catalogue record for this publication is available from the British Library

Library of Congress Cataloguing in Publication data

Auslander, Maurice
Representation theory of Artin algebras / Maurice Auslander, Idun
Reiten, Sverre O. Smalø.
p. cm. – (Cambridge studies in advanced mathematics: 36)
Includes bibliographical references.
ISBN 0 521 41134 3
1. Artin rings. 2. Artin algebras. 3. Representation of algebras.
I. Reiten, Idun, 1942– . II. Smalø. Sverre O.
III. Title. IV. Series.
QA251.5.A87 1994
512′.4–dc20 93-43326 CIP

ISBN-13 978-0-521-59923-8 paperback
ISBN-10 0-521-59923-7 paperback

Transferred to digital printing 2006

To our parents

Charles and Ida

Ivar and Alma

Olaf and Svanhild

Contents

Contents

Introduction

A major concern of elementary linear algebra is the description of how one linear transformation can act on a finite dimensional vector space over a field. Stated in simplest terms, the central problem of this book is to describe how a finite number of linear transformations can act simultaneously on a finite dimensional vector space. While the language of linear algebra suffices in dealing with one transformation on a vector space, it is inadequate to the more general task of dealing with several linear transformations acting simultaneously. As so often happens, when straightforward approaches to seemingly simple problems fail, one adopts a more devious strategy, usually involving a more abstract approach to the problem. In our case, this more abstract approach is called the representation theory of finite dimensional algebras, which in its broadest terms is the study of modules over finite dimensional algebras. One of the advantages of the module theoretic approach is that the language and machinery of both category theory and homological algebra become available. While these theories play a central role in this book, no extensive knowledge of these subjects is required, since only the most elementary concepts and results, as contained in most introductory courses or books on homological algebra, are assumed.

Although many of the concepts and results presented here are of recent origin, having been developed for the most part over the past twenty-five years, the subject itself dates from the middle part of the nineteenth century with the discovery of the quaternions, the first noncommutative field, and the subsequent development, during the first part of this century, of the theory of semisimple finite dimensional algebras over fields. This well developed theory has played an important role in such classical subjects as the representation theory of finite groups over the complex numbers and the Brauer groups of fields, the first of which has proven to be a powerful tool in finite group theory and algebraic number theory, while the second is an object of deep significance also in algebraic number theory as well as in algebraic geometry and abstract field theory. However, since the theory of semisimple finite dimensional algebras is not only well developed but also easily accessible either through elementary algebra courses or textbooks, we assume the reader is familiar with this theory which we use freely in discussing non-semisimple finite dimensional algebras, the algebras of primary concern to us in this book.

While the interest in nonsemisimple finite dimensional algebras goes back to the latter part of the nineteenth century, the development of a general theory of these algebras has been much slower and more sporadic than the semisimple theory. Until recently much of the work has been concentrated on studying specific types of algebras such as modular group algebras, the Kronecker algebra and Nakayama algebras, to name a few. This tradition continues to this day. For instance, algebras of finite representation types have been studied extensively, as have hereditary algebras of finite and tame representation type. But there is at least one respect in which the more recent work differs sharply from the earlier work. There is now a much more highly developed theoretical framework, which has made a more systematic, less *ad hoc* approach to the subject possible. It is our purpose in this book to give an introduction to the part of the theory built around almost split sequences. While this necessitates discussing other aspects of the general theory such as categories of modules modulo various subcategories and the dual of the transpose, other important topics such as coverings, tilting, bocses, vector space categories, posets, derived categories, homologically finite subcategories and finitely presented functors are not dealt with. We do not discuss tame algebras, except for one example, and we do not deal with quantum groups, perverse sheaves or quasihereditary algebras. Some of the topics which are omitted are basic to representation theory, and our original plan, and even first manuscript, included many of them. Since we wanted to include enough preliminary material to make the book accessible to graduate students, space requirements made it necessary to modify our original ambition and leave out many developments, including some of our own favorite ones, from this volume. In particular we postponed the treatment of finitely presented functors since we felt there would not be enough space in the present volume to illustrate their use. It is hoped that there will be a forthcoming volume dealing with other aspects of the subject. Also some other aspects are dealt with in the books [GaRo], [Hap2], [JL], [Pr], [Rin3], [Si].

Besides personal taste, our reason for concentrating on the theory centered around almost split sequences is that these invariants of indecomposable modules appear either explicitly or implicitly in much of the recent work on the subject. We illustrate this point by giving applications to Grothendieck groups, criteria for finite representation type, hereditary algebras of finite representation type and the Kronecker algebra, which is of tame but not finite representation type.

Our proof of the existence theorem for almost split sequences has not appeared before in the literature. It is based on an easily derived, but remarkably useful, relationship between the dimensions of vector spaces of homomorphisms between modules. Amongst other things, this formula comes up naturally in studying cycles of morphisms and their impact on the question of when modules are determined by their composition factors, as well as in the theory of morphisms determined by modules, which is in essence a method for classifying homomorphisms. In fact, one of the important features of the present day representation theory of finite dimensional algebras is this concern with morphisms between modules in addition to the modules themselves.

Although we have been pretending that this book is about finite dimensional algebras over fields, it is for the most part concerned with the slightly more general class of rings called artin algebras which are algebras Λ over commutative artin rings R with Λ a finitely generated R-module. The reason for this is that while the added generality considerably widens the applicability of the theory, there is little added complication in developing the theory once one has established the duality theory for finitely generated modules over commutative artin rings. Of course, we have not hesitated to specialize to fields or algebraically closed fields when this is necessary or convenient.

The book is divided into eleven chapters, each of which is subdivided into sections. The first two chapters contain relevant background material on artin rings and algebras. Chapter III provides a large source of examples of artin algebras and their module categories, especially through the discussion of quivers and their representations. The next four chapters contain basic material centered around almost split sequences and Auslander–Reiten quivers. The first seven chapters together with Chapter VIII on hereditary algebras form the core of the book. The last four chapters are more or less independent of each other. Following each chapter is a set of exercises of various degrees of depth and complexity. Some are superficial "finger exercises" while others are outlines of proofs of significant theories not covered in the text. These exercises are followed by notes containing brief historical and bibliographical comments as well as suggested further readings. There has been no attempt made to give a comprehensive list of references. Many important papers related to the material presented in this book do not appear in our reference list. The reader can consult the books and papers quoted for further references. We provide no historical comments or specific references on standard facts on ring theory and homological algebra, but give references to appropriate textbooks.

At the end of the book we list conjectures and open problems, some of which are well known questions in the area. We give some background and references for what is already known.

Finally we would like to thank various people for making helpful comments on parts of the book, especially Dieter Happel and Svein Arne Sikko, and also Øyvind Bakke, Bill Crawley-Boevey, Wei Du, Otto Kerner, Henning Krause, Shiping Liu, Brit Rohnes, Claus Michael Ringel, Øyvind Solberg, Gordana Todorov, Stig Venås and Dan Zacharia. In addition we are grateful to students at Brandeis, Düsseldorf, Syracuse and Trondheim for trying out various versions of the book.

Our thanks go especially to Jo Torsmyr for the excellent typing of the manuscript.

Finally, we thank the Cambridge University Press, in particular David Tranah, for their help and patience in the preparation of this book.

Some minor corrections are made in the paperback edition. We would like to thank the people who sent us comments, in addition to our own local students Aslak Bakke Buan, Ole Enge, Dag Madsen and Inger Heidi Slungård.

Maurice Auslander died on November 18, 1994. We deeply regret that he did not live to see the book in print.

I
Artin rings

While we are assuming that the reader is familiar with general concepts of ring theory, such as the radical of a ring, and of module theory, such as projective, injective and simple modules, we are not assuming that the reader, except for semisimple modules and semisimple rings, is necessarily familiar with the special features of the structure of artin algebras and their finitely generated modules. This chapter is devoted to presenting background material valid for left artin rings, and the next chapter deals with special features of artin algebras. All rings considered in this book will be assumed to have an identity and all modules are unitary, and unless otherwise stated all modules are left modules.

We start with a discussion of finite length modules over arbitrary rings. After proving the Jordan–Hölder theorem, we introduce the notions of right minimal morphisms and left minimal morphisms and show their relationship to arbitrary morphisms between finite length modules. When applied to finitely generated modules over left artin rings, these results give the existence of projective covers which in turn gives the structure theorem for projective modules as well as the theory of idempotents in left artin rings. We also include some results from homological algebra which we will need in this book.

1 Finite length modules

In this section we introduce the composition series and composition factors for modules of finite length. We prove the Jordan–Hölder theorem and give an interpretation of it in terms of Grothendieck groups.

Let Λ be an arbitrary ring. Given a family of Λ-modules $\{A_i\}_{i \in I}$ we denote by $\coprod_{i \in I} A_i$ the **sum** of the A_i in the category of Λ-modules. The reader should note that direct sum is another commonly used terminology

1

for what we call sum, and another notation is $\oplus_{i \in I} A_i$. We recall that a Λ-module A is **semisimple** if A is a sum of simple Λ-modules and that Λ is a **semisimple ring** if Λ is a semisimple Λ-module.

A basic characterization of such modules is that A is semisimple if and only if every submodule of A is a summand of A. As a consequence, every submodule and every factor module of a semisimple module are again semisimple. But in general the category of semisimple modules, or finitely generated semisimple modules, is not closed under extensions. This leads to the study of modules of finite length, which is the smallest category closed under extensions which contains the simple modules.

A module A is said to be of **finite length** if there is a finite filtration of submodules $A = A_0 \supset A_1 \supset \cdots \supset A_n = 0$ such that A_i/A_{i+1} is either zero or simple for $i = 0, \ldots, n-1$. We call such a filtration F of A a **generalized composition series**, and the nonzero factor modules A_i/A_{i+1} the **composition factors** of the filtration F. If no factor module A_i/A_{i+1} is zero for $i = 0, \ldots, n-1$, then F is a **composition series** for A. For a simple Λ-module S we then define $m_S^F(A)$ to be the number of composition factors of F which are isomorphic to S, and we define the length $l_F(A)$ to be $\sum m_S^F(A)$, where the sum is taken over all the simple Λ-modules. We define the **length** $l(A)$ of A to be the minimum of $l_F(A)$ for composition series F of A, and $m_S(A)$ to be the minimum of the $m_S^F(A)$. Note that $l(0) = 0$. Our aim is to prove that the numbers $m_S^F(A)$ and $l_F(A)$ are independent of the choice of composition series F.

Let $0 \to A \xrightarrow{f} B \xrightarrow{g} C \to 0$ be an exact sequence, and let F be a generalized composition series $B = B_0 \supset B_1 \supset \cdots \supset B_n = 0$ of B. This filtration induces filtrations F' of A given by $A = f^{-1}(B) \supset f^{-1}(B_1) \supset \cdots \supset f^{-1}(B_n) = 0$ and F'' of C given by $C = g(B) \supset g(B_1) \supset \cdots \supset g(B_n) = 0$. We write $f^{-1}(B_i) = A_i$ and $g(B_i) = C_i$. Then we have the following preliminary result.

Proposition 1.1 *Let the notation be as above.*

(a) *The filtrations F' of A and F'' of C are generalized composition series.*

(b) *For each simple module S we have*

$$m_S^{F'}(A) + m_S^{F''}(C) = m_S^F(B).$$

(c) $l_{F'}(A) + l_{F''}(C) = l_F(B).$

Proof For each $i = 0, \ldots, n$ we have an exact sequence $0 \to A_i \to B_i \to$

$C_i \to 0$ and for each $i = 0, \ldots, n-1$ an exact commutative diagram

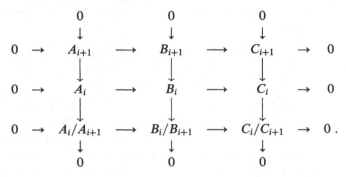

Hence we have that if $B_i/B_{i+1} = 0$, then $A_i/A_{i+1} = 0 = C_i/C_{i+1}$. If B_i/B_{i+1} is simple, then either $A_i/A_{i+1} \simeq B_i/B_{i+1}$ and $C_i/C_{i+1} = 0$, or $B_i/B_{i+1} \simeq C_i/C_{i+1}$ and $A_i/A_{i+1} = 0$. Parts (a), (b) and (c) now follow easily. □

We can now prove the Jordan–Hölder theorem.

Theorem 1.2 *Let B be a Λ-module of finite length, and F and G two composition series for B. Then for each simple Λ-module S we have $m_S^F(B) = m_S^G(B) = m_S(B)$, and hence $l_F(B) = l_G(B) = l(B)$.*

Proof We prove this by induction on $l(B)$. Our claim clearly holds if $l(B) \le 1$. Assume now that $l(B) > 1$. Then B contains a nonzero submodule $A \neq B$. Since it follows by Proposition 1.1 that $l(A) + l(B/A) \le l(B)$, we have $l(A) < l(B)$ and $l(B/A) < l(B)$, using that $l(A)$ and $l(B/A)$ are nonzero. Let F and G be two composition series for B and let F' and G' denote the induced filtrations on A and F'' and G'' the induced ones on $C = B/A$. For each simple Λ-module S we have by induction that $m_S^{F'}(A) = m_S^{G'}(A)$ and $m_S^{F''}(C) = m_S^{G''}(C)$. Since $m_S^F(B) = m_S^{F'}(A) + m_S^{F''}(C)$ and $m_S^G(B) = m_S^{G'}(A) + m_S^{G''}(C)$ by Proposition 1.1, we get $m_S^F(B) = m_S^G(B)$, and hence also $l_F(B) = l_G(B)$. □

Let $0 \to A \xrightarrow{f} B \xrightarrow{g} C \to 0$ be an exact sequence. If F' is a generalized composition series $A = A_0 \supset A_1 \supset \ldots \supset A_s = 0$ of A and F'' is a generalized composition series $C = C_0 \supset C_1 \supset \ldots \supset C_t = 0$ of C, then it follows as above that we get a generalized composition series $B = B_0 \supset g^{-1}(C_1) \supset \ldots \supset g^{-1}(C_{t-1}) \supset f(A) \supset f(A_1) \ldots \supset f(A_s) = 0$ of B. Using Proposition 1.1 and Theorem 1.2 we then have the following.

Corollary 1.3 *Let* $0 \rightarrow A \xrightarrow{f} B \xrightarrow{g} C \rightarrow 0$ *be an exact sequence of* Λ-*modules where A and C have finite length. Then B has finite length and* $l(A) + l(C) = l(B)$. $\qquad\qquad\square$

A semisimple module of finite length is clearly uniquely determined by its composition factors, but this does not hold in general for modules with composition series. For example if B has finite length, then B and $\coprod m_S(B)S$, where the sum is taken over all nonisomorphic simple Λ-modules S, have the same composition factors. Hence all finitely generated Λ-modules are determined by their composition factors if and only if Λ is a semisimple ring. It is however an interesting question when indecomposable modules are determined by their composition factors, and this will be discussed in Chapters VIII and IX.

Before giving the following useful consequence of Corollary 1.3 we recall that a morphism of modules is called a **monomorphism** if it is a one to one map and an **epimorphism** if it is an onto map.

Proposition 1.4 *Let A be a* Λ-*module of finite length and* $f : A \rightarrow A$ *a* Λ-*homomorphism. Then the following are equivalent.*

(a) *f is an isomorphism.*
(b) *f is a monomorphism.*
(c) *f is an epimorphism.*

Proof This follows directly from the fact that $l(f(A)) + l(A/f(A)) = l(A)$. $\qquad\qquad\square$

For a ring Λ we denote by $\operatorname{Mod}\Lambda$ the category of left Λ-modules. A subcategory $\mathscr{C}$ of $\operatorname{Mod}\Lambda$ is **closed under extensions** if B is in $\mathscr{C}$ for any exact sequence $0 \rightarrow A \rightarrow B \rightarrow C \rightarrow 0$ with A and C in $\mathscr{C}$.

The following characterization of the category of finite length modules which we denote by $\mathrm{f.l.}\,\Lambda$ is useful.

Proposition 1.5

(a) *The category* $\mathrm{f.l.}\,\Lambda$ *is the smallest subcategory of* $\operatorname{Mod}\Lambda$ *closed under extensions and containing the simple modules.*
(b) *A* Λ-*module A is of finite length if and only if A is both artin and noetherian.*

Proof (a) The category $\mathrm{f.l.}\,\Lambda$ contains the simple modules, and is closed under extensions by Corollary 1.3. It is also clear that any subcategory of $\operatorname{Mod}\Lambda$ closed under extensions and containing the simple modules must contain $\mathrm{f.l.}\,\Lambda$.

(b) Since artin and noetherian modules are closed under extensions and simple modules are both artin and noetherian, we have that any module of finite length is artin and noetherian.

Suppose now that a module B is both noetherian and artin. Clearly every submodule and every factor module of B have these properties. Since B is noetherian, there exists a submodule A of B maximal with respect to being of finite length. If $A \neq B$, then B/A has a simple submodule C since B/A is artin. Let A' be the submodule of B containing A such that $A'/A \simeq C$. Then A' is of finite length, which contradicts the maximality of A. Hence we get $A = B$ and so B is of finite length. □

For semisimple modules it is easy to see that any one of the chain conditions implies the other one. Hence we have the following.

Proposition 1.6 *For a semisimple Λ-module B the following are equivalent.*

(a) *B has finite length.*
(b) *B is noetherian.*
(c) *B is artin.* □

A useful point of view concerning the composition factors of a module of finite length is to study a special group associated with the finite length modules. Since the finite length modules are finitely generated by Proposition 1.5, the collection of isomorphism classes of modules of finite length is a set. Hence we can associate with the category of finite length modules f.l. Λ the free abelian group $F(\text{f.l.}\,\Lambda)$ with basis the isomorphism classes $[A]$ of finite length modules A. Denote by $R(\text{f.l.}\,\Lambda)$ the subgroup of $F(\text{f.l.}\,\Lambda)$ generated by expressions $[A] + [C] - [B]$ for each exact sequence $0 \to A \to B \to C \to 0$ in f.l. Λ. Then the **Grothendieck group** $K_0(\text{f.l.}\,\Lambda)$ of f.l. Λ is defined to be the factor group $F(\text{f.l.}\,\Lambda)/R(\text{f.l.}\,\Lambda)$. Associated with a finite length module A is the coset of the isomorphism class $[A]$ in the Grothendieck group $K_0(\text{f.l.}\,\Lambda)$, which we also denote by $[A]$. It turns out that this element $[A]$ in $K_0(\text{f.l.}\,\Lambda)$ contains all information on the composition factors of A. It follows directly that $K_0(\text{f.l.}\,\Lambda)$ is generated by elements $[S]$ where S is a simple Λ-module. Using the Jordan–Hölder theorem we get the following stronger result.

Theorem 1.7 *$K_0(\text{f.l.}\,\Lambda)$ is a free abelian group with basis $\{[S_i]\}_{i \in I}$, where the S_i are the simple Λ-modules and for each finite length module A we have that $[A] = \sum_{i \in I} m_{S_i}(A)[S_i]$ in $K_0(\text{f.l.}\,\Lambda)$.*

Proof Let $F(\text{s.s.}\,\Lambda)$ be the subgroup of $F(\text{f.l.}\,\Lambda)$ generated by the $[S_i]$, where the S_i are a complete set of simple Λ-modules up to isomorphism. Define $\alpha: F(\text{s.s.} \Lambda) \rightarrow K_0(\text{f.l.} \Lambda)$ by $\alpha([S_i]) = [S_i]$ in $K_0(\text{f.l.} \Lambda)$. We have seen that if $0 \rightarrow A \rightarrow B \rightarrow C \rightarrow 0$ is an exact sequence in f.l. Λ, then $m_S(B) = m_S(A) + m_S(C)$ for all simple Λ-modules S. Therefore if for each A in f.l. Λ we let $\beta([A])$ be the element $\sum_{i \in I} m_{S_i}(A)[S_i]$ in $F(\text{s.s.} \Lambda)$, we obtain a morphism $\beta: K_0(\text{f.l.} \Lambda) \rightarrow F(\text{s.s.} \Lambda)$. It is now not difficult to see that $\beta\alpha = 1_{F(\text{s.s.} \Lambda)}$ and $\alpha\beta = 1_{K_0(\text{f.l.} \Lambda)}$. Hence $\alpha: F(\text{s.s.} \Lambda) \rightarrow K_0(\text{f.l.} \Lambda)$ is an isomorphism, giving our desired result. □

2 Right and left minimal morphisms

In this section we introduce the concepts of right minimal and left minimal morphisms between modules. These notions are especially interesting for modules of finite length, and they also specialize to the concepts of projective covers and injective envelopes.

Let Λ be an arbitrary ring. For a fixed Λ-module C, consider the category $\text{Mod}\,\Lambda/C$ whose objects are the Λ-morphisms $f: B \rightarrow C$, and where a morphism $g: f \rightarrow f'$ from $f: B \rightarrow C$ to $f': B' \rightarrow C$ is a Λ-morphism $g: B \rightarrow B'$ such that

$$
\begin{array}{ccc}
B & \xrightarrow{f} & C \\
{\scriptstyle g}\big\downarrow & \nearrow_{\scriptstyle f'} & \\
B' & &
\end{array}
$$

commutes. It follows that $g: f \rightarrow f'$ is an isomorphism in $\text{Mod}\,\Lambda/C$ if and only if the associated morphism $g: B \rightarrow B'$ is an isomorphism in $\text{Mod}\,\Lambda$. We say that $f: B \rightarrow C$ is **right minimal** if every morphism $g: f \rightarrow f$ is an automorphism. We introduce an equivalence relation on the objects of $\text{Mod}\,\Lambda/C$ by defining $f \sim f'$ if $\text{Hom}(f, f') \neq \emptyset$ and $\text{Hom}(f', f) \neq \emptyset$. We now show that for modules of finite length, each equivalence class contains a right minimal morphism.

Proposition 2.1 *Let Λ be a ring and C a Λ-module. Every equivalence class in $\text{Mod}\,\Lambda/C$ containing some $f: B \rightarrow C$ with B of finite length contains a right minimal morphism, which is unique up to isomorphism.*

Proof Choose $f: B \rightarrow C$ in the given equivalence class with $l(B)$ smallest possible, and let $g: f \rightarrow f$ be a morphism in $\text{Mod}\,\Lambda/C$. We then have a commutative diagram

$$
\begin{array}{ccc}
B & \xrightarrow{\;f\;} & C \\
\downarrow & \nearrow{\scriptstyle f|_{g(B)}} & \uparrow{\scriptstyle f} \\
g(B) & \hookrightarrow & B
\end{array}
$$

which shows that $g(B) = B$ by minimality of $l(B)$. Then $g: f \to f$ must be an isomorphism, so that $f: B \to C$ is right minimal.

Assume that $f': B' \to C$ is a right minimal morphism which is equivalent to $f: B \to C$. We then have morphisms $g: f \to f'$ and $h: f' \to f$. Using that both f and f' are right minimal, we get that hg and gh are isomorphisms. Hence h and g are isomorphisms. □

Let $f: B \to C$ be a morphism with B of finite length. Then the unique, up to isomorphism in $\operatorname{Mod}\Lambda/C$, right minimal morphisms in the equivalence class in $\operatorname{Mod}\Lambda/C$ of f are called **right minimal versions of** f.

Whenever there is a morphism of Λ-modules $f: M \to N$ and M' is a submodule of M then $f|_{M'}: M' \to N$ denotes the restriction of f to M'. The next result gives a reduction to right minimal morphisms.

Theorem 2.2 *Let Λ be a ring and C a Λ-module. Let $g: X \to C$ be an object in $\operatorname{Mod}\Lambda/C$ with X of finite length. Then there is a decomposition $X = X' \coprod X''$ such that $g|_{X'}: X' \to C$ is right minimal and $g|_{X''} = 0$. Moreover, the morphism $g|_{X'}$ is a right minimal version of g.*

Proof Choose $f: B \to C$ minimal and equivalent to $g: X \to C$, as is possible by Proposition 2.1. We then have a commutative diagram

$$
\begin{array}{ccc}
B & \xrightarrow{\;f\;} & C \\
\downarrow{\scriptstyle s} & & \| \\
X & \xrightarrow{\;g\;} & C \\
\downarrow{\scriptstyle t} & & \| \\
B & \xrightarrow{\;f\;} & C .
\end{array}
$$

Then $f = fts$, so that ts is an isomorphism. Letting $\operatorname{Im} s$ denote the image of s and $\operatorname{Ker} t$ the kernel of t, we get $X = \operatorname{Im} s \coprod \operatorname{Ker} t$, and $g|_{\operatorname{Im} s}$ is right minimal and $g|_{\operatorname{Ker} t} = 0$. It is easy to see that $g|_{\operatorname{Im} s}$ is in the same equivalence class as g in $\operatorname{Mod}\Lambda/C$. □

We state the following easy consequence.

Corollary 2.3 *The following are equivalent for a morphism* $f: B \to C$ *with* B *of finite length.*

(a) *f is right minimal.*

(b) *If B' is a nonzero summand of B, then $f|_{B'} \neq 0$.* $\square$

For a fixed Λ-module A consider the category $\text{Mod}\,\Lambda \setminus A$ whose objects are the Λ-morphisms $f: A \to B$ and where a morphism $g: f \to f'$ from $f: A \to B$ to $f': A \to B'$ is a Λ-morphism $g: B \to B'$ such that $gf = f'$. Dual to the notion of right minimal morphism we define a morphism $f: A \to B$ of Λ-modules to be **left minimal** if whenever $g: B \to B$ has the property that $gf = f$, then g is an automorphism. We also have the following dual version of Proposition 2.1. In each equivalence class in $\text{Mod}\,\Lambda \setminus A$ of a morphism $h: A \to Y$ with Y of finite length there are unique, up to isomorphism in $\text{Mod}\,\Lambda \setminus A$, left minimal morphisms called the **left minimal versions** of h. For the convenience of the reader we state the following dual result.

Theorem 2.4 *Let Λ be a ring and A a Λ-module. Let $f: A \to Y$ be an object in $\text{Mod}\,\Lambda \setminus A$ with Y of finite length. Then there is a decomposition $Y = Y' \coprod Y''$ such that $p'f: A \to Y'$ is left minimal and $p''f: A \to Y''$ is zero, where $p': Y \to Y'$ and $p'': Y \to Y''$ are the projections according to the decomposition $Y = Y' \coprod Y''$. Moreover, $p'f$ is a left minimal version of f.* $\square$

3 Radical of rings and modules

In this section we give the definition and basic properties of the radical of rings and modules for left artin rings. We are mainly interested in rings Λ where all finitely generated Λ-modules have finite length. This clearly holds for semisimple rings. Actually, we prove that Λ has this property if and only if the Λ-module Λ has finite length, or equivalently, Λ is both left noetherian and left artin. We show that it is superfluous to assume left noetherian.

First we recall that the **radical** of a ring Λ, which we denote by r_Λ, or simply r, is the intersection of the maximal left ideals of Λ, as well as the intersection of the maximal right ideals of Λ, and is hence an ideal, where an ideal of Λ always means a two-sided ideal. The radical plays a central role in the theory of left artin rings. We recall Nakayama's lemma which states that a left ideal $\mathfrak{a}$ is contained in $\mathfrak{r}$ if and only if $\mathfrak{a}M = M$ implies $M = 0$ when M is a finitely generated Λ-module.

We now prove that left artin rings are left noetherian.

Proposition 3.1 *Assume that Λ is a left artin ring. Then we have the following.*

(a) *The radical $\mathfrak{r}$ of Λ is nilpotent.*

(b) *$\Lambda/\mathfrak{r}$ is a semisimple ring.*

(c) *A Λ-module A is semisimple if and only if $\mathfrak{r}A = 0$.*

(d) *There is only a finite number of nonisomorphic simple Λ-modules.*

(e) *Λ is left noetherian.*

Proof (a) Since Λ is left artin and $\Lambda \supset \mathfrak{r} \supset \mathfrak{r}^2 \supset \cdots \supset \mathfrak{r}^n \supset \cdots$ is a descending sequence of left ideals, there is some n such that $\mathfrak{r}^n = \mathfrak{r}^{n+1}$. Suppose $\mathfrak{r}^n \neq 0$. Then $\mathfrak{r}^{n+1} = \mathfrak{r}^n\mathfrak{r} = \mathfrak{r}^n \neq 0$, so the class $\mathscr{F}$ of all left ideals $\mathfrak{a}$ with $\mathfrak{r}^n\mathfrak{a} \neq 0$ is nonempty. Choose a left ideal $\mathfrak{a}$ in Λ which is minimal in $\mathscr{F}$. Then there is some x in $\mathfrak{a}$ with $\mathfrak{r}^n x \neq 0$ and therefore $\mathfrak{r}^n(\Lambda x) \neq 0$. By the minimality of $\mathfrak{a}$ we have $\mathfrak{a} = \Lambda x$, so $\mathfrak{a}$ is a finitely generated left ideal. Now $0 \neq \mathfrak{r}^n\mathfrak{a} = \mathfrak{r}^{n+1}\mathfrak{a} = \mathfrak{r}^n\mathfrak{r}\mathfrak{a}$, so $\mathfrak{r}\mathfrak{a}$ is also in $\mathscr{F}$ and therefore $\mathfrak{a} = \mathfrak{r}\mathfrak{a}$. This is a contradiction by Nakayama's lemma, and hence $\mathfrak{r}^n = 0$.

(b) Let I be an ideal in Λ containing $\mathfrak{r}$ such that $I/\mathfrak{r}$ is nilpotent in $\Lambda/\mathfrak{r}$. Then there is an integer t with $I^t \subset \mathfrak{r}$. Since $\mathfrak{r}^n = 0$, we have $I^s = 0$ for $s = nt$. Let $\mathfrak{m}$ be a maximal left ideal in Λ, and consider the natural map $p: \Lambda \to \Lambda/\mathfrak{m}$. If $I \not\subset \mathfrak{m}$, then $p(I) \neq 0$, and hence $p(I) = \Lambda/\mathfrak{m}$ since $\Lambda/\mathfrak{m}$ is a simple Λ-module. Then we get $p(I^2) = Ip(I) = I(\Lambda/\mathfrak{m}) = \Lambda/\mathfrak{m}$, and further $0 = p(I^s) = \Lambda/\mathfrak{m}$, a contradiction. This shows that $I \subset \mathfrak{m}$, and hence $I \subset \mathfrak{r}$, so that $I/\mathfrak{r}$ is 0 in $\Lambda/\mathfrak{r}$. Since $\Lambda/\mathfrak{r}$ has no nonzero nilpotent ideals, and is left artin since Λ is left artin, we conclude that $\Lambda/\mathfrak{r}$ is a semisimple ring.

(c) If for a Λ-module A we have that $\mathfrak{r}A = 0$, then A is a $(\Lambda/\mathfrak{r})$-module and hence a semisimple $(\Lambda/\mathfrak{r})$-module. Hence A is also a semisimple Λ-module.

Conversely it is clear by the definition of $\mathfrak{r}$ that if A is a semisimple Λ-module, then $\mathfrak{r}A = 0$.

(d) Since there is only a finite number of nonisomorphic simple $(\Lambda/\mathfrak{r})$-modules and every simple Λ-module is a $(\Lambda/\mathfrak{r})$-module, there is only a finite number of nonisomorphic simple Λ-modules.

(e) It follows from (a) that Λ has a finite filtration $\Lambda \supset \mathfrak{r} \supset \mathfrak{r}^2 \supset \cdots \supset \mathfrak{r}^n = 0$. We write $\Lambda = \mathfrak{r}^0$. Each $\mathfrak{r}^i/\mathfrak{r}^{i+1}$ is a semisimple Λ-module by (c) for $i = 0, 1, \ldots, n-1$ and is artin since Λ is a left artin ring. Hence $\mathfrak{r}^i/\mathfrak{r}^{i+1}$

is noetherian by Proposition 1.6, and consequently Λ is a left noetherian ring.   □

We now have the following description of the rings where all finitely generated modules have finite length.

Corollary 3.2 *For a ring Λ the following are equivalent.*
(a) *Every finitely generated Λ-module has finite length.*
(b) Λ *is left artin.*
(c) *The radical $\mathfrak{r}$ of Λ is nilpotent and $\mathfrak{r}^i/\mathfrak{r}^{i+1}$ is a finitely generated semisimple module for all $i \geq 0$.* □

In general it may be difficult to compute the radical of a left artin ring Λ by first finding the maximal left ideals. The following criterion is usually easy to apply. However, before giving this result, it is convenient to introduce the following notation. If A and B are submodules of a module C, we denote by $A + B$ the submodule of C generated by A and B.

Proposition 3.3 *Let Λ be a left artin ring and $\mathfrak{a}$ an ideal in Λ such that $\mathfrak{a}$ is nilpotent and $\Lambda/\mathfrak{a}$ is semisimple. Then we have $\mathfrak{a} = \mathfrak{r}$.*

Proof Let $\mathfrak{a}$ be a nilpotent ideal with $\Lambda/\mathfrak{a}$ semisimple. To show that $\mathfrak{a} \subset \mathfrak{r}$, assume to the contrary that there is a maximal ideal $\mathfrak{m}$ in Λ with $\mathfrak{a} \not\subset \mathfrak{m}$. Then $\mathfrak{a} + \mathfrak{m} = \Lambda$, where $\mathfrak{a} + \mathfrak{m}$ denotes the smallest left ideal containing $\mathfrak{a}$ and $\mathfrak{m}$. Hence $\mathfrak{a}^2 + \mathfrak{a}\,\mathfrak{m} = \mathfrak{a}$, so that $\mathfrak{a}^2 + \mathfrak{m} = \Lambda$. Continuing this way, we get $\mathfrak{a}^n + \mathfrak{m} = \Lambda$ for all n, which gives a contradiction since $\mathfrak{a}$ is nilpotent. This shows $\mathfrak{a} \subset \mathfrak{m}$, and consequently $\mathfrak{a} \subset \mathfrak{r}$.

Clearly the radical of $\Lambda/\mathfrak{a}$ is equal to $\mathfrak{r}/\mathfrak{a}$, so that $\mathfrak{r} = \mathfrak{a}$ since $\Lambda/\mathfrak{a}$ is semisimple.   □

For left artin rings the radical of a module also plays an important role. The **radical** rad$\,A$ of a Λ-module A over an arbitrary ring Λ is the intersection of the maximal submodules. We have the following useful characterization of the radical of a module. Recall first that a submodule B of a Λ-module A is **small** in A if $B + X = A$ for a submodule X of A implies $X = A$.

Lemma 3.4 *Let A be a finitely generated module over an arbitrary ring Λ. Then a submodule B of A is small in A if and only if $B \subset \mathrm{rad}\,A$.*

Proof Suppose $B \subset \text{rad}\,A$ and let X be a submodule of A with $B+X = A$. Since A is finitely generated, every proper submodule of A is contained in a maximal submodule. We can then conclude that $X = A$ since B is contained in all maximal submodules of A.

Assume that B is a submodule of A which is not contained in $\text{rad}\,A$, and let X be a maximal submodule of A not containing B. Then B is not small in A since $B + X = A$, but $X \neq A$. □

The following gives a useful description of the radical of a module for left artin rings.

Proposition 3.5 *Let A be a finitely generated module over a left artin ring Λ. Then we have $\text{rad}\,A = \mathfrak{r}A$.*

Proof We first show $\mathfrak{r}A \subset \text{rad}\,A$. Assume $\mathfrak{r}A + X = A$ for a submodule X of A. Then we get that $\mathfrak{r}^n A + X = A$ for all $n \geq 1$, so that $X = A$ since $\mathfrak{r}$ is nilpotent. Hence $\mathfrak{r}A$ is small in A and is then contained in $\text{rad}\,A$ by Lemma 3.4.

Since $A/\mathfrak{r}A$ is a semisimple Λ-module, it is easy to see that $\text{rad}(A/\mathfrak{r}A)$ is zero. On the other hand, $\text{rad}(A/\mathfrak{r}A) = (\text{rad}\,A)/\mathfrak{r}A$, and we are done. □

We end this section by connecting the radical of modules with the notion of an essential epimorphism. An epimorphism $f: A \to B$ is called an **essential epimorphism** if a morphism $g: X \to A$ is an epimorphism whenever $fg: X \to B$ is an epimorphism. The following result is an easy consequence of Proposition 3.5.

Proposition 3.6 *The following are equivalent for an epimorphism $f: A \to B$, where A and B are finitely generated modules over a left artin ring.*

(a) *f is an essential epimorphism.*
(b) *$\text{Ker}\,f \subset \mathfrak{r}A$.*
(c) *The induced epimorphism $A/\mathfrak{r}A \to B/\mathfrak{r}B$ is an isomorphism.* □

We also have the following connection with right minimal morphisms.

Proposition 3.7 *Let Λ be a left artin ring. If A is a finitely generated Λ-module and $f: A \to B$ is an essential epimorphism, then f is right minimal.*

Proof Since $\text{Ker}\,f \subset \mathfrak{r}A$ by Proposition 3.6, it is easy to see that no nonzero summand A' of A can be contained in Ker f. Therefore f is right minimal by Theorem 2.2.   □

For a module A of finite length over a left artin ring Λ the smallest integer i with $\mathfrak{r}^i A = 0$ is called the **radical length** of A, denoted by rl(A), and $0 \subset \mathfrak{r}^{i-1}A \subset \cdots \subset \mathfrak{r}A \subset A$ is the **radical series** of A. Sometimes the radical length of A is called the **Loewy length** of A.

4 Structure of projective modules

In this section we give the structure of projective modules over left artin rings, and their connection with simple modules. For this the notion of projective cover is important. All rings will be left artin and we will deal with the category mod Λ of finitely generated left Λ-modules unless otherwise stated.

Let Λ be a left artin ring and A a Λ-module. A **projective cover** of A is an essential epimorphism $f: P \to A$ with P a projective Λ-module. Our proof of the existence of projective covers for A in mod Λ is based on the following characterization of projective covers.

Proposition 4.1 *Let A be in* mod Λ *where Λ is a left artin ring and let $f: P \to A$ be an epimorphism with P projective. Then f is a projective cover if and only if f is right minimal.*

Proof If f is a projective cover, then f is right minimal by Proposition 3.7.

Suppose f is right minimal and let $g: X \to P$ be such that $fg: X \to A$ is an epimorphism. We want to show that g is an epimorphism. Since $fg: X \to A$ is an epimorphism and P is projective, we have the following commutative diagram.

$$
\begin{array}{ccc}
P & \xrightarrow{f} & A \\
\downarrow{\scriptstyle h} & & \| \\
X & \xrightarrow{fg} & A \\
\downarrow{\scriptstyle g} & & \| \\
P & \xrightarrow{f} & A
\end{array}
$$

Using that f is right minimal, we get that gh is an isomorphism, which shows that g is an epimorphism.   □

As an immediate consequence of this proposition we have the following existence and uniqueness theorem for projective covers.

Theorem 4.2 *Let A be in* mod Λ *where Λ is a left artin ring. Then we have the following.*

(a) *There is a projective cover $f: P \to A$ in* mod Λ.

(b) *Any two projective covers $f_1: P_1 \to A$ and $f_2: P_2 \to A$ are isomorphic in* mod Λ/A.

Proof (a) Since A is in mod Λ, there is an epimorphism $f: P \to A$ in mod Λ with P projective. But we know by Theorem 2.2 that $P = P' \coprod P''$ where $f|_{P'}$ is right minimal and $f|_{P''} = 0$. Therefore $f|_{P'}: P' \to A$ is a right minimal epimorphism with P' projective and is hence a projective cover by Proposition 4.1.

(b) We leave the proof to the reader. $\qquad\square$

We now point out some easily verified properties of projective covers.

Proposition 4.3

(a) *An epimorphism $f: P \to A$ with P projective is a projective cover if and only if the induced epimorphism $P/\mathfrak{r}P \to A/\mathfrak{r}A$ is an isomorphism.*

(b) *Let $\{f_i: P_i \to A_i\}_{i \in I}$ be a finite family of epimorphisms with the P_i projective modules. Then the induced epimorphism $\coprod_{i \in I} P_i \to \coprod_{i \in I} A_i$ is a projective cover if and only if each $f_i: P_i \to A_i$ is a projective cover.*

Proof Part (a) follows from Proposition 3.6, and (b) is an easy consequence of (a). $\qquad\square$

We now apply these results to obtain the basic structure theorem for projective modules in mod Λ. We denote the full subcategory of mod Λ consisting of the projective modules by $\mathscr{P}(\Lambda)$. We use projective covers and the uniqueness of decomposition of a semisimple module into a sum of simple modules to get a corresponding decomposition result for projective modules. Recall that a Λ-module A is called **indecomposable** if A cannot be decomposed as a sum of proper submodules.

Theorem 4.4

(a) *For each P in $\mathscr{P}(\Lambda)$ the natural epimorphism $P \to P/\mathfrak{r}P$ is a projective cover.*

(b) *If P and Q are in $\mathscr{P}(\Lambda)$, then $P \simeq Q$ if and only if $P/\mathfrak{r}P \simeq Q/\mathfrak{r}Q$.*

(c) *P in $\mathscr{P}(\Lambda)$ is indecomposable if and only if $P/\mathfrak{r}P$ is simple.*

(d) *Suppose P is in $\mathscr{P}(\Lambda)$ and $P = \coprod_{i=1}^{n} P_i \simeq \coprod_{j=1}^{m} Q_j$ where the P_i and Q_j are indecomposable modules. Then $m = n$ and there is a permutation σ of $\{1, \ldots, n\}$ such that $P_i \simeq Q_{\sigma(i)}$ for all $i = 1, \ldots, n$.*

Proof Part (a) follows from Proposition 4.3 and (b) is a consequence of Theorem 4.2 and Proposition 4.3.

(c) Clearly $P/\mathfrak{r}P$ being simple implies that P is indecomposable. Suppose $P/\mathfrak{r}P$ is not simple. Then $P/\mathfrak{r}P \simeq U \coprod V$ with U and V nonzero semisimple modules. Let $P(U) \to U$ and $P(V) \to V$ be projective covers. Then $P \simeq P(U) \coprod P(V)$ by (b) and so P is not indecomposable.

(d) Suppose $P = \coprod_{i=1}^{n} P_i \simeq \coprod_{j=1}^{m} Q_j$ with P_i and Q_j indecomposable modules. Then we have $P/\mathfrak{r}P \simeq \coprod_{i=1}^{n}(P_i/\mathfrak{r}P_i) \simeq \coprod_{j=1}^{m}(Q_j/\mathfrak{r}Q_j)$. Since the $P_i/\mathfrak{r}P_i$ and the $Q_j/\mathfrak{r}Q_j$ are simple modules, it follows that $n = m$ and there is a permutation σ of $\{1, \ldots, n\}$ such that $P_i/\mathfrak{r}P_i \simeq Q_{\sigma(i)}/\mathfrak{r}Q_{\sigma(i)}$ for $i = 1, \ldots, n$. Hence by (b) we have $P_i \simeq Q_{\sigma(i)}$ for all $i = 1, \ldots, n$, giving the desired result. □

As an immediate consequence of this theorem we have the following.

Corollary 4.5 *Let $S_1, \ldots, S_n$ be a complete list of nonisomorphic simple Λ-modules. Then their projective covers $P_1, \ldots, P_n$ are a complete list of nonisomorphic indecomposable projective Λ-modules. Moreover each P_i is isomorphic to a summand of Λ as a left Λ-module.* □

Before giving our final statement on the structure of projective modules, we recall the definition of a local ring.

A (not necessarily left artin) ring Λ is **local** if the nonunits of Λ form an ideal in Λ. We shall need the following property.

Proposition 4.6 *If Λ is a (not necessarily left artin) local ring, then 0 and 1 are the only idempotents in Λ.*

Proof Assume that the idempotent e is a unit. Then there is some f in Λ such that $ef = 1$, and hence $e = eef = ef = 1$. So if e is an idempotent different from 0 or 1, then e and $1 - e$ are not units. But their sum is a unit, so that Λ would not be local. □

We now give the following useful characterizations of indecomposable projective modules.

Proposition 4.7 *The following are equivalent for a projective module P over a left artin ring* Λ.

(a) P *is indecomposable.*
(b) $\mathfrak{r}P$ *is the unique maximal submodule of* P.
(c) $\text{End}_\Lambda(P)$ *is a local ring.*

Proof (a)$\Rightarrow$(b) We know that P being indecomposable implies that $P/\mathfrak{r}P$ is simple. But this implies that $\mathfrak{r}P$ is a maximal submodule of P, and hence the only one since $\mathfrak{r}P$ is the intersection of the maximal submodules of P.

(b)$\Rightarrow$(c) Let f be in $\text{End}_\Lambda(P)$. Since $\mathfrak{r}P$ is the unique maximal submodule of P, we know that $\text{Im} f \not\subset \mathfrak{r}P$ if and only if $f: P \to P$ is onto. But $f: P \to P$ being an epimorphism means that it is an isomorphism. Therefore f is a unit in $\text{End}_\Lambda(P)$ if and only if $\text{Im} f \not\subset \mathfrak{r}P$. It follows that the nonunits of $\text{End}_\Lambda(P)$ are precisely the f in $\text{End}_\Lambda(P)$ such that $\text{Im} f \subset \mathfrak{r}P$, which are easily seen to be an ideal in $\text{End}_\Lambda(P)$. Hence $\text{End}_\Lambda(P)$ is a local ring.

(c)$\Rightarrow$(a) This follows from the fact that 1 and 0 are the only idempotents in $\text{End}_\Lambda(P)$ when $\text{End}_\Lambda(P)$ is local. $\square$

We now show how we can find the indecomposable projective modules by using idempotents. A set $\{e_1, \ldots, e_n\}$ of idempotents in Λ is **orthogonal** if $e_i e_j = 0$ when $i \neq j$. A nonzero idempotent e is **primitive** if e cannot be written as a sum of two nonzero orthogonal idempotents.

There is the following connection between decompositions of Λ as a left Λ-module and idempotent elements in Λ.

Proposition 4.8 *Let* Λ *be a left artin ring and* e *a nonzero idempotent in* Λ.

(a) *Suppose* $\Lambda e = P_1 \coprod \cdots \coprod P_n$ *with the* $P_i \neq 0$ *and let* e_i *in* P_i *for all* $i = 1, \ldots, n$ *be such that* $e = e_1 + \cdots + e_n$. *Then* $\{e_1, \ldots, e_n\}$ *is a set of nonzero orthogonal idempotents with the property* $\Lambda e_i = P_i$ *for all* $i = 1, \ldots, n$.
(b) *Suppose* $\{e_1, \ldots, e_n\}$ *is a set of nonzero orthogonal idempotents such that* $e = e_1 + \cdots + e_n$. *Then* Λe_i *is a submodule of* Λe *for all* $i = 1, \ldots, n$ *and* $\Lambda e = \Lambda e_1 \coprod \cdots \coprod \Lambda e_n$.
(c) e *in* Λ *is a primitive idempotent if and only if* Λe *is an indecomposable projective* Λ-*module.*
(d) 1 *can be written as a sum of primitive orthogonal idempotents.*

Proof (a) Since e is in Λe, it follows that there are unique elements e_i in P_i for $i = 1,\ldots,n$ such that $e = e_1 + \cdots + e_n$. Now each x in Λe can be written as λe, so $xe = (\lambda e)e = \lambda e = x$ since $e^2 = e$. Now let x_i be in P_i for some i. Then $x_i = x_i e_1 + \cdots + x_i e_i + \cdots + x_i e_n$. Since $x_i e_j$ is in P_j for all $j = 1,\ldots,n$ and there is only one way of writing x_i as a sum of elements in the P_j, it follows that $x_i e_j = 0$ for $i \neq j$ and $x_i e_i = x_i$. Hence $P_i = \Lambda e_i$, which implies $e_i \neq 0$ for all $i = 1,\ldots,n$ since all the P_i are nonzero. We have also shown that $e_i e_j = 0$ if $j \neq i$ and that $e_i^2 = e_i$. Thus $\{e_1,\ldots,e_n\}$ is a set of nonzero orthogonal idempotents.

(b) We have that $e_i e = e_i e_1 + \cdots + e_i e_i + \cdots + e_i e_n = e_i^2 = e_i$ since $e_i e_j = 0$ if $j \neq i$. Hence each e_i is in Λe and so each $\Lambda e_i \subset \Lambda e$. We now want to show that each x in Λe can be written uniquely as a sum $x_1 + \cdots + x_n$ with the x_i in Λe_i for all $i = 1,\ldots,n$. It is clear that every element in Λe_i can be written as such a sum, so we only have to show the uniqueness or equivalently if $0 = x_1 + \cdots + x_n$ with the x_i in Λe_i, then $x_i = 0$ for all i. Suppose $0 = \lambda_1 e_1 + \lambda_2 e_2 + \cdots + \lambda_n e_n$ with the λ_i in Λ. Then $0 = \lambda_1 e_1 e_i + \lambda_2 e_2 e_i + \cdots + \lambda_i e_i e_i + \cdots + \lambda_n e_n e_i$. Since $e_j e_i = 0$ if $j \neq i$, we have that $0 = \lambda_i e_i^2 = \lambda_i e_i$. So all the $\lambda_i e_i$ are 0, which is our desired result.

(c) and (d) These follow easily from (a) and (b). $\qquad\square$

This result shows that there is a close connection between the decomposition of Λ into a sum of indecomposable Λ-modules and the decomposition of 1 into a sum of primitive orthogonal idempotents.

We end this section by showing the connection between subgroups of Λ of the form $e\Lambda f$ and $e\mathfrak{r}f$ for idempotents e and f and the groups of Λ-morphisms from Λe to Λf and from Λe to $\mathfrak{r}f$ respectively.

Proposition 4.9 *Let Λ be a left artin ring and let e and f be idempotents. The morphism $\phi: e\Lambda f \to \mathrm{Hom}_\Lambda(\Lambda e, \Lambda f)$ given by $\phi(e\lambda f)(\lambda' e) = \lambda' e\lambda f$ is an isomorphism and $\phi|_{e\mathfrak{r}^m f}$ gives an isomorphism from $e\mathfrak{r}^m f$ to $\mathrm{Hom}_\Lambda(\Lambda e, \mathfrak{r}^m f)$ for all m.* $\qquad\square$

Proof The proof of this is straightforward and left to the reader. $\qquad\square$

5 Some homological facts

This section is devoted to recalling some facts from homological algebra which will be used freely (without reference) in the rest of the book. For the most part, no proofs are given. The reader is referred to the notes for references for this material.

We begin with the following way of computing the left global dimension l.gl.dim Λ of a left artin ring Λ. For a Λ-module X we denote by $\mathrm{pd}_\Lambda X$ the projective dimension of X.

Proposition 5.1 *Let $\mathfrak{r}$ be the radical of a left artin ring Λ. Then we have* l.gl.dim $\Lambda = \mathrm{pd}_\Lambda(\Lambda/\mathfrak{r})$.

Proof By definition we have l.gl.dim $\Lambda \geq \mathrm{pd}_\Lambda(\Lambda/\mathfrak{r})$. To prove the reverse inequality it suffices to show that $\mathrm{pd}_\Lambda(\Lambda/\mathfrak{r}) \geq$ l.gl.dim Λ when $\mathrm{pd}_\Lambda(\Lambda/\mathfrak{r}) = n < \infty$. Since $\Lambda/\mathfrak{r}$ is a semisimple Λ-module containing every simple Λ-module as a summand, it follows that $\mathrm{pd}_\Lambda(\Lambda/\mathfrak{r}) = \max\{\mathrm{pd}_\Lambda S|S$ a simple Λ-module$\}$. Therefore we have that $\mathrm{pd}_\Lambda S \leq n$ for all simple Λ-modules S. We now prove by induction on $l(M)$ that this implies that $\mathrm{pd}_\Lambda M \leq n$ for all finitely generated Λ-modules M. If $l(M) = 0$ we are done. If $l(M) = t > 0$, then there is an exact sequence $0 \to S \to M \to M' \to 0$ with S a simple Λ-module and $l(M') = t - 1$. Since $\mathrm{pd}_\Lambda M \leq \max\{\mathrm{pd}_\Lambda S, \mathrm{pd}_\Lambda M'\}$, we have by the inductive hypothesis that $\mathrm{pd}_\Lambda M \leq n$. But it is a standard fact that l.gl.dim $\Lambda = \sup\{\mathrm{pd}_\Lambda M|M$ a finitely generated Λ-module $\}$, and hence l.gl.dim $\Lambda \leq n$. This completes the proof of the proposition. $\square$

A ring Λ is called **left hereditary** if all left ideals are projective. It is well known that if Λ is left hereditary then submodules of projective Λ-modules are projective. A left artin ring Λ will be called **hereditary** if Λ is left hereditary.

The following characterization of hereditary left artin rings is an immediate consequence of Proposition 5.1.

Corollary 5.2 *The following are equivalent for a left artin ring Λ.*

(a) Λ *is hereditary.*
(b) $\mathfrak{r}$ *is a projective Λ-module.*
(c) $\mathrm{pd}_\Lambda(\Lambda/\mathfrak{r}) \leq 1$.
(d) l.gl.dim $\Lambda \leq 1$. $\square$

The rest of this section is devoted to a discussion of some basic facts concerning short exact sequences. Since these concepts and results hold for arbitrary modules over arbitrary rings we make no assumptions on our ring Λ or on the Λ-modules we are considering.

Let A and B be two fixed Λ-modules. Two short exact sequences $0 \to B \to E \to A \to 0$ and $0 \to B \to E' \to A \to 0$ are said to be equivalent if there is a commutative diagram

$$
\begin{array}{ccccccccc}
0 & \to & B & \to & E & \to & A & \to & 0 \\
  &   & \| &   & \downarrow & & \| & & \\
0 & \to & B & \to & E' & \to & A & \to & 0.
\end{array}
$$

Clearly in such a diagram the morphism $E \to E'$ is an isomorphism. This shows that two sequences being equivalent is an equivalence relation. We denote by $E(A, B)$ the set of equivalence classes of short exact sequences $0 \to B \to E \to A \to 0$. The equivalence class of a short exact sequence $\xi : 0 \to B \to E \to A \to 0$ is denoted by $[\xi]$.

Now suppose we are given an exact sequence $\xi : 0 \to B \to E \to A \to 0$ and a morphism $f : A' \to A$. Then we have the exact commutative pullback diagram

$$
\begin{array}{ccccccccc}
\xi_f: & 0 & \to & B & \to & E \times_A A' & \to & A' & \to & 0 \\
       &   &   & \| &   & \downarrow & & \downarrow f & & \\
\xi: & 0 & \to & B & \to & E & \to & A & \to & 0.
\end{array}
$$

It can be shown that if $[\xi] = [\tau]$ in $E(A, B)$ then $[\xi_f] = [\tau_f]$ in $E(A', B)$. Then we obtain the map $E(f, B) : E(A, B) \to E(A', B)$ given by $E(f, B)[\xi] = [\xi_f]$. It is easily seen that $E(1_A, B) : E(A, B) \to E(A, B)$ is the identity and $E(fg, B) = E(g, B)E(f, B)$ for morphisms $g : A'' \to A$ and $f : A' \to A$.

Dually, given a morphism $g : B \to B'$, we have the exact commutative pushout diagram

$$
\begin{array}{ccccccccc}
\xi: & 0 & \to & B & \to & E & \to & A & \to & 0 \\
        &   &   & \downarrow g & & \downarrow & & \| & & \\
\xi^g: & 0 & \to & B' & \to & B' \times^B E & \to & A & \to & 0.
\end{array}
$$

It can be shown that if $[\xi] = [\tau]$ in $E(A, B)$, then $[\xi^g] = [\tau^g]$ in $E(A, B')$. Thus we obtain the map $E(A, g) : E(A, B) \to E(A, B')$ given by $E(A, g)([\xi]) = [\xi^g]$. It is easily seen that $E(A, 1_B) : E(A, B) \to E(A, B)$ is the identity and $E(A, fg) = E(A, f)E(A, g)$ for morphisms $g : B \to B'$ and $f : B' \to B''$.

We now point out the following important fact about these maps.

Proposition 5.3 *Let* $f : A' \to A$ *and* $g : B \to B'$ *be morphisms of Λ-modules. Then we have* $E(f, B')E(A, g)[\xi] = E(A', g)E(f, B)[\xi]$ *in* $E(A', B')$ *for all* $[\xi]$ *in* $E(A, B)$. $\square$

This proposition suggests the following definition. Let $f: A' \to A$ and $g: B \to B'$ be morphisms of Λ-modules. Then define $E(f, g): E(A, B) \to E(A', B')$ by $E(f, g) = E(f, B')E(A, g) = E(A', g)E(f, B)$.

Using these observations we define an addition in $E(A, B)$ called the **Baer sum**, which makes $E(A, B)$ an abelian group.

Let $\xi_1 : 0 \to B \xrightarrow{\beta_1} E_1 \xrightarrow{\alpha_1} A \to 0$ and $\xi_2 : 0 \to B \xrightarrow{\beta_2} E_2 \xrightarrow{\alpha_2} A \to 0$ be exact sequences. We define $[\xi_1] + [\xi_2]$ in $E(A, B)$ as follows. Let $f: A \to A \coprod A$ be the map given by $f(a) = (a, a)$ for all a in A and let $g: B \coprod B \to B$ be given by $g(b_1, b_2) = b_1 + b_2$ for all b_1, b_2 in B. Let $\xi_1 \coprod \xi_2$ be the sum

$$0 \to B \amalg B \xrightarrow{\beta_1 \amalg \beta_2} E_1 \amalg E_2 \xrightarrow{\alpha_1 \amalg \alpha_2} A \amalg A \to 0$$

in $E(A \coprod A, B \coprod B)$ and define $[\xi_1] + [\xi_2] = E(f, g)(\xi_1 \coprod \xi_2)$ in $E(A, B)$. Then $E(A, B)$ with this addition is an abelian group with the class of the split exact sequence $0 \to B \to B \coprod A \to A \to 0$ the zero element and the inverse of the class of $0 \to B \xrightarrow{\beta} E \xrightarrow{\alpha} A \to 0$ being the class of $0 \to B \xrightarrow{\beta} E \xrightarrow{-\alpha} A \to 0$. Also for all morphisms $f: A' \to A$ and $g: B \to B'$, the map $E(f, g): E(A, B) \to E(A', B')$ is a homomorphism of abelian groups.

We also make the following useful observation. Let $\xi : 0 \to B \xrightarrow{g} E \xrightarrow{h} A \to 0$ be an exact sequence of Λ-modules and $f: A' \to A$ a Λ-morphism. Then $E(f, B)([\xi]) = 0$ in $E(E', B)$ if and only if there is some $t: A' \to E$ such that $ht = f$. To see this we use that an element in $E(A', B)$ is zero if and only if a corresponding exact sequence splits, together with the universal property of pullbacks.

Before continuing with our discussion of the abelian groups $E(A, B)$ we pause to review the notion of a bimodule.

In concrete situations where one has two rings Λ and Γ operating on an abelian group M, M is in a natural way either a Λ-module or a Λ^{op}-module and M is either a Γ-module or a Γ^{op}-module. In all such situations we will call M a Λ-Γ-bimodule if for each λ in Λ the operation of λ on M commutes with the operation of each γ in Γ on M. If M is a Λ-module and a Γ-module this is expressed by the relation $\lambda(\gamma(m)) = \gamma(\lambda(m))$ for all $\lambda \in \Lambda$, $\gamma \in \Gamma$ and m in M. If M is a Λ-module and a right Γ-module this commutativity relation is expressed by the "associative" relation $\lambda(m\gamma) = (\lambda m)\gamma$ for all $\lambda \in \Lambda$, $\gamma \in \Gamma$ and m in M. For brevity, if $\Gamma = \Lambda$, we refer to a Λ-Γ-module as a Λ-bimodule.

For the rest of this section we shall mean that M is a (left) Λ-module and a (left) Γ-module when M is a Λ-Γ-bimodule. For example, suppose M is a Λ-module and $\Gamma = \mathrm{End}_\Lambda(M)$. Then we consider M as a Γ-module by means of the action $f \cdot m = f(m)$ for all f in $\text{End}_\Lambda(M)$ and m in M. Then M is a Λ-Γ-bimodule since $\lambda(f(m)) = f(\lambda m)$ for all λ in Λ, all f in $\text{End}_\Lambda(M)$ and m in M. This is the only way we consider M as a Λ-Γ-bimodule.

Suppose now that A and B are Λ-modules and A is a Λ-Γ-bimodule. Then for each γ in Γ the map $f_\gamma: A \rightarrow A$ given by $f_\gamma(a) = \gamma a$ for all a in A is a Λ-module homomorphism. Also we have that $f_{\gamma_1\gamma_2}(a) = (\gamma_1\gamma_2)(a) = \gamma_1(\gamma_2 a) = f_{\gamma_1}(f_{\gamma_2}a)$ for all γ_1 and γ_2 in Γ and a in A. Hence we have $f_{\gamma_1\gamma_2} = f_{\gamma_1}f_{\gamma_2}$ for all γ_1 and γ_2 in Γ. Then for each γ in Γ we can define the operation of γ on the abelian group $E(A, B)$ by $\gamma[\xi] = E(f_\gamma, B)[\xi]$. It is not difficult to check that this operation makes $E(A, B)$ a Γ^{op}-module. It is a Γ^{op}-module and not a Γ-module since $\gamma_1(\gamma_2[\xi]) = E(f_{\gamma_1}, B)(E(f_{\gamma_2}, B)[\xi]) = E(f_{\gamma_2}f_{\gamma_1}, B)[\xi] = (\gamma_2\gamma_1)[\xi]$ for all γ_1 and γ_2 in Γ and $[\xi]$ in $E(A, B)$. This is the way we consider $E(A, B)$ a Γ^{op}-module when A is a Λ-Γ-bimodule. In particular, since A is always a Λ-$\text{End}_\Lambda(A)$-bimodule, we can always consider $E(A, B)$ as an $\text{End}_\Lambda(A)^{\text{op}}$-module by means of the operation of $\text{End}_\Lambda(A)^{\text{op}}$ on $E(A, B)$ given above. This is the only way we consider $E(A, B)$ as an $\text{End}_\Lambda(A)^{\text{op}}$-module.

Similarly, suppose A and B are Λ-modules and B is a Λ-Σ-bimodule. Then for each σ in Σ, the map $g_\sigma: B \rightarrow B$ given by $g_\sigma(b) = \sigma b$ for all b in B is a Λ-module morphism. Thus for each σ in Σ we can define the operation of σ on $E(A, B)$ by $\sigma \cdot [\xi] = E(A, g_\sigma)[\xi]$ for all $[\xi]$ in $E(A, B)$. It is then straightforward to check that this operation makes $E(A, B)$ a Σ-module. This is the way we will consider $E(A, B)$ as a Σ-module when B is a Λ-Σ-bimodule. Since B is always a Λ-$\text{End}_\Lambda(B)$-bimodule, we can always consider $E(A, B)$ as an $\text{End}_\Lambda(B)$-module and this is the only way we will consider $E(A, B)$ as an $\text{End}_\Lambda(B)$-module.

Suppose now that A is a Λ-Γ-bimodule and B is a Λ-Σ-bimodule. Then the abelian group $E(A, B)$ is a Γ^{op}-module and a Σ-module. In fact, it is a Γ^{op}-Σ-bimodule. For let γ be in Γ and σ in Σ and let $f_\gamma: A \rightarrow A$ be the Λ-module morphism given by $f_\gamma(a) = \gamma a$ for all a in A. Let $g_\sigma: B \rightarrow B$ be the Λ-module morphism given by $g_\sigma(b) = \sigma b$ for all b in B. Then for $[\xi]$ in $E(A, B)$ we have that $\gamma(\sigma[\xi]) = E(f_\gamma, B)(E(A, g_\sigma)[\xi])$ which by Proposition 5.3 is the same as $E(A, g_\sigma)(E(f_\gamma, B)[\xi]) = \sigma(\gamma[\xi])$. Therefore $E(A, B)$ is a Γ^{op}-Σ-bimodule. In particular $E(A, B)$ is an $\text{End}_\Lambda(A)^{\text{op}}$-$\text{End}_\Lambda(B)$-bimodule.

Associated with each pair of Λ-modules (A, B) is also the abelian group $\text{Ext}_\Lambda^1(A, B)$. We now describe for each pair of Λ-modules (A, B) isomorphisms of abelian groups $\Theta_{(A,B)}: E(A, B) \rightarrow \text{Ext}_\Lambda^1(A, B)$ which are functorial in A and B.

Let $0 \to K \to P \to A \to 0$ be an exact sequence with P a projective Λ-module. Then for each Λ-module B we have the exact sequence $\mathrm{Hom}_\Lambda(P, B) \to \mathrm{Hom}_\Lambda(K, B) \xrightarrow{\delta_B} \mathrm{Ext}^1_\Lambda(A, B) \to 0$. Now let $\xi : 0 \to B \to E \to A \to 0$ be an exact sequence. Then we have a commutative exact diagram

$$
\begin{array}{ccccccccc}
0 & \to & K & \to & P & \to & A & \to & 0 \\
 & & \downarrow h_\xi & & \downarrow g & & \| & & \\
0 & \to & B & \to & E & \to & A & \to & 0.
\end{array}
$$

Now $\delta_B(h_\xi)$ in $\mathrm{Ext}^1_\Lambda(A, B)$ does not depend on the particular choice of $g : P \to E$. Also if $\xi' : 0 \to B \to E' \to A \to 0$ is in the same equivalence class as $\xi : 0 \to B \to E \to A \to 0$ and we have a commutative diagram

$$
\begin{array}{ccccccccc}
0 & \to & K & \to & P & \to & A & \to & 0 \\
 & & \downarrow h'_{\xi'} & & \downarrow g' & & \| & & \\
0 & \to & B & \to & E' & \to & A & \to & 0,
\end{array}
$$

then $\delta_B(h'_{\xi'}) = \delta_B(h_\xi)$ in $\mathrm{Ext}^1_\Lambda(A, B)$. Therefore we obtain a well defined map $\Theta_{(A,B)} : E(A, B) \to \mathrm{Ext}^1_\Lambda(A, B)$ by setting $\Theta([\xi]) = \delta_B(h_\xi)$ for all $[\xi]$ in $E(A, B)$. One can prove that these maps $\Theta_{(A,B)}$ have the following properties.

Theorem 5.4 *For each pair (A, B) of Λ-modules the maps $\Theta_{(A,B)} : E(A, B) \to \mathrm{Ext}^1_\Lambda(A, B)$ are isomorphisms of abelian groups which are functorial in A and B.* □

Using the functorial isomorphism in Theorem 5.4 we get the following important consequence of the equality $E(f, B')E(A, g) = E(A', g)E(f, B)$ in Proposition 5.3.

Corollary 5.5 *Let $f : A' \to A$ and $g : B \to B'$ be Λ-morphisms. Then $\mathrm{Ext}^1_\Lambda(f, B')\mathrm{Ext}^1_\Lambda(A, g) = \mathrm{Ext}^1_\Lambda(A', g)\mathrm{Ext}^1_\Lambda(f, B)$.* □

Suppose now that A is a Λ-Γ-bimodule and B is a Λ-module. Then we define a Γ^{op}-structure on $\mathrm{Ext}^1_\Lambda(A, B)$ just as we did for $E(A, B)$. Similarly if B is a Λ-Σ-bimodule, then we define a Σ-module structure on $\mathrm{Ext}^1_\Lambda(A, B)$ just as we did for $E(A, B)$. Finally if A is a Λ-Γ-bimodule and B is a Λ-Σ-bimodule, then $\mathrm{Ext}^1_\Lambda(A, B)$ is a Γ^{op}-Σ-bimodule and the isomorphism $\Theta_{(A,B)} : E(A, B) \to \mathrm{Ext}^1_\Lambda(A, B)$ is an isomorphism of Γ^{op}-Σ-bimodules. In particular $\mathrm{Ext}^1_\Lambda(A, B)$ is an $\mathrm{End}_\Lambda(A)^{\mathrm{op}}$-$\mathrm{End}_\Lambda(B)$-bimodule, and $\Theta_{(A,B)}$ an $\mathrm{End}_\Lambda(A)^{\mathrm{op}}$-$\mathrm{End}_\Lambda(B)$-bimodule isomorphism.

We end this section by pointing out connections between pullback diagram, pushout diagram and decomposition of the middle term B in a short exact sequence $0 \to A \to B \to C \to 0$.

Proposition 5.6 *Let*

$$
(*) \quad
\begin{array}{ccc}
A & \overset{f}{\to} & B \\
\downarrow f' & & \downarrow g \\
B' & \overset{g'}{\to} & C
\end{array}
$$

be a commutative diagram of morphisms between Λ-modules.

(a) *The following are equivalent.*

(i) *The diagram $(*)$ is a pushout diagram.*

(ii) *The induced sequence $A \overset{\binom{f}{-f'}}{\to} B \coprod B' \overset{(g,g')}{\to} C \to 0$ is exact.*

(b) *If $(*)$ is a pushout diagram, then in the induced exact commutative diagram*

$$
\begin{array}{ccccccc}
A & \overset{f}{\to} & B & \to & \mathrm{Coker}\, f & \to & 0 \\
\downarrow f' & & \downarrow g & & \downarrow h & & \\
B' & \overset{g'}{\to} & C & \to & \mathrm{Coker}\, g' & \to & 0
\end{array}
$$

h is an isomorphism.

(c) *The following are equivalent.*

(i) *The diagram $(*)$ is a pullback diagram.*

(ii) *The induced sequence $0 \to A \overset{\binom{f}{-f'}}{\to} B \coprod B' \overset{(g,g')}{\to} C$ is exact.*

(d) *If $(*)$ is a pullback diagram, then in the induced exact commutative diagram*

$$
\begin{array}{ccccccc}
0 & \to & \mathrm{Ker}\, f & \to & A & \overset{f}{\to} & B \\
& & \downarrow h & & \downarrow f' & & \downarrow g \\
0 & \to & \mathrm{Ker}\, g' & \to & B' & \overset{g'}{\to} & C
\end{array}
$$

h is an isomorphism. $\square$

We have the following consequence.

Corollary 5.7 *Let $0 \to A \overset{\binom{f}{-f'}}{\to} B \coprod B' \overset{(g,g')}{\to} C \to 0$ be an exact sequence of Λ-modules. Then the following hold.*

(a) *The diagram*

$$
\begin{array}{ccc}
A & \xrightarrow{f} & B \\
\downarrow{\scriptstyle -f'} & & \downarrow{\scriptstyle g} \\
B' & \xrightarrow{g'} & C
\end{array}
$$

  is both a pushout and a pullback diagram.
(b) *f is an epimorphism if and only if g' is an epimorphism.*
(c) *f is a monomorphism if and only if g' is a monomorphism.* ☐

Exercises

1. For each nonzero integer n let $\mathbb{Z}_n$ denote the cyclic abelian group of order n.

(a) Find all five composition series for $\mathbb{Z}_4 \times \mathbb{Z}_2$.
(b) How many composition series exist for $\mathbb{Z}_4 \times \mathbb{Z}_4$ and how many exist for $\mathbb{Z}_2 \times \mathbb{Z}_2 \times \mathbb{Z}_2$?

2. Let k be a field, $\Lambda = k[x]/(x^3)$ and let $f:\Lambda \coprod k \to \Lambda/(x^2)$ be the Λ-homomorphism given by $f(\lambda, a) = (\lambda + ax)\bmod(x^2)$, where k is the trivial Λ-module. Prove that f is right minimal but not an essential epimorphism.

3. In each of the three cases below give examples of exact sequences $0 \to A \xrightarrow{f} B \xrightarrow{g} C \to 0$ and $0 \to A' \xrightarrow{f'} B' \xrightarrow{g'} C' \to 0$ of finite abelian groups A, B, C, A', B' and C'.

(i) $A \simeq A'$, $B \simeq B'$, $C \not\simeq C'$.
(ii) $A \simeq A'$, $C \simeq C'$, $B \not\simeq B'$.
(iii) $B \simeq B'$, $C \simeq C'$, $A \not\simeq A'$.

4. Let Λ be a hereditary left artin ring. Let $f:A \to P$ be a nonzero Λ-homomorphism with P a projective Λ-module. Show that there exists a nonzero left ideal I in Λ such that $A \simeq I \coprod N$ for some Λ-module N.

5. Let k be a field and let Λ be a finite dimensional k-algebra. Show that Λ is isomorphic to a k-subalgebra of a full matrix algebra over k.

6. Let $A = \left\{ \left(\begin{smallmatrix} a & b & c \\ c & a & b \\ b & c & a \end{smallmatrix} \right) \mid a, b, c \in \mathbb{R} \right\}$. Show that $A \simeq \mathbb{R} \times \mathbb{C}$ where $\mathbb{R}$ is the real numbers and $\mathbb{C}$ is the complex numbers.

7. (Nakayama's lemma) Let Λ be a ring with radical $\mathfrak{r}$ and $\mathfrak{a}$ a left ideal of Λ. Prove that the following statements are equivalent.

(i) $\mathfrak{a} \subset \mathfrak{r}$.

(ii)   For every finitely generated Λ-module M we have that $\mathfrak{a}M = M$ implies $M = 0$.

(iii) For every finitely generated Λ-module M we have that $\mathfrak{a}M$ is small in M.

8. Let Λ be a left artin ring and P a finitely generated indecomposable projective Λ-module. Prove that P/M is an indecomposable Λ-module for each submodule M of P with $M \neq P$.

9. Let Λ be a left artin ring. Show that every finitely generated Λ-module is isomorphic to a finite sum of indecomposable Λ-modules.

10. Show that a ring Λ has a composition series when viewed as a module over itself if and only if every finitely generated Λ-module has a composition series.

11. Let $\mathfrak{r}$ be the radical of a left artin ring Λ.

(a) Show that for each Λ-module M we have $(\Lambda/\mathfrak{r}) \otimes_\Lambda M \simeq M/\mathfrak{r}M$.

(b) Show that a finitely generated Λ-module M is projective if and only if $\mathrm{Tor}_1^\Lambda(\Lambda/\mathfrak{r}, M) = 0$.

(c) Show that a finitely generated Λ-module M is projective if and only if $\mathrm{Ext}_\Lambda^1(M, \Lambda/\mathfrak{r}) = 0$.

(d) Let $0 \to K \to P \to M \to 0$ be an exact sequence of finitely generated Λ-modules with P projective. Prove that the following are equivalent.

   (i) M is projective
   (ii)   The induced morphism $K/\mathfrak{r}K \to P/\mathfrak{r}P$ is a monomorphism
   (iii) The induced sequence $0 \to K/\mathfrak{r}K \to P/\mathfrak{r}P \to M/\mathfrak{r}M \to 0$ is exact.

12. Let Λ be any ring and $\mathfrak{a}$ a left ideal. Let Σ be the idealizer of $\mathfrak{a}$ in Λ, which is $\Sigma = \{\lambda \in \Lambda | \mathfrak{a}\lambda \subset \mathfrak{a}\}$.

(a) Prove that Σ is a subring of Λ containing $\mathfrak{a}$ as a two-sided ideal.

(b) Prove that $\phi : \Sigma \to \mathrm{End}_\Lambda(\Lambda/\mathfrak{a})^{\mathrm{op}}$, given by $\phi(\gamma)(\lambda + \mathfrak{a}) = \lambda\gamma + \mathfrak{a}$ for $\gamma \in \Sigma$ and $\lambda + \mathfrak{a} \in \Lambda/\mathfrak{a}$, is a surjective ring morphism with kernel $\mathfrak{a}$.

13. Let p be a prime number. Prove that $f:(\mathbb{Z}/p\mathbb{Z})\coprod(\mathbb{Z}/p^2\mathbb{Z}) \to \mathbb{Z}/p^3\mathbb{Z}$ given by $f(x + (p), y + (p^2)) = p^2x + py + (p^3)$ is not right minimal in f.l. $\mathbb{Z}$.

Notes

The basic material on rings and modules can be found for example in [AnF]. Note that our proof of the Jordan–Hölder theorem differs slightly from the standard proofs through the use of generalized composition series.

Right and left minimal morphisms were introduced in [AuS1] in connection with work on preprojective partitions.

For the homological facts in the last section we refer to standard texts on homological algebra, like [HiS], [No], and [Rot].

II
Artin algebras

In this chapter we turn our attention to artin algebras and their finitely generated modules, the main subject of this book. One important feature of the theory of artin algebras as opposed to left artin rings is that endomorphism rings of finitely generated modules are again artin algebras. In principle, this enables one to convert problems involving only a finite number of modules over one artin algebra to problems about finitely generated projective modules over some other artin algebra. This procedure, which we call projectivization, is illustrated by our proofs of the Krull–Schmidt theorem and other results. Another important property of artin algebras is that there is a duality between finitely generated left and finitely generated right modules. It is convenient to start the chapter with a section on categories over a commutative artin ring R, and study equivalences of such categories.

1 Artin algebras and categories

Generalizing the category $\mathrm{mod}\,\Lambda$ for an artin R-algebra Λ we introduce the notion of R-categories, and study equivalences between such categories.

Let R be a commutative artin ring. We recall that an R-algebra Λ is a ring together with a ring morphism $\phi: R \to \Lambda$ whose image is in the center of Λ. For an R-algebra $\phi: R \to \Lambda$ we usually write $r\lambda$ for $\phi(r)\lambda$ where r is in R and λ is in Λ. If $\phi_1: R \to \Lambda_1$ and $\phi_2: R \to \Lambda_2$ make Λ_1 and Λ_2 R-algebras, then Λ_1 is an R-subalgebra of Λ_2 if it is a subring of Λ_2 via $i: \Lambda_1 \to \Lambda_2$ and $i\phi_1 = \phi_2$. We say that Λ is an **artin R-algebra**, or **artin algebra** for short, if Λ is finitely generated as an R-module. It is clear that an artin algebra is both a left and a right artin ring. It can be

26

shown that the center $Z(\Lambda)$ of an artin R-algebra is a commutative artin ring and that Λ is an artin $Z(\Lambda)$-algebra.

Important examples of artin algebras are finite dimensional algebras over a field. Examples which are not of this type are found amongst proper factor rings of principal ideal domains and endomorphism rings of finite abelian groups.

It is clear that if Λ is an artin R-algebra via a ring morphism $R \to \Lambda$, then the same ring morphism $R \to \Lambda^{op}$ makes Λ^{op} an artin R-algebra. We give another way of associating new artin algebras with a given one. If A and B are in $\text{mod}\,\Lambda$ for an artin R-algebra Λ, then A and B are in $\text{mod}\,R$. For $r \in R$ and $f \in \text{Hom}_R(A, B)$ we have $f(ra) = rf(a)$ for all $a \in A$. We define rf by $(rf)(a) = rf(a)$ for $a \in A$. Then $\text{Hom}_R(A, B)$ is an R-module. $\text{Hom}_\Lambda(A, B)$ is contained in $\text{Hom}_R(A, B)$ and is clearly a subgroup. Since the image of R is in the center of Λ, we have $(rf)(\lambda a) = r(f(\lambda a)) = r(\lambda f(a)) = \lambda r f(a) = \lambda((rf)(a))$ for $\lambda \in \Lambda$ and $f \in \text{Hom}_\Lambda(A, B)$. Hence $\text{Hom}_\Lambda(A, B)$ is an R-submodule of $\text{Hom}_R(A, B)$, and we shall always consider $\text{Hom}_\Lambda(A, B)$ as an R-module this way. Similarly for A in $\text{mod}\,\Lambda$ we define for $r \in R$ a map $\phi(r): A \to A$ by $\phi(r)(a) = ra$ for $a \in A$. Since the image of R is in the center of Λ, we have $\phi(r)(\lambda a) = r(\lambda a) = \lambda(ra) = \lambda\phi(r)(a)$ for $\lambda \in \Lambda$. Hence we have the map $\phi: R \to \text{End}_\Lambda(A) \subset \text{End}_R(A)$ which is clearly a ring morphism. For $g \in \text{End}_R(A)$ and $r \in R$ and $a \in A$ we have $(\phi(r)g)(a) = r(g(a)) = g(ra) = (g\phi(r))(a)$, so that $\text{Im}\,\phi \subset Z(\text{End}_R(A)) \cap \text{End}_\Lambda(A) \subset Z(\text{End}_\Lambda(A))$. This makes $\text{End}_R(A)$ and $\text{End}_\Lambda(A)$ R-algebras, with $\text{End}_\Lambda(A)$ an R-subalgebra of $\text{End}_R(A)$, and we shall always consider $\text{End}_R(A)$ and $\text{End}_\Lambda(A)$ as R-algebras this way. Note that the R-module structure on $\text{End}_R(A)$ given by $\phi: R \to \text{End}_R(A)$ is the same as the R-module structure for $\text{Hom}_R(A, B)$ with $B = A$.

We can now give the following result.

Proposition 1.1 *Let Λ be an artin R-algebra.*

(a) *If A and B are in $\text{mod}\,\Lambda$, then $\text{Hom}_\Lambda(A, B)$ is a finitely generated R-submodule of $\text{Hom}_R(A, B)$.*

(b) *If A is in $\text{mod}\,\Lambda$, then $\text{End}_\Lambda(A)$ is an artin R-algebra which is an R-subalgebra of the artin R-algebra $\text{End}_R(A)$.*

Proof (a) Let A and B be in $\text{mod}\,\Lambda$. Since A is in $\text{mod}\,\Lambda$, and hence in $\text{mod}\,R$, we have an epimorphism $nR \to A$ for some $n > 0$, where nR denotes the sum of n copies of R. Hence we get a monomorphism $\text{Hom}_R(A, B) \to \text{Hom}_R(nR, B) \xrightarrow{\sim} nB$. Since R is noetherian and nB is a finitely generated R-module, we get that the submodule $\mathrm{Hom}_\Lambda(A, B)$ of $\mathrm{Hom}_R(A, B)$ is a finitely generated R-module.

(b) We have seen that $\mathrm{End}_\Lambda(A)$ is an R-subalgebra of $\mathrm{End}_R(A)$. Now $\mathrm{End}_R(A)$ and $\mathrm{End}_\Lambda(A)$ are finitely generated R-modules by (a), and are hence artin R-algebras. $\qquad\square$

Recall that a category $\mathscr{C}$ is **preadditive** if for each pair of objects A, B in $\mathscr{C}$ the morphism set $\mathrm{Hom}_\mathscr{C}(A, B)$ is an abelian group and for A, B, C in $\mathscr{C}$ the composition morphism $\mathrm{Hom}_\mathscr{C}(A, B) \times \mathrm{Hom}_\mathscr{C}(B, C) \to \mathrm{Hom}_\mathscr{C}(A, C)$ is bilinear. If in addition $\mathscr{C}$ has finite sums, then $\mathscr{C}$ is **additive**. For a commutative artin ring R we say that a preadditive category $\mathscr{C}$ is an R-**category** if each $\mathrm{Hom}_\mathscr{C}(A, B)$ is an R-module and the composition morphism $\mathrm{Hom}_\mathscr{C}(A, B) \times \mathrm{Hom}_\mathscr{C}(B, C) \to \mathrm{Hom}_\mathscr{C}(A, C)$ is R-bilinear. Further we recall that a covariant (respectively contravariant) functor $F:\mathscr{C} \to \mathscr{D}$ between two preadditive categories is **additive** if for each pair of objects A, B in $\mathscr{C}$, the induced morphisms $\mathrm{Hom}_\mathscr{C}(A, B) \to \mathrm{Hom}_\mathscr{D}(F(A), F(B))$ (respectively $\mathrm{Hom}_\mathscr{C}(A, B) \to \mathrm{Hom}_\mathscr{D}(F(B), F(A)))$ are group homomorphisms. We say that an additive functor $F:\mathscr{C} \to \mathscr{D}$ between two R-categories is an R-**functor** if the induced morphisms above are R-homomorphisms. Unless stated to the contrary, all our categories will be R-categories and all our functors are R-functors for some commutative artin ring R.

When Λ is an artin R-algebra we have seen in Proposition 1.1 that $\mathrm{Hom}_\Lambda(A, B)$ is an R-module for A and B in $\mathrm{mod}\,\Lambda$. For $f:A \to B$ and $g:B \to C$ in $\mathrm{mod}\,\Lambda$ and $r \in R$ we have $(g(rf))(a) = g(rf(a)) = rg(f(a)) = ((rg)f)(a)$ for $a \in A$. Hence the composition morphism $\mathrm{Hom}_\Lambda(A, B) \times \mathrm{Hom}_\Lambda(B, C) \to \mathrm{Hom}_\Lambda(A, C)$ is R-bilinear. Since $\mathrm{mod}\,\Lambda$ is clearly an additive category, it follows that $\mathrm{mod}\,\Lambda$ is an additive R-category.

For A in $\mathrm{mod}\,\Lambda$, $\mathrm{add}\,A$ denotes the full subcategory of $\mathrm{mod}\,\Lambda$ whose objects are summands of finite sums of copies of A. Then it is easy to see that $\mathrm{add}\,A$ is an R-category.

We now give some examples of R-functors. Various well known functors from $\mathrm{mod}\,\Lambda$ to the category Ab of abelian groups are in fact R-functors when Λ is an artin R-algebra.

First we note that the forgetful functor $F:\mathrm{mod}\,\Lambda \to \mathrm{mod}\,R$ is easily seen to be an R-functor, by using Proposition 1.1.

For A in $\mathrm{mod}\,\Lambda$ it follows from Proposition 1.1 that $\mathrm{End}_\Lambda(A)$ and $\mathrm{End}_\Lambda(A)^{\mathrm{op}}$ are artin R-algebras. Then A is an $\mathrm{End}_\Lambda(A)$-module, when for $a \in A$ and $f \in \mathrm{End}_\Lambda(A)$ we define $fa = f(a)$, and hence A is a right $\mathrm{End}_\Lambda(A)^{\mathrm{op}}$-module. If X is also in $\mathrm{mod}\,\Lambda$, then $\mathrm{Hom}_\Lambda(A, X)$ has a natural structure as a module over $\Gamma = \mathrm{End}_\Lambda(A)^{\mathrm{op}}$ given as follows. For $t \in \Gamma$ and $f \in \mathrm{Hom}_\Lambda(A, X)$, define $(tf)(a) = f(t(a))$ for $a \in A$. Since $\mathrm{Hom}_\Lambda(A, X)$ is in $\mathrm{mod}\,R$ by Proposition 1.1, it is also in $\mathrm{mod}\,\Gamma$. It is then easy to see that the functor $\mathrm{Hom}_\Lambda(A, \ ) : \mathrm{mod}\,\Lambda \to \mathrm{mod}\,\Gamma$ is an R-functor.

If B is in $\mathrm{mod}(\Lambda^{\mathrm{op}})$ for an artin algebra Λ, then B is a module over the artin algebra $\Gamma = \mathrm{End}_{\Lambda^{\mathrm{op}}}(B)$. For X in $\mathrm{mod}\,\Lambda$ we have that $B \otimes_\Lambda X$ is then a Γ-module, where $f(b \otimes x) = f(b) \otimes x$ for $f \in \Gamma$, $b \in B$ and $x \in X$. It is easy to see that the functors $B \otimes_\Lambda - : \mathrm{mod}\,\Lambda \to \mathrm{mod}\,\Gamma$ and $B \otimes_\Lambda - : \mathrm{mod}\,\Lambda \to \mathrm{mod}\,R$ are R-functors.

Similarly $\mathrm{Ext}^1_\Lambda(A, \ )$ is an R-functor from $\mathrm{mod}\,\Lambda$ to $\mathrm{mod}\,\mathrm{End}_\Lambda(A)^{\mathrm{op}}$ or to $\mathrm{mod}\,R$, and $\mathrm{Tor}^\Lambda_1(B, \ )$ is an R-functor from $\mathrm{mod}\,\Lambda$ to $\mathrm{mod}\,\mathrm{End}_{\Lambda^{\mathrm{op}}}(B)$ or to $\mathrm{mod}\,R$, when A is in $\mathrm{mod}\,\Lambda$ and B in $\mathrm{mod}(\Lambda^{\mathrm{op}})$.

We shall now see that there is a general procedure for associating with any additive functor $F : \mathscr{C} \to Ab$, where $\mathscr{C}$ is an R-category, an R-functor $\widetilde{F} : \mathscr{C} \to \mathrm{Mod}\,R$, in a natural way. To see this we first note that for each A in $\mathscr{C}$ there is a ring morphism $R \to \mathrm{End}_\mathscr{C}(A)$ obtained by sending r to $r1_A$. For since composition of morphisms in $\mathscr{C}$ is R-bilinear, we have $(r_1 1_A)(r_2 1_A) = 1_A(r_1(r_2 1_A)) = (r_1 r_2)1_A$. Then we define $\widetilde{F}(A)$ in $\mathrm{Mod}\,R$ to be the abelian group $F(A)$ with R-module structure given by the composite ring morphism $R \to \mathrm{End}_\mathscr{C}(A) \to \mathrm{End}_{Ab}(F(A))$. It is then easy to see that for $f : A \to B$ in $\mathscr{C}$ the induced group morphism $F(f) : \widetilde{F}(A) \to \widetilde{F}(B)$ is an R-morphism, and it is easily seen that we get an R-functor $\widetilde{F} : \mathscr{C} \to \mathrm{Mod}\,R$ this way.

To formally express that two R-categories $\mathscr{C}$ and $\mathscr{D}$ are essentially the same we say that a covariant R-functor $F : \mathscr{C} \to \mathscr{D}$ is an **isomorphism** if there is an R-functor $G : \mathscr{D} \to \mathscr{C}$ such that $GF = 1_\mathscr{C}$ and $FG = 1_\mathscr{D}$.

We illustrate this concept through the following. For an R-algebra Λ and a unit α in Λ we have a ring automorphism $g_\alpha : \Lambda \to \Lambda$ defined by $g_\alpha(\lambda) = \alpha^{-1}\lambda\alpha$ for $\lambda \in \Lambda$. There is an induced R-functor $F_\alpha : \mathrm{mod}\,\Lambda \to \mathrm{mod}\,\Lambda$ sending a Λ-module C to a Λ-module ${}^\alpha C$ which is equal to C as an abelian group and where the new Λ-structure is defined by $\lambda \cdot c = g_\alpha(\lambda)c$ for $\lambda \in \Lambda$ and $c \in {}^\alpha C$. We clearly have $F_{\alpha^{-1}}F_\alpha = 1_{\mathrm{mod}\Lambda} = F_\alpha F_{\alpha^{-1}}$, so that $F_\alpha : \mathrm{mod}\,\Lambda \to \mathrm{mod}\,\Lambda$ is an isomorphism.

Isomorphisms of R-categories are, however, rare, and in practice it is sufficient to have a weaker relationship between R-categories $\mathscr{C}$ and $\mathscr{D}$ in order to ensure that their algebraic structures are essentially the same. We say that a covariant R-functor $F : \mathscr{C} \to \mathscr{D}$ between two R-categories is an **equivalence** of R-categories if there is an R-functor $G : \mathscr{D} \to \mathscr{C}$ such that the composite functors GF and FG are isomorphic to the identity functors.

It is useful to have another way of describing when two categories are equivalent. We say that an R-functor $F:\mathscr{C} \to \mathscr{D}$ between R-categories is **faithful** if the morphism $F_{A,B}:\operatorname{Hom}_{\mathscr{C}}(A, B) \to \operatorname{Hom}_{\mathscr{D}}(F(A), F(B))$ given by F is a monomorphism for all A, B in $\mathscr{C}$, and **full** if this morphism is an epimorphism. The functor F is **dense** if for each M in $\mathscr{D}$ there is some C in $\mathscr{C}$ with $F(C) \simeq M$.

We have the following characterization of an equivalence of categories.

Theorem 1.2 *A covariant functor* $F:\mathscr{C} \to \mathscr{D}$ *between R-categories is an equivalence if and only if it is full, faithful and dense.*

Proof Assume first that $F:\mathscr{C} \to \mathscr{D}$ is an equivalence. We then have an R-functor $G:\mathscr{D} \to \mathscr{C}$ and isomorphisms $\phi:GF \to 1_{\mathscr{C}}$ and $\psi:FG \to 1_{\mathscr{D}}$. For each A and B in $\mathscr{C}$ we then have an isomorphism $\bar{\phi}_{A,B}:\operatorname{Hom}_{\mathscr{C}}(A, B) \to \operatorname{Hom}_{\mathscr{C}}(GF(A), GF(B))$ given by $\bar{\phi}_{A,B}(f) = \phi_B^{-1} f \phi_A$. But we clearly have $\bar{\phi}_{A,B} = G_{F(A),F(B)}F_{A,B}$. Hence $F_{A,B}:\operatorname{Hom}_{\mathscr{C}}(A, B) \to \operatorname{Hom}_{\mathscr{D}}(F(A), F(B))$ is a monomorphism and $G_{F(A),F(B)}$ is an epimorphism. Using the isomorphism ψ we get that $F_{A,B}$ is an epimorphism. Hence $F_{A,B}$ is an isomorphism and therefore F is full and faithful. It is clear that F is dense.

Assume conversely that the R-functor F is full, faithful and dense. Since F is dense, we can for each M in $\mathscr{D}$ choose an isomorphism $\eta_M:M \to F(A)$ for some A in $\mathscr{C}$. We define $G(M) = A$, so that we have a morphism $\eta_M:M \to FG(M)$ for each M in $\mathscr{D}$. If $f:M \to N$ is a morphism in $\mathscr{D}$, let $h:FG(M) \to FG(N)$ be given by $h = \eta_N f \eta_M^{-1}$, i.e. the diagram

$$
\begin{array}{ccc}
M & \xrightarrow{f} & N \\
\downarrow{\eta_M} & & \downarrow{\eta_N} \\
FG(M) & \xrightarrow{h} & FG(N)
\end{array}
$$

commutes. Since F is full and faithful, there is a unique morphism $t:G(M) \to G(N)$ such that $F(t) = h$. We define $G(f) = t$ so that $h = FG(t)$. It is obvious that $G(1_M) = 1_{GM}$ for each M in $\mathscr{D}$ and $G(g'g) = G(g')G(g)$ for morphisms $g:M \to N$ and $g':N \to L$ in $\mathscr{D}$. Hence G is a covariant functor and the η_M give an isomorphism between $1_{\mathscr{D}}$ and FG.

The functor FG is an R-functor since it is isomorphic to $1_{\mathscr{D}}$ by the R-isomorphism η_M. Let f and g be in $\operatorname{Hom}_{\mathscr{D}}(M, N)$. Because F is also an R-functor, we have $FG(f + g) = FG(f) + FG(g) = F(G(f) + G(g))$ and

$FG(rf) = rFG(f) = F(rG(f))$ for r in R. Using that F is full and faithful we get $G(f + g) = G(f) + G(g)$ and $(G(rf) = rG(f)$, and consequently G is an R-functor.

It remains to show that we have an isomorphism $\epsilon : 1_{\mathscr{D}} \to GF$. For A in $\mathscr{C}$ we have the isomorphism $\eta_{F(A)} : F(A) \to FGF(A)$. Since F is full and faithful, there is a unique morphism $\epsilon_A : A \to GF(A)$ in $\mathscr{C}$ such that $F(\epsilon_A) = \eta_{F(A)}$. Since $\eta_{F(A)}$ is an isomorphism, there is a morphism $g : FGF(A) \to F(A)$ such that $\eta_{F(A)}g = 1_{FGF(A)}$ and $g\eta_{F(A)} = 1_{F(A)}$. Choosing $t : GF(A) \to A$ such that $F(t) = g$ we have $F(\epsilon_A t) = F(1_{GF(A)})$ and $F(t\epsilon_A) = F(1_A)$. Since F is faithful we see that $\epsilon_A : A \to GF(A)$ is an isomorphism. Then for $f : A \to B$ in $\mathscr{C}$, the diagram

$$
\begin{array}{ccc}
A & \overset{f}{\to} & B \\
\downarrow{\scriptstyle \epsilon_A} & & \downarrow{\scriptstyle \epsilon_B} \\
GF(A) & \overset{GF(f)}{\to} & GF(B)
\end{array}
$$

has the property that after applying F it is commutative. Since F is faithful, the diagram is then itself commutative, and we are done. □

Theorem 1.2 can be used to show that many homological properties of R-categories are preserved under equivalence. We illustrate this on projective and injective objects. To cover more general categories than module categories, we generalize the following concepts. A morphism $g : B \to C$ in an R-category $\mathscr{C}$ is an **epimorphism** if for any nonzero morphism $h : C \to X$ in $\mathscr{C}$ we have $hg \neq 0$. And $g : B \to C$ is a **monomorphism** if for any nonzero morphism $f : Y \to B$ we have $gf \neq 0$. An object P in $\mathscr{C}$ is **projective** if for any epimorphism $g : B \to C$ and morphism $h : P \to C$ there is a morphism $s : P \to B$ such that $gs = h$. Similarly an object I in $\mathscr{C}$ is **injective** if for any monomorphism $g : B \to C$ and morphism $h : B \to I$ there is a morphism $s : C \to I$ such that $sg = h$. It is not difficult to check that in the case $\mathscr{C}$ is Mod Λ for some R-algebra Λ, then the concepts just introduced coincide with the usual module theoretic concepts of epimorphism, monomorphism and projective and injective modules.

We then have the following direct consequence of the definitions and Theorem 1.2.

Proposition 1.3 *Let $F : \mathscr{C} \to \mathscr{D}$ be an R-functor which is an equivalence of R-categories.*

(a) *A morphism $g : B \to C$ in $\mathscr{C}$ is an epimorphism (respectively mono-*

morphism) if and only if $F(g):F(B) \to F(C)$ is an epimorphism (re-spectively monomorphism) in $\mathscr{D}$.

(b) *An object C in $\mathscr{C}$ is projective (respectively injective) if and only if $F(C)$ is projective (respectively injective) in $\mathscr{D}$.*   □

We shall see examples of equivalences of R-categories in the next section and in later chapters. Here we just point out that if $\mathscr{C} = \mathscr{D} = \text{mod}\,k$ for a field k, then the natural functor $F:\text{mod}\,k \to \text{mod}\,k$ defined on objects by $F(C) = C^{**}$ is an equivalence of k-categories where $C^* = \text{Hom}_k(C,k)$ is the dual of the vector space C. Even though for every finite dimensional vector space C we have a natural isomorphism $C \xrightarrow{\sim} C^{**}$, not every C in $\text{mod}\,k$ is actually a dual vector space. Hence the functor $F:\text{mod}\,k \to \text{mod}\,k$ is not an isomorphism of k-categories.

If $F:\mathscr{C} \to \mathscr{D}$ is a contravariant R-functor between R-categories, there is an induced covariant R-functor $\overline{F}:\mathscr{C}^{\text{op}} \to \mathscr{D}$ and F is a full, faithful or dense R-functor if and only if $\overline{F}$ is a full, faithful or dense R-functor respectively. We say that the R-functor $F:\mathscr{C} \to \mathscr{D}$ is a **duality** if $\overline{F}:\mathscr{C}^{\text{op}} \to \mathscr{D}$ is an equivalence. Then we have the following analogues of Theorem 1.2 and Proposition 1.3.

Theorem 1.4 *A contravariant R-functor $F:\mathscr{C} \to \mathscr{D}$ between two R-categories is a duality if and only if F is full, faithful and dense.*   □

Proposition 1.5 *Let $F:\mathscr{C} \to \mathscr{D}$ be an R-functor which is a duality of R-categories.*

(a) *A morphism $f:A \to B$ in $\mathscr{C}$ is a monomorphism if and only if $F(f):F(B) \to F(A)$ is an epimorphism.*

(b) *An object C in $\mathscr{C}$ is projective (respectively injective) if and only if $F(C)$ is injective (respectively projective) in $\mathscr{D}$.*   □

The functor $F:\text{mod}\,k \to \text{mod}\,k$ which sends a finite dimensional vector space to its dual space is an example of a duality of k-categories. More examples will be discussed later.

2 Projectivization

In this section we show that passing from an artin algebra Λ to the endomorphism algebra $\Gamma = \text{End}_\Lambda(A)^{\text{op}}$ for A in $\text{mod}\,\Lambda$ provides a technique for reducing questions about the module A to questions about projective modules. We prove the Krull–Schmidt theorem using this procedure. Throughout this section we assume that all rings are artin algebras.

For A in mod Λ we denote the R-functor $\text{Hom}_\Lambda(A, \ ): \text{mod } \Lambda \to \text{mod } \Gamma$ by e_A and call it the **evaluation functor** at A. Note that e_A is a left exact functor which commutes with finite sums. The following result gives the basis for our reduction.

Proposition 2.1 *Let A be in* mod Λ *where Λ is an artin algebra. Then the evaluation functor* $e_A: \text{mod } \Lambda \to \text{mod } \Gamma$ *has the following properties.*

(a) $e_A: \text{Hom}_\Lambda(Z, X) \to \text{Hom}_\Gamma(e_A(Z), e_A(X))$ *is an isomorphism for Z in* add A *and X in* mod Λ.

(b) *If X is in* add A, *then $e_A(X)$ is in* $\mathcal{P}(\Gamma)$.

(c) $e_A|_{\text{add } A}: \text{add } A \to \mathcal{P}(\Gamma)$ *is an equivalence of R-categories.*

Proof (a) It is clear that $e_A: \text{Hom}_\Lambda(A, X) \to \text{Hom}_\Gamma(e_A(A), e_A(X))$ is an isomorphism. Our desired result follows by the additivity of e_A.

(b) Since $e_A(A) \simeq \Gamma$, it is clear that $e_A(nA) \simeq n\Gamma$ for all positive integers n. Since e_A commutes with sums, it follows that $e_A(X)$ is in $\mathcal{P}(\Gamma)$ for all X in add A.

(c) By (a) we know that $e_A|_{\text{add } A}$ is full and faithful. To see that it is dense let P be in $\mathcal{P}(\Gamma)$. Then we have that $P \coprod Q = n\Gamma$ for some n. Hence there is an idempotent $f \in \text{End}_\Gamma(n\Gamma)$ such that $P = \text{Ker } f$. Therefore there is an idempotent $u \in \text{End}_\Lambda(nA)$ such that $e_A(u) = f$. Then $\text{Ker } u$ is in add A and $e_A(\text{Ker } u) \simeq P$ since e_A is left exact. $\quad\square$

Our first application of this proposition will be the Krull–Schmidt theorem for mod Λ.

Theorem 2.2

(a) *A module A in* mod Λ *where Λ is an artin algebra is indecomposable if and only if* $\text{End}_\Lambda(A)$ *is a local ring.*

(b) *Let $\{A_i\}_{i \in I}$ and $\{B_j\}_{j \in J}$ be two finite families of finitely generated indecomposable Λ-modules. If $\coprod_{i \in I} A_i \simeq \coprod_{j \in J} B_j$, then there is a bijection $\sigma: I \to J$ such that $A_i \simeq B_{\sigma(i)}$ for all $i \in I$.*

Proof (a) Let A be in mod Λ and let $\Gamma = \text{End}_\Lambda(A)^{\text{op}}$. Since e_A induces an equivalence add $A \to \mathcal{P}(\Gamma)$ we have that A is indecomposable if and only if $e_A(A)$ is an indecomposable projective Γ-module. But we have already seen in I Proposition 4.7 that since $e_A(A)$ is a finitely generated projective module over the artin algebra Γ, then $e_A(A)$ is indecomposable if and only if $\text{End}_\Gamma(e_A(A))$ is local. But $e_A: \text{End}_\Lambda(A) \to \text{End}_\Gamma(e_A(A))$ is an isomorphism. Hence A is indecomposable if and only if $\text{End}_\Lambda(A)$ is a local ring.

(b) Let $C = \coprod_{i \in I} A_i$. Then C is a finitely generated Λ-module and B_j and A_i are in $\text{add } C$ for all i and j. Let $\Gamma = \text{End}_\Lambda(C)^{\text{op}}$ and let $e_C : \text{mod }\Lambda \to \text{mod }\Gamma$ be the evaluation functor. Since e_C induces an equivalence $\text{add } C \to \mathscr{P}(\Gamma)$, we use the corresponding result about projective modules in I Theorem 4.4 to complete the proof. $\square$

We now show how the evaluation functor can be used to connect up right minimal morphisms of arbitrary modules with projective covers, and use this to show that a direct sum of right minimal morphisms is right minimal.

Proposition 2.3 *Let A be in $\text{mod }\Lambda$ and $f : X \to Y$ a morphism in $\text{add } A$ where Λ is an artin algebra. Then f is right minimal if and only if the induced morphism*

$$\text{Hom}_\Lambda(A, X) \to \text{Im}(\text{Hom}_\Lambda(A, f))$$

is a projective cover in $\text{mod End}_\Lambda(A)^{\text{op}}$.

Proof Let $\Gamma = \text{End}_\Lambda(A)^{\text{op}}$ and let $e_A : \text{mod }\Lambda \to \text{mod }\Gamma$ be the evaluation functor. Since $e_A : \text{add } A \to \mathscr{P}(\Gamma)$ is an equivalence of categories, it follows from the definition of right minimal morphisms that $f : X \to Y$ is right minimal if and only if $\text{Hom}_\Lambda(A, f) : \text{Hom}_\Lambda(A, X) \to \text{Hom}_\Lambda(A, Y)$ is right minimal. But $\text{Hom}_\Lambda(A, f) : \text{Hom}_\Lambda(A, X) \to \text{Hom}_\Lambda(A, Y)$ is right minimal if and only if the induced morphism $\text{Hom}_\Lambda(A, X) \to \text{Im}(\text{Hom}_\Lambda(A, f))$ is a projective cover by I Proposition 4.1. $\square$

As an easy consequence of this characterization of right minimal morphisms we have the following.

Corollary 2.4 *Let $\{f_i : A_i \to B_i\}_{i \in I}$ be a finite family of morphisms in $\text{mod }\Lambda$ where Λ is an artin algebra. Then $\coprod_{i \in I} f_i : \coprod_{i \in I} A_i \to \coprod_{i \in I} B_i$ is right minimal if and only if each $f_i : A_i \to B_i$ is right minimal.*

Proof Let A be a finitely generated Λ-module such that the A_i and B_i are in $\text{add } A$ for all i in I. Let $e_A : \text{mod }\Lambda \to \text{mod }\Gamma$ be the usual evaluation functor. Now we have already seen that $\coprod_{i \in I} f_i : \coprod_{i \in I} A_i \to \coprod_{i \in I} B_i$ is right minimal if and only if $e_A(\coprod_{i \in I} f_i) : e_A(\coprod_{i \in I} A_i) \to \text{Im } e_A(\coprod_{i \in I} f_i)$ is a projective cover. But we know that $e_A(\coprod_{i \in I} f_i)$ is a projective cover if and only if each $e_A(f_i)$ is a projective cover by I Proposition 4.3. Therefore each $f_i: A_i \to B_i$ is right minimal if and only if $\coprod_{i \in I} f_i: \coprod_{i \in I} A_i \to \coprod_{i \in I} B_i$ is right minimal. $\square$

In many instances it is technically easier to deal with modules over artin algebras Λ which have the property that if $\Lambda = \coprod_{i=1}^{n} P_i$ with the P_i indecomposable projective modules, then $P_i \not\simeq P_j$ for $i \neq j$. Such artin algebras are called **basic** artin algebras.

Suppose now that Λ is an arbitrary artin algebra and $\Lambda = \coprod_{i=1}^{m} n_i P_i$ with the $n_i > 0$ and the P_i nonisomorphic indecomposable projective modules. Let $P = \coprod_{i=1}^{m} P_i$ and let $\Gamma = \operatorname{End}_{\Lambda}(P)^{\mathrm{op}}$. Consider the evaluation functor $e_P: \operatorname{mod} \Lambda \to \operatorname{mod} \Gamma$. Then $e_P(P) = \Gamma$ and the decomposition $P = \coprod_{i=1}^{m} P_i$ with nonisomorphic indecomposable P_i gives the decomposition $\Gamma = \coprod_{i=1}^{m} e_P(P_i)$ where the $e_P(P_i)$ are nonisomorphic indecomposable projective Γ-modules. Thus Γ is a basic artin algebra which is called the **reduced form of** Λ. Our next aim is to show that $e_P: \operatorname{mod} \Lambda \to \operatorname{mod} \Gamma$ is an equivalence of categories, which means that Λ and its reduced form Γ have essentially the same module theory. This result will be a trivial consequence of a more general result which will have other applications later on. It is convenient to first give some definitions.

Let X be a Λ-module. A **projective presentation** for X is an exact sequence $P_1 \xrightarrow{f_1} P_0 \xrightarrow{f_0} X \to 0$ with the P_i projective modules for $i = 0, 1$. A projective presentation $P_1 \xrightarrow{f_1} P_0 \xrightarrow{f_0} X \to 0$ for X is called a **minimal projective presentation** if $f_0: P_0 \to X$ and $f_1: P_1 \to \operatorname{Ker} f_0$ are projective covers.

Suppose P is a projective module. Then we denote by $\operatorname{mod} P$ the full subcategory of $\operatorname{mod} \Lambda$ whose objects are those A in $\operatorname{mod} \Lambda$ which have projective presentations $P_1 \to P_0 \to A \to 0$ with the P_i in $\operatorname{add} P$ for $i = 0, 1$.

Proposition 2.5 *Let P be a projective Λ-module and let $\Gamma = \operatorname{End}_{\Lambda}(P)^{\mathrm{op}}$. Then the restriction $e_P|_{\operatorname{mod} P}: \operatorname{mod} P \to \operatorname{mod} \Gamma$ of the evaluation functor $e_P: \operatorname{mod} \Lambda \to \operatorname{mod} \Gamma$ is an equivalence of categories.*

Proof We first show that $e_P|_{\operatorname{mod} P}$ is dense. Suppose X is a Γ-module and $Q_1 \xrightarrow{f} Q_0 \to X \to 0$ a projective Γ-presentation of X. Then there is some P_i in $\operatorname{add} P$ with $e_P(P_i) \simeq Q_i$ for $i = 0, 1$. Hence there is a morphism

$P_1 \xrightarrow{g} P_0$ in add P such that the diagram

$$
\begin{array}{ccc}
e_P(P_1) & \xrightarrow{e_P(g)} & e_P(P_0) \\
\wr\wr & & \wr\wr \\
Q_1 & \xrightarrow{f} & Q_0
\end{array}
$$

commutes. Therefore $e_P(\text{Coker } g) \simeq \text{Coker } f = X$ since P being projective implies that e_P is exact. This shows that $e_P|_{\text{mod } P}$ is dense.

We now want to show that $e_P|_{\text{mod } P}$ is full and faithful. Let A and B be in mod P and let $P_1 \to P_0 \to A \to 0$ be a projective presentation for A. Then we have the following commutative exact diagram.

$$
\begin{array}{ccccccc}
0 & \to & \text{Hom}_\Lambda(A,B) & \to & \text{Hom}_\Lambda(P_0,B) & \to & \text{Hom}_\Lambda(P_1,B) \\
& & \downarrow & & \downarrow & & \downarrow \\
0 & \to & \text{Hom}_\Gamma(e_P(A),e_P(B)) & \to & \text{Hom}_\Gamma(e_P(P_0),e_P(B)) & \to & \text{Hom}_\Gamma(e_P(P_1),e_P(B))
\end{array}
$$

Since the P_i are in add P, we know that the second and third vertical morphisms are isomorphisms. Hence $e_P : \text{Hom}_\Lambda(A, B) \to \text{Hom}_\Gamma(e_P(A), e_P(B))$ is an isomorphism for all A and B in mod P. Therefore $e_P|_{\text{mod } P}$ is full and faithful and hence an equivalence of categories. $\square$

As an immediate consequence of this result we have our desired result about the reduced form Γ of an artin algebra Λ.

Corollary 2.6 *Suppose* $\{P_1, \ldots, P_n\}$ *is a complete set of nonisomorphic indecomposable projective* Λ-*modules. Let* $P = \coprod_{i=1}^n P_i$ *and let* $\Gamma = \text{End}_\Lambda(P)^{\text{op}}$. *Then* $e_P : \text{mod } \Lambda \to \text{mod } \Gamma$ *is an equivalence of categories.*

Proof Clearly add $P = \mathscr{P}(\Lambda)$, so mod $P = \text{mod } \Lambda$. Therefore by our previous result, $e_P : \text{mod } \Lambda \to \text{mod } \Gamma$ is an equivalence of categories. $\square$

Finally it is not difficult to see that an artin algebra Λ is basic if and only if Λ^{op} is basic. Using I Theorem 4.4 (b) it is not hard to obtain the following result.

Proposition 2.7 *The following statements are equivalent.*

(a) Λ *is basic.*
(b) $\Lambda/\mathfrak{r}$ *is basic.*
(c) $\Lambda/\mathfrak{r} \simeq \coprod_{i=1}^k D_i$ *where the* D_i *are division rings.*
(d) Λ^{op} *is basic.* $\square$

3 Duality

Another important reason for restricting our attention to artin algebras is the existence of a duality between $\mathrm{mod}\,\Lambda$ and $\mathrm{mod}(\Lambda^{\mathrm{op}})$. We begin by describing a duality $D:\mathrm{mod}\,R \to \mathrm{mod}\,R$ for a commutative artin ring R.

Recall that for an arbitrary ring Λ and arbitrary modules A and B with $B \subset A$ we say that A is an **essential extension** of B if $X \cap B \neq 0$ for all submodules $X \neq 0$ of A. A monomorphism $i:A \to I$ with I injective is called an **injective envelope** of A if I is an essential extension of $i(A)$. Also recall that for an arbitrary ring Λ and any module A in $\mathrm{Mod}\,\Lambda$ there exists an injective envelope $i:A \to I$ in $\mathrm{Mod}\,\Lambda$ unique up to isomorphism. Let R be a commutative artin ring. Since R is an artin ring, it has only a finite number of nonisomorphic simple modules $S_1, \ldots, S_n$. Let $I(S_i)$ be the injective envelope of S_i and let $J = \coprod_{i=1}^{n} I(S_i)$. Then J is the injective envelope of $\coprod_{i=1}^{n} S_i$ and for each A in $\mathrm{mod}\,R$ there is a monomorphism $A \to J'$ with J' in $\mathrm{add}\,J$. We have $S_i \simeq R/\mathfrak{m}_i$ where $\mathfrak{m}_i$ is a maximal ideal of R, and we get R-isomorphisms $\mathrm{Hom}_R(S_i, S_i) \simeq \mathrm{Hom}_R(R/\mathfrak{m}_i, S_i) \simeq \mathrm{Hom}_R(R, S_i) \simeq S_i$. For $i \neq j$ we have $\mathrm{Hom}_R(S_i, S_j) = 0$. As a consequence of these observations, using the notation for the multiplicities of the simple composition factors from Chapter I, we have the following.

Theorem 3.1 *Let X be in $\mathrm{mod}\,R$. Then we have the following.*

(a) $\mathrm{Hom}_R(X, J)$ *is of finite length and* $m_{S_i}(\mathrm{Hom}_R(X, J)) = m_{S_i}(X)$ *for* $i = 1, \ldots, n$.

(b) *The natural morphism* $X \to \mathrm{Hom}_R(\mathrm{Hom}_R(X, J), J)$ *is an isomorphism.*

(c) *The contravariant R-functor $D:\mathrm{mod}\,R \to \mathrm{mod}\,R$ defined by $D = \mathrm{Hom}_R(\ , J)$ is a duality.*

Proof (a) We prove this by induction on $l(X)$. If $X = 0$ there is nothing to prove. Suppose $l(X) = 1$. Then $X \simeq S_j$ for some j. But $\mathrm{Hom}_R(S_j, J) \simeq \mathrm{Hom}_R(S_j, \coprod_{i=1}^{n} S_i) \simeq \mathrm{Hom}_R(S_j, S_j) \simeq S_j$. So $m_{S_i}(X) = m_{S_i}(\mathrm{Hom}_R(X, J))$. Now suppose that $l(X) > 1$ and let $0 \to X' \to X \to X'' \to 0$ be exact with $X'' \simeq S_j$ for some j. Then $0 \to \mathrm{Hom}_R(X'', J) \to \mathrm{Hom}_R(X, J) \to \mathrm{Hom}_R(X', J) \to 0$ is exact. Since $l(X') < l(X)$ and $l(X'') < l(X)$, the result now follows from the induction hypothesis.

(b) Since by (a) we have that X and $\mathrm{Hom}_R(\mathrm{Hom}_R(X, J), J)$ have the same length, it suffices to show that $\phi:X \to \mathrm{Hom}_R(\mathrm{Hom}_R(X, J), J)$ is a monomorphism. Because $\phi(x)(f) = f(x)$ for all f in $\mathrm{Hom}_R(X, J)$ and x in X, we have $\phi(x) \neq 0$ if there is an $f:X \to J$ such that $f(x) \neq 0$. Suppose x is a nonzero element of X and let Rx be the submodule of

X generated by x. But then $Rx/(\text{rad } R)Rx \neq 0$ by Nakayama's lemma and so there is a nonzero morphism $Rx/(\text{rad } R)Rx \to \coprod_{i=1}^{n} S_i$. Hence the induced map $g: Rx \to J$ is not zero, in particular $g(x) \neq 0$. Since J is injective, we can extend g to a map $f: X \to J$ which does not vanish on x. It follows that $\phi: X \to \text{Hom}_R(\text{Hom}_R(X, J), J)$ is an isomorphism.

(c) This is a direct consequence of (b) and the definition of duality. $\square$

As an immediate consequence of this we have the following facts about J, by choosing X to be R in Theorem 3.1.

Corollary 3.2
(a) *J and R have the same length, and $m_{S_i}(J) = m_{S_i}(R)$ for all i.*
(b) *$R \simeq \text{End}_R(J)$.* $\square$

We now want to show that the duality $D: \text{mod } R \to \text{mod } R$ for a commutative artin ring R induces a duality $D: \text{mod } \Lambda \to \text{mod}(\Lambda^{\text{op}})$ for an artin R-algebra Λ. To see this we first show that for X in $\text{mod } \Lambda$ the R-module $\text{Hom}_R(X, J)$ has the following natural structure as a Λ^{op}-module. For f in $\text{Hom}_R(X, J)$ and λ in Λ we define $(f\lambda)(x) = f(\lambda x)$ for all $x \in X$. This is the way we will consider DX as a Λ^{op}-module for each Λ-module X. It is clear that $\text{Hom}_R(X, J)$ is a finitely generated Λ^{op}-module since it is a finitely generated R-module by Proposition 1.1. If $f: X \to Y$ is a morphism in $\text{mod } \Lambda$, it is easy to see that $\text{Hom}_R(f, J): \text{Hom}_R(Y, J) \to \text{Hom}_R(X, J)$ is a morphism in $\text{mod}(\Lambda^{\text{op}})$. This shows that D gives a contravariant R-functor $D: \text{mod } \Lambda \to \text{mod}(\Lambda^{\text{op}})$. In fact we have the following.

Theorem 3.3 *If Λ is an artin R-algebra, then the contravariant R-functor $D: \text{mod } \Lambda \to \text{mod}(\Lambda^{\text{op}})$ is a duality.*

Proof If X is in $\text{mod } \Lambda$, then we have seen that $\text{Hom}_R(X, J)$ is in $\text{mod}(\Lambda^{\text{op}})$, and similarly $\text{Hom}_R(\text{Hom}_R(X, J), J)$ is in $\text{mod } \Lambda$. The natural R-isomorphism $\phi_X: X \to \text{Hom}_R(\text{Hom}_R(X, J), J)$ is then also a Λ-isomorphism. For if λ is in Λ and x is in X, we have $\phi(\lambda x)(f) = f(\lambda x) = (f\lambda)(x) = \phi(x)(f\lambda) = (\lambda\phi(x))(f)$ for all f in $\text{Hom}_R(X, J)$, so that $\phi(\lambda x) = \lambda\phi(x)$. Hence we have an isomorphism $\phi: 1_{\text{mod } \Lambda} \to D^2$, and similarly an isomorphism $\phi: 1_{\text{mod}(\Lambda^{\text{op}})} \to D^2$. It follows that $D: \text{mod } \Lambda \to \text{mod}(\Lambda^{\text{op}})$ is a duality. $\square$

For any left artin ring Λ the category $\text{mod } \Lambda$ has **enough projectives**, that is, for each C in mod Λ there is an epimorphism $f: P \to C$ in mod Λ with P projective. The category mod Λ is said to have **enough injectives** if for any C in mod Λ there is a monomorphism $g: C \to I$ with I injective. For an artin ring the category mod Λ does not necessarily have enough injectives, but for artin algebras we have the following consequence of the above.

Corollary 3.4 *For an artin R-algebra Λ the category* mod Λ *has enough injectives.*

Proof For C in mod Λ we have an epimorphism $h: P \to D(C)$ in mod(Λ^{op}) with P projective. So we have a monomorphism $D(h): D^2(C) \to D(P)$, and $D(P)$ is injective by Proposition 1.5. Since $D^2(C) \simeq C$, this finishes the proof. $\square$

We end this section by showing the following alternative description of the duality D which we will use freely in the rest of the book.

Proposition 3.5 *The functors* $\mathrm{Hom}_R(\ ,J)$ *and* $\mathrm{Hom}_\Lambda(\ ,D\Lambda)$ *from* mod Λ *to* mod(Λ^{op}) *are isomorphic.*

Proof Considering the $_R\Lambda_\Lambda$ we have by using adjointness functorial isomorphisms $\mathrm{Hom}_R(M,J) \xrightarrow{\sim} \mathrm{Hom}_R(\Lambda \otimes_\Lambda M, J) \xrightarrow{\sim} \mathrm{Hom}_\Lambda(M, \mathrm{Hom}_R(\Lambda, J)) = \mathrm{Hom}_\Lambda(M, D\Lambda)$. $\square$

4 Structure of injective modules

Throughout this section Λ will always denote an artin algebra. We have seen that for an artin algebra Λ the category mod Λ has enough injectives. In this section we investigate the structure of the injective modules in mod Λ for an artin algebra Λ. In this connection we describe the injective envelope of a finitely generated Λ-module and introduce the notion of the socle of a module.

For A in mod Λ the **socle** of A, denoted by soc A, is defined to be the submodule of A generated by all semisimple submodules of A. We have the following basic properties of the socle of a module for an artin algebra Λ.

Proposition 4.1 *For A in* $\operatorname{mod}\Lambda$ *where Λ is an artin algebra we have the following.*

(a) $A = 0$ *if and only if* $\operatorname{soc}A = 0$.

(b) *A is an essential extension of* $\operatorname{soc}A$.

(c) *$A \to I$ is an injective envelope of A if and only if the induced morphism* $\operatorname{soc}A \to I$ *is an injective envelope.*

(d) *An injective module I in* $\operatorname{mod}\Lambda$ *is indecomposable if and only if* $\operatorname{soc}I$ *is simple.*

Proof (a) Clearly if $A = 0$ then $\operatorname{soc}A = 0$. Let A be a nonzero module. Since the radical r of Λ is nilpotent, there is a largest integer t such that $\mathrm{r}^t A \neq 0$. Since $\mathrm{r}^t A$ is semisimple, it follows that $\operatorname{soc}A \neq 0$.

(b), (c) and (d) are easily verified consequences of (a). □

It follows from the above proposition that one way to describe the injective envelope of a module A is to describe $\operatorname{soc}A$ and the injective envelopes of simple modules. To describe the injective envelopes of the simple modules it is first useful to investigate a duality between projective left and right modules.

Given a Λ-module A we consider the abelian group $\operatorname{Hom}_\Lambda(A,\Lambda)$ to be a Λ^{op}-module by the operation $(f\lambda)(a) = f(a)\lambda$ for all λ in Λ, for all f in $\operatorname{Hom}_\Lambda(A,\Lambda)$ and a in A. It is clear that if A is a finitely generated Λ-module, then $\operatorname{Hom}_\Lambda(A,\Lambda)$, which we denote by A^*, is a finitely generated Λ^{op}-module. Also it is easily checked that if $f:A \to B$ is a morphism of Λ-modules, then $\operatorname{Hom}_\Lambda(f,\Lambda):B^* \to A^*$ given by $\operatorname{Hom}_\Lambda(f,\Lambda)(g) = gf$ for all g in B^* is a morphism of Λ^{op}-modules, which we denote by f^*. It is clear that the contravariant functor $T = \operatorname{Hom}_\Lambda(\ ,\Lambda):\operatorname{mod}\Lambda \to \operatorname{mod}\Lambda^{\mathrm{op}}$ given by $A \mapsto A^*$ for all A in $\operatorname{mod}\Lambda$ and $f \mapsto f^*$ for all morphisms f in $\operatorname{mod}\Lambda$ is an R-functor.

As in the case of vector spaces we define $\phi_A:A \to A^{**}$ by $\phi_A(a)(f) = f(a)$ for all a in A and all f in A^*. This is easily seen to be a Λ-morphism which is functorial in A, i.e. we have a morphism of functors $\phi:1_{\operatorname{mod}\Lambda} \to T^2$.

We now want to consider what happens to the category $\mathscr{P}(\Lambda)$ of finitely generated projective Λ-modules under the functor T. We begin with the following easily verified result.

Lemma 4.2

(a) *The map $t:\Lambda \to \Lambda^*$ given by $t(\lambda)(x) = x\lambda$ for all λ in Λ and x in Λ is a Λ^{op}-isomorphism which we view as an identification.*

(b) $\phi_\Lambda : \Lambda \to \Lambda^{**}$ *as defined above is a Λ-isomorphism.* $\square$

This readily gives the following.

Proposition 4.3

(a) *Let Λ be an artin algebra. For each P in $\mathscr{P}(\Lambda)$ we have that P^* is in $\mathscr{P}(\Lambda^{op})$.*

(b) *For each P in $\mathscr{P}(\Lambda)$, the morphism $\phi_P : P \to P^{**}$ is an isomorphism which we will view as an identification.*

(c) *The functor $\mathrm{Hom}_\Lambda(\ , \Lambda)|_{\mathscr{P}(\Lambda)} : \mathscr{P}(\Lambda) \to \mathscr{P}(\Lambda^{op})$ is a duality.*

Proof For $P = \Lambda$ we have by Lemma 4.2 that P^* is in $\mathscr{P}(\Lambda^{op})$ and that $\phi_P : P \to P^{**}$ is an isomorphism. Since $\mathrm{Hom}_\Lambda(\ , \Lambda)$ is an R-functor, the same holds for $\phi_{n\Lambda}$ when n is a positive integer. For an arbitrary P in $\mathscr{P}(\Lambda)$ there is some Q in $\mathscr{P}(\Lambda)$ and some integer n such that $n\Lambda \simeq P \coprod Q$. Hence (a) and (b) follow by using the additivity of $\mathrm{Hom}_\Lambda(\ , \Lambda)$ again.

It follows from (b) that the morphism $\phi : 1_{\mathscr{P}(\Lambda)} \to \mathrm{Hom}_\Lambda(\ , \Lambda^{op})|_{\mathscr{P}(\Lambda^{op})} \mathrm{Hom}_\Lambda(\ , \Lambda)|_{\mathscr{P}(\Lambda)}$ is an isomorphism. Since also the morphism $\psi : 1_{\mathscr{P}(\Lambda^{op})} \to \mathrm{Hom}_\Lambda(\ , \Lambda)|_{\mathscr{P}(\Lambda)} \mathrm{Hom}_{\Lambda^{op}}(\ , \Lambda)|_{\mathscr{P}(\Lambda^{op})}$ is an isomorphism, it follows that $\mathrm{Hom}_\Lambda(\ , \Lambda)|_{\mathscr{P}(\Lambda)} : \mathscr{P}(\Lambda) \to \mathscr{P}(\Lambda^{op})$ is a duality. $\square$

We now give the final preliminary observation about finitely generated projective Λ-modules we need in studying finitely generated injective Λ-modules.

Proposition 4.4 *Let A and B be in $\mathrm{mod}\,\Lambda$, where Λ is an artin algebra.*

(a) *The morphism $\psi : \mathrm{Hom}_\Lambda(A, \Lambda) \otimes_\Lambda B \to \mathrm{Hom}_\Lambda(A, B)$, defined by $\psi(f \otimes b)(a) = f(a)b$ for all f in $\mathrm{Hom}_\Lambda(A, \Lambda)$, for all b in B and all a in A, is functorial in A and B.*

(b) *If A is in $\mathscr{P}(\Lambda)$, then ψ is an isomorphism for all B in $\mathrm{mod}\,\Lambda$.*

Proof The first part of the proposition is a routine calculation. Also it is easily seen that $\psi : \mathrm{Hom}_\Lambda(\Lambda, \Lambda) \otimes_\Lambda B \to \mathrm{Hom}_\Lambda(\Lambda, B)$ is an isomorphism for all B in $\mathrm{mod}\,\Lambda$. Since $\mathscr{P}(\Lambda) = \mathrm{add}\,\Lambda$, the rest follows from the functoriality of ψ in A. $\square$

We now make some observations concerning simple modules over semisimple rings which we then generalize to projective Λ-modules.

Lemma 4.5 *Let* $\Delta = \Delta_1 \times \cdots \times \Delta_n$ *be the unique decomposition of a semisimple ring* Δ *into a product of simple rings* Δ_i *for* $i = 1, \ldots, n$. *Let* $\mathfrak{m}_i = \prod_{j \neq i} \Delta_j$ *for all* $i = 1, \ldots, n$.

(a) *The* $\mathfrak{m}_i$ *are the unique maximal ideals in* Δ.
(b) *If* S *is a simple module, then* $\text{ann} \, S$, *the annihilator of* S, *is* $\mathfrak{m}_i$ *for some* i.
(c) *Two simple* Δ-*modules* S *and* T *are isomorphic if and only if* $\text{ann} \, S = \text{ann} \, T$.
(d) *If* S *is a simple* Δ-*module, then* $\text{Hom}_\Delta(S, \Delta)$ *is a simple* Δ^{op}-*module and* $\text{ann} \, S = \text{ann} \, \text{Hom}_\Delta(S, \Delta)$.

Proof We only discuss (d), and leave the rest as an exercise. Since Δ is semisimple, S is an indecomposable projective Δ-module, so $\text{Hom}_\Delta(S, \Delta)$ is indecomposable and hence simple. Let $\mathfrak{m}_i = \text{ann} \, S$. Then the epimorphism $\Delta \to \Delta_i$ gives an isomorphism $\text{Hom}_\Delta(S, \Delta) \simeq \text{Hom}_\Delta(S, \Delta_i)$ of Δ^{op}-modules, which shows that $\text{ann} \, \text{Hom}_\Delta(S, \Delta) = \mathfrak{m}_i$, finishing the proof. $\square$

We now have our main result concerning injective envelopes.

Proposition 4.6 *Let* A *be a semisimple module over the artin algebra* Λ *and let* $P \to A$ *be a projective cover for* A *in* $\text{mod} \, \Lambda$. *Then* $D(P^*)$ *is an injective envelope of* A.

Proof It clearly suffices to prove this in the case where A is a simple Λ-module and P its projective cover. Hence P is indecomposable and so P^* is also an indecomposable projective Λ^{op}-module. Now $P^*/\mathfrak{r}P^* \simeq P^* \otimes_\Lambda (\Lambda/\mathfrak{r}) = \text{Hom}_\Lambda(P, \Lambda/\mathfrak{r}) = \text{Hom}_{\Lambda/\mathfrak{r}}(P/\mathfrak{r}P, \Lambda/\mathfrak{r})$. Therefore, by Lemma 4.5, we have that $\text{ann}(P/\mathfrak{r}P) = \text{ann}(P^*/\mathfrak{r}P^*)$ and so $A \simeq D(P^*/\mathfrak{r}P^*)$. Since $P^* \to P^*/\mathfrak{r}P^*$ is a projective cover, it follows that $D(P^*/\mathfrak{r}P^*) \to D(P^*)$ is an injective envelope. $\square$

We now have a one to one correspondence between the nonisomorphic simple Λ-modules, the nonisomorphic indecomposable projective Λ-modules and the nonisomorphic indecomposable injective Λ-modules. More explicitly, if S is a simple Λ-module, we associate with S its projective cover P and its injective envelope $D(P^*)$.

Let A be in $\text{mod} \, \Lambda$ for an artin algebra Λ. For $j > 1$ we define by induction $\text{soc}^j A \subset A$ to be the preimage of $\text{soc}(A/\text{soc}^{j-1} A)$ in A. The smallest integer t with $\text{soc}^t A = A$ is called the **socle length** of A, denoted by sl(A), and $0 \subset \text{soc}\,A \subset \text{soc}^2 A \subset \cdots \subset \text{soc}^{t-1} A \subset A$ is the **socle series** of A.

The radical series defined in I Section 3 and the socle series do not necessarily define the same series of submodules of a module A. But using that $\text{r}^m A = 0$ if and only if $(DA)\text{r}^m = 0$ and that the duality D takes radical series to socle series and socle series to radical series, we have the following.

Proposition 4.7 *Let Λ be an artin algebra and let A be in $\text{mod}\,\Lambda$. Then we have* rl(A) = sl(A). $\qquad\qquad\qquad\qquad\qquad\qquad\qquad\qquad\qquad\quad$ □

5 Blocks

In this section Λ will be an artin algebra and we study the decomposition of Λ into a product of indecomposable algebras, which are called blocks. We shall see that the study of the module theory of Λ reduces to the module theory for the corresponding blocks.

We first consider the case of a commutative artin ring R, and we show that there is a close connection between ring summands and idempotents. Let $1 = e_1 + \cdots + e_n$ be a decomposition of 1 into a sum of primitive orthogonal idempotents. We first observe that $e_1, \ldots, e_n$ are then the only primitive idempotents in R. For let e be a primitive idempotent in R. Since $e = ee_1 + \cdots + ee_n$ and each ee_j is clearly an idempotent, there is some i with $ee_i \neq 0$ and $ee_j = 0$ for $i \neq j$ since e is primitive. We further have $e_i = (1 - e)e_i + ee_i$, which implies that $(1 - e)e_i = 0$ since e_i is primitive. This shows that $e_i = ee_i$ and hence $e = e_i$.

The decomposition $R = Re_1 \coprod \ldots \coprod Re_n$ of R as a sum of indecomposable projective modules gives in this case a decomposition of R into a product of indecomposable rings Re_i, with identity e_i. Such a ring decomposition clearly has to be given by a decomposition of 1 into a sum of primitive orthogonal idempotents, and must hence be unique. We also point out that $Re_i = e_iRe_i \simeq \text{End}_R(Re_i)$ is a local ring for all $i = 1, \ldots, n$, since Re_i is indecomposable.

Now let Λ be an artin R-algebra, where R is the center of Λ. We then get the following.

Proposition 5.1 *Let Λ be an artin algebra with center R, and $1 = e_1 + \cdots + e_n$ the decomposition of 1 into a sum of primitive orthogonal idempotents in R. Then $\Lambda = \Lambda e_1 \times \cdots \times \Lambda e_n$ is a decomposition of Λ into a finite product of* *indecomposable algebras* $\Lambda_i = \Lambda e_i$ *with identity* e_i, *and this decomposition is unique.*

Proof We first have a decomposition $\Lambda = \Lambda e_1 \coprod \cdots \coprod \Lambda e_n$ of left Λ-modules. But since the e_i are in the center of Λ, this is clearly an algebra decomposition and e_i is the identity of Λe_i. We want to show that Λe is an indecomposable algebra when e is a primitive idempotent of R. For if $\Lambda e = \Lambda_1 \times \Lambda_2$, then $e = f_1 + f_2$, where f_1 and f_2 are idempotents in Λ_1 and Λ_2 respectively, and they clearly lie in the center R of Λ. This shows that f_1 or f_2 must be zero, and hence Λe is an indecomposable algebra. Similarly any decomposition $\Lambda = \Lambda_1 \times \cdots \times \Lambda_m$ into a product of indecomposable algebras is given by a decomposition of 1 into a sum of primitive orthogonal idempotents in R, and is hence unique. $\square$

Let now $\Lambda = \Lambda_1 \times \cdots \times \Lambda_n$ be a product of indecomposable algebras and $1 = e_1 + \cdots + e_n$ the corresponding decomposition of 1. The Λ_i are called the **blocks** of Λ. We have natural ring maps $\Lambda \to \Lambda_i$ for each i, and in this way Λ_i has a natural structure as a Λ-module. If A is a Λ-module, then $A = e_1 A \coprod \cdots \coprod e_n A$, where $e_i A$ is a Λ_i-module and hence a Λ-module. If A is an indecomposable Λ-module there is a unique i such that $A = e_i A$ and $e_j A = 0$ for $j \neq i$. We then say that A belongs to the block Λ_i. If $B = e_1 B \coprod \cdots \coprod e_n B$ is another Λ-module, we have an isomorphism $\mathrm{Hom}_\Lambda(A, B) \simeq \prod_{i=1}^n \mathrm{Hom}_{\Lambda_i}(e_i A, e_i B)$, functorial in A and B, since clearly $\mathrm{Hom}_\Lambda(e_i A, e_j B) = 0$ for $i \neq j$. This shows that the study of $\mathrm{mod}\,\Lambda$ can be reduced to the study of $\mathrm{mod}\,\Lambda_i$ for $i = 1, \ldots, n$. Hence we can often without loss of generality assume that our algebras are indecomposable.

Since the decomposition of Λ into blocks gives a partition of the indecomposable modules according to the blocks to which they belong, we get in particular a partition of the indecomposable projective Λ-modules. We now show how to define this partition of the indecomposable projective Λ-modules without reference to the block decomposition of Λ.

Denote by $\mathscr{P}$ the set of indecomposable projective Λ-modules. We say that a partition $\mathscr{P} = \mathscr{P}_1 \cup \cdots \cup \mathscr{P}_n$ of the indecomposable projective Λ-modules is a **block partition** if

(a) $\mathrm{Hom}_\Lambda(P, Q) = 0$ when P and Q belong to different $\mathscr{P}_i$ and
(b) if P and Q are in $\mathscr{P}_i$, we have a chain

$$P = Q_1 - Q_2 - \cdots - Q_{n-1} - Q_n = Q$$

in $\mathscr{P}_i$, with nonzero maps from Q_i to Q_{i+1} or from Q_{i+1} to Q_i, for each $i = 1, \ldots, n-1$.

We leave the proof of the following result to the reader.

Proposition 5.2 *Let Λ be an artin algebra and $\mathscr{P} = \mathscr{P}_1 \cup \cdots \cup \mathscr{P}_n$ a block partition of the indecomposable projective modules.*

(a) *If we write $\Lambda = P_1 \coprod \cdots \coprod P_n$, where P_i is a sum of indecomposable modules from $\mathscr{P}_i$, then $\Lambda \simeq \operatorname{End}_\Lambda(\Lambda)^{\mathrm{op}} = \operatorname{End}_\Lambda(P_1)^{\mathrm{op}} \times \cdots \times \operatorname{End}_\Lambda(P_n)^{\mathrm{op}}$ gives the block decomposition for Λ.*

(b) *If P and Q are in $\mathscr{P}_i$ with $P \not\simeq Q$ there is a chain $P = Q_1 - Q_2 - \cdots - Q_{n-1} - Q_n = Q$ in $\mathscr{P}_i$ such that $\operatorname{Hom}_\Lambda(Q_i, \mathfrak{r}Q_{i+1}/\mathfrak{r}^2 Q_{i+1}) \neq 0$ or $\operatorname{Hom}_\Lambda(Q_{i+1}, \mathfrak{r}Q/\mathfrak{r}^2 Q_i) \neq 0$.*

Exercises

1. Let Λ be an artin algebra, P an indecomposable projective module, I an indecomposable injective module in $\operatorname{mod}\Lambda$ and M arbitrary in $\operatorname{mod}\Lambda$.

(a) Prove that $\operatorname{Hom}_\Lambda(P, M) \neq 0$ if and only if $P/\mathfrak{r}P$ is a composition factor of M.

(b) Show that the length of $\operatorname{Hom}_\Lambda(P, M)$ as an $\operatorname{End}_\Lambda(P)^{\mathrm{op}}$-module is the same as the multiplicity of $P/\mathfrak{r}P$ as a composition factor of M.

(c) Prove that $\operatorname{Hom}_\Lambda(M, I) \neq 0$ if and only if $\operatorname{soc} I$ is a composition factor of M.

(d) Show that the length of $\operatorname{Hom}_\Lambda(M, I)$ as an $\operatorname{End}_\Lambda(I)$-module is the same as the multiplicity of $\operatorname{soc} I$ as a composition factor of M.

2. Let P be a finitely generated projective Λ-module. Prove that $P \otimes_{\operatorname{End}(P)^{\mathrm{op}}} : \operatorname{mod}\operatorname{End}(P)^{\mathrm{op}} \to \operatorname{mod}\Lambda$ induces an equivalence from $\operatorname{mod}\operatorname{End}(P)^{\mathrm{op}}$ to $\operatorname{mod}P$ which is an inverse of $\operatorname{Hom}_\Lambda(P, \ )|_{\operatorname{mod}P} : \operatorname{mod}P \to \operatorname{mod}\operatorname{End}(P)^{\mathrm{op}}$. (Hint: Define $\phi_M : P \otimes_{\operatorname{End}(P)^{\mathrm{op}}} \operatorname{Hom}_\Lambda(P, M) \to M$ by $\phi_M(p \otimes f) = f(p)$ for M in $\operatorname{mod}\Lambda$, for p in P and $f \in \operatorname{Hom}_\Lambda(P, M)$, and define $\psi_X : X \to \operatorname{Hom}_\Lambda(P, P \otimes_{\operatorname{End}(P)^{\mathrm{op}}} X)$ for each X in $\operatorname{mod}\operatorname{End}(P)^{\mathrm{op}}$ by $\psi_X(x)(p) = p \otimes x$ for all $x \in X$ and $p \in P$).

3. Let $\mathbb{Z}$ be the integers and for each integer n let $\mathbb{Z}_n = \mathbb{Z}/(n)$.

(a) Show that $\mathbb{Z}_n$ is an injective $\mathbb{Z}_n$-module for all $n \neq 0$.

(b) Let A be a finite abelian group. Show that there is a monomorphism $\phi : \mathbb{Z}/(\operatorname{ann} A) \to A$ where $\operatorname{ann} A$ is the annihilator of A.

(c) Let ϕ be as in (b). Show that $\operatorname{ann} A \subset \operatorname{ann}(\operatorname{coker}\phi)$.

(d) Show that $A \simeq (\mathbb{Z}/(n_1)) \coprod (\mathbb{Z}/(n_2)) \coprod \cdots \coprod (\mathbb{Z}/(n_j))$ for integers $n_1, n_2, \ldots, n_j$ such that $n_{i+1} | n_i$ for $i = 1, \ldots, j - 1$ (rational canonical form of finite abelian groups).

(e) Let $n = p_1^{t_1} p_2^{t_2} \cdots p_m^{t_m}$ be the prime factorization of an integer $n > 0$. Show that $\mathbb{Z}/(n) \simeq (\mathbb{Z}/(p_1^{t_1})) \coprod \cdots \coprod (\mathbb{Z}/(p_m^{t_m}))$ (Chinese remainder theorem).

(f) Let A be a finite abelian group with $(n) = \operatorname{ann} A$ with prime decomposition $n = p_1^{t_1} p_2^{t_2} \cdots p_m^{t_m}$. Show that there are integers s_{ij}, with $A \simeq \coprod_{i=1}^{n} \coprod_{j=0}^{t_i - 1} s_{ij}(\mathbb{Z}/(p_i^{t_i - j}))$.

4. Let Λ be an artin algebra and S a simple Λ-module. Let e be a primitive idempotent in Λ.

(a) Show that there is a projective cover $\Lambda e \to S$ if and only if $eS \neq 0$.

(b) Show that $eS \neq 0 \Leftrightarrow (DS)e \neq 0$.

(c) Show that there is a projective cover $\Lambda e \to S$ if and only if there is a projective cover $e\Lambda \to DS$.

(d) Show that the morphism $\phi : \operatorname{Hom}_\Lambda(\Lambda e, \Lambda) \to e\Lambda$ given by $\phi(f) = f(e)$ is a Λ^{op}-isomorphism for each idempotent e in Λ.

5. Let k be a field and let $\Lambda = k[X_1, \ldots, X_n]$ be the polynomial ring in n indeterminates $X_1, \cdots, X_n$. Let $\mathscr{A}$ be the category of finitely generated graded Λ-modules, i.e. the objects M of $\mathscr{A}$ are the finitely generated Λ-modules M together with a decomposition of M as a k-vector space $M = \coprod_{i \in \mathbb{Z}} M_i$ where $X_j m_i \in M_{i+1}$ for $m_i \in M_i$ and $j = 1, \ldots, n$. We denote the objects by $\widetilde{M} = (M, \coprod_{i \in \mathbb{Z}} M_i)$.

The morphisms in $\mathscr{A}$ are the degree zero morphisms, i.e. if $\widetilde{M}$ and $\widetilde{N}$ are two objects in $\mathscr{A}$ then $\operatorname{Hom}_{\mathscr{A}}(\widetilde{M}, \widetilde{N})$ are the Λ-homomorphisms $f : M \to N$ such that $f(M_i) \subset N_i$.

(a) Let $\widetilde{M} = (M, \coprod_{i \in \mathbb{Z}} M_i)$ be in $\mathscr{A}$. Show that there exists $n \in \mathbb{Z}$ with $M_i = 0$ for $i \leq n$.

(b) Prove that $\mathscr{A}$ is a k-category in which $\dim_k(\operatorname{Hom}_{\mathscr{A}}(\widetilde{M}, \widetilde{N})) < \infty$ for each pair of objects $\widetilde{M}$ and $\widetilde{N}$.

(c) Prove that $\mathscr{A}$ is not equivalent to $\operatorname{mod} \Lambda$ for any artin algebra Λ.

6. Let Λ be any ring and M a Λ-module of finite length. Let $0 = M_0 \subset M_1 \subset M_2 \subset \cdots \subset M_n = M$ be a filtration such that M_{i+1}/M_i is semisimple for $i = 0, \ldots, n - 1$ and such that M_{i+1}/M_i' is not semisimple if M_i' is a proper submodule of M_i.

(a) Prove that for each i we have that $M_i \subset \mathrm{soc}^i M$ and consequently $n \geq \mathrm{sl}(M)$.

(b) Prove that for each i we have that $M_i \supset \mathrm{rad}^{n-i} M$ and consequently $n \geq \mathrm{rl}(M)$.

(c) Prove that $M_i \not\subset \mathrm{soc}^{i-1} M$ for any $i \geq 1$.

(d) Use (a) and (b) to prove $n = \mathrm{sl}(M) = \mathrm{rl}(M)$. (Note that this generalizes Proposition 4.7.)

7. Let Λ be a ring and let C be in $\mathrm{Mod}\,\Lambda$.

(a) Show that set of morphisms from f to f' in $\mathrm{Mod}\,\Lambda/C$ is not closed under sums so that $\mathrm{Mod}\,\Lambda/C$ is not a preadditive category.

(b) Let $f : X \to C$ and $g : Y \to C$ be in $\mathrm{Mod}\,\Lambda/C$.

   (a) Show that the pullback $X \times^C Y \to C$ is the product of f and g in the category $\mathrm{Mod}\,\Lambda/C$.

   (b) Show that $(f, g) : X \coprod Y \to C$ is the sum of f and g in $\mathrm{Mod}\,\Lambda/C$.

8. Let Λ be an artin algebra and $\mathscr{P}(\Lambda)$ the category of projective Λ-modules.

(a) Prove that for an indecomposable object P in $\mathscr{P}(\Lambda)$ we have that $P/\mathrm{r}P$ is isomorphic to a submodule of Λ as a left Λ-module if and only if for each epimorphism $f : Q \to P$ in $\mathscr{P}(\Lambda)$ there is some $g : P \to Q$ with $fg = 1_P$.

(b) Prove that if an indecomposable object P in $\mathscr{P}(\Lambda)$ is projective in $\mathscr{P}(\Lambda)$ then $P/\mathrm{r}P$ is isomorphic to a submodule of Λ.

(c) Prove that all indecomposable objects P in $\mathscr{P}(\Lambda)$ are projective in $\mathscr{P}(\Lambda)$ if and only if $P/\mathrm{r}P$ is isomorphic to a submodule of Λ for all indecomposable objects P in $\mathscr{P}(\Lambda)$.

(d) Prove that all objects in $\mathscr{P}(\Lambda)$ are injective as objects of $\mathscr{P}(\Lambda)$ if and only if for each simple right Λ-module S we have that Λ contains a right Λ-submodule isomorphic to S.

(e) Prove that $\mathrm{mod}\,\Lambda$ has no module of projective dimension 1 if and only if all objects of $\mathscr{P}(\Lambda)$ are injective as objects of $\mathscr{P}(\Lambda)$.

Notes

The basic theory of equivalences of categories can be found in for example [No]. Our approach follows [No] closely, formulated in terms of

R-categories and *R*-functors. Most of the standard material on rings and modules is found in [AnF]. Note that our proof of the Krull–Schmidt theorem is essentially the standard proof, but is organized differently since we first prove it for projective modules and then reduce the general case to the projective case. For a generalization to the case where there is an infinite number of summands with local endomorphism rings (Azumaya's Theorem) we refer to [AnF].

III

Examples of algebras and modules

The main object of study in this book is the finitely generated modules over artin algebras. A central role is played by the simple, projective and injective modules studied in the previous chapters. In this chapter we study some classes of algebras where the module categories have an alternative description which is sometimes easier to work with. The algebras we investigate are path algebras of quivers with or without relations, triangular matrix algebras, group algebras over a field and skew group algebras over artin algebras. These examples of algebras and their module categories are used to illustrate various concepts and results discussed in the first two chapters.

1 Quivers and their representations

In this section we introduce quivers and their representations over a field k. The notion of quiver and the associated path algebra come up naturally in the study of (not necessarily finite dimensional) tensor algebras of a bimodule over a semisimple k-algebra. The representations of a quiver with relations correspond to modules over a factor algebra of the associated path algebra. This way we get a concrete description of the modules in terms of vector spaces together with linear transformations. This is particularly effective in describing the simple, projective and injective modules. We show that any finite dimensional basic k-algebra is given by a quiver with relations when k is algebraically closed.

We start with the basic definitions. A **quiver** $\Gamma = (\Gamma_0, \Gamma_1)$ is an oriented graph, where Γ_0 is the set of vertices and Γ_1 the set of arrows between vertices. We assume in this section that Γ is a finite quiver, that is, Γ_0 and Γ_1 are both finite sets. We denote by $s: \Gamma_1 \to \Gamma_0$ and $e: \Gamma_1 \to \Gamma_0$ the maps where $s(\alpha) = i$ and $e(\alpha) = j$ when $\alpha: i \to j$ is an arrow from

49

the vertex i to the vertex j. A **path** in the quiver Γ is either an ordered sequence of arrows $p = \alpha_n \cdots \alpha_1$ with $e(\alpha_t) = s(\alpha_{t+1})$ for $1 \le t < n$, or the symbol e_i for $i \in \Gamma_0$. We call the paths e_i **trivial paths** and we define $s(e_i) = e(e_i) = i$. For a nontrivial path $p = \alpha_n \cdots \alpha_1$ we define $s(p) = s(\alpha_1)$ and $e(p) = e(\alpha_n)$. A nontrivial path p is said to be an **oriented cycle** if $s(p) = e(p)$.

For a field k, let $k\Gamma$ be the k-vector space with the paths of Γ as basis. To see that $k\Gamma$ has a natural structure as a k-algebra, we define a k-linear map $f : k\Gamma \to \mathrm{End}_k(k\Gamma)$ as follows. It is enough to define $f(p)$ for any path p in Γ, and it is sufficient to define $f(p)(q)$ for any path q in Γ, since the paths form a basis for $k\Gamma$ as k-vector space. We then define, for the trivial paths e_i

$$f(e_i)(q) = \begin{cases} q & \text{if } e(q) = i, \\ 0 & \text{otherwise,} \end{cases}$$

and for an arrow $\alpha \in \Gamma_1$,

$$f(\alpha)(q) = \begin{cases} \alpha q & \text{if } e(q) = s(\alpha) \text{ and } q \text{ is a nontrivial path,} \\ \alpha & \text{if } q = e_{s(\alpha)}, \\ 0 & \text{otherwise.} \end{cases}$$

If $p = \alpha_n \cdots \alpha_1$ is a nontrivial path in Γ, we define $f(p) = f(\alpha_n) \cdots f(\alpha_1)$. For an element $\sigma = \sum_{i=1}^t a_i p_i$ in $k\Gamma$, with $a_i \in k$ and p_i a path for $i = 1, \ldots, t$, we then have $f(\sigma) = \sum_{i=1}^t a_i f(p_i)$. We see that $f(\sum_{i \in \Gamma_0} e_i) = 1$. If $\sigma = \sum_{i=1}^t a_i p_i \ne 0$, then $f(\sigma)(\sum_{i \in \Gamma_0} e_i) = \sigma \ne 0$, so that $f(\sigma) \ne 0$. Hence f is an injective map, so that $f : k\Gamma \to \mathrm{Im} f$ is an isomorphism of k-vector spaces. Since $1 \in \mathrm{Im} f$ and clearly $f(p)f(p') = f(pp') \in \mathrm{Im} f$ for paths p and p' in Γ, it is easy to see that $\mathrm{Im} f$ is a k-subalgebra of $\mathrm{End}_k(k\Gamma)$. Hence $\mathrm{Im} f$ is a k-algebra, and there is induced via f a k-algebra structure on $k\Gamma$, by defining, for paths p and q in Γ,

$$(p)(q) = f^{-1}(f(p)f(q)) = \begin{cases} pq & \text{if } e(q) = s(p) \text{ and } p, q \text{ nontrivial,} \\ p & \text{if } q = e_{s(p)}, \\ q & \text{if } p = e_{e(q)}, \\ 0 & \text{otherwise.} \end{cases}$$

This k-algebra $k\Gamma$ is called the **path algebra** of Γ over k. Note that $\sum_{i \in \Gamma_0} e_i$ is the identity element of $k\Gamma$.

The following is an immediate consequence of the definition.

Proposition 1.1 *If k is a field and Γ a finite quiver, then $k\Gamma$ is a finite dimensional k-algebra if and only if Γ has no oriented cycles.* $\square$

We illustrate with some examples.

Example Let k be a field and Γ the quiver $\underset{1}{\cdot} \overset{\alpha}{\to} \underset{2}{\cdot} \overset{\beta}{\to} \underset{3}{\cdot}$. Then $\{e_1, e_2, e_3, \alpha,$ $\beta, \beta\alpha\}$ is a basis for the path algebra $k\Gamma$ over k.

As illustration of multiplication of paths we have $(e_1)(\alpha) = 0$, $(\alpha)(e_1) = \alpha$, $(\alpha)(\beta) = 0$ and $(\beta)(\alpha) = \beta\alpha$.

Example Let k be a field and Γ the quiver $1 \cdot \circlearrowright^{\alpha}$. Then $\{e_1, \alpha, \alpha^2, \ldots, \alpha^i, \ldots\}$ is a k-basis for $k\Gamma$, and $(\alpha^i)(\alpha^j) = \alpha^{i+j}$ for $i > 0$ and $j > 0$. Clearly $k\Gamma$ is isomorphic to the polynomial ring $k[X]$ in one variable X over k, by identifying α^i with X^i for $i > 0$.

Example Let k be a field and Γ the quiver $1 \overset{\alpha}{\underset{\gamma}{\rightleftarrows}} \overset{2}{\nwarrow}{}^{\beta} \cdot 3$. Write

$p_1 = \gamma\beta\alpha$, $p_2 = \alpha\gamma\beta$ and $p_3 = \beta\alpha\gamma$. Then $\{p_1^i, p_2^i, p_3^i, \alpha p_1^i, \beta p_2^i, \gamma p_3^i, \beta\alpha p_1^i, \gamma\beta p_2^i, \alpha\gamma p_3^i \mid i \geq 0\}$ is a k-basis for $k\Gamma$, where $p_i^0 = e_i$.

Example Let k be a field and Γ the quiver $2 \cdot \overset{\alpha}{\underset{\gamma}{\begin{smallmatrix}\nearrow\\\searrow\end{smallmatrix}}} \overset{1}{\underset{4}{\cdot}} \overset{\beta}{\underset{\delta}{\begin{smallmatrix}\searrow\\\nearrow\end{smallmatrix}}} \cdot 3$. Then

$\{e_1, e_2, e_3, e_4, \alpha, \beta, \gamma, \delta, \gamma\alpha, \delta\beta\}$ is a k-basis for $k\Gamma$.

We now show how path algebras of quivers over a field come up naturally in connection with tensor algebras. Associated with the pair $(\Sigma, {}_\Sigma V_\Sigma)$ where Σ is a ring and V a Σ-bimodule is the **tensor ring** $T(\Sigma, V)$. If we write the n-fold tensor product $V \otimes_\Sigma V \otimes \cdots \otimes_\Sigma V$ as V^n, then $T(\Sigma, V) = \Sigma \coprod V \coprod V^2 \coprod \cdots \coprod V^i \coprod \cdots$ as an abelian group. Writing $V^0 = \Sigma$, multiplication is induced by the natural Σ-bilinear maps $V^i \times V^j \to V^{i+j}$ for $i \geq 0$ and $j \geq 0$.

We shall use the following criterion for constructing ring morphisms from tensor rings to other rings.

Lemma 1.2 *Let Σ be a ring and V a Σ-bimodule. Let Λ be a ring and $f : \Sigma \coprod V \to \Lambda$ a map such that the following two conditions are satisfied.*

(i) $f|_\Sigma : \Sigma \to \Lambda$ *is a ring morphism.*

(ii)   *Viewing Λ as a Σ-bimodule via $f|_\Sigma : \Sigma \to \Lambda$ then $f|_V : V \to \Lambda$ is a Σ-bimodule map.*

Then there is a unique ring morphism $\tilde{f} : T(\Sigma, V) \to \Lambda$ such that $\tilde{f}|_{\Sigma \coprod V} = f$.

Proof Consider the map $\phi: V \times V \to \Lambda$ defined by $\phi(v_1, v_2) = f(v_1)f(v_2)$ for v_1 and v_2 in V. We have for $r \in \Sigma$ that $\phi(v_1 r, v_2) = f(v_1 r)f(v_2) = f(v_1)rf(v_2) = f(v_1)f(rv_2) = \phi(v_1, rv_2)$, using that $f|_V:$ $V \to \Lambda$ is a Σ-bimodule map. Hence there is a unique group morphism $f_2: V \otimes_\Sigma V \to \Lambda$ such that $f_2(v_1 \otimes v_2) = f(v_1)f(v_2)$. Considering $V \otimes_\Sigma V$ as a Σ-bimodule in the natural way, it is easy to see that $f_2: V \otimes_\Sigma V \to \Lambda$ is a Σ-bimodule map. By induction we get a unique Σ-bimodule map $f_n: V^n \to \Lambda$ such that $f_n(v_1 \otimes \cdots \otimes v_n) = f(v_1) \cdots f(v_n)$. Then we define $\tilde{f}: T(\Sigma, V) \to \Lambda$ by $\tilde{f}(\sum_{n=0}^\infty w_n) = \sum_{n=0}^\infty f_n(w_n)$ for $\sum_{n=0}^\infty w_n \in T(\Sigma, V)$ with $w_n \in V^n$. It is an easy calculation to check that $\tilde{f}$ is a ring morphism, and $\tilde{f}$ is clearly uniquely determined by f.   □

For an artin R-algebra $\phi: R \to \Sigma$ we only consider those Σ-bimodules V such that $rv = vr$ for all r in R and v in V. Then the ring morphism $\psi: R \to \Sigma \coprod V \coprod \cdots \coprod V^i \coprod \cdots$ given by $\psi(r) = (\phi(r), 0, \ldots, 0, \ldots)$ for all r in R has its image in the center of $T(\Sigma, V)$ because R acts centrally on Σ and V and hence on V^i for all $i = 0, 1, \ldots$, i.e. $rv = vr$ for all r in R and v in V^i for all $i \geq 0$. Thus the ring morphism $\psi: R \to T(\Sigma, V)$ makes $T(\Sigma, V)$ an R-algebra and this is the only way we consider $T(\Sigma, V)$ as an R-algebra. Further if in Lemma 1.2 the rings Σ and Λ are R-algebras and $f: \Sigma \coprod V \to \Lambda$ is such that $f|_\Sigma$ is a morphism of R-algebras, then the unique ring morphism $\tilde{f}: T(\Sigma, V) \to \Lambda$ such that $\tilde{f}|_{\Sigma \coprod V} = f$ is a morphism of R-algebras.

Before beginning our discussion of the connection between path algebras and tensor algebras, it is convenient to make some definitions.

Let k be a field. For each positive integer n we denote by $\prod_n(k)$ the k-algebra which as a ring is $k \times \cdots \times k$, the product of k with itself n times, and has the k-algebra structure given by the ring morphism $\phi: k \to \prod_n(k)$ where $\phi(x) = (x, \ldots, x)$ for all x in k. Let $\Sigma = \prod_n(k)$ and let V be a Σ-bimodule where k acts centrally, that is $av = va$ for $a \in k$ and $v \in V$, and assume that V is finite dimensional over k. Then the tensor ring $T(\Sigma, V)$ is a k-algebra, and we can associate with $T(\Sigma, V)$ a quiver $\Gamma = (\Gamma_0, \Gamma_1)$ in the following way. The set of vertices Γ_0 is $\{1, \ldots, n\}$. Let ϵ_i for $i = 1, \ldots, n$ be the idempotent of Σ with the ith coordinate equal to 1 and the other coordinates zero. Then $\epsilon_j V \epsilon_i$ is a k-subspace of V and there will be $\dim_k \epsilon_j V \epsilon_i$ arrows from i to j in Γ. The quiver $\Gamma = (\Gamma_0, \Gamma_1)$ constructed in this way is called the **quiver of** $T(\Sigma, V)$.

For a path algebra $k\Gamma$ denote by J the ideal in $k\Gamma$ generated by all the arrows in Γ. We then have the following connection between tensor algebras and path algebras.

Proposition 1.3 *Let k be a field, and $\Sigma = \prod_n(k)$. Let V be a Σ-bimodule where k acts centrally and which is finite dimensional over k. If Γ is the quiver of the tensor algebra $T(\Sigma, V)$, there is a k-algebra isomorphism $\phi: T(\Sigma, V) \to k\Gamma$ such that $\phi(\coprod_{j \geq t} V^j) = J^t$.*

Proof We have a k-algebra homomorphism $f: \Sigma \to k\Gamma$ with image $k\Gamma_0$ defined by $f(a_1, a_2, \ldots, a_n) = \sum_{i=1}^n a_i e_i$ for a_i in k and a k-isomorphism $f: V \to k\Gamma_1$ defined by giving a bijection between a chosen basis for each $\epsilon_j V \epsilon_i$ and the set of arrows from i to j. Then it is easy to see that $k\Gamma_1$ is a Σ-subbimodule of $k\Gamma$ and that $f: V \to k\Gamma_1$ is an isomorphism of Σ-bimodules when $k\Gamma$ is viewed as a Σ-bimodule via the isomorphism $f: \Sigma \to k\Gamma_0$. Hence there is by Lemma 1.2 a ring morphism $\tilde{f}: T(\Sigma, V) \to k\Gamma$ extending $f: \Sigma \coprod V \to k\Gamma$. It is clear that $\tilde{f}$ is a surjective k-algebra morphism with $\tilde{f}(\coprod_{j \geq t} V^j) = J^t$. We obtain a k-basis for V^t formed by elements $v_1 \otimes \cdots \otimes v_t$ where there is some path $\underset{i_t}{\cdot} \to \underset{i_{t-1}}{\cdot} \to \cdots \to \underset{i_1}{\cdot} \to \underset{i_0}{\cdot}$ in Γ such that v_j is amongst the chosen basis elements in $\epsilon_{i_{j-1}} V \epsilon_{i_j}$ for $j = 1, \ldots, t$. Then $\tilde{f}(v_1 \otimes \cdots \otimes v_t)$ is a path from i_t to i_0, and distinct basis elements are mapped to distinct paths. This shows that $\tilde{f}$ is injective, and hence $\phi = \tilde{f}$ is a k-algebra isomorphism with the desired properties. $\square$

We now investigate the indecomposable projective modules for a finite dimensional path algebra $k\Gamma$ over a field k. Since $k\Gamma$ is finite dimensional it follows that Γ has no oriented cycles by Proposition 1.1. It is then easy to see that the ideal J generated by the arrows is nilpotent. Clearly $k\Gamma/J \simeq ke_1 \times \cdots \times ke_n$ as k-algebras, where $\Gamma_0 = \{1, \ldots, n\}$ is the set of vertices of Γ. Since $k\Gamma/J$ is semisimple, it follows from I Proposition 3.3 that $J = \mathfrak{r}$, the radical of $\Lambda = k\Gamma$. Hence Λ is a basic k-algebra by II Proposition 2.7. Since $1 = e_1 + \cdots + e_n$ is a decomposition of 1 into a sum of orthogonal idempotents, we have $\Lambda = \Lambda e_1 \coprod \cdots \coprod \Lambda e_n$, where Λe_i is a projective module for $i = 1, \ldots, n$, which is indecomposable since $\Lambda e_i / \mathfrak{r} e_i$ is one-dimensional over k. Clearly the paths p with $s(p) = i$ constitute a k-basis for the indecomposable projective Λ-module Λe_i.

A vertex i in Γ is called a **sink** if there is no arrow α with $s(\alpha) = i$ and a **source** if there is no arrow α with $e(\alpha) = i$. Then Λe_i is simple if and only if i is a sink.

The nontrivial paths p with $s(p) = i$ are obviously a basis for re_i. Denote by $\alpha_1, \ldots, \alpha_t$ the arrows with $s(\alpha_j) = i$. Then any nontrivial path p with $s(p) = i$ is of the form $q\alpha_j$ for some j, where q is a path with $s(q) = e(\alpha_j)$. Hence we have $re_i = \coprod_{j=1}^t \Lambda e_{e(\alpha_j)} \alpha_j \simeq \coprod_{j=1}^t \Lambda e_{e(\alpha_j)}$, so we have proved that r is a projective Λ-module. Therefore we have the following by using I Corollary 5.2.

Proposition 1.4 *Let k be a field and Γ a quiver without oriented cycles. Then the finite dimensional k-algebra $k\Gamma$ is hereditary.*   □

Actually, one can show that $k\Gamma$ is (left) hereditary even if Γ has oriented cycles.

For a finite dimensional path algebra $k\Gamma$ with vertex set $\Gamma_0 = \{1, \ldots, n\}$ the indecomposable injective $k\Gamma$-modules are up to isomorphism of the form $I_i = \mathrm{Hom}_k(e_i(k\Gamma), k)$ for $i = 1, \ldots, n$. Clearly the paths p with $e(p) = i$ form a k-basis for the projective $(k\Gamma)^{op}$-module $e_i(k\Gamma)$. We take the dual basis consisting of elements p^* for paths p with $e(p) = i$ to get a k-basis for I_i. For such a basis element p^* and a path q in Γ we have $(qp^*)(u) = p^*(uq)$ when u is a path with $e(u) = i$. Hence we have

$$qp^* = \begin{cases} u^* & \text{if } p = uq, \\ 0 & \text{otherwise.} \end{cases}$$

We illustrate the description of the projective modules and their radicals and of the injective modules with some examples.

Example Let k be a field and Γ the quiver $\underset{1}{\cdot} \overset{\alpha}{\to} \underset{2}{\cdot} \overset{\beta}{\to} \underset{3}{\cdot}$. Then $(k\Gamma)e_1$ has k-basis $\{e_1, \alpha, \beta\alpha\}$, $(k\Gamma)e_2$ has k-basis $\{e_2, \beta\}$ and $(k\Gamma)e_3$ has k-basis $\{e_3\}$. We have $re_1 = (k\Gamma)\alpha = (k\Gamma)e_2\alpha \simeq (k\Gamma)e_2$ and $re_2 = (k\Gamma)\beta = (k\Gamma)e_3\beta \simeq (k\Gamma)e_3$ and $re_3 = 0$.

The projective $(k\Gamma)^{op}$-module $e_1(k\Gamma)$ has k-basis $\{e_1\}$, so that $\{e_1^*\}$ is a k-basis for the corresponding injective $k\Gamma$-module I_1. Further the injective $k\Gamma$-module I_2 has k-basis $\{e_2^*, \alpha^*\}$ and I_3 has k-basis $\{e_3^*, \beta^*, (\beta\alpha)^*\}$. The operation of Λ on I_3 is illustrated by $e_3\beta^* = 0$, $e_2\beta^* = \beta^*$, $\alpha(\beta\alpha)^* = \beta^*$ and $\beta\beta^* = e_3^*$.

Example Let k be a field and Γ the quiver $\begin{smallmatrix} & 1 & \\ \alpha \swarrow & & \searrow \beta \\ 2 & & 3 \\ \gamma \searrow & & \swarrow \delta \\ & 4 & \end{smallmatrix}$. Then $(k\Gamma)e_1$

has k-basis $\{e_1, \alpha, \beta, \gamma\alpha, \delta\beta\}$, $(k\Gamma)e_2$ has k-basis $\{e_2, \gamma\}$, $(k\Gamma)e_3$ has k-basis $\{e_3, \delta\}$ and $(k\Gamma)e_4$ has k-basis $\{e_4\}$.

We have $\mathfrak{r}e_1 \simeq (k\Gamma)e_2 \coprod (k\Gamma)e_3$ and $\mathfrak{r}e_2 \simeq (k\Gamma)e_4 \simeq \mathfrak{r}e_3$ and $\mathfrak{r}e_4 = 0$.

The injective $k\Gamma$-module I_1 has k-basis $\{e_1^*\}$, I_2 has k-basis $\{e_2^*, \alpha^*\}$, I_3 has k-basis $\{e_3^*, \beta^*\}$ and I_4 has k-basis $\{e_4^*, \gamma^*, \delta^*, (\gamma\alpha)^*, (\delta\beta)^*\}$. The operation of Λ on I_4 is illustrated by $\alpha(\gamma\alpha)^* = \gamma^*$ and $\beta(\gamma\alpha)^* = 0$.

We next want to study the connection between modules over path algebras and representations of the quivers.

Let C be a module over the path algebra $k\Gamma$. Then $C = \coprod_{i \in \Gamma_0} e_i C$ gives a decomposition of C into a finite sum of vector spaces over k. And if α is an arrow from i to j, then left multiplication by α induces a k-linear map from $e_i C$ to $e_j C$. This motivates the following definition, which gives us a concrete way of viewing the modules over path algebras.

A **representation** (V, f) of a quiver Γ over a field k is a set of vector spaces $\{V(i) | i \in \Gamma_0\}$ together with k-linear maps $f_\alpha : V(i) \to V(j)$ for each arrow $\alpha : i \to j$. We here assume that the representations are finite dimensional, that is, each $V(i)$ has finite dimension over k.

A morphism $h : (V, f) \to (V', f')$ between two representations of Γ over k is a collection $\{h_i : V(i) \to V'(i)\}_{i \in \Gamma_0}$ of k-linear maps such that for each arrow $\alpha : i \to j$ in Γ the diagram

$$\begin{array}{ccc} V(i) & \overset{h_i}{\to} & V'(i) \\ \downarrow f_\alpha & & \downarrow f'_\alpha \\ V(j) & \overset{h_j}{\to} & V'(j) \end{array}$$

commutes. If $h : (V, f) \to (V', f')$ and $g : (V', f') \to (V'', f'')$ are two morphisms between representations then the composition gh is defined to be the collection of maps $\{g_i h_i : V(i) \to V''(i)\}$. In this way we get **the category of** (finite dimensional) **representations** of Γ over k, which we denote by $\text{Rep}\,\Gamma$. We introduce some basic terminology in $\text{Rep}\,\Gamma$.

We say that an object (V, f) is a **subobject** of an object (V', f') in $\text{Rep}\,\Gamma$ if $V(i) \subset V'(i)$ for all $i \in \Gamma_0$ and $f_\alpha = f'_\alpha |_{V(i)}$ for each arrow $\alpha : i \to j$. For a morphism $h : (V, f) \to (V', f')$ we define the kernel $\text{Ker}\,h$ to be the subobject (W, f'') of (V, f) defined by $W(i) = \text{Ker}\,h_i$ for $i \in \Gamma_0$ and $f''_\alpha = f_\alpha |_{W(i)}$ for each arrow $\alpha : i \to j$. We here use that $f_\alpha(\text{Ker}\,h_i) \subset \text{Ker}\,h_j$ for each arrow $\alpha : i \to j$, which is easy to see. Further we define the image

Im h of h to be the subobject (U, g) of (V', f') defined by $U(i) = \text{Im } h_i$ and $g_\alpha = f'_\alpha|_{\text{Im } h_i}$ for each arrow $\alpha: i \to j$. The object (V, f) where $V(i) = 0$ for all $i \in \Gamma_0$ and $f_\alpha = 0$ for all $\alpha \in \Gamma_1$ is the zero object in Rep Γ. It is easy to see that a morphism $h:(V, f) \to (V', f')$ is a monomorphism in Rep Γ if and only if Ker $h = 0$ and that h is an epimorphism if and only if Im $h = (V', f')$. Clearly h is an isomorphism if and only if $h_i: V(i) \to V'(i)$ is an isomorphism for all $i \in \Gamma_0$. We say that a sequence of morphisms $(V, f) \xrightarrow{g} (V', f') \xrightarrow{h} (V'', f'')$ is **exact** if Im $g = $ Ker h. This is clearly the case if and only if the induced sequences $V(i) \xrightarrow{g_i} V'(i) \xrightarrow{h_i} V''(i)$ are exact for all $i \in \Gamma_0$.

A **sum** of two objects (V, f) and (V', f') in Rep Γ is the object (W, g) where $W(i) = V(i) \coprod V'(i)$ for each $i \in \Gamma_0$ and $g_\alpha = f_\alpha \coprod f'_\alpha$ for all $\alpha \in \Gamma_1$. An object (V, f) is said to be **indecomposable** if it is not isomorphic to the sum of two nonzero representations. An object (V, f) is **simple** if it has no proper nonzero subobjects. A simple object is clearly indecomposable. For each vertex $i \in \Gamma_0$ we have a simple object (S_i, f) given by $S_i(i) = k$ and $S_i(j) = 0$ for $j \neq i$, and by $f_\alpha = 0$ for each arrow $\alpha \in \Gamma_1$. We leave to the reader to check that Rep Γ is a k-category.

We illustrate with some examples.

Example Let k be a field and Γ the quiver $1 \cdot \overset{\alpha}{\underset{\beta}{\rightrightarrows}} \cdot 2$. Let V be the representation $k \overset{1}{\underset{a}{\rightrightarrows}} k$ and W the representation $k \overset{a^{-1}}{\underset{1}{\rightrightarrows}} k$, where $a \neq 0$ in k. Then we have an isomorphism $h: V \to W$ given by $h_1 = 1_k$ and $h_2: k \to k$ being multiplication by a^{-1}.

Example Let k be a field and Γ the quiver $\cdot \circlearrowleft^\alpha$. For each $\lambda \in k$ there is a simple representation $(S_\lambda, f^\lambda_\alpha)$ given by $S_\lambda(1) = k$ and $f^\lambda_\alpha: k \to k$ being multiplication by λ. It is easy to see that if $\lambda \neq \lambda'$ then $(S_\lambda, f^\lambda_\alpha)$ and $(S_{\lambda'}, f^{\lambda'}_\alpha)$ are not isomorphic in Rep Γ.

We now want to show that the categories Rep Γ and f.d.$(k\Gamma)$ are equivalent, where f.d.$(k\Gamma)$ denotes the **category of $k\Gamma$-modules of finite k-dimension**. We start by defining functors $F: \text{Rep } \Gamma \to \text{f.d.}(k\Gamma)$ and $H: \text{f.d.}(k\Gamma) \to \text{Rep } \Gamma$.

For (V, f) in Rep Γ we define $F(V, f)$ to be $\coprod_{i \in \Gamma_0} V(i)$ as a k-vector space. For each arrow $\alpha: i \to j$ we have a k-linear map $f_\alpha: V(i) \to V(j)$. If we denote by $\pi_i: F(V, f) \to V(i)$ the projection and by $\xi_i: V(i) \to F(V, f)$ the inclusion according to the decomposition of $F(V, f)$, there is induced a map $\bar{f}_\alpha = \xi_j f_\alpha \pi_i : F(V,f) \to F(V,f)$. For the trivial path e_i we have the induced map $\bar{f}_{e_i} = \xi_i f_{e_i} \pi_i : F(V,f) \to F(V,f)$ where $f_{e_i} = 1_{V(i)} : V(i) \to V(i)$. Then $\bar{f} : k\Gamma_0 \to \text{End}_k(F(V,f))$ is a k-algebra morphism and $\bar{f} : k\Gamma_1 \to \text{End}_k(F(V,f))$ a $k\Gamma_0$-bimodule morphism. Hence there is by Lemma 1.2 and Proposition 1.3 a unique k-algebra morphism $\tilde{f} : k\Gamma \to \text{End}_k(F(V,f))$ extending $\bar{f}$, so that $F(V,f)$ becomes a $k\Gamma$-module.

Let $h : (V,f) \to (V',f')$ be a morphism in $\text{Rep}\,\Gamma$. Then we have a k-linear map $h_i : V(i) \to V'(i)$ for each $i \in \Gamma_0$, and hence an induced map of vector spaces $\tilde{h} : F(V,f) \to F(V',f')$. For each arrow $\alpha : i \to j$ in Γ_1 we have $h_j f_\alpha = f'_\alpha h_i : V(i) \to V'(j)$. Hence we have $\widetilde{hf_\alpha} = \widetilde{f'_\alpha h}$, so that we get $\widetilde{hf_\sigma} = \widetilde{f'_\sigma h}$ for each $\sigma \in k\Gamma$. In other words, $\tilde{h}(\sigma v) = \sigma \tilde{h}(v)$ for $v \in F(V,f)$, so that $\tilde{h}$ is a $k\Gamma$-map. Then we define $F(h) = \tilde{h}$. It is now straightforward to see that $F : \text{Rep}\,\Gamma \to \text{f.d.}(k\Gamma)$ is a k-functor.

We next want to define a functor $H : \text{f.d.}(k\Gamma) \to \text{Rep}\,\Gamma$. Let C be in $\text{f.d.}(k\Gamma)$. Since $1 = e_1 + \cdots + e_n$ is a sum of orthogonal idempotents in $k\Gamma$, we get a sum of vector spaces $C = \coprod_{i \in \Gamma_0} e_i C$. For each $\sigma \in k\Gamma$ we have a map $\tilde{f}_\sigma : C \to C$ defined by $\tilde{f}_\sigma(c) = \sigma c$. If $\alpha : i \to j$ is an arrow in Γ we have $\alpha(e_i C) = e_j \alpha C \subset e_j C$, so that α induces by restriction a k-linear map $f_\alpha : e_i C \to e_j C$. We now define $H(C)$ to be the representation given by the k-vector spaces $e_i C$ for each $i \in \Gamma_0$ together with the maps $f_\alpha : e_i C \to e_j C$ for each arrow $\alpha : i \to j$.

If $h : B \to C$ is a morphism in $\text{f.d.}(k\Gamma)$, we have $h(e_i B) \subset e_i h(B) \subset e_i C$, so we get by restriction a k-linear map $h_i : e_i B \to e_i C$. For an arrow $\alpha : i \to j$ we have $\alpha h(b) = h(\alpha b)$ for $b \in B$, and hence $\alpha h_i(b) = h_j(\alpha b)$ for $b \in e_i B$. In other words we have $f'_\alpha h_i(b) = h_j f_\alpha(b)$, so by letting $H(h) = \{h_i\}$ we get that $H(h) : H(B) \to H(C)$ is a map in $\text{Rep}\,\Gamma$. It is straightforward to show that H is a functor from $\text{f.d.}\,k\Gamma$ to $\text{Rep}\,\Gamma$.

We can now prove the following.

Theorem 1.5 *Let k be a field and Γ a finite quiver. Then the functors $F : \text{Rep}\,\Gamma \to \text{f.d.}(k\Gamma)$ and $H : \text{f.d.}(k\Gamma) \to \text{Rep}\,\Gamma$ are inverse equivalences of k-categories.*

Proof Let (V,f) be in $\text{Rep}\,\Gamma$. Then $F(V,f) = \coprod_{i \in \Gamma_0} V(i)$ and $e_i F(V,f) = \tilde{f}_{e_i}(F(V,f)) = \xi_i(V(i))$. If $\alpha : i \to j$ is an arrow in Γ, the k-linear map $f_\alpha : V(i) \to V(j)$ induces the k-linear map $f_\alpha : F(V,f) \to F(V,f)$. The restriction of f_α to $\xi_i(V(i))$ gives a k-linear map $f'_\alpha : \xi_i(V(i)) \to \xi_j(V(j))$.

For each arrow $\alpha : i \to j$ we have the commutative diagram

$$
\begin{array}{ccc}
V(i) & \overset{\xi_i}{\to} & \xi_i(V(i)) \\
\downarrow f_\alpha & & \downarrow f'_\alpha \\
V(j) & \overset{\xi_i}{\to} & \xi_j(V(j)).
\end{array}
$$

Using that $HF(V, f)$ is the representation given by the collection $\{\xi_i(V(i))\}$, we get that $\xi = \{\xi_i\}$ gives an isomorphism $\xi : (V, f) \to HF(V, f)$. It is not hard to verify that ξ is an isomorphism of functors from $1_{\mathrm{Rep}\,\Gamma}$ to HF.

Let now B and C be in f.d.$(k\Gamma)$, and $f : B \to C$ a $k\Gamma$-map. Then we have the commutative diagram

$$
\begin{array}{ccc}
B & \overset{f}{\to} & C \\
\downarrow \wr & & \downarrow \wr \\
\coprod_{i \in \Gamma_0} e_i B & \overset{f_1 \amalg \cdots \amalg f_n}{\to} & \coprod_{i \in \Gamma_0} e_i C,
\end{array}
$$

where $f_i : e_i B \to e_i C$ are the restriction maps. From this it follows that we have an isomorphism of functors from $1_{\mathrm{f.d.}(k\Gamma)}$ to FH. $\square$

Since there are finite dimensional k-algebras which are not hereditary, not every basic finite dimensional k-algebra, even when k is algebraically closed, is isomorphic to the path algebra of a quiver. But as we shall see later on in this section, if Λ is a basic finite dimensional algebra over an algebraically closed field k, then Λ is a factor of a path algebra $k\Gamma$ for some finite quiver Γ. Therefore mod Λ is a full subcategory of f.d.$(k\Gamma)$, and is hence equivalent to some subcategory of Rep Γ. In order to describe this subcategory we are led to the following definitions.

A **relation** σ on a quiver Γ over a field k is a k-linear combination of paths $\sigma = a_1 p_1 + \cdots + a_n p_n$ with $a_i \in k$ and $e(p_1) = \cdots = e(p_n)$ and $s(p_1) = \cdots = s(p_n)$. We here assume that the length $l(p_i)$ of each p_i, that is the number of arrows in each path, is at least 2. If $\rho = \{\sigma_t\}_{t \in T}$ is a set of relations on Γ over k, the pair (Γ, ρ) is called a **quiver with relations** over k. Associated with (Γ, ρ) is the k-algebra $k(\Gamma, \rho) = k\Gamma/\langle \rho \rangle$, where $\langle \rho \rangle$ denotes the ideal in $k\Gamma$ generated by the set of relations ρ. We have by assumption $\langle \rho \rangle \subset J^2$, where J is the ideal of $k\Gamma$ generated by all the arrows in Γ.

We are mainly interested in the algebras $k(\Gamma, \rho)$ where $J^t \subset \langle \rho \rangle \subset J^2$ for some integer t. These are clearly finite dimensional, and we have the following description of the radical, where for an element x in $k\Gamma$ we denote by $\bar{x}$ the corresponding element in $k(\Gamma, \rho)$.

Proposition 1.6 *Let k be a field and (Γ, ρ) a quiver with relations over k. Assume that $J^t \subset \langle \rho \rangle \subset J^2$ for some t. Then the image $\bar{J}$ of J in $k(\Gamma, \rho)$ is* $\operatorname{rad} k(\Gamma, \rho)$.

Proof We have $\bar{J}^t = 0$ and $k(\Gamma, \rho)/\bar{J} \simeq \prod_n(k)$, where n is the number of vertices of Γ. In particular $k(\Gamma, \rho)/\bar{J}$ is semisimple. It then follows from I Proposition 3.3 that $\bar{J} = \operatorname{rad} k(\Gamma, \rho)$. $\qquad\qquad \square$

Note that $k(\Gamma, \rho)$ may be a finite dimensional algebra without $\bar{J}$ being the radical as we now show.

Example Let k be a field and Γ the quiver $\cdot \circlearrowleft^\alpha$ and $\rho = \{\alpha^3 - \alpha^2\}$. Then $k(\Gamma, \rho) = k\Gamma/\langle \alpha^3 - \alpha^2 \rangle \simeq k[X]/(X^3 - X^2) \simeq (k[X]/(X^2)) \times k$ and $\bar{J}$ has k-basis $\{\bar{\alpha}, \bar{\alpha}^2\}$. But $\bar{\alpha}$ is not in the radical of $k(\Gamma, \rho)$.

We now want to describe the indecomposable projective Λ-modules when $\Lambda = k(\Gamma, \rho)$ for a quiver with relations (Γ, ρ) such that $J^t \subset \langle \rho \rangle$ for some t. We have that $1 = \sum_{i \in \Gamma_0} \bar{e}_i$ is a decomposition of 1 into a sum of orthogonal idempotents, and $\Lambda = \coprod_{i \in \Gamma_0} \Lambda \bar{e}_i$. We see that $\Lambda \bar{e}_i$ is a projective cover of the simple module associated with the vertex i, and is hence indecomposable. Since the elements $\bar{p}$ in Λ where p is a path with $s(p) = i$ generate $\Lambda \bar{e}_i$ as k-vector space, we can always find a k-basis for $\Lambda \bar{e}_i$ consisting of elements $\bar{p}$ where p is a path with $s(p) = i$.

We illustrate with some examples.

Example Let k be a field and Γ the quiver $\underset{1}{\cdot} \xrightarrow{\alpha} \underset{2}{\cdot} \xrightarrow{\beta} \underset{3}{\cdot} \xrightarrow{\gamma} \underset{4}{\cdot}$. Let ρ be the set of relations $\{\beta\alpha, \gamma\beta\}$ for Γ over k. Then $\{\beta\alpha, \gamma\beta, \gamma\beta\alpha\}$ is a k-basis for the ideal $\langle \rho \rangle$, and $\{\bar{e}_1, \bar{e}_2, \bar{e}_3, \bar{e}_4, \bar{\alpha}, \bar{\beta}, \bar{\gamma}\}$ is a k-basis for the algebra $\Lambda = k(\Gamma, \rho)$. The module $\Lambda \bar{e}_1$ has k-basis $\{\bar{e}_1, \bar{\alpha}\}$, $\Lambda \bar{e}_2$ has k-basis $\{\bar{e}_2, \bar{\beta}\}$, $\Lambda \bar{e}_3$ has k-basis $\{\bar{e}_3, \bar{\gamma}\}$ and $\Lambda \bar{e}_4$ has k-basis $\{\bar{e}_4\}$.

Example Let k be a field and Γ the quiver

$$\begin{array}{ccc} & \underset{}{\overset{1}{\cdot}} & \\ {}^\alpha\swarrow & & \searrow^\beta \\ \underset{2}{\cdot} & & \underset{3}{\cdot} \\ {}_\gamma\searrow & & \swarrow_\delta \\ & \underset{4}{\cdot} & \end{array}$$

. Let ρ consist

of the relation $\gamma\alpha - \delta\beta$ on Γ over k. Then for $\Lambda = k(\Gamma, \rho)$ we see that $\Lambda \bar{e}_1$ has k-basis $\{\bar{e}_1, \bar{\alpha}, \bar{\beta}, \overline{\gamma\alpha}\}$. Further, $\Lambda \bar{e}_2$ has k-basis $\{\bar{e}_2, \bar{\gamma}\}$, $\Lambda \bar{e}_3$ has k-basis $\{\bar{e}_3, \bar{\delta}\}$ and $\Lambda \bar{e}_4$ has k-basis $\{\bar{e}_4\}$.

Example Let k be a field and Γ the quiver $1 \xleftarrow{\alpha} \overset{2}{\underset{\gamma}{\nearrow}} \xrightarrow{\beta} 3$ and $\rho =$ $\{\beta\alpha, \gamma\beta, \alpha\gamma\}$ a set of relations on Γ over k. Then $\Lambda = k(\Gamma, \rho)$ has k-basis $\{\bar{e}_1, \bar{e}_2, \bar{e}_3, \bar{\alpha}, \bar{\beta}, \bar{\gamma}\}$ and $\Lambda\bar{e}_1$ has k-basis $\{\bar{e}_1, \bar{\alpha}\}$, $\Lambda\bar{e}_2$ has k-basis $\{\bar{e}_2, \bar{\beta}\}$ and $\Lambda\bar{e}_3$ has k-basis $\{\bar{e}_3, \bar{\gamma}\}$.

For a quiver with relations (Γ, ρ) over a field k we define the category $\mathrm{Rep}(\Gamma, \rho)$ of representations to be the full subcategory of $\mathrm{Rep}\,\Gamma$ whose objects are the (V, f) with $f_\sigma = 0$ for each relation σ in ρ. Then we get the following.

Proposition 1.7 *Let k be a field and (Γ, ρ) a quiver with relations over k. Then the functor $F: \mathrm{Rep}\,\Gamma \to \mathrm{f.d.}(k\Gamma)$ induces an equivalence of k-categories between $\mathrm{Rep}(\Gamma, \rho)$ and $\mathrm{f.d.}(k(\Gamma, \rho))$.*

Proof If (V, f) is in $\mathrm{Rep}(\Gamma, \rho)$, then by definition the map f_σ is zero for all σ in ρ. Hence $\sigma F(V, f) = 0$, so that $F(V, f)$ is a $k(\Gamma, \rho)$-module.

If conversely $F(V, f)$ is a $k(\Gamma, \rho)$-module, then $\sigma F(V, f) = 0$ for all σ in ρ, so that $f_\sigma = 0$ for all σ in ρ. Hence (V, f) is in $\mathrm{Rep}(\Gamma, \rho)$. In view of Theorem 1.5, this finishes our claim. $\square$

Note that we have $\mathrm{f.d.}\,k(\Gamma, \rho) = \mathrm{mod}\,k(\Gamma, \rho)$ when $J^t \subset \langle \rho \rangle \subset J^2$ for some t. We then get the following consequence of F being an equivalence of categories by using II Proposition 1.3.

Proposition 1.8 *Let k be a field and (Γ, ρ) a quiver with relations over k such that $J^t \subset \langle \rho \rangle \subset J^2$ for some t, and let $F: \mathrm{Rep}(\Gamma, \rho) \to \mathrm{mod}\,k(\Gamma, \rho)$ be the above equivalence. Then we have the following.*

(a) *An object (V, f) in $\mathrm{Rep}(\Gamma, \rho)$ is projective (respectively injective, simple, indecomposable) if and only if $F(V, f)$ is projective (respectively injective, simple, indecomposable) in $\mathrm{mod}\,k(\Gamma, \rho)$.*

(b) *A sequence $(U, f) \to (V, g) \to (W, h)$ in $\mathrm{Rep}(\Gamma, \rho)$ is exact if and only if the induced sequence $F(U, f) \to F(V, g) \to F(W, h)$ is exact in $\mathrm{mod}\,k(\Gamma, \rho)$.*

Proof Since F is an equivalence of k-categories we know by II Proposition 1.3 that (V, f) is projective (respectively injective) in $\mathrm{Rep}(\Gamma, \rho)$ if and only if $F(V, f)$ is projective (respectively injective) in $\mathrm{mod}\,k(\Gamma, \rho)$. The rest is an easy consequence of the definitions. $\square$

Whenever it is convenient we shall view the equivalence between Rep(Γ, ρ) and mod $k(\Gamma, \rho)$ as an identification. It is useful to describe the projective, injective and simple objects in Rep(Γ, ρ) directly. We also interpret the ordinary duality D and the duality Hom$_\Lambda(\ , \Lambda)$ on projectives in the context of representations when $\Lambda = k(\Gamma, \rho)$. In addition we interpret the radical and the socle of a module in the category Rep(Γ, ρ), and also the Grothendieck group. For these interpretations we assume that $J^t \subset \langle \rho \rangle \subset J^2$ for some t.

We have considered simple objects S_i in Rep(Γ, ρ), for each $i \in \Gamma_0$, which are in one to one correspondence with Γ_0 and hence with the simple $k(\Gamma, \rho)$-modules. Therefore there are no other simple representations of (Γ, ρ).

Let $P_i = \Lambda \bar{e}_i$ be an indecomposable projective Λ-module where $\Lambda = k(\Gamma, \rho)$, and let $(V, f) = H(P_i)$ be the corresponding representation of (Γ, ρ). Then we have by definition that $V(j) = e_j P_i$. If we start with a k-basis for P_i consisting of elements $\bar{p}$ where p is a path in Γ with $s(p) = i$, we get a k-basis for $V(j)$ by picking the basis elements $\bar{p}$ where $e(p) = j$. If $\alpha : i' \to j'$ is an arrow, we get a map $f_\alpha : V(i') \to V(j')$ by defining $f_\alpha(\bar{p}) = \overline{\alpha p}$ when $e(p) = i'$.

For a quiver Γ with set of vertices $\{1, \ldots, n\}$ we denote by Γ^{op} the quiver having the same set of vertices. For each arrow $\alpha : i \to j$ in Γ there is an arrow $\alpha^{\mathrm{op}} : j \to i$ in Γ^{op}. For a path $p = \alpha_n \cdots \alpha_1$ in Γ let p^{op} be the path $\alpha_1^{\mathrm{op}} \cdots \alpha_n^{\mathrm{op}}$ in Γ^{op}. For a relation σ in Γ over k, denote by σ^{op} the induced relation in Γ^{op}. Then we have $k(\Gamma, \rho)^{\mathrm{op}} \simeq k(\Gamma^{\mathrm{op}}, \rho^{\mathrm{op}})$. For the finite dimensional algebra $\Lambda = k(\Gamma, \rho)$ we have an isomorphism $(\Lambda \bar{e}_i)^* = \mathrm{Hom}_\Lambda(\Lambda \bar{e}_i, \Lambda) \simeq \bar{e}_i \Lambda = \Lambda^{\mathrm{op}} \bar{e}_i^{\mathrm{op}}$ given by sending $g : \Lambda \bar{e}_i \to \Lambda$ to $g(\bar{e}_i)$.

When (V, f) is in Rep(Γ, ρ) we want to describe the representation of $(\Gamma^{\mathrm{op}}, \rho^{\mathrm{op}})$ which corresponds to the $k(\Gamma, \rho)^{\mathrm{op}}$-module $DF(V, f)$. We denote this representation by $(D(V), D(f))$. As a k-vector space $DF(V, f)$ is $D(\coprod_{i \in \Gamma_0} V(i)) = \mathrm{Hom}_k(\coprod_{i \in \Gamma_0} V(i), k)$. For the vertex i the vector space $D(V)(i)$ is $e_i^{\mathrm{op}} DF(V, f) = D(e_i F(V, f)) = D(V(i)) = \mathrm{Hom}_k(V(i), k)$, where we identify $e_i F(V, f)$ with $V(i)$. This describes the vector space of the dual representation as the dual spaces at each vertex. Next consider an arrow $\alpha : i \to j$ and the corresponding linear map $f_\alpha : V(i) \to V(j)$. Applying the duality with respect to the field we get a map $D(f_\alpha) : D(V(j)) \to D(V(i))$ given by $D(f_\alpha)(t)(v) = t(f_\alpha(v))$ for t in $D(V(j))$ and v in $V(i)$. Now let $D(f)_{\alpha^{\mathrm{op}}} = D(f_\alpha)$. It is straightforward to verify that if (V, f) is in Rep(Γ, ρ) we get that $(D(V), D(f))$ is in Rep$(\Gamma^{\mathrm{op}}, \rho^{\mathrm{op}})$ and that $DF(V, f) \simeq F(D(V), D(f))$.

We now get the following way of interpreting the duality for the simple, projective and injective representations. If (S_i, f) is the simple representation in $\text{Rep}(\Gamma, \rho)$ corresponding to the vertex i we know that $S_i(j) = 0$ for $j \neq i$ and $S_i(i)$ is a one-dimensional vector space and $f_\alpha = 0$ for all arrows α. Hence $(D(S_i), D(f))$ is obtained by taking the dual space $D(S_i(i))$ at the vertex i in Γ^{op} which is also one-dimensional and with all maps being zero.

Consider the projective representation (V, f) in $\text{Rep}(\Gamma, \rho)$ corresponding to the vertex i of Γ. We have vector spaces $V(i)$ for $i \in \Gamma_0$ and morphisms f_α for $\alpha \in \Gamma_1$. Then $(D(V), D(f))$ is obtained by $D(V)(i) = D(V(i))$ for $i \in \Gamma_0^{\text{op}}$ and $D(f)_{\alpha^{\text{op}}} = D(f_\alpha)$ for $\alpha^{\text{op}} \in \Gamma_1^{\text{op}}$. But we have that (V, f) is a projective representation if and only if $F(V, f)$ is a projective $k(\Gamma, \rho)$-module. It then follows that $DF(V, f)$ is an injective $k(\Gamma, \rho)^{\text{op}}$-module. But $k(\Gamma, \rho)^{\text{op}} = k(\Gamma^{\text{op}}, \rho^{\text{op}})$ and hence $(D(V), D(f))$ is an injective representation in $\text{Rep}(\Gamma^{\text{op}}, \rho^{\text{op}})$.

Conversely starting with the projective representations in $\text{Rep}(\Gamma^{\text{op}}, \rho^{\text{op}})$ and taking their duals we get the injective representations in $\text{Rep}(\Gamma, \rho)$. From linear algebra it is well known and easy to see that if $f : V \to W$ is a linear map between finite dimensional k-vector spaces represented by a matrix M relative to bases $\mathcal{B}$ and $\mathcal{C}$ in V and W respectively, then using the dual bases $\mathcal{B}^*$ of $D(V)$ and $\mathcal{C}^*$ of $D(W)$ respectively one gets that the matrix representing $D(f)$ with respect to these bases is the transpose of the matrix M. In this way one gets an easy way of describing the injective representations of (Γ, ρ) as we will show on examples.

Identifying modules with representations, the duality $(\ )^* : \mathscr{P}(\Lambda) \to \mathscr{P}(\Lambda^{\text{op}})$ between projective left and right Λ-modules induces a duality also denoted by $(\ )^*$ between the projective representations of (Γ, ρ) and of $(\Gamma^{\text{op}}, \rho^{\text{op}})$, when $J^t \subset \langle \rho \rangle \subset J^2$ for some t. If P_i is the indecomposable projective representation corresponding to a vertex i in Γ, then P_i^* is the projective representation of $(\Gamma^{\text{op}}, \rho^{\text{op}})$ corresponding to i.

If (V, f) is a representation of (Γ, ρ) we want to describe the subrepresentation $\mathfrak{r}(V, f) = (U, f')$ of (V, f) corresponding to the $k(\Gamma, \rho)$-module $\mathfrak{r}F(V, f)$ and the subrepresentation $\text{soc}(V, f) = (W, f'')$ of (V, f) corresponding to the $k(\Gamma, \rho)$-module $\text{soc}\, F(V, f)$. Since $\mathfrak{r}$ is generated by the arrows, we have $U(i) = \sum_{e(\alpha)=i} f_\alpha(V(s(\alpha)))$, the subspace generated by the images of the maps f_α to $V(i)$. For each arrow $\alpha : i \to j$ let f_α' be $f_\alpha|_{U(i)}$ which clearly is a k-linear map from $U(i)$ with image in $U(j)$. Further we have $W(i) = \bigcap_{s(\alpha)=i} \text{Ker}\, f_\alpha$, consisting of the elements which go to 0 by all maps f_α from $V(i)$. Clearly $f_\alpha(W(i)) = 0$ for each arrow $\alpha : i \to j$ so we let $f_\alpha'' = 0$.

We also note that the associated elements in the Grothendieck group have a simple description when we start with a representation (V, f) of (Γ, ρ). If S_i denotes the simple $k(\Gamma, \rho)$-module corresponding to the vertex i, then the associated element $[F(V, f)]$ of (V, f) in the Grothendieck group $K_0(\operatorname{mod} k(\Gamma, \rho))$ is $\sum_{i \in \Gamma_0} \dim_k V(i)[S_i]$.

We now illustrate the various interpretations on concrete examples.

Example Let k be a field and Γ the quiver $\underset{1}{\cdot} \overset{\alpha}{\to} \underset{2}{\cdot} \overset{\beta}{\to} \underset{3}{\cdot} \overset{\gamma}{\to} \underset{4}{\cdot}$. Then Γ^{op} is the quiver $\underset{1}{\cdot} \overset{\alpha^{\mathrm{op}}}{\leftarrow} \underset{2}{\cdot} \overset{\beta^{\mathrm{op}}}{\leftarrow} \underset{3}{\cdot} \overset{\gamma^{\mathrm{op}}}{\leftarrow} \underset{4}{\cdot}$. The projective $k\Gamma$-module $P_1 = \Lambda e_1$ which has the k-basis $\{e_1, \alpha, \beta\alpha, \gamma\beta\alpha\}$ corresponds to the representation $k \overset{1}{\to} k \overset{1}{\to} k \overset{1}{\to} k$ and P_1^* corresponds to the representation $k \leftarrow 0 \leftarrow 0 \leftarrow 0$ of Γ^{op}. The module $P_2 = \Lambda e_2$ has k-basis $\{e_2, \beta, \gamma\beta\}$ and corresponds to the representation $0 \to k \overset{1}{\to} k \overset{1}{\to} k$, whereas P_2^* corresponds to the representation $k \overset{1}{\leftarrow} k \leftarrow 0 \leftarrow 0$ of Γ^{op}. Then $I_1 = D(P_1^*)$ corresponds to the representation $k \to 0 \to 0 \to 0$ and $I_2 = D(P_2^*)$ to the representation $k \overset{1}{\to} k \to 0 \to 0$ of Γ. Note that we could here also use the explicit description of the injective $k\Gamma$-modules given earlier in this section.

If we consider the relation $\rho = \{\gamma\beta\}$ on Γ, then the representation corresponding to the projective $k(\Gamma, \rho)$-module $\Lambda \bar{e}_2$ with k-basis $\{\bar{e}_2, \bar{\beta}\}$ is $0 \to k \overset{1}{\to} k \to 0$. The representation corresponding to the projective $k(\Gamma, \rho)^{\mathrm{op}}$-module $(\Lambda \bar{e}_2)^*$ is $k \overset{1}{\leftarrow} k \leftarrow 0 \leftarrow 0$. The representation corresponding to the projective $k(\Gamma, \rho)$-module $\Lambda \bar{e}_4$ is $0 \to 0 \to 0 \to k$ and to the injective module I_4 is $0 \to 0 \to k \overset{1}{\to} k$.

Example Let k be a field and Γ the quiver $1 \cdot \underset{\beta}{\overset{\alpha}{\rightrightarrows}} \cdot 2$. Then Γ^{op} is the quiver $1 \cdot \underset{\beta^{\mathrm{op}}}{\overset{\alpha^{\mathrm{op}}}{\leftleftarrows}} \cdot 2$. Let (V, f) be the representation $k \underset{a}{\overset{1}{\rightrightarrows}} k$ of Γ, where $V(1)$ has k-basis $\{u\}$, $V(2)$ has k-basis $\{v\}$ and $f_\alpha(u) = v$ and $f_\beta(u) = av$, where $a \in k$. Let $\{u^*\}$ and $\{v^*\}$ be the dual basis for $DV(1) = V(1)^*$ and $DV(2) = V(2)^*$ respectively. If we write $D(V, f) = (DV, Df)$, then $(Df)_{\alpha^{\mathrm{op}}}(v^*)(u) = v^*(f_\alpha(u)) = v^*(v) = 1$, so that $(Df)_{\alpha^{\mathrm{op}}}(v^*) = u^*$ and consequently $(Df)_{\alpha^{\mathrm{op}}} = f_\alpha^*$. Further we have $(Df)_{\beta^{\mathrm{op}}}(v^*)(u) = v^*(f_\beta(u)) = v^*(av) = a$ so that $(Df)_{\beta^{\mathrm{op}}}(v^*) = au^*$. Hence $D(V, f)$ is the representation $k \underset{a}{\overset{1}{\leftleftarrows}} k$ of Γ^{op}.

The projective $k(\Gamma)$-module P_1 has k-basis $\{e_1, \alpha, \beta\}$ and corresponds to the representation $k \overset{f_\alpha}{\underset{f_\beta}{\rightrightarrows}} k \coprod k$ where $f_\alpha(b) = (b,0)$ and $f_\beta(b) = (0,b)$ for $b \in k$. The projective $k(\Gamma)$-module P_2 has k-basis $\{e_2\}$ and corresponds to the representation $0 \rightrightarrows k$. The projective $k\Gamma^{\mathrm{op}}$-module P_1^* corresponds to the representation $k \leftleftarrows 0$ of Γ^{op} and P_2^* to $k \coprod k \overset{g_{\alpha^{\mathrm{op}}}}{\underset{g_{\beta^{\mathrm{op}}}}{\leftleftarrows}} k$, where $g_{\alpha^{\mathrm{op}}}(c) = (c,0)$ and $g_{\beta^{\mathrm{op}}}(c) = (0,c)$ for $c \in k$. The injective $k(\Gamma)$-module $I_1 = DP_1^*$ corresponds to $k \rightrightarrows 0$ and $I_2 = DP_2^*$ to $k \coprod k \overset{f_\alpha}{\underset{f_\beta}{\rightrightarrows}} k$ where $f_\alpha(b,c) = b$ and $f_\beta(b,c) = c$ for b, c in k.

Let (V,f) be the representation $k \coprod k \coprod k \overset{f_\alpha}{\underset{f_\beta}{\rightrightarrows}} k \coprod k$ where $f_\alpha = \left(\begin{smallmatrix} 1 & 0 & 0 \\ 0 & 1 & 0 \end{smallmatrix}\right)$ and $f_\beta = \left(\begin{smallmatrix} 0 & 1 & 0 \\ 0 & 0 & 1 \end{smallmatrix}\right)$ relative to the standard bases for $k \coprod k \coprod k$ and $k \coprod k$. Then an easy calculation shows that $D(V,f) = (DV, Df)$ is the representation $k \coprod k \coprod k \overset{(Df)_{\alpha^{\mathrm{op}}}}{\underset{(Df)_{\beta^{\mathrm{op}}}}{\leftleftarrows}} k \coprod k$ when $D(nk)$ is identified with nk through dual bases, where $(Df)_{\alpha^{\mathrm{op}}} = \left(\begin{smallmatrix} 1 & 0 \\ 0 & 1 \\ 0 & 0 \end{smallmatrix}\right)$ and $(Df)_{\beta^{\mathrm{op}}} = \left(\begin{smallmatrix} 0 & 0 \\ 1 & 0 \\ 0 & 1 \end{smallmatrix}\right)$ relative to the standard bases. We see that the dual is given by the transposed matrices, when we are using dual bases.

Example Let Γ be the quiver

$$1 \cdot \overset{\alpha}{\longrightarrow} \overset{2}{\cdot} \overset{\beta}{\longrightarrow} \cdot 3$$
$$\downarrow \gamma$$
$$\overset{\cdot}{4}$$

and (V,f) the representation

$$k \overset{f_\alpha}{\rightarrow} k \amalg k \overset{f_\beta}{\rightarrow} k$$
$$\downarrow f_\gamma$$
$$k \quad ,$$

where $f_\alpha(a) = (a,0)$, $f_\beta(a,b) = b$ and $f_\gamma(a,b) = a - b$. If $\mathfrak{r}(V,f) = (U, f')$, then $U(1) = 0$ since no nontrivial path ends at 1, $U(2) = \mathrm{Im} f_\alpha = k \coprod 0$, $U(3) = k$, and $U(4) = k$. Hence the radical $\mathfrak{r}(V,f)$ is the representation

$$0 \rightarrow k \overset{g_\beta}{\rightarrow} k$$
$$\downarrow g_\gamma$$
$$k \quad ,$$

where clearly $g_\gamma = 1$ and $g_\beta = 0$.

We now want to compute $\mathrm{soc}(V,f) = (W, f'')$. Using that $\mathrm{soc}\, F(V,f) =$

$\{v \in \coprod_{i=1}^{n} V(i) | \mathfrak{r}v = 0\}$, we have $W(3) = k = W(4)$ since no arrows start at 3 or 4. $W(2) = \operatorname{Ker} f_\beta \cap \operatorname{Ker} f_\gamma = 0$ and $W(1) = \operatorname{Ker} f_\alpha = 0$.

The element $[F(V,f)]$ corresponding to (V,f) in the Grothendieck group is $[S_1] + 2[S_2] + [S_3] + [S_4]$, where S_i is the simple $k\Gamma$-module corresponding to the vertex i.

Our aim now is to show that every basic finite dimensional algebra over an algebraically closed field k is isomorphic to some $k(\Gamma, \rho)$. To this end it is convenient to have the following definitions. A finite dimensional algebra Λ over a field k is said to be **elementary** if $\Lambda/\mathfrak{r} \simeq \prod_n(k)$ for some n as k-algebras. When Λ is an elementary k-algebra we call the quiver of the tensor algebra $T(\Lambda/\mathfrak{r}, \mathfrak{r}/\mathfrak{r}^2)$ the **quiver of** Λ. We first show that every elementary k-algebra is isomorphic to some $k(\Gamma, \rho)$ and then show that every finite dimensional algebra over an algebraically closed field k is elementary, completing the proof of our desired result.

Theorem 1.9 *Let Λ be a finite dimensional elementary k-algebra.*

(a) *Let $\{e_1, \ldots, e_n\}$ be a complete set of primitive orthogonal idempotents in Λ, and $\{r_1, \ldots, r_t\}$ a set of elements in $\mathfrak{r}$ such that the images $\bar{r}_1, \ldots, \bar{r}_t$ in $\mathfrak{r}/\mathfrak{r}^2$ generate $\mathfrak{r}/\mathfrak{r}^2$ as a $\Lambda/\mathfrak{r}$-module. Then $\{e_1, \ldots, e_n, r_1, \ldots, r_t\}$ generate Λ as a k-algebra.*

(b) *There is a surjective ring homomorphism $\tilde{f}: T(\Lambda/\mathfrak{r}, \mathfrak{r}/\mathfrak{r}^2) \to \Lambda$ with $\coprod_{j \geq rl(\Lambda)} (\mathfrak{r}/\mathfrak{r}^2)^j \subset \operatorname{Ker} \tilde{f} \subset \coprod_{j \geq 2} (\mathfrak{r}/\mathfrak{r}^2)^j$.*

(c) *$\Lambda \simeq k(\Gamma, \rho)$ with $J^s \subset \langle \rho \rangle \subset J^2$ for some s, where Γ is the quiver of Λ and ρ is a set of relations of Γ over k.*

(d) *If $\Lambda \simeq k(\Gamma, \rho)$ with $J^t \subset \langle \rho \rangle \subset J^2$ for some t, then Γ is the quiver of Λ.*

Proof (a) We prove this by induction on the Loewy length $rl(\Lambda)$. For $rl(\Lambda) = 1$ we have $\Lambda \simeq \prod_n(k)$ and Λ is hence generated as a k-algebra by the idempotents e_i for $i = 1, \ldots, n$. For $rl(\Lambda) = 2$ the result is then easy by the assumption on $r_1, \ldots, r_t$. Assume now that the claim holds for $rl(\Lambda) = m \geq 2$ and assume that $rl(\Lambda) = m + 1$. Let A be the k-subalgebra of Λ generated by $e_1, \ldots, e_n, r_1, \ldots, r_t$, and let $x \in \Lambda$. By the induction assumption we have $A/(A \cap \mathfrak{r}^m) = \Lambda/\mathfrak{r}^m$. Hence there is some $y \in A$ such that $x - y$ is in $\mathfrak{r}^m$, and therefore there are α_i in $\mathfrak{r}^{m-1}$ and β_i in $\mathfrak{r}$ such that $x - y = \sum_{i=1}^{s} \alpha_i \beta_i$. Again by the induction assumption we have $\alpha_i = a_i + a_i'$ with $a_i \in A \cap \mathfrak{r}^m$ and $a_i' \in \mathfrak{r}^m$ and $\beta_i = b_i + b_i'$ with b_i in $A \cap \mathfrak{r}$ and b_i' in $\mathfrak{r}^m$. We then get $\alpha_i \beta_i = a_i b_i$ for all $i = 1, \ldots, s$, so that $x - y$ is in A. Since y is in A, we conclude that x is in A, and this finishes our claim.

(b) Let $e_1, \ldots, e_n$ be a complete set of primitive orthogonal idempotents in Λ, and denote by $\bar{e}_i$ the image of e_i in $\Lambda/\mathfrak{r}$. For each pair of integers i, j with $1 \leq i, j \leq n$, choose elements $\{y_s\}$ in $e_j \mathfrak{r} e_i$ such that if $\bar{y}_s$ denotes the image in $\mathfrak{r}/\mathfrak{r}^2$, then $\{\bar{y}_s\}$ is a k-basis for $\bar{e}_j (\mathfrak{r}/\mathfrak{r}^2) \bar{e}_i$. Define $f : (\Lambda/\mathfrak{r}) \coprod (\mathfrak{r}/\mathfrak{r}^2) \to \Lambda$ by $f(\bar{e}_i) = e_i$ for $i = 1, \ldots, n$ and $f(\bar{y}_s) = y_s$ for each chosen element y_s in $\mathfrak{r}$. Then $f|_{\Lambda/\mathfrak{r}} : \Lambda/\mathfrak{r} \to f(\Lambda/\mathfrak{r})$ is a ring isomorphism and $f|_{\mathfrak{r}/\mathfrak{r}^2} : \mathfrak{r}/\mathfrak{r}^2 \to f(\mathfrak{r}/\mathfrak{r}^2)$ is an isomorphism of $(\Lambda/\mathfrak{r})$-bimodules. By Lemma 1.2 there is then a ring homomorphism $\tilde{f} : T(\Lambda/\mathfrak{r}, \mathfrak{r}/\mathfrak{r}^2) \to \Lambda$ such that $\tilde{f}|_{(\Lambda/\mathfrak{r}) \coprod (\mathfrak{r}/\mathfrak{r}^2)} = f$. It follows from (a) that $\tilde{f}$ is surjective. Since $\tilde{f}((\mathfrak{r}/\mathfrak{r}^2)^j) \subset \mathfrak{r}^j \subset \mathfrak{r}^2$ for $j \geq 2$ and $\tilde{f}|_{(\Lambda/\mathfrak{r}) \coprod (\mathfrak{r}/\mathfrak{r}^2)} : (\Lambda/\mathfrak{r}) \coprod (\mathfrak{r}/\mathfrak{r}^2) \to \Lambda$ is a monomorphism with image intersecting $\mathfrak{r}^2$ trivially, it follows that $\operatorname{Ker}\tilde{f} \subset \coprod_{j \geq 2} (\mathfrak{r}/\mathfrak{r}^2)^j$. Since $\tilde{f}((\mathfrak{r}/\mathfrak{r}^2)^j) = 0$ for $j \geq \operatorname{rl}(\Lambda)$, we have $\coprod_{j \geq \operatorname{rl}(\Lambda)} (\mathfrak{r}/\mathfrak{r}^2)^j \subset \operatorname{Ker}\tilde{f}$.

(c) Let Γ be the associated quiver of $T(\Lambda/\mathfrak{r}, \mathfrak{r}/\mathfrak{r}^2)$. From (b) we have the surjective k-algebra morphism $\tilde{f} : T(\Lambda/\mathfrak{r}, r/\mathfrak{r}^2) \to \Lambda$ with $\coprod_{j \geq \operatorname{rl}(\Lambda)} (r/\mathfrak{r}^2)^j \subset \operatorname{Ker}\tilde{f} \subset \coprod_{j \geq 2} (r/\mathfrak{r}^2)^j$. Since by Proposition 1.3 there is a k-algebra isomorphism $\phi : T(\Lambda/\mathfrak{r}, r/\mathfrak{r}^2) \to k\Gamma$ with $\phi(\coprod_{j \geq n} (r/\mathfrak{r}^2)^j) = J^n$ we get that $\tilde{f}\phi^{-1} : k\Gamma \to \Lambda$ is a surjective k-algebra morphism where $I = \operatorname{Ker}(\tilde{f}\phi^{-1})$ has the property that $J^s \subset I \subset J^2$ for some integer s. Then I is a finitely generated ideal in $k\Gamma$ since J^s is finitely generated in $k\Gamma$ and I/J^s is a finitely generated ideal in the artin algebra $k\Gamma/J^s$. For each σ from a finite set of generators for the ideal I, write $\sigma = \sum_{1 \leq i, j \leq m} e_j \sigma e_i$ and replace σ by the $e_j \sigma e_i$ for $1 \leq i, j \leq m$, which are relations on Γ over k. This gives us a finite set of relations ρ on Γ with $\langle \rho \rangle = I$.

(d) If $\Lambda \simeq k(\Gamma, \rho)$ with $J^t \subset \langle \rho \rangle \subset J^2$ for some t, then it follows from (b) that $\Lambda/\mathfrak{r}^2 \simeq k\Gamma/J^2$ as k-algebras. Since it is easy to see that Γ is the quiver of $k\Gamma/J^2$, we get that Γ is the quiver of $\Lambda/\mathfrak{r}^2$ and hence of Λ. □

Corollary 1.10 *Let Λ be a basic finite dimensional algebra over an algebraically closed field k.*

(a) Λ *is an elementary k-algebra.*

(b) Λ *is isomorphic to $k(\Gamma, \rho)$ where Γ is the quiver of Λ.*

Proof (a) In order to show that Λ is elementary, it suffices to show that the basic semisimple algebras over the algebraically closed field k are elementary. So let k be an algebraically closed field and Σ a semisimple basic k-algebra. Then $\Sigma = k_1 \times \cdots \times k_n$ where the k_i are division rings.

Let $\pi_i: \Sigma \to k_i$ be the ith projection and let $\phi: k \to \Sigma$ be the inclusion making Σ a k-algebra. Then $\pi_i\phi: k \to k_i$ makes k_i a finite dimensional division ring extension of k. But k being algebraically closed implies that $\pi_i\phi$ is an isomorphism. Now identifying each k_i with k through the isomorphism $\pi_i\phi$ gives that Σ is isomorphic to $\prod_n(k)$ as a k-algebra.

(b) This is an immediate consequence of part (a) and Theorem 1.9. $\square$

It is worth noting that if Λ is an elementary k-algebra and Λ is isomorphic to $k(\Gamma, \rho)$, then Γ is, up to isomorphism, determined by Λ but neither ρ nor the ideal $\langle \rho \rangle$ in $k\Gamma$ is determined by Λ.

We have seen that the path algebra $k\Gamma$ of a quiver without oriented cycles over a field k is hereditary. Using Theorem 1.9 we can show that any elementary hereditary algebra over a field k is isomorphic to a path algebra $k\Gamma$. To do this we need the following results on hereditary artin algebras.

Lemma 1.11 *If Λ is a hereditary artin algebra and $\mathfrak{a}$ a nonzero ideal of Λ contained in $\mathfrak{r}^2$, then $\Lambda/\mathfrak{a}$ is not hereditary.*

Proof Let Λ be a hereditary artin algebra and $\mathfrak{a} \subset \mathfrak{r}^2$ a nonzero ideal of Λ. Consider the exact sequence $0 \to \mathfrak{a}/(\mathfrak{a}\,\mathfrak{r}) \to \mathfrak{r}/(\mathfrak{a}\,\mathfrak{r}) \to \mathfrak{r}/\mathfrak{a} \to 0$ of $(\Lambda/\mathfrak{a})$-modules. Since $\mathfrak{r}$ is a projective Λ-module we have that $\mathfrak{r}/(\mathfrak{a}\,\mathfrak{r})$ is a projective $(\Lambda/\mathfrak{a})$-module. Also $\mathfrak{a} \neq 0$ implies that $\mathfrak{a}/(\mathfrak{a}\,\mathfrak{r}) \neq 0$ by Nakayama's lemma, and hence $\mathfrak{r}/(\mathfrak{a}\,\mathfrak{r}) \not\simeq \mathfrak{r}/\mathfrak{a}$. Now using that $\mathfrak{a} \subset \mathfrak{r}^2$ we get $\mathfrak{a}/(\mathfrak{a}\,\mathfrak{r}) \subset \mathfrak{r}_{\Lambda/\mathfrak{a}}(\mathfrak{r}/(\mathfrak{a}\,\mathfrak{r}))$ and therefore $\mathfrak{r}/\mathfrak{a}\,\mathfrak{r} \to \mathfrak{r}/\mathfrak{a}$ is a projective cover which is not an isomorphism. Hence $\mathfrak{r}/\mathfrak{a}$ is not a projective $(\Lambda/\mathfrak{a})$-module and therefore $\Lambda/\mathfrak{a}$ is not hereditary. $\square$

Lemma 1.12 *Let Λ be a hereditary artin algebra and $f: P \to Q$ a nonzero morphism in $\operatorname{mod}\Lambda$ with P and Q indecomposable Λ-modules and Q a projective module. Then f is a monomorphism and P is a projective module.*

Proof Since f is nonzero $\operatorname{Im} f$ is a nonzero submodule of Q. Since Q is projective and Λ is hereditary, it follows that $\operatorname{Im} f$ is projective. Hence $f: P \to \operatorname{Im} f$ is a split epimorphism. Therefore $f: P \to \operatorname{Im} f$ is an isomorphism since P is indecomposable. Thus $f: P \to Q$ is a monomorphism with P projective. $\square$

We now have the following.

Proposition 1.13 *Let Λ be a finite dimensional elementary hereditary algebra over a field k. Then the associated quiver Γ of Λ has no oriented cycles and Λ is isomorphic to $k\Gamma$.*

Proof If there is an arrow $\alpha: i \to j$ in the quiver Γ of Λ, then by definition $e_j(\mathfrak{r}/\mathfrak{r}^2)e_i$ is not zero, where i and j are the vertices corresponding to the idempotents e_i and e_j. Then $e_j\mathfrak{r}e_i$ is not zero, and a nonzero element x in $e_j\mathfrak{r}e_i$ gives by right multiplication a nonzero Λ-homomorphism from $P_j = \Lambda e_j$ to $P_i = \Lambda e_i$ which is not an isomorphism but which must be a monomorphism by Lemma 1.12. Hence an oriented cycle in the quiver would give rise to a sequence of proper monomorphisms from some P_i to itself. Since this is impossible, Γ has no oriented cycles.

Since Λ is assumed to be elementary, we have by Theorem 1.9 that $\Lambda \simeq k\Gamma/\langle\rho\rangle$ with $\langle\rho\rangle \subset J^2$. Since Λ is hereditary, it follows from Lemma 1.11 that $\langle\rho\rangle$ must be 0, and consequently Λ is isomorphic to $k\Gamma$.
□

Since any basic finite dimensional algebra over an algebraically closed field k is the factor algebra of the path algebra of the associated quiver, it is useful to have various descriptions of the associated quiver.

Proposition 1.14 *Let Λ be a basic finite dimensional algebra over an algebraically closed field k and $1 = \epsilon_1 + \cdots + \epsilon_n$ a decomposition of 1 into a sum of primitive orthogonal idempotents. Let $P_i = \Lambda\epsilon_i$ and $S_i = P_i/\mathfrak{r}P_i$ for $i = 1, \ldots, n$. Then for a given pair of numbers i, j in $\{1, \ldots, n\}$ the following numbers are the same.*

(a) $\dim_k(\epsilon_j\mathfrak{r}/\mathfrak{r}^2\epsilon_i)$.

(b) *The multiplicity of the simple module S_j in $\mathfrak{r}P_i/\mathfrak{r}^2P_i$.*

(c) *The multiplicity of P_j in P where $P \to P_i \to S_i \to 0$ is a minimal projective presentation of S_i.*

(d) $\dim_k \operatorname{Ext}^1_\Lambda(S_i, S_j)$.

Proof For each $i \in \{1, \ldots, n\}$ we have the exact sequence $0 \to \mathfrak{r}P_i \to P_i \to S_i \to 0$ where $P_i \to S_i$ is a projective cover. Applying $\operatorname{Hom}_\Lambda(\ , S_j)$ we obtain the exact sequence of k-vector spaces $0 \to \operatorname{Hom}_\Lambda(S_i, S_j) \to \operatorname{Hom}_\Lambda(P_i, S_j) \overset{h}{\to} \operatorname{Hom}_\Lambda(\mathfrak{r}P_i, S_j) \to \operatorname{Ext}^1_\Lambda(S_i, S_j) \to 0$. Since h must be zero, we have $\operatorname{Hom}_\Lambda(\mathfrak{r}P_i, S_j) \simeq \operatorname{Ext}^1_\Lambda(S_i, S_j)$. But now $\operatorname{Hom}_\Lambda(\mathfrak{r}P_i, S_j) = \operatorname{Hom}_\Lambda(\mathfrak{r}P_i/\mathfrak{r}^2P_i, S_j)$ and $\dim_k \operatorname{Hom}_\Lambda(S_p, S_q)$ is 1 if $p = q$ and 0 otherwise. Hence we obtain $\dim_k \operatorname{Ext}^1_\Lambda(S_i, S_j) = \dim_k \operatorname{Hom}_\Lambda(\mathfrak{r}P_i, S_j) =$

$\dim_k \operatorname{Hom}_\Lambda(\mathfrak{r}P_i/\mathfrak{r}^2 P_i, S_j)$. But the last number is clearly the multiplicity of S_j as a summand of $\mathfrak{r}P_i/\mathfrak{r}^2 P_i$ which again is the multiplicity of P_j as a summand of the projective cover P of $\mathfrak{r}P_i/\mathfrak{r}^2 P_i$. However the projective cover of $\mathfrak{r}P_i/\mathfrak{r}^2 P_i$ is the same as the projective cover of $\mathfrak{r}P_i$. Hence we have established the equality of the numbers in (b), (c) and (d). Further we have that $\dim_k(\operatorname{Hom}_\Lambda(\mathfrak{r}P_i/\mathfrak{r}^2 P_i, S_j)) = \dim_k(\operatorname{Hom}_{\Lambda/\mathfrak{r}}(\mathfrak{r}P_i/\mathfrak{r}^2 P_i, S_j)) = \dim_k(\operatorname{Hom}_{\Lambda/\mathfrak{r}}(S_j, \mathfrak{r}P_i/\mathfrak{r}^2 P_i)) = \dim_k(\operatorname{Hom}_\Lambda(P_j, \mathfrak{r}P_i/\mathfrak{r}^2 P_i))$. But we have $\operatorname{Hom}_\Lambda(P_j, \mathfrak{r}^m P_i) \simeq \epsilon_j \mathfrak{r}^m \epsilon_i$ (see I Proposition 4.9), and hence $\dim_k(\operatorname{Hom}_\Lambda(\mathfrak{r}P_i/\mathfrak{r}^2 P_i, S_j)) = \dim_k(\epsilon_j \mathfrak{r}/\mathfrak{r}^2 \epsilon_i)$. This finishes the proof of the proposition. $\qquad\square$

Motivated by Proposition 1.13 we associate with any artin algebra Λ a **valued quiver**, that is, a quiver with at most one arrow from a vertex i to a vertex j, and with an ordered pair of positive integers associated with each arrow. This is done by writing an arrow from i to j if $\operatorname{Ext}^1_\Lambda(S_i, S_j) \neq 0$, and assigning to this arrow the pair of integers $(\dim_{\operatorname{End}_\Lambda(S_j)} \operatorname{Ext}^1_\Lambda(S_i, S_j), \dim_{\operatorname{End}_\Lambda(S_i)^{op}} \operatorname{Ext}^1_\Lambda(S_i, S_j))$. When Λ is elementary this corresponds to replacing m arrows from i to j by one arrow with valuation (m, m).

Another way of interpreting these numbers is given in the following proposition which we leave to the reader to verify.

Proposition 1.15 *Let Λ be an artin R-algebra.*

(a) *Let $P_1 \to S_1$ and $P_2 \to S_2$ be projective covers of the simple modules S_1 and S_2 respectively. Then the following numbers are the same.*

    (i) $\dim_{\operatorname{End}_\Lambda(S_2)} \operatorname{Ext}^1_\Lambda(S_1, S_2)$.

    (ii) $l_R(\operatorname{Ext}^1_\Lambda(S_1, S_2))/l_R(\operatorname{End}_\Lambda(S_2))$.

    (iii)   *The multiplicity of the simple module S_2 as a summand of $\mathfrak{r}P_1/\mathfrak{r}^2 P_1$.*

    (iv)   *The multiplicity of P_2 as a summand of the projective cover of $\mathfrak{r}P_1$.*

(b) *Let $S_1 \to I_1$ and $S_2 \to I_2$ be injective envelopes of the simple modules S_1 and S_2 respectively. Then the following numbers are the same.*

    (i) $\dim_{\operatorname{End}_\Lambda(S_1)^{op}} \operatorname{Ext}^1_\Lambda(S_1, S_2)$.

    (ii)   *The multiplicity of I_1 as a summand of the injective envelope of $I_2/\operatorname{soc} I_2$.* $\qquad\square$

Note that the valued quiver of the opposite algebra of Λ has the opposite underlying quiver but the valuations are the same. This is easily seen from the fact that $\mathrm{Ext}^1_\Lambda(S_i, S_j) \neq 0$ if and only if $\mathrm{Ext}^1_{\Lambda^{\mathrm{op}}}(DS_j, DS_i) \neq 0$ and that $\dim_{\mathrm{End}_\Lambda(S_i)^{\mathrm{op}}} \mathrm{Ext}^1_\Lambda(S_i, S_j) = \dim_{\mathrm{End}_\Lambda(DS_i)} \mathrm{Ext}^1_{\Lambda^{\mathrm{op}}}(DS_i, DS_j)$.

We illustrate with the following.

Example Let k be a field and $T_3(k) = \left(\begin{smallmatrix} k & 0 & 0 \\ k & k & 0 \\ k & k & k \end{smallmatrix} \right)$ the k-subalgebra of the full 3×3 matrix algebra $M_3(k)$ over k consisting of 3×3 matrices where all entries above the main diagonal are 0. Let I be the ideal consisting of elements of the form $\left(\begin{smallmatrix} 0 & 0 & 0 \\ 0 & 0 & 0 \\ a & 0 & 0 \end{smallmatrix} \right)$ for $a \in k$, let $\Lambda = T_3(k)/I$ and let $e_1 = \left(\begin{smallmatrix} 1 & 0 & 0 \\ 0 & 0 & 0 \\ 0 & 0 & 0 \end{smallmatrix} \right) + I$, $e_2 = \left(\begin{smallmatrix} 0 & 0 & 0 \\ 0 & 1 & 0 \\ 0 & 0 & 0 \end{smallmatrix} \right) + I$ and $e_3 = \left(\begin{smallmatrix} 0 & 0 & 0 \\ 0 & 0 & 0 \\ 0 & 0 & 1 \end{smallmatrix} \right) + I$. Then $\mathrm{r}(\Lambda e_1) \simeq \Lambda e_2/\mathrm{r}(\Lambda e_2)$, $\mathrm{r}(\Lambda e_2) \simeq \Lambda e_3$ and $\mathrm{r}(\Lambda e_3) = 0$. Hence the associated quiver Γ is $\underset{1}{\cdot} \overset{\alpha}{\to} \underset{2}{\cdot} \overset{\beta}{\to} \underset{3}{\cdot}$ and $\Lambda \simeq k(\Gamma, \rho)$ where $\rho = \{\beta\alpha\}$.

Example Let $\mathbb{R}$ be the real numbers and $\mathbb{C}$ the complex numbers and let $\Lambda = \left(\begin{smallmatrix} \mathbb{C} & 0 \\ \mathbb{C} & \mathbb{R} \end{smallmatrix} \right)$ be the $\mathbb{R}$-subalgebra of the 2×2 matrices over $\mathbb{C}$. Let $e_1 = \left(\begin{smallmatrix} 1 & 0 \\ 0 & 0 \end{smallmatrix} \right)$ and $e_2 = \left(\begin{smallmatrix} 0 & 0 \\ 0 & 1 \end{smallmatrix} \right)$. Writing $\underset{i}{\cdot} \overset{(a,b)}{\to} \underset{j}{\cdot}$ when the ordered pair (a, b) is assigned to the arrow $\alpha: i \to j$, we get that the associated valued quiver is $\underset{1}{\cdot} \overset{(2,1)}{\to} \underset{2}{\cdot}$.

We end this section by pointing out that viewing the modules as representations is closely related to viewing modules as functors. Associated with a quiver Γ (without oriented cycles) and a field k, we have the path category $\mathscr{C}$ whose objects are the vertices of Γ. For i and j in Γ_0 we have that $\mathrm{Hom}(i, j)$ is the vector space over k spanned by all paths from i to j in Γ, and composition of morphisms is induced by composition of paths. Then clearly $\mathrm{Rep}\,\Gamma$ is equivalent to the category of covariant functors from $\mathscr{C}$ to $\mathrm{mod}\,k$. It is easy to see that $\mathscr{C}$ is equivalent to $\mathrm{ind}\,\mathscr{P}(\Lambda)^{\mathrm{op}}$. Actually, for any artin R-algebra Λ we have that $\mathrm{mod}\,\Lambda$ is equivalent to the category of contravariant functors from $\mathrm{ind}\,\mathscr{P}(\Lambda)$ to $\mathrm{mod}\,R$.

2 Triangular matrix rings

Writing the identity as a sum of two orthogonal idempotents $1 = e + (1-e)$ gives a coarser way of dividing up an artin algebra than writing 1 as a sum of primitive orthogonal idempotents. Writing $1 = e + (1 - e)$ with e a nontrivial idempotent gives rise to the algebras $e\Lambda e$ and $(1-e)\Lambda(1-e)$ and the two bimodules $e\Lambda(1 - e)$ and $(1 - e)\Lambda e$. This decomposition takes on an especially nice form if $e\Lambda(1 - e) = 0$. In this case we obtain an isomorphism between Λ and a triangular matrix algebra, a class of algebras we introduce and investigate in this section.

Let T and U be rings and $_U M_T$ a U-T-bimodule. In this section this means that M is a left U-module and right T-module such that $(um)t = u(mt)$ for all $u \in U$, $t \in T$ and $m \in M$. Then we can construct the **triangular matrix ring** $\Lambda = \begin{pmatrix} T & 0 \\ M & U \end{pmatrix}$. The elements of Λ are 2×2 matrices $\begin{pmatrix} t & 0 \\ m & u \end{pmatrix}$ with $t \in T$, $u \in U$ and $m \in M$. Addition and multiplication are given by the ordinary operation on matrices, as $\begin{pmatrix} t_1 & 0 \\ m_1 & u_1 \end{pmatrix} + \begin{pmatrix} t_2 & 0 \\ m_2 & u_2 \end{pmatrix} = \begin{pmatrix} t_1+t_2 & 0 \\ m_1+m_2 & u_1+u_2 \end{pmatrix}$ and $\begin{pmatrix} t_1 & 0 \\ m_1 & u_1 \end{pmatrix} \begin{pmatrix} t_2 & 0 \\ m_2 & u_2 \end{pmatrix} = \begin{pmatrix} t_1 t_2 & 0 \\ m_1 t_2 + u_1 m_2 & u_1 u_2 \end{pmatrix}$. We give a description of the Λ-modules in terms of triples $(_T A, _U B, f)$ where A is a T-module, B a U-module and $f: M \otimes_T A \to B$ a U-morphism. In particular, in the important special case $T = U = M$, we just have morphisms between T-modules. We use this point of view to give a description of projective and injective Λ-modules in terms of A, B and f, and we illustrate how this description is convenient for establishing connections between the homological properties of Λ and those of U, T and M.

Triangular matrix rings also come up as endomorphism rings of a sum $M \coprod N$ of two Λ-modules M and N with $\mathrm{Hom}_\Lambda(N, M) = 0$. In this case we have a ring isomorphism $\mathrm{End}_\Lambda(M \coprod N) \simeq \begin{pmatrix} \mathrm{End}_\Lambda(M) & 0 \\ \mathrm{Hom}_\Lambda(M,N) & \mathrm{End}_\Lambda(N) \end{pmatrix}$. This happens for example when $\Lambda = \mathbb{Z}$ and M is a torsionfree $\mathbb{Z}$-module and N is a torsion $\mathbb{Z}$-module.

If T is a division ring, a triangular matrix ring $\Lambda = \begin{pmatrix} T & 0 \\ M & U \end{pmatrix}$ is called a **one-point extension** of U by the bimodule $_U M_T$. The reason for this terminology comes from the following.

Let $\Lambda = k(\Gamma, \rho)$ be a finite dimensional path algebra over a field k of the quiver (Γ, ρ) with relations. Let i be a source in Γ and $\bar{e}_i$ the corresponding idempotent in Λ. Since there are no nontrivial paths ending in i, we have $\bar{e}_i \Lambda \bar{e}_i \simeq k$ and $\bar{e}_i \Lambda (1 - \bar{e}_i) = 0$. If (Γ', ρ') denotes the quiver with relations we obtain by removing the vertex i and the relations starting at i, then $(1 - \bar{e}_i)\Lambda(1 - \bar{e}_i) \simeq k(\Gamma', \rho')$. So $k(\Gamma, \rho)$ is obtained from $\Lambda' = k(\Gamma', \rho')$ by adding one vertex i, together with arrows and relations starting at i. We then have $\Lambda \simeq \begin{pmatrix} k & 0 \\ (1-\bar{e}_i)\Lambda\bar{e}_i & \Lambda' \end{pmatrix}$ so that Λ is a one-point extension of Λ'.

The special case $\begin{pmatrix} T & 0 \\ T & T \end{pmatrix}$ of a triangular matrix ring is also related to another general ring construction. Let T and U be k-algebras over a field k. Then the vector space $U \otimes_k T$ becomes a ring by defining $(t \otimes u)(t' \otimes u') =$

$tt' \otimes uu'$. We then have a ring isomorphism $g: \left(\begin{smallmatrix} k & 0 \\ k & k \end{smallmatrix} \right) \otimes_k T \rightarrow \left(\begin{smallmatrix} T & 0 \\ T & T \end{smallmatrix} \right)$ defined by $g \left(\left(\begin{smallmatrix} a & 0 \\ b & c \end{smallmatrix} \right) \otimes t \right) = \left(\begin{smallmatrix} at & 0 \\ bt & ct \end{smallmatrix} \right)$ on generators.

When we want to construct a triangular matrix ring starting with rings T and U, we are interested in natural choices of U-T-bimodules. When T and U are artin R-algebras we have ring homomorphisms $\phi_1: R \rightarrow T$ and $\phi_2: R \rightarrow U$ inducing an R-bimodule structure on a U-T-bimodule M. Then we are especially interested in the case where R acts centrally on M, that is, $rm = mr$ for all $r \in R$ and $m \in M$. Whenever we have a ring homomorphism $f: T \rightarrow U$, then U has a natural structure as U-T-bimodule. In the case $f = 1_T: T \rightarrow T$, this specializes to $\left(\begin{smallmatrix} T & 0 \\ T & T \end{smallmatrix} \right)$. If $T = U$, then two-sided ideals can be used for M, and also the bimodule $D(T)$ when T is an artin algebra and D is the ordinary duality. When T and U are algebras over a field k, and A is a left U-module and B a right T-module, then $A \otimes_k B$ is a U-T-bimodule in a natural way.

Since we are mainly interested in artin rings, especially artin algebras, we investigate when a triangular matrix ring is an artin ring or an artin algebra.

Proposition 2.1 *Let* $\Lambda = \left(\begin{smallmatrix} T & 0 \\ {}_U M_T & U \end{smallmatrix} \right)$ *where* T *and* U *are rings and* ${}_U M_T$ *is a* U-T-*bimodule.*

(a) Λ *is left artin if and only if* T *and* U *are left artin and* M *is a finitely generated* U-*module.*

(b) Λ *is right artin if and only if* T *and* U *are right artin and* M *is a finitely generated* T-*module.*

(c) Λ *is an artin algebra if and only if there is a commutative ring* R *such that* T *and* U *are artin* R-*algebras, and* M *is finitely generated over* R *which acts centrally on* M.

Proof (a) and (b) We have the exact sequence of left as well as right Λ-modules $0 \rightarrow \left(\begin{smallmatrix} 0 & 0 \\ M & 0 \end{smallmatrix} \right) \rightarrow \Lambda \rightarrow T \times U \rightarrow 0$. The left Λ-submodules of $\left(\begin{smallmatrix} 0 & 0 \\ M & 0 \end{smallmatrix} \right)$ correspond to the U-submodules of M and the right Λ-submodules correspond to T-submodules of M. Hence Λ is left artin if and only if T and U are left artin and M is an artin U-module, or equivalently, a finitely generated U-module. Similarly Λ is right artin if and only if T and U are right artin and M is a finitely generated T-module.

(c) Assume there is a commutative artin ring R such that $\phi_T: R \rightarrow T$ makes T an artin R-algebra and $\phi_U: R \rightarrow U$ makes U an artin R-algebra and such that M is finitely generated over R, which acts centrally on

M. Then the morphism $\phi: R \to \Lambda$ given by $\phi(r) = \begin{pmatrix} \phi_T(r) & 0 \\ 0 & \phi_U(r) \end{pmatrix}$ is a ring homomorphism. Since T and U are R-algebras and R acts centrally on M, the image of ϕ is in the center of Λ. Further Λ is a finitely generated R-module since T, U and M are finitely generated R-modules. In other words, Λ is an artin R-algebra.

Assume conversely that Λ is an artin R-algebra, so that we have a ring homomorphism $\phi: R \to \Lambda$ with $\text{Im}\,\phi$ in the center of Λ. Considering the natural composite maps $\phi_T: R \to \Lambda \to T$ and $\phi_U: R \to \Lambda \to U$, we see that T and U are finitely generated R-modules. Since the image of R is clearly in the center in both cases, T and U must be artin R-algebras. Since Λ is a finitely generated R-module and M is an R-submodule, M is a finitely generated R-module. Since $\text{Im}\,\phi$ is in the center of Λ, R acts centrally on M. □

Using this result it is easy to provide examples of left artin rings which are not right artin. For example if $\Lambda = \begin{pmatrix} \mathbb{Q} & 0 \\ \mathbb{R} & \mathbb{R} \end{pmatrix}$ where $\mathbb{Q}$ are the rational numbers and $\mathbb{R}$ the real numbers, then Λ is left artin since $\mathbb{R}$ is a finitely generated $\mathbb{R}$-module, but not right artin since $\mathbb{R}$ is not a finitely generated $\mathbb{Q}$-module. This shows that the class of left artin rings considered in Chapter I is much wider than the class of artin algebras.

We assume from now on that $\Lambda = \begin{pmatrix} T & 0 \\ M & U \end{pmatrix}$ is an artin algebra. When C is a module over a triangular matrix ring $\Lambda = \begin{pmatrix} T & 0 \\ M & U \end{pmatrix}$, then the idempotents $e_1 = \begin{pmatrix} 1 & 0 \\ 0 & 0 \end{pmatrix}$ and $e_2 = \begin{pmatrix} 0 & 0 \\ 0 & 1 \end{pmatrix}$ give rise to a decomposition $C = e_1 C \coprod e_2 C$ into a sum of abelian groups. Here $e_1 C$ is in a natural way a T-module and $e_2 C$ a U-module. Since for each $m \in M$ we have that $\begin{pmatrix} 0 & 0 \\ m & 0 \end{pmatrix} e_1 = e_2 \begin{pmatrix} 0 & 0 \\ m & 0 \end{pmatrix}$ and $\begin{pmatrix} 0 & 0 \\ m & 0 \end{pmatrix} e_2 = e_1 \begin{pmatrix} 0 & 0 \\ m & 0 \end{pmatrix} = 0$ we get that multiplying an element of C with an element of the form $\tilde{m} = \begin{pmatrix} 0 & 0 \\ m & 0 \end{pmatrix}$ gives that $\tilde{m} e_1 C \subset e_2 C$ and $\tilde{m} e_2 C = 0$. Hence we have a map $\psi: M \times e_1 C \to e_2 C$, given by $\psi(m, e_1 c) = e_2 \begin{pmatrix} 0 & 0 \\ m & 0 \end{pmatrix} c$ which is easily seen to be T-bilinear. One then obtains a unique map $M \otimes_T e_1 C \to e_2 C$ which is a U-morphism. This gives a convenient way of viewing the modules over a triangular matrix ring Λ. To make this point of view precise we introduce the following category which we shall show is equivalent to $\text{mod}\,\Lambda$.

Let $\Lambda = \begin{pmatrix} T & 0 \\ M & U \end{pmatrix}$ be a triangular matrix R-algebra. Let $_\Lambda\mathscr{C}$ be the category whose objects are the triples (A, B, f) with A in $\text{mod}\,T$, B in $\text{mod}\,U$ and $f: M \otimes_T A \to B$ a U-morphism. The morphisms between two objects (A, B, f) and (A', B', f') are pairs of morphisms (α, β) where $\alpha: A \rightarrow A'$ is a T-morphism and $\beta: B \rightarrow B'$ is a U-morphism, such that the diagram

$$
\begin{array}{ccc}
M \otimes_T A & \xrightarrow{M \otimes \alpha} & M \otimes_T A' \\
\downarrow f & & \downarrow f' \\
B & \xrightarrow{\beta} & B'
\end{array}
$$

commutes. If (α_1, β_1) and (α_2, β_2) are morphisms from (A, B, f) to (A', B', f'), then their sum is defined by $(\alpha_1, \beta_1) + (\alpha_2, \beta_2) = (\alpha_1 + \alpha_2, \beta_1 + \beta_2)$. It is easy to check that $_\Lambda \mathscr{C}$ is an R-category.

The relationship between $_\Lambda \mathscr{C}$ and $\mathrm{mod}\, \Lambda$ is given via the functor $F: {_\Lambda \mathscr{C}} \rightarrow \mathrm{mod}\, \Lambda$ which is defined as follows. For (A, B, f) in $_\Lambda \mathscr{C}$ we define $F(A, B, f) = A \coprod B$ as abelian group, and the Λ-module structure is given by $\begin{pmatrix} t & 0 \\ m & u \end{pmatrix} (a, b) = (ta, f(m \otimes a) + ub)$ for $t \in T$, $u \in U$, $m \in M$, $a \in A$ and $b \in B$. If we have $(\alpha, \beta): (A, B, f) \rightarrow (A', B', f')$ in $_\Lambda \mathscr{C}$, then $F(\alpha, \beta) = \alpha \coprod \beta: A \coprod B \rightarrow A' \coprod B'$. The following result gives the formal connection between $_\Lambda \mathscr{C}$ and $\mathrm{mod}\, \Lambda$.

Proposition 2.2 *Let* $\Lambda = \begin{pmatrix} T & 0 \\ M & U \end{pmatrix}$ *be a triangular matrix algebra. Then the functor* $F: {_\Lambda \mathscr{C}} \rightarrow \mathrm{mod}\, \Lambda$ *defined above is an equivalence of categories.*

Proof It is easy to see that F is an R-functor. Let $(\alpha, \beta): (A, B, f) \rightarrow (A', B', f')$ be a map in $_\Lambda \mathscr{C}$, and assume that $F(\alpha, \beta) = \alpha \coprod \beta: A \coprod B \rightarrow A' \coprod B'$ is 0. Then $\alpha = 0$ and $\beta = 0$ so that $(\alpha, \beta) = (0, 0)$, and hence F is faithful.

Let (A, B, f) and (A', B', f') be in $_\Lambda \mathscr{C}$, and let $s: F(A, B, f) \rightarrow F(A', B', f')$ be a morphism in $\mathrm{mod}\, \Lambda$. Hence we have a map $s: A \coprod B \rightarrow A' \coprod B'$, and for $a \in A$ and $b \in B$ let $s(a, b) = (a', b')$. Write $e_1 = \begin{pmatrix} 1 & 0 \\ 0 & 0 \end{pmatrix}$ and $e_2 = \begin{pmatrix} 0 & 0 \\ 0 & 1 \end{pmatrix}$. Then $s(a, 0) = s(e_1(a, b)) = e_1 s(a, b) = e_1(a', b') = (a', 0)$, so that there is induced a map $s_1: A \rightarrow A'$ given by $s_1(a) = a'$. Similarly there is induced a map $s_2: B \rightarrow B'$. For $m \in M$, write as before $\widetilde{m} = \begin{pmatrix} 0 & 0 \\ m & 0 \end{pmatrix}$. Then $s(\widetilde{m}(a, 0)) = s(0, f(m \otimes a)) = (0, s_2 f(m \otimes a))$ and $\widetilde{m} s(a, 0) = \widetilde{m}(s_1(a), 0) = (0, f'(m \otimes s_1(a)))$. It follows that $s_2 f(m \otimes a) = f'(m \otimes s_1(a))$, so that $(s_1, s_2): (A, B, f) \rightarrow (A', B', f')$ is a map in $_\Lambda \mathscr{C}$. Then we have $F(s_1, s_2) = s$, and this shows that F is full.

To see that F is dense, let C be in $\mathrm{mod}\, \Lambda$ and consider $C = e_1 C \coprod e_2 C$. We have seen that $e_1 C$ is in $\mathrm{mod}\, T$ and $e_2 C$ is in $\mathrm{mod}\, U$ and that there is a T-bilinear map $M \times e_1 C \rightarrow e_2 C$ inducing a U-homomorphism $f: M \otimes_T e_1 C \rightarrow e_2 C$. It is then easy to check that $F(e_1 C, e_2 C, f) \simeq C$.

It now follows from II Theorem 1.2 that F is an equivalence of categories. □

From now on we identify $_\Lambda\mathscr{C}$ with $\mathrm{mod}\,\Lambda$ by means of the functor $F\colon{}_\Lambda\mathscr{C}\to\mathrm{mod}\,\Lambda$. It is in many respects more convenient to deal with the category $_\Lambda\mathscr{C}$ than $\mathrm{mod}\,\Lambda$, especially with respect to homological properties. In particular we shall give a description of the projective and the injective objects in $_\Lambda\mathscr{C}$.

We say that a sequence of maps $(A,B,f)\overset{(\alpha,\beta)}{\to}(A',B',f')\overset{(\alpha',\beta')}{\to}(A'',B'',f'')$ is exact if the sequences $A\overset{\alpha}{\to}A'\overset{\alpha'}{\to}A''$ and $B\overset{\beta}{\to}B'\overset{\beta'}{\to}B''$ are exact. And an object (A,B,f) in $_\Lambda\mathscr{C}$ is simple if it has no proper nonzero subobject. Here we say that (A',B',f') is a subobject of (A,B,f) if $A'\subset A$, $B'\subset B$ and $f'=f|_{M\otimes_T A'}$. We then have the following.

Proposition 2.3 *Let Λ be the triangular matrix algebra $\left(\begin{smallmatrix} T & 0 \\ M & U \end{smallmatrix}\right)$ and $F\colon$ $_\Lambda\mathscr{C}\to\mathrm{mod}\,\Lambda$ the equivalence of categories defined above.*

(a) *X in $_\Lambda\mathscr{C}$ is projective if and only if FX is projective in $\mathrm{mod}\,\Lambda$.*

(b) *X in $_\Lambda\mathscr{C}$ is injective if and only if FX is injective in $\mathrm{mod}\,\Lambda$.*

(c) *$0\to X\to Y\to Z\to 0$ is an exact sequence in $_\Lambda\mathscr{C}$ if and only if $0\to FX\to FY\to FZ\to 0$ is an exact sequence in $\mathrm{mod}\,\Lambda$.*

(d) *X in $_\Lambda\mathscr{C}$ is simple if and only if FX is simple in $\mathrm{mod}\,\Lambda$.*

Proof (a) and (b) follow from II Proposition 1.3 and (c) is a direct consequence of the definitions. Part (d) follows easily from (c). □

In order to describe the injective objects in $_\Lambda\mathscr{C}$ for a triangular matrix R-algebra $\left(\begin{smallmatrix} T & 0 \\ M & U \end{smallmatrix}\right)$ we use the adjoint isomorphism $\phi\colon\mathrm{Hom}_U(M\otimes_T A,B)\to\mathrm{Hom}_T(A,\mathrm{Hom}_U(M,B))$ to give a description of a category isomorphic to $_\Lambda\mathscr{C}$.

Let $_\Lambda\widetilde{\mathscr{C}}$ be the category whose objects are triples (A,B,g) where A is a T-module, B is a U-module and $g\colon A\to\mathrm{Hom}_U(M,B)$ is a T-morphism. A morphism from (A,B,g) to (A',B',g') is a pair of morphisms (α,β) with $\alpha\colon A\to A'$ a T-morphism and $\beta\colon B\to B'$ a U-morphism such that the following diagram commutes:

$$
\begin{array}{ccc}
A & \overset{\alpha}{\to} & A' \\
{\scriptstyle g}\downarrow & & \downarrow{\scriptstyle g'} \\
\mathrm{Hom}_U(M,B) & \overset{\mathrm{Hom}_U(M,\beta)}{\to} & \mathrm{Hom}_U(M,B').
\end{array}
$$

Now define a functor $H: {}_\Lambda\mathscr{C} \to {}_\Lambda\widetilde{\mathscr{C}}$ by $H(A, B, f) = (A, B, \phi(f))$ on objects and $H(\alpha, \beta) = (\alpha, \beta)$ on morphisms. Clearly the functor $G: {}_\Lambda\widetilde{\mathscr{C}} \to {}_\Lambda\mathscr{C}$ given by $G(A, B, g) = (A, B, \phi^{-1}(g))$ on objects and $G(\alpha, \beta) = (\alpha, \beta)$ on morphisms is an inverse of H. Hence we have the rare occasion of an isomorphism of categories.

Proposition 2.4 *Let* Λ, ${}_\Lambda\mathscr{C}$, ${}_\Lambda\widetilde{\mathscr{C}}$ *and* $H: {}_\Lambda\mathscr{C} \to {}_\Lambda\widetilde{\mathscr{C}}$ *be as above. Then* H *is an isomorphism of categories with* G *as an inverse.*   □

Using the equivalence between ${}_\Lambda\widetilde{\mathscr{C}}$ and $\mathrm{mod}\,\Lambda$ obtained by composing the functors $G: {}_\Lambda\widetilde{\mathscr{C}} \to {}_\Lambda\mathscr{C}$ and $F: {}_\Lambda\mathscr{C} \to \mathrm{mod}\,\Lambda$ we can now give a direct description of how the duality acts on ${}_\Lambda\mathscr{C}$. In order to do this we identify the R-algebra Λ^{op} with the triangular matrix algebra $\begin{pmatrix} U^{\mathrm{op}} & 0 \\ {}_{T^{\mathrm{op}}}M_{U^{\mathrm{op}}} & T^{\mathrm{op}} \end{pmatrix}$ when Λ is the R-algebra $\begin{pmatrix} T & 0 \\ {}_UM_T & U \end{pmatrix}$, by sending $\begin{pmatrix} t & 0 \\ m & u \end{pmatrix}^{\mathrm{op}}$ to $\begin{pmatrix} u^{\mathrm{op}} & 0 \\ m & t^{\mathrm{op}} \end{pmatrix}$.

If (A, B, f) is in ${}_\Lambda\mathscr{C}$, then $H(A, B, f) = (A, B, \phi(f))$ is in ${}_\Lambda\widetilde{\mathscr{C}}$ where $\phi: \mathrm{Hom}_U(M \otimes_T A, B) \to \mathrm{Hom}_T(A, \mathrm{Hom}_U(M, B))$ is the adjoint isomorphism. Applying the duality we obtain the T^{op}-morphism $D\phi(f)$: $D\,\mathrm{Hom}_U(M, B) \to DA$. Now it is easy to see that $\psi: M \otimes_{U^{\mathrm{op}}} DB \to D\,\mathrm{Hom}_U(M, B)$ given by $\psi(m \otimes g)(f) = gf(m)$ for $m \in M$, $g \in DB$ and $f \in \mathrm{Hom}_U(M, B)$ is a T^{op}-isomorphism functorial in B by first considering $B = D(U_U)$ and using that $M \otimes_{U^{\mathrm{op}}} D(\ )$ and $D\,\mathrm{Hom}_U(M, \ )$ are right exact functors from $\mathrm{mod}\,U$ to $\mathrm{mod}\,T^{\mathrm{op}}$. Hence we obtain the object $(DB, DA, D\phi(f)\psi)$ in ${}_{\Lambda^{\mathrm{op}}}\mathscr{C}$. For a morphism $(\alpha, \beta): (A, B, f) \to (A', B', f')$ in ${}_\Lambda\mathscr{C}$ it is now straightforward to check that $(D\beta, D\alpha): (DB', DA', D\phi(f')\psi) \to (DB, DA, D\phi(f)\psi)$ is a morphism in ${}_{\Lambda^{\mathrm{op}}}\mathscr{C}$ and that we get a contravariant functor $\widetilde{D}: {}_\Lambda\mathscr{C} \to {}_{\Lambda^{\mathrm{op}}}\mathscr{C}$ this way. Let $F: {}_\Lambda\mathscr{C} \to \mathrm{mod}\,\Lambda$ and $F': {}_{\Lambda^{\mathrm{op}}}\mathscr{C} \to \mathrm{mod}\,\Lambda^{\mathrm{op}}$ be the equivalences described in Proposition 2.2 for Λ and Λ^{op} respectively. Then an easy calculation shows that the following diagram of functors commutes.

$$\begin{array}{ccc} {}_\Lambda\mathscr{C} & \xrightarrow{F} & \mathrm{mod}\,\Lambda \\ \downarrow \widetilde{D} & & \downarrow D \\ {}_{\Lambda^{\mathrm{op}}}\mathscr{C} & \xrightarrow{F'} & \mathrm{mod}(\Lambda^{\mathrm{op}}) \end{array}$$

where D is the ordinary duality from $\mathrm{mod}\,\Lambda$ to $\mathrm{mod}(\Lambda^{\mathrm{op}})$. Hence $\widetilde{D}$ is a duality which we shall also denote by D.

We now give a description of simple, projective and injective objects in ${}_\Lambda\mathscr{C}$.

Proposition 2.5 *Let* $\Lambda = \begin{pmatrix} T & 0 \\ {}_UM_T & U \end{pmatrix}$ *be a triangular matrix algebra.*

(a) $\operatorname{rad}\Lambda = \begin{pmatrix} \operatorname{rad} T & 0 \\ M & \operatorname{rad} U \end{pmatrix}$, *and the simple objects in* $_\Lambda\mathscr{C}$ *are of the form* $(S, 0, 0)$ *where S is a simple T-module and $(0, S', 0)$ where S' is a simple U-module.*

(b) *The indecomposable projective objects in* $_\Lambda\mathscr{C}$ *are objects isomorphic to objects of the form* $(P, M \otimes_T P, 1_{M\otimes P})$ *where P is an indecomposable projective T-module and $(0, Q, 0)$ where Q is an indecomposable projective U-module.*

(c) *The indecomposable injective objects in* $_\Lambda\mathscr{C}$ *are objects of the form* $(I, 0, 0)$ *where I is an indecomposable injective T-module and objects isomorphic to objects of the form* $(\operatorname{Hom}_U(M, J), J, \phi)$ *where J is an indecomposable injective U-module and $\phi\colon M \otimes_T \operatorname{Hom}_U(M, J) \to J$ is given by $\phi(m \otimes f) = f(m)$ for $m \in M$ and $f \in \operatorname{Hom}_U(M, J)$.*

Proof (a) Since T and U are artin rings, we have $(\operatorname{rad} T)^i = 0 = (\operatorname{rad} U)^i$ for some i. Then an easy calculation shows that $A^{2i} = 0$ when A is the ideal $\begin{pmatrix} \operatorname{rad} T & 0 \\ M & \operatorname{rad} U \end{pmatrix}$ of Λ. Since clearly $\Lambda/A \simeq (T/\operatorname{rad} T) \times (U/\operatorname{rad} U)$ and is hence a semisimple ring, it follows from I Proposition 3.3 that $A = \operatorname{rad}\Lambda$.

The simple Λ-modules are hence given by the simple $(T/\operatorname{rad} T)$-modules and the simple $(U/\operatorname{rad} U)$-modules. Therefore it is easy to see that the simple objects in $_\Lambda\mathscr{C}$ are $(S, 0, 0)$ where S is a simple T-module and $(0, S', 0)$ where S' is a simple U-module.

(b) We have $\Lambda = P \coprod Q$ where $P = \begin{pmatrix} T & 0 \\ M & 0 \end{pmatrix}$ and $Q = \begin{pmatrix} 0 & 0 \\ 0 & U \end{pmatrix}$ as left Λ-modules. The object of $_\Lambda\mathscr{C}$ corresponding to P is clearly $(T, M, 1_M)$, where we identify M with $M \otimes_T T$ via the natural isomorphism and the object corresponding to Q is $(0, U, 0)$. The indecomposable summands of $(0, U, 0)$ are of the form $(0, Q, 0)$ where Q is an indecomposable projective U-module. If $T = P_1 \coprod \cdots \coprod P_n$ is a decomposition of T into a sum of indecomposable projective T-modules, then $(T, M, 1_M) = (P_1, M \otimes_T P_1, 1_{M\otimes_T P_1}) \coprod \cdots \coprod (P_n, M \otimes_T P_n, 1_{M\otimes_T P_n})$. Since $\operatorname{End}_{\Lambda\mathscr{C}}(P_i, M \otimes_T P_i, 1_{M\otimes P_i}) \simeq \operatorname{End}_T(P_i)$, it follows that the $(P_i, M \otimes_T P_i, 1_{M\otimes_T P_i})$ are indecomposable. In this way we get all indecomposable projective objects up to isomorphism.

(c) This follows from (b) and the description of the duality between $_\Lambda\mathscr{C}$ and $_{\Lambda^{\operatorname{op}}}\mathscr{C}$. □

We also interpret the duality $\operatorname{Hom}_\Lambda(\ ,\Lambda)\colon \mathscr{P}(\Lambda) \to \mathscr{P}(\Lambda^{\operatorname{op}})$, which we denote by $(\ )^*$, on objects in $_\Lambda\mathscr{C}$. Let $X = (P, M \otimes_T P, 1_{M\otimes_T P})$ be an indecomposable projective object in $_\Lambda\mathscr{C}$, and let $(S, 0, 0)$ be the corresponding simple object. We define X^* to be the object in $_{\Lambda^{op}}\mathscr{C}$ corresponding to the Λ^{op}-module $(FX)^*$. Recall that Λ^{op} is identified with $\begin{pmatrix} U^{op} & 0 \\ _{T^{op}}M_{U^{op}} & T^{op} \end{pmatrix}$. Then we have $D(S, 0, 0) = (0, D(S), 0)$ and the projective cover is $(0, Q, 0)$, where Q is the projective cover of the T^{op}-module $D(S)$. Then we have $Q = P^*$, and $X^* = (0, P^*, 0)$.

When $X = (0, P, 0)$, it follows similarly that $X^* = (P^*, M \otimes P^*, 1_{M \otimes P^*})$.

We now illustrate how to use the category $_{\Lambda}\mathscr{C}$ in computations, where we identify $_{\Lambda}\mathscr{C}$ with its image in mod Λ by the functor F.

Proposition 2.6 *Let* $\Lambda = \begin{pmatrix} T & 0 \\ T & T \end{pmatrix}$ *for an artin algebra* T. *Then we have* gl.dim $\Lambda =$ gl.dim $T + 1$.

Proof It is easy to see from the description of the projective objects in $_{\Lambda}\mathscr{C}$ that for any T-module A we have $\mathrm{pd}_{\Lambda}(A, A, 1_A) = \mathrm{pd}_T A = \mathrm{pd}_{\Lambda}(0, A, 0)$. It follows that if gl.dim $T = \infty$, then gl.dim $\Lambda = \infty$. Assume then that gl.dim $T = n < \infty$. Then there are simple T-modules S and S' with $\mathrm{Ext}_T^n(S, S') \neq 0$. The exact sequence $0 \to (0, S, 0) \to (S, S, 1_S) \to (S, 0, 0) \to 0$ implies that $\mathrm{pd}_{\Lambda}(S, 0, 0) \leq n + 1$ and gives rise to the exact sequence $\mathrm{Ext}_{\Lambda}^n((S, S, 1_S), (0, S', 0)) \to \mathrm{Ext}_{\Lambda}^n((0, S, 0), (0, S', 0)) \to \mathrm{Ext}_{\Lambda}^{n+1}((S, 0, 0), (0, S', 0)) \to 0$. Since $\mathrm{Hom}_{\Lambda}((P, P, 1_P), (0, S', 0)) = 0$ for each projective T-module P, we have $\mathrm{Ext}_{\Lambda}^n((S, S, 1_S), (0, S', 0)) = 0$. Because $\mathrm{Ext}_{\Lambda}^n((0, S, 0), (0, S', 0)) \simeq \mathrm{Ext}_T^n(S, S') \neq 0$, it follows that $\mathrm{Ext}_{\Lambda}^{n+1}((S, 0, 0), (0, S', 0)) \neq 0$, so that $\mathrm{pd}_{\Lambda}(S, 0, 0) = n + 1$. This shows that gl.dim $\Lambda = n + 1$. $\quad\square$

Proposition 2.7 *Suppose* T *and* U *are artin* R-*algebras with* T *semisimple and let* $_UM_T$ *be a nonzero bimodule where* R *acts centrally and* M *is finitely generated over* R. *If* $\Lambda = \begin{pmatrix} T & 0 \\ _UM_T & U \end{pmatrix}$, *then* gl.dim $\Lambda =$ max(gl.dim U, $\mathrm{pd}_U M + 1$).

Proof For a simple U-module S we have as above that $\mathrm{pd}_{\Lambda}(0, S, 0) = \mathrm{pd}_U S$. Further we have an exact sequence $0 \to (0, M, 0) \to (T, M, 1_M) \to (T, 0, 0) \to 0$ showing that $\mathrm{pd}_{\Lambda}(T, 0, 0) = \mathrm{pd}_U M + 1$, and we are done. $\quad\square$

The triangular matrix ring construction is a special case of the more general construction of **trivial extensions**. If V is a ring and $_VM_V$ a V-bimodule, we define the ring $V \ltimes M$ as follows. The elements are pairs (v, m) with $v \in V$ and $m \in M$, addition is componentwise and multiplication is given by $(v, m)(v', m') = (vv', vm' + mv')$. In other words,

$V \ltimes M$ is isomorphic to the tensor algebra $T(V,M)$ modulo the ideal generated by $M \otimes_V M$. For example the ring $k[X]/(X^2)$ for a field k is isomorphic to the trivial extension $k \ltimes k$. If $\Lambda = \begin{pmatrix} T & 0 \\ M & U \end{pmatrix}$ is a triangular matrix ring, let $V = T \times U$ and consider $_U M_T$ as a V-bimodule by $(t,u)m = um$ and $m(t,u) = mt$ for $t \in T$, $u \in U$ and $m \in M$. Then it is easy to see that $\Lambda \simeq V \ltimes M$.

3 Group algebras

Throughout this section G denotes a finite group, k a field and kG the group algebra of G over k. In this section we point out some other features of group algebras not shared by arbitrary artin algebras, which come from the fact that kG-modules can be viewed as group representations. In particular, we describe how for two kG-modules A and B, the k-vector spaces $A \otimes_k B$ and $\mathrm{Hom}_k(A,B)$ are considered as kG-modules. These observations are applied to obtain the classical result that kG is semisimple if and only if the characteristic of k does not divide the order of G. In Chapter V these considerations will be used to give a way of constructing almost split sequences when kG is not semisimple.

Before describing how $A \otimes_k B$ is considered as a kG-module when A and B are kG-modules, it is convenient to point out the following description of kG-modules and kG-morphisms.

Clearly G is a subgroup of the group of units in kG and k is a subring of kG. Therefore associated with a kG-module A is the structure of A as a k-vector space together with the operation of G on A given by $\sigma(a) = \sigma a$ for all σ in G and a in A having the following properties.

(i)   For each σ in G, the map $A \to A$ given by $a \mapsto \sigma a$ for all a in A is k-linear.

(ii) $(\sigma_1 \sigma_2)(a) = \sigma_1(\sigma_2 a)$ for all σ_1 and σ_2 in G and a in A.

(iii) $1a = a$ for all a in A where 1 is the identity in G.

Also it is not difficult to see that if A is a k-vector space and $G \times A \to A$ is an operation of G on A satisfying the above properties (i), (ii) and (iii), then the operation $kG \times A \to A$ given by $(\sum_{\sigma \in G} t_\sigma \sigma)(a) = \sum_{\sigma \in G} t_\sigma(\sigma(a))$, for all t_σ in k, σ in G and a in A, is the only way of considering A as a kG-module so that the induced operations of k and G on A are the ones we started with. Thus we see that the kG-modules can be considered as k-vector spaces A together with an action of G on A satisfying (i), (ii) and (iii).

Suppose A and B are kG-modules. Then it is easily seen that $\text{Hom}_{kG}(A, B)$ is the k-subspace of $\text{Hom}_k(A, B)$ consisting of all k-linear maps $f : A \to B$ satisfying $\sigma(f(a)) = f(\sigma a)$ for all σ in G and a in A.

Suppose now that A and B are kG-modules. Then in order to define a kG-module structure on the k-vector space $A \otimes_k B$ it suffices to describe an operation $G \times (A \otimes_k B) \to A \otimes_k B$ satisfying conditions (i), (ii) and (iii). We leave it to the reader to check that the operation $G \times (A \otimes_k B) \to A \otimes_k B$ given by $\sigma(a \otimes b) = \sigma a \otimes \sigma b$ for all σ in G and a in A and b in B satisfies conditions (i), (ii) and (iii). Unless stated to the contrary, this is the only way we consider $A \otimes_k B$ as a kG-module. It is also easily seen that if $f : B \to B'$ is a morphism of kG-modules, then $A \otimes f : A \otimes_k B \to A \otimes_k B'$ given by $(A \otimes f)(a \otimes b) = a \otimes f(b)$ for all a in A and b in B is a kG-morphism. Thus with each A in $\text{mod}\,kG$ is associated the exact functor $A \otimes_k - : \text{mod}\,kG \to \text{mod}\,kG$ given by $B \mapsto A \otimes_k B$ for all B in $\text{mod}\,kG$, and each $f : B \to B'$ in $\text{mod}\,kG$ goes to $A \otimes f : A \otimes_k B \to A \otimes_k B'$ in $\text{mod}\,kG$. Similarly we have the exact functor $- \otimes_k B : \text{mod}\,kG \to \text{mod}\,kG$. We now point out two important properties of these functors which are used constantly when dealing with the representation theory of kG.

We say that G **operates trivially** on a k-vector space V if $\sigma v = v$ for all σ in G and v in V. We denote by k the kG-module which is the k-vector space k together with the trivial action of G on k. This module is called the **trivial** kG-module. Then for each kG-module A the morphism $A \to k \otimes_k A$ given by $a \mapsto 1 \otimes a$ for all a in A is a kG-isomorphism functorial in A which we will usually consider an identification. We leave it to the reader to verify that this is indeed the case. We record the other important property of tensor products we need in the following.

Proposition 3.1 *Let A be a projective kG-module. Then $A \otimes_k B$ and $B \otimes_k A$ are projective kG-modules for all kG-modules B.*

Before giving the proof of this result we give an important application.

Corollary 3.2 *The following statements are equivalent for kG.*

(a) *The group algebra kG is semisimple.*

(b) *The trivial kG-module k is projective.*

Proof We know that kG is semisimple if and only if every kG-module is projective. Therefore (a) implies (b). Suppose k is projective. Then $A \simeq A \otimes_k k$ is projective for each kG-module A by Proposition 3.1. Hence (b) implies (a). $\square$

In order to determine precisely when the trivial kG-module is projective, it is convenient to make the following observations. For a kG-module A we denote by A^G the kG-submodule of A consisting of all a in A such that $\sigma a = a$ for all σ in G. Moreover, if $f : A \to B$ is a morphism in $\text{mod}\, kG$, then $f(A^G) \subset B^G$. Hence we obtain the fixed point functor $(\ )^G : \text{mod}\, kG \to \text{mod}\, kG$ given by $(\ )^G(A) = A^G$ and for $f : A \to B$ in $\text{mod}\, kG$, $(\ )^G(f) = f|_{A^G} : A^G \to B^G$. We now give another description of the fixed point functor.

Let A be a kG-module. Then it is easily seen that an element a of A is in A^G if and only if there is a kG-morphism $f : k \to A$ such that $f(1) = a$, where k is the trivial kG-module. Therefore for each A in $\text{mod}\, kG$ we have isomorphisms $\text{Hom}_{kG}(k, A) \to A^G$ functorial in A given by $f \mapsto f(1)$ for all f in $\text{Hom}_{kG}(k, A)$. Hence we have an isomorphism between $\text{Hom}_{kG}(k, \ )$ and the fixed point functor $(\ )^G$. We now apply these considerations to determine when k is a projective kG-module.

We have the kG-epimorphism $\epsilon : kG \to k$ given by $\epsilon(\sum_{\sigma \in G} t_\sigma \sigma) = \sum_{\sigma \in G} t_\sigma$. Therefore k is a projective kG-module if and only if there is a kG-morphism $f : k \to kG$ such that $\epsilon f = 1_k$. By our previous remarks, this is equivalent to saying that there is an element z in $(kG)^G$ such that $\epsilon(z) = 1$. Straightforward calculations show that $\sum_{\sigma \in G} t_\sigma \sigma$ is in $(kG)^G$ if and only if all the t_σ are the same element of k. For if $\sum_{\sigma \in G} t_\sigma \sigma$ is in $(kG)^G$, then $\sum_{\sigma \in G} t_\sigma \tau \sigma = \sum_{\sigma \in G} t_\sigma \sigma$ for all τ in G. Hence $t_1 = t_\sigma$ for all σ in G which means that all the t_σ are the same. On the other hand it is obvious that $\sum_{\sigma \in G} t_\sigma \sigma$ is in $(kG)^G$ when all the t_σ are the same. Hence $(kG)^G = k \sum_{\sigma \in G} \sigma$, which means $\epsilon((kG)^G) = |G|k$, where $|G|$ is the order of G. Therefore the trivial kG-module is projective if and only if $|G|$ is not zero in k, or equivalently, the characteristic of k does not divide $|G|$.

Summarizing, we have the following.

Theorem 3.3 *The following are equivalent for the group algebra kG.*

(a) *kG is semisimple.*
(b) *The trivial kG-module is projective.*
(c) *The characteristic of k does not divide $|G|$.*

We now finish the proof of Theorem 3.3 by giving a proof of Proposition 3.1. This will require some preliminary definitions and results. In these considerations we are not making any assumptions about the characteristic of k or the order of G.

Let A and B be kG-modules. Then $\text{Hom}_k(A, B)$ is a k-vector space on which we define the following operation of G. Given σ in G and f in $\mathrm{Hom}_k(A, B)$ we define σf in $\mathrm{Hom}_k(A, B)$ by $(\sigma f)(a) = \sigma(f(\sigma^{-1}a))$ for $a \in A$. We leave it to the reader to check that this operation makes $\mathrm{Hom}_k(A, B)$ a kG-module. Now it is not difficult to see that if $f : B \to B'$ is a morphism in $\mathrm{mod}\,kG$, then the induced morphism $\mathrm{Hom}_k(A, B) \overset{\mathrm{Hom}_k(A,f)}{\longrightarrow} \mathrm{Hom}_k(A, B')$ is a kG-morphism. Thus for A in $\mathrm{mod}\,kG$ we get an exact functor $\mathrm{Hom}_k(A, \ ) : \mathrm{mod}\,kG \to \mathrm{mod}\,kG$ given by $B \mapsto \mathrm{Hom}_k(A, B)$ for all B in $\mathrm{mod}\,kG$ and a morphism $f : B \to B'$ in $\mathrm{mod}\,kG$ goes to $\mathrm{Hom}_k(A, f) : \mathrm{Hom}_k(A, B) \to \mathrm{Hom}_k(A, B')$.

Similarly, if $g : A \to A'$ is in $\mathrm{mod}\,kG$ then $\mathrm{Hom}_k(g, B) : \mathrm{Hom}_k(A', B) \to \mathrm{Hom}_k(A, B)$ is a kG-morphism for each B in $\mathrm{mod}\,kG$. Therefore for each B in $\mathrm{mod}\,kG$ we get the contravariant exact functor $\mathrm{Hom}_k(\ , B) : \mathrm{mod}\,kG \to \mathrm{mod}\,kG$ given by $A \mapsto \mathrm{Hom}_k(A, B)$ and $g : A' \to A$ goes to $\mathrm{Hom}_k(g, B) : \mathrm{Hom}_k(A', B) \to \mathrm{Hom}_k(A, B)$.

Another important feature of the operation of G on $\mathrm{Hom}_k(A, B)$ is that $\mathrm{Hom}_{kG}(A, B) = \mathrm{Hom}_k(A, B)^G$ for all kG-modules A and B. For we have that an f in $\mathrm{Hom}_k(A, B)$ is in $\mathrm{Hom}_k(A, B)^G$ if and only if $\sigma f(\sigma^{-1}a) = (\sigma f)(a) = f(a)$ for all σ in G and a in A. But $\sigma f(\sigma^{-1}a) = f(a)$ if and only if $f(\sigma^{-1}a) = \sigma^{-1}f(a)$. Therefore f is in $\mathrm{Hom}_k(A, B)^G$ if and only if f is in $\mathrm{Hom}_{kG}(A, B)$.

We now point out the following basic connection between the functors $- \otimes_k -$ and $\mathrm{Hom}_k(\ ,)$.

Proposition 3.4 *Let A, B and C be in* $\mathrm{mod}\,kG$. *Then the morphism* α: $\mathrm{Hom}_{kG}(A, \mathrm{Hom}_k(B, C)) \to \mathrm{Hom}_{kG}(A \otimes_k B, C)$ *given by* $\alpha(f)(a \otimes b) = f(a)(b)$ *for all f in* $\mathrm{Hom}_{kG}(A, \mathrm{Hom}_k(B, C))$ *and all a in A and b in B is an isomorphism functorial in A, B and C.*

Proof We leave it to the reader to check that the isomorphism $\alpha' : \mathrm{Hom}_k(A, \mathrm{Hom}_k(B, C)) \to \mathrm{Hom}_k(A \otimes_k B, C)$ of k-vector spaces given by $\alpha'(f)(a \otimes b) = f(a)(b)$ for all f in $\mathrm{Hom}_k(A, \mathrm{Hom}_k(B, C))$ and all a in A and b in B is a kG-isomorphism functorial in A, B and C in $\mathrm{mod}\,kG$. Therefore α' induces an isomorphism α on fixed points, which gives our desired result.   □

As a consequence of this result we obtain the following proof of Proposition 3.1.

Proof Let A be a projective kG-module and let B be an arbitrary kG-module. We want to show that $A \otimes_k B$ is projective. Let $0 \to C' \to C \to C'' \to 0$ be an exact sequence of kG-modules. Since $\mathrm{Hom}_k(B, \ )$ is an exact functor and A is a projective kG-module, we obtain by applying Proposition 3.4 the following commutative exact diagram.

$$
\begin{array}{ccccccc}
\operatorname{Hom}_{kG}(A,\operatorname{Hom}_k(B,C')) & \to & \operatorname{Hom}_{kG}(A,\operatorname{Hom}_k(B,C)) & \to & \operatorname{Hom}_{kG}(A,\operatorname{Hom}_k(B,C'')) & \to & 0 \\
\downarrow \wr & & \downarrow \wr & & \downarrow \wr & & \\
\operatorname{Hom}_{kG}(A \otimes_k B,C') & \to & \operatorname{Hom}_{kG}(A \otimes_k B,C) & \to & \operatorname{Hom}_{kG}(A \otimes_k B,C'') & &
\end{array}
$$

This shows that $\operatorname{Hom}_{kG}(A \otimes_k B, C) \to \operatorname{Hom}_{kG}(A \otimes_k B, C'') \to 0$ is exact, which proves that $A \otimes_k B$ is projective. This finishes the proof of Proposition 3.1 as well as the proof of Theorem 3.3. $\qquad\square$

4 Skew group algebras

In this section we introduce the artin algebras known as skew group algebras, which are a natural generalization of group algebras of finite groups over fields discussed in the previous section. As in the case of group algebras, we also discuss some elementary features of the module theory of skew group algebras.

Throughout this section all groups considered are finite. Let Λ be an artin R-algebra. By an R-algebra automorphism $\sigma : \Lambda \to \Lambda$ we mean a ring automorphism of Λ with the additional property that $\sigma(r\lambda) = r\sigma(\lambda)$ for all r in R and λ in Λ.

Suppose now that G is a finite group. Then an R-algebra operation of G on Λ is a function $\varphi : G \times \Lambda \to \Lambda$ satisfying the following, where we write $\varphi(\sigma, \lambda) = \sigma(\lambda)$ for σ in G and λ in Λ:

(i) $\sigma : \Lambda \to \Lambda$ is an R-algebra automorphism for all σ in G.
(ii) $(\sigma_1 \sigma_2)(\lambda) = \sigma_1(\sigma_2(\lambda))$ for all σ_1 and σ_2 in G and λ in Λ.
(iii) $1\lambda = \lambda$ for all λ in Λ, where 1 is the identity element in G.

Suppose $G \times \Lambda \to \Lambda$ is an R-algebra operation of G on Λ. Then the skew group algebra of G over Λ, which we denote by ΛG, is given by the following data.

(a)   As an abelian group ΛG is the free left Λ-module with the elements of G as a basis.
(b)   The multiplication in ΛG is defined by the rule $(\lambda_\sigma \sigma)(\lambda_\tau \tau) = (\lambda_\sigma \sigma(\lambda_\tau))\sigma\tau$ for all λ_σ and λ_τ in Λ and σ and τ in G.

We leave it to the reader to check that ΛG is indeed a ring with identity element the formal product of the identity of Λ with the identity in G, also denoted by 1. It is also easily checked that the map $R \to \Lambda G$ given by $r \mapsto r1$ for all r in R makes ΛG an artin R-algebra. We follow the usual convention of identifying each σ in G with 1σ in ΛG. In this way G is a subgroup of the group of units in ΛG. Similarly, we identify each element λ in Λ with $\lambda 1$ in ΛG, so Λ becomes an R-subalgebra of ΛG.

We give some examples of skew group algebras.

Example The group algebra of a finite group G over a field k is an example of a skew group algebra over k where the operation of G on k is the trivial operation given by $\sigma(x) = x$ for all σ in G and x in k.

Example Suppose N and H are groups together with an operation of H on N satisfying $h(n_1n_2) = (hn_1)(hn_2)$ for all n_1 and n_2 in N and h in H. Let G be the semidirect product of N by H which as a set is the cartesian product $N \times H$ together with multiplication given by $(n_1, h_1)(n_2, h_2) = (n_1h_1(n_2), h_1h_2)$ for all n_1 and n_2 in N and all h_1 and h_2 in H. Let R be a commutative artin ring. Define an action of H on the group algebra RN by $h\sum_{x\in N} r_x x = \sum_{x\in N} r_x hx$ for all h in H, r_x in R and x in N. If we denote RN by Λ, it is easily seen that the action of H on Λ is an R-algebra action, so we can form the skew group algebra ΛH. Define the map $\alpha: RG \to \Lambda H$ by $\alpha(\sum_{(x,h)\in G} r_{(x,h)}(x,h)) = \sum_{h\in H}(\sum_{x\in N} r_{(x,h)}x)h$ where RG is the group algebra of G over R. It is not difficult to check that α is an R-algebra isomorphism. Thus the group algebra RG can also be viewed as the skew group algebra ΛH.

In particular, let N be the cyclic group $\mathbb{Z}/n\mathbb{Z}$ and let $H = \mathbb{Z}/2\mathbb{Z}$ with generator σ. Define the action of H on N by $\sigma(x) = -x$ for all x in N. Then $G = H \ltimes N$ is D_n, the nth dihedral group which is the group of symmetries of a regular n-gon in the plane. Therefore $RD_n \simeq \Lambda H$ where Λ is the group algebra RN.

Example Let k be a field of characteristic different from 2 and let Γ be the quiver

$$3 \xleftarrow{\beta} 2 \xleftarrow{\alpha} 1 \xrightarrow{\alpha'} 2' \xrightarrow{\beta'} 3'.$$

Denote $k\Gamma$ by Λ and let $G = \langle \sigma \rangle$ be the group of order 2. Consider the elements e_1, e_2, e_3, α and β in Λ and let $\sigma e_1 = e_1$, $\sigma e_2 = e_{2'}$, $\sigma e_3 = e_{3'}$, $\sigma\alpha = \alpha'$ and $\sigma\beta = \beta'$. Then there is only one way of extending σ to a k-algebra automorphism of Λ and this is the way we will consider G as a group of automorphisms of Λ. We now want to compute ΛG and we start by computing $(\Lambda/\mathfrak{r})G$.

We clearly have a decomposition $(\Lambda/\mathfrak{r})G = (ke_1)G \times (ke_2 \times ke_{2'})G \times (ke_3 \times ke_{3'})G$. Let $\widetilde{e}_1 = \frac{1}{2}(e_1 + e_1\sigma)$ and $\widetilde{\widetilde{e}}_1 = \frac{1}{2}(e_1 - e_1\sigma)$ which are nonzero idempotents in $(ke_1)G$, and we have $e_1 = \widetilde{e}_1 + \widetilde{\widetilde{e}}_1$. This gives a decomposition $(ke_1)G \simeq (kG)\widetilde{e}_1 \times (kG)\widetilde{\widetilde{e}}_1$. Further we obtain an isomorphism $(ke_2 \times ke_{2'})G \xrightarrow{\sim} \left(\begin{smallmatrix} k & k \\ k & k \end{smallmatrix}\right)$ by sending e_2 to $\left(\begin{smallmatrix} 1 & 0 \\ 0 & 0 \end{smallmatrix}\right)$, $e_{2'}$ to $\left(\begin{smallmatrix} 0 & 0 \\ 0 & 1 \end{smallmatrix}\right)$, $e_2\sigma$ to $\left(\begin{smallmatrix} 0 & 1 \\ 0 & 0 \end{smallmatrix}\right)$ and $e_{2'}\sigma$ to $\left(\begin{smallmatrix} 0 & 0 \\ 1 & 0 \end{smallmatrix}\right)$. Similarly we get an isomorphism $(ke_3 \times ke_{3'})G \simeq \left(\begin{smallmatrix} k & k \\ k & k \end{smallmatrix}\right)$ by sending e_3 to $\left(\begin{smallmatrix} 1 & 0 \\ 0 & 0 \end{smallmatrix}\right)$, $e_{3'}$ to $\left(\begin{smallmatrix} 0 & 0 \\ 0 & 1 \end{smallmatrix}\right)$, $e_3\sigma$ to $\left(\begin{smallmatrix} 0 & 1 \\ 0 & 0 \end{smallmatrix}\right)$ and $e_{3'}\sigma$ to $\left(\begin{smallmatrix} 0 & 0 \\ 1 & 0 \end{smallmatrix}\right)$. Hence $(\Lambda/\mathfrak{r})G$ is isomorphic to the k-algebra $k \times k \times \left(\begin{smallmatrix} k & k \\ k & k \end{smallmatrix}\right) \times \left(\begin{smallmatrix} k & k \\ k & k \end{smallmatrix}\right)$. Since the simple $\left(\begin{smallmatrix} k & k \\ k & k \end{smallmatrix}\right)$-modules $\left(\begin{smallmatrix} k & 0 \\ k & 0 \end{smallmatrix}\right)$ and $\left(\begin{smallmatrix} 0 & k \\ 0 & k \end{smallmatrix}\right)$ are isomorphic, we see that $e(\Lambda G)e$ is a reduced form of ΛG, where $e = \widetilde{e}_1 + \widetilde{\widetilde{e}}_1 + e_2 + e_3$. This means that $e(\Lambda G)e$ is a basic algebra which is Morita equivalent to ΛG.

We have that $\{e_1, e_2, e_{2'}, e_3, e_{3'}, \alpha, \alpha', \beta, \beta', \beta\alpha, \beta'\alpha', e_1\sigma, e_2\sigma, e_{2'}\sigma, e_3\sigma,$ $e_{3'}\sigma, \alpha\sigma, \alpha'\sigma, \beta\sigma, \beta'\sigma, \beta\alpha\sigma, \beta'\alpha'\sigma\}$ is a k-basis for ΛG. By multiplying on the left and right it is then easy to see that $\{\widetilde{e}_1, \widetilde{\widetilde{e}}_1, e_2, e_3, \frac{1}{2}(\alpha + \alpha\sigma), \frac{1}{2}(\alpha - \alpha\sigma), \beta, \frac{1}{2}\beta(\alpha+\alpha\sigma), \frac{1}{2}\beta(\alpha-\alpha\sigma)\}$ is a k-basis for $e\Lambda Ge$. Writing $\widetilde{\alpha} = \frac{1}{2}(\alpha+\alpha\sigma)$ and $\widetilde{\widetilde{\alpha}} = \frac{1}{2}(\alpha - \alpha\sigma)$, we see that $e\Lambda Ge$ is isomorphic to the path algebra of the quiver

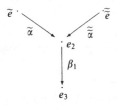

This last example illustrates that even if we only want to deal with basic artin algebras the skew group algebra construction may lead to an algebra which is not basic. We have here seen how to construct the associated basic algebra in a concrete situation.

We now give a description of the modules over a skew group algebra ΛG over an artin R-algebra Λ, which closely parallels the description given in Section 3 of the modules over a group algebra.

Suppose X is a ΛG-module. Since Λ is a subring of ΛG, the ΛG-module X is also a Λ-module. The fact that G is a subgroup of the group of units in ΛG commuting with the action of R gives an operation $G \times X \to X$ satisfying the following.

(i) $\sigma(\lambda x) = \sigma(\lambda)\sigma(x)$ for all σ in G, λ in Λ and x in X.

(ii) $\sigma(rx) = r\sigma(x)$ for all σ in G, r in R and x in X.

(iii) $(\sigma_1\sigma_2)(x) = \sigma_1(\sigma_2 x)$ for all σ_1, σ_2 in G and x in X.

(iv) $1x = x$ for all x in X when 1 is the identity element in G.

Thus associated with a ΛG-module X is the Λ-module X together with an operation of G on X satisfying conditions (i), (ii), (iii) and (iv) above.

On the other hand, suppose we are given a Λ-module Y and an operation of G on Y satisfying (i), (ii), (iii) and (iv) above. It is easily checked that Y is a ΛG-module by means of the operation $(\sum_{\sigma \in G} \lambda_\sigma \sigma)y = \sum_{\sigma \in G} \lambda_\sigma(\sigma(y))$ for all $\sum_{\sigma \in G} \lambda_\sigma \sigma$ in ΛG and y in Y. Hence we see that the ΛG-modules where the operation of R commutes with the operation of G can be viewed as Λ-modules together with an operation of G satisfying (i), (ii), (iii) and (iv). We will use these points of view interchangeably. As an illustration, we now describe the morphisms between ΛG-modules from this alternative point of view.

Let X and Y be ΛG-modules. Since Λ is an R-subalgebra of ΛG, we have that each f in $\text{Hom}_{\Lambda G}(X, Y)$ is a Λ-morphism from X to Y. Because G is contained in ΛG, it follows that each f in $\text{Hom}_{\Lambda G}(X, Y)$ also satisfies $f(\sigma x) = \sigma f(x)$ for all σ in G and x in X. Now it can be easily checked that if f in $\text{Hom}_\Lambda(X, Y)$ satisfies $f(\sigma x) = \sigma f(x)$ for all σ in G and x in X, then f is in $\text{Hom}_{\Lambda G}(X, Y)$. Therefore $\text{Hom}_{\Lambda G}(X, Y) = \{f \in \text{Hom}_\Lambda(X, Y) | f(\sigma x) = \sigma f(x)$ for all σ in G and x in $X\}$.

We now illustrate some of these points by considering some special, but important, modules.

As for group algebras, associated with each ΛG-module X is the R-submodule X^G of X consisting of the fixed points in X, i.e. all elements x in X such that $\sigma(x) = x$ for all σ in G. Also it is clear that if $f : X \to Y$ is a ΛG-module morphism, then $f(X^G) \subset Y^G$. It is further easily seen that $f|_{X^G} : X^G \to Y^G$ is an R-module morphism. We then obtain the fixed point functor $(\ )^G : \text{mod} \Lambda G \to \text{mod} R$ given by $X \mapsto X^G$ and $f \mapsto f|_{X^G}$ for all objects X and morphisms f in $\text{mod} \Lambda G$. As in the case of group algebras, this important functor has an alternative description which we now give.

Unless stated to the contrary, we always consider the Λ-module Λ as a ΛG-module by means of the operation $(\sum_{\sigma \in G} \lambda_\sigma \sigma)(x) = \sum_{\sigma \in G} \lambda_\sigma \sigma(x)$ for each $x \in \Lambda$ and $\sum_{\sigma \in G} \lambda_\sigma \sigma \in \Lambda G$. Next we consider the map $\epsilon : \Lambda G \to \Lambda$ given by $\epsilon(\sum_{\sigma \in G} \lambda_\sigma \sigma) = \sum_{\sigma \in G} \lambda_\sigma$. This is easily seen to be a ΛG-epimorphism with $\text{Ker}\,\epsilon$ the left ideal of ΛG generated over Λ by $\{\sigma - 1 | \sigma \in G - \{1\}\}$. For each ΛG-module X it follows that $\text{Hom}_{\Lambda G}(\Lambda, X)$ consists precisely of those Λ-morphisms $f : \Lambda \to X$ such that $f(1)$ is in X^G. This implies that for each ΛG-module X, the map $\text{Hom}_{\Lambda G}(\Lambda, X) \to X^G$ given by $f \mapsto f(1)$ is an R-module isomorphism functorial in X. Thus we obtain a canonical isomorphism between the functors $\text{Hom}_{\Lambda G}(\Lambda, \ )$ and $(\ )^G$ showing that our ΛG-module Λ is a natural generalization of the trivial module for group algebras. For this reason we will call the ΛG-module Λ the trivial ΛG-module. Having described in Section 3 when the trivial module over a group algebra is a projective module, it is natural to wonder more generally when the trivial module over a skew group algebra is a projective module. One such criterion is given in the following.

Proposition 4.1 *The trivial ΛG-module Λ is a projective ΛG-module if and only if there is some λ in Λ such that $\sum_{\sigma \in G} \sigma(\lambda) = 1$.*

Proof Since $\epsilon \colon \Lambda G \to \Lambda$ is an epimorphism, Λ is a projective ΛG-module if and only if there is a ΛG-morphism $g \colon \Lambda \to \Lambda G$ such that $\epsilon g = 1_\Lambda$. But $\epsilon g = 1_\Lambda$ if and only if $\epsilon g(1) = 1$. In view of the isomorphism $\text{Hom}_{\Lambda G}(\Lambda, \Lambda G) \to (\Lambda G)^G$ given by $f \mapsto f(1)$, we know there is some $g \colon \Lambda \to \Lambda G$ such that $\epsilon g(1) = 1$ if and only if there is some $\sum_{\sigma \in G} \lambda_\sigma \sigma$ in $(\Lambda G)^G$ such that $\epsilon(\sum_{\sigma \in G} \lambda_\sigma \sigma) = \sum_{\sigma \in G} \lambda_\sigma = 1$. Therefore it is essential to have a description of $(\Lambda G)^G$. But it is fairly straightforward to show that $(\Lambda G)^G = \{\sum_{\sigma \in G} \sigma(\lambda)\sigma | \lambda \in \Lambda\}$, as we shall see in the next result. Therefore Λ is ΛG-projective if and only if there is an element $\sum_{\sigma \in G} \sigma(\lambda)\sigma$ in ΛG such that $1 = \epsilon(\sum_{\sigma \in G} \sigma(\lambda)\sigma) = \sum_{\sigma \in G} \sigma(\lambda)$, which is our desired result. $\square$

We now prove our required lemma.

Lemma 4.2 $(\Lambda G)^G = \{\sum_{\sigma \in G} \sigma(\lambda)\sigma | \lambda \in \Lambda\}$.

Proof We first show that the elements of the form $\sum_{\sigma \in G} \sigma(\lambda)\sigma$ for each λ in Λ are in $(\Lambda G)^G$. This follows from the fact that $\tau(\sum_{\sigma \in G} \sigma(\lambda)\sigma) = \sum_{\sigma \in G} \tau(\sigma(\lambda)\sigma) = \sum_{\sigma \in G} \tau\sigma(\lambda)\tau\sigma = \sum_{\sigma \in G} \sigma(\lambda)\sigma$ for all τ in G.

Suppose now that $\sum_{\sigma \in G} \lambda_\sigma \sigma$ is in $(\Lambda G)^G$. Then for each τ in G we have that $\sum_{\sigma \in G} \lambda_\sigma \sigma = \tau \sum_{\sigma \in G} \lambda_\sigma \sigma = \sum_{\sigma \in G} \tau(\lambda_\sigma)\tau\sigma$. But this implies that $\tau(\lambda_1) = \lambda_\tau$ for all τ in G, or in other words, $\sum_{\sigma \in G} \lambda_\sigma \sigma = \sum_{\tau \in G} \tau(\lambda_1)\tau$, which is our desired result. $\square$

As an application of this criterion for the trivial ΛG-module to be a projective module, we have the following.

Corollary 4.3 *Let ΛG be a skew group algebra with G a group of order n.*

(a) *If n is invertible in Λ, then the trivial ΛG-module Λ is a projective module.*

(b) *Suppose G operates trivially on Λ, i.e. $\sigma\lambda = \lambda$ for all σ in G and λ in Λ. Then Λ is a projective ΛG-module if and only if n is invertible in Λ.*

Proof (a) Suppose $1/n$ in Λ is the inverse of n. Then for each σ in G we have $1 = \sigma(1) = \sigma(n \cdot 1/n) = n\sigma(1/n)$. Hence $\sigma(1/n) = 1/n$ for all σ in G. Therefore $\sum_{\sigma \in G} \sigma(1/n) = 1$, which by Proposition 4.1 implies that Λ is a projective ΛG-module.

(b) In view of part (a) we only have to show that if G operates trivially on Λ and Λ is a projective ΛG-module, then n is invertible in Λ. By Proposition 4.1, since Λ is a projective ΛG-module, there is some λ in Λ such that $\sum_{\sigma \in G} \sigma(\lambda) = 1$. Since G operates trivially on Λ, we have that $n\lambda = 1$ so that λ is the inverse of n in Λ.   □

The following example shows that it is possible for Λ to be a projective ΛG-module without the order of G being invertible in Λ.

Example Let k be a field and let Λ be the subalgebra of the 4×4 matrix algebra over k consisting of all matrices of the form $\begin{pmatrix} a & 0 & 0 & 0 \\ 0 & b & 0 & 0 \\ 0 & c & a & 0 \\ d & 0 & 0 & b \end{pmatrix}$ with a, b, c, d in k. Now the invertible matrix

$$\begin{pmatrix} 0 & 1 & 0 & 0 \\ 1 & 0 & 0 & 0 \\ 0 & 0 & 0 & 1 \\ 0 & 0 & 1 & 0 \end{pmatrix}$$

is of order 2 and acts as a k-automorphism ϕ of Λ by conjugation. Let $G = \{1, \phi\}$. From the equality

$$\begin{pmatrix} 1 & 0 & 0 & 0 \\ 0 & 1 & 0 & 0 \\ 0 & 0 & 1 & 0 \\ 0 & 0 & 0 & 1 \end{pmatrix} = \begin{pmatrix} 1 & 0 & 0 & 0 \\ 0 & 0 & 0 & 0 \\ 0 & 0 & 1 & 0 \\ 0 & 0 & 0 & 0 \end{pmatrix} + \phi \begin{pmatrix} 1 & 0 & 0 & 0 \\ 0 & 0 & 0 & 0 \\ 0 & 0 & 1 & 0 \\ 0 & 0 & 0 & 0 \end{pmatrix}$$

we see that Λ is a projective ΛG-module. In particular, Λ is a projective ΛG-module even when char $k = 2$. Therefore Λ is a projective ΛG-module even though 2, the order of G, is zero in Λ.

For the rest of this section we will be mainly concerned with skew group R-algebras ΛG with the order of G invertible in Λ. This is not because arbitrary skew group R-algebras are not of interest. Rather, it is because this additional hypothesis gives a simpler and better understood module theory. Our aim now is to prove the following.

Theorem 4.4 *Let ΛG be a skew group R-algebra where the order of G is invertible in Λ. Then* gl.dim Λ = gl.dim ΛG.

The proof of this result will take several steps and involves general concepts of independent interest. We begin by pointing out the following relationships between the subalgebra Λ of a skew group algebra ΛG, and ΛG. In particular, since Λ is a subalgebra of ΛG, ΛG is both a left Λ-module and a right Λ-module. It is these structures which are of particular interest to us.

Lemma 4.5

(a) *The group $G \subset \Lambda G$ is a basis for ΛG as a right as well as a left Λ-module.*

(b) *Let C be the subset of ΛG consisting of all elements $\sum_{\sigma \in G} \lambda_\sigma \sigma$ satisfying $\lambda_1 = 0$. Then C is a Λ-subbimodule of ΛG and we have that $\Lambda G \simeq \Lambda \coprod C$ as Λ-bimodules.*

Proof (a) By definition ΛG is a free left Λ-module with basis G. We now show that G is also a basis for ΛG as a right Λ-module. Since $\sum_{\sigma \in G} \lambda_\sigma \sigma = \sum_{\sigma \in G} \sigma(\sigma^{-1}\lambda_\sigma)$, it follows that G generates ΛG as a right Λ-module. Also if $\sum_{\sigma \in G} \sigma \lambda_\sigma = 0$, then $\sum_{\sigma \in G} \sigma(\lambda_\sigma)\sigma = 0$ which means that each $\sigma(\lambda_\sigma) = 0$ or equivalently, $\lambda_\sigma = 0$ for each σ in G. This shows that G is a basis for ΛG as a right Λ-module.

(b) It is clear that C is a left Λ-submodule of ΛG. Now for each λ in Λ and $\sum_{\sigma \in G} \lambda_\sigma \sigma$ in ΛG we have that $(\sum_{\sigma \in G} \lambda_\sigma \sigma)\lambda = \sum_{\sigma \in G} \lambda_\sigma \sigma(\lambda)\sigma$. From this it follows that if $\sum_{\sigma \in G} \lambda_\sigma \sigma$ is in C, then $(\sum_{\sigma \in G} \lambda_\sigma \sigma)\lambda$ is in C for each λ in Λ. The fact that $\Lambda \coprod C \simeq \Lambda G$ as both left and right Λ-modules now follows trivially from (a) and hence $\Lambda \coprod C \simeq \Lambda G$ as Λ-bimodules. $\square$

We now use these observations to show that gl.dim $\Lambda G \geq$ gl.dim Λ. In fact, we prove the following more general result.

Proposition 4.6 *Let Λ be a subalgebra of the R-algebra Γ satisfying the following two conditions.*

(i) *Γ is a projective left Λ-module.*

(ii) *There is a subgroup C of Γ which is a Λ-subbimodule of Γ such that $\Gamma = \Lambda \coprod C$.*

Then we have gl.dim $\Gamma \geq$ gl.dim Λ.

Proof Let X be a Λ-module. Then $\Gamma \otimes_\Lambda X$ is a Γ-module by means of the operation $\gamma(\gamma_1 \otimes x) = \gamma\gamma_1 \otimes x$. Let

$$(*) \qquad \cdots \to P_t \to \cdots \to P_1 \to P_0 \to \Gamma \otimes_\Lambda X \to 0$$

be a minimal projective Γ-resolution of $\Gamma \otimes_\Lambda X$. Since Γ is a projective left Λ-module, the exact sequence $(*)$ when viewed as an exact sequence of Λ-modules is a Λ-projective resolution of $\Gamma \otimes_\Lambda X$ viewed as a Λ-module, and so $\mathrm{pd}_\Gamma(\Gamma \otimes_\Lambda X) \geq \mathrm{pd}_\Lambda(\Gamma \otimes_\Lambda X)$. The fact that $\Gamma = \Lambda \coprod C$ as Λ-bimodules gives that $\Gamma \otimes_\Lambda X = (\Lambda \otimes_\Lambda X) \coprod (C \otimes_\Lambda X)$ as left Λ-modules, where the action of Λ on $C \otimes_\Lambda X$ is given by $\lambda(c \otimes x) = \lambda c \otimes x$ for all x in Λ, c in C and x in X. Because $\Lambda \otimes_\Lambda X \simeq X$, we have that X is a Λ-summand of $\Gamma \otimes_\Lambda X$. Therefore $\mathrm{pd}_\Lambda X \leq \mathrm{pd}_\Lambda(\Gamma \otimes_\Lambda X)$ and so $\mathrm{pd}_\Lambda X \leq \mathrm{pd}_\Gamma(\Gamma \otimes_\Lambda X)$. The fact that this holds for all Λ-modules X implies that gl.dim $\Lambda \leq$ gl.dim Γ. $\qquad \square$

Combining this proposition with Lemma 4.5, we have the following.

Corollary 4.7 *The skew group algebra ΛG has the property that* gl.dim $\Lambda \leq$ gl.dim ΛG. $\qquad \square$

Thus in order to finish the proof of Theorem 4.4, it suffices to show that gl.dim $\Lambda G \leq$ gl.dim Λ when the order of G is invertible in Λ. Our proof of this result is based on the following where we consider $\Lambda G \otimes_\Lambda Y$ as a ΛG-module by $\gamma(v \otimes y) = \gamma v \otimes y$ for $\gamma \in G$ and $v \otimes y \in \Lambda G \otimes_\Lambda Y$.

Lemma 4.8 *Suppose ΛG is a skew group algebra with the order of G invertible in Λ. Then Y is a ΛG-summand of $\Lambda G \otimes_\Lambda Y$ for all ΛG-modules Y.*

Proof For each ΛG-module Y define the multiplication map m_Y : $\Lambda G \otimes_\Lambda Y \to Y$ by $m_Y(\sum_{\sigma \in G} \lambda_\sigma \sigma \otimes y) = (\sum_{\sigma \in G} \lambda_\sigma \sigma)y$. It is easily seen that m_Y is a ΛG-epimorphism. We now describe a ΛG-morphism $h_Y : Y \to \Lambda G \otimes_\Lambda Y$ with the property that $m_Y h_Y = 1_Y$, which implies that Y is a ΛG-summand of $\Lambda G \otimes_\Lambda Y$.

Define $h_Y : Y \to \Lambda G \otimes_\Lambda Y$ by $h_Y(y) = \sum_{\sigma \in G} \sigma \otimes \sigma^{-1}(1/n)y$ for all y in Y where $1/n$ is the inverse of n, the order of G in Λ. It is clear that h_Y is additive. Also for each λ in Λ we have that $h_Y(\lambda y) = \sum_{\sigma \in G} \sigma \otimes \sigma^{-1}(\lambda y/n) = \sum_{\sigma \in G}(\sigma \otimes \sigma^{-1}(\lambda)\sigma^{-1}(y/n)) = \sum_{\sigma \in G}(\sigma \cdot \sigma^{-1}(\lambda) \otimes \sigma^{-1}(y)) = \sum_{\sigma \in G}(\lambda\sigma \otimes \sigma^{-1}(y/n)) = \lambda h_Y(y)$. So h_Y is a Λ-morphism. The reader may check that $h_Y(\sigma y) = \sigma h_Y(y)$ for all

σ in G and y in Y. So h_Y is in fact a ΛG-morphism. Now the composition $m_Y h_Y : Y \to Y$ is 1_Y since $m_Y h_Y(y) = m_Y \sum_{\sigma \in G} \sigma \otimes \sigma^{-1}(y/n) = \sum_{\sigma \in G} \sigma \sigma^{-1}(1/n)y = n(1/n)y = y$ for all y in Y. Therefore Y is isomorphic to a ΛG-summand of $\Lambda G \otimes_\Lambda Y$. □

We now obtain our desired result as a special case of the following more general result.

Proposition 4.9 *Suppose* Λ *is an R-subalgebra of an artin R-algebra* Γ *having the following properties.*

(i) Γ *is a projective right* Λ-*module.*
(ii) *Each* Γ-*module* Y *is isomorphic over* Γ *to a summand of* $\Gamma \otimes_\Lambda Y$. *Then we have* gl.dim $\Gamma \leq$ gl.dim Λ.

Proof Let Y be a Γ-module. Then viewing Y as a Λ-module, we have a Λ-projective resolution

$$\cdots \to P_t \to \cdots \to P_0 \to Y \to 0.$$

Since Γ is right Λ-projective, the sequence

$$\cdots \to \Gamma \underset{\Lambda}{\otimes} P_t \to \cdots \to \Gamma \underset{\Lambda}{\otimes} P_0 \to \Gamma \underset{\Lambda}{\otimes} Y \to 0$$

is exact and is therefore a projective Γ-resolution of $\Gamma \otimes_\Lambda Y$. Hence we get $\text{pd}_\Gamma \Gamma \otimes_\Lambda Y \leq \text{pd}_\Lambda Y$ for all Γ-modules Y. But $\text{pd}_\Gamma Y \leq \text{pd}_\Gamma \Gamma \otimes_\Lambda Y$ since Y is a summand of $\Gamma \otimes_\Lambda Y$, and hence gl.dim $\Gamma \leq$ gl.dim Λ. □

Combining this proposition with Lemma 4.8 we obtain the following result which finishes the proof of Theorem 4.4.

Corollary 4.10 *Let* ΛG *be a skew group algebra with the order of* G *invertible in* Λ. *Then we have* gl.dim $\Lambda G \leq$ gl.dim Λ. □

As an application of Theorem 4.4, we give a description of the radical of ΛG when the order of G is invertible in Λ. To this end it is convenient to make the following general remarks.

We say that an ideal $\mathfrak{a}$ in Λ is G-invariant if $\sigma \mathfrak{a} \subset \mathfrak{a}$ for all σ in G. Clearly this is the case if and only if $\mathfrak{a}$ is a ΛG-submodule of Λ. If $\mathfrak{a}$ is a G-invariant ideal of Λ, then we can define an R-algebra operation of G on $\Lambda/\mathfrak{a}$ by $\sigma(\lambda + \mathfrak{a}) = \sigma(\lambda) + \mathfrak{a}$ for all σ in G and λ in Λ. We also have the natural surjective R-algebra morphism $h : \Lambda \to \Lambda/\mathfrak{a}$

which induces the natural R-algebra morphism $t: \Lambda G \to (\Lambda/\mathfrak{a})G$ given by $t(\sum_{\sigma \in G} \lambda_\sigma \sigma) = \sum_{\sigma \in G} h(\lambda_\sigma)\sigma$ for all λ_σ in Λ and σ in G. It is also trivial to check that $\operatorname{Ker} t = \mathfrak{a}\Lambda G$. In particular, the fact that $\sigma: \Lambda \to \Lambda$ is an R-algebra automorphism implies that the radical $\mathfrak{r}$ of Λ is a G-invariant ideal of Λ. Since $(\mathfrak{r}\Lambda G)^i = \mathfrak{r}^i \Lambda G$ for all i, we have that $\mathfrak{r}\Lambda G$ is a nilpotent ideal and therefore $\mathfrak{r}\Lambda G$ is contained in $\operatorname{rad}(\Lambda G)$. Combining these remarks with Theorem 4.4 we have the following.

Proposition 4.11 *Suppose ΛG is a skew group algebra with the order of G invertible in Λ. Then $\mathfrak{r}\Lambda G = \operatorname{rad}(\Lambda G)$ where $\mathfrak{r}$ is the radical of Λ.*

Proof We have already established that $\mathfrak{r}\Lambda G$ is contained in $\operatorname{rad}(\Lambda G)$. Therefore we will have our desired equality if we show that $\Lambda G/\mathfrak{r}\Lambda G$ is semisimple. We have already observed that $\Lambda G/\mathfrak{r}\Lambda G \simeq (\Lambda/\mathfrak{r})G$. The fact that the order of G is invertible in Λ implies that the order of G is invertible in the semisimple algebra $\Lambda/\mathfrak{r}$. Therefore by Theorem 4.4, we have that $\operatorname{gl.dim}(\Lambda/\mathfrak{r})G = \operatorname{gl.dim}(\Lambda/\mathfrak{r}) = 0$. □

We have already seen that the functor $\Lambda G \otimes_\Lambda -: \operatorname{mod}\Lambda \to \operatorname{mod}\Lambda G$ given by $X \mapsto \Lambda G \otimes_\Lambda X$ for all X in $\operatorname{mod}\Lambda$ plays an important role in studying the module theory of ΛG. Another equally natural functor from $\operatorname{mod}\Lambda$ to $\operatorname{mod}\Lambda G$ is given as follows. For each X in $\operatorname{mod}\Lambda$ we consider $\operatorname{Hom}_\Lambda(\Lambda G, X)$ a ΛG-module by means of the operation $(xf)(y) = f(yx)$ for all x and y in ΛG and f in $\operatorname{Hom}_\Lambda(\Lambda G, X)$. It is not difficult to see that if $f: X \to Y$ is a morphism in $\operatorname{mod}\Lambda$, then $\operatorname{Hom}_\Lambda(\Lambda G, f): \operatorname{Hom}_\Lambda(\Lambda G, X) \to \operatorname{Hom}_\Lambda(\Lambda G, Y)$ is a ΛG-morphism. This data defines the functor $\operatorname{Hom}_\Lambda(\Lambda G, \ ): \operatorname{mod}\Lambda \to \operatorname{mod}\Lambda G$ given by $X \mapsto \operatorname{Hom}_\Lambda(\Lambda G, X)$. Our aim now is to show that $\Lambda G \otimes_\Lambda -$ and $\operatorname{Hom}_\Lambda(\Lambda G, \ )$ are isomorphic functors.

We first consider the following more general situation. Let Λ be an R-subalgebra of the artin R-algebra Γ. Clearly we can define the functors $\Gamma \otimes_\Lambda -: \operatorname{mod}\Lambda \to \operatorname{mod}\Gamma$ and $\operatorname{Hom}_\Lambda(\Gamma, \ ): \operatorname{mod}\Lambda \to \operatorname{mod}\Gamma$ as we did above in the special case $\Gamma = \Lambda G$. We then have the following criterion for when these functors are isomorphic, where it is understood that $\operatorname{Hom}_\Lambda(\Gamma, \Lambda)$ is considered as a Γ-Λ-bimodule by means of the operations $(\gamma f)(x) = f(x\gamma)$ for all f in $\operatorname{Hom}_\Lambda(\Gamma, \Lambda)$ and γ and x in Γ and $(f\lambda)(x) = f(x)\lambda$ for all λ in Λ, x in Γ and f in $\operatorname{Hom}_\Lambda(\Gamma, \Lambda)$.

Proposition 4.12 *Let Λ be an R-subalgebra of the artin R-algebra Γ. Then the following are equivalent.*

(a) *The functors* $\text{Hom}_\Lambda(\Gamma, \ )$ *and* $\Gamma \otimes_\Lambda -$ *from* $\text{mod}\,\Lambda$ *to* $\text{mod}\,\Gamma$ *are isomorphic.*

(b) *Both of the following conditions are satisfied.*

   (i) Γ *is a projective left* Λ-*module.*

   (ii) Γ *and* $\text{Hom}_\Lambda(\Gamma, \Lambda)$ *are isomorphic as* Γ-Λ-*bimodules.*

Proof (a) $\Rightarrow$ (b) Since $\Gamma \otimes_\Lambda -$ is right exact and $\Gamma \otimes_\Lambda -$ is isomorphic to $\text{Hom}_\Lambda(\Gamma, \ )$, it follows that $\text{Hom}_\Lambda(\Gamma, \ )$ is right exact. Hence $\text{Hom}_\Lambda(\Gamma, \ )$ is exact, which is equivalent to Γ being a projective left Λ-module.

In order to prove that (ii) is satisfied let $\alpha_X : \Gamma \otimes_\Lambda X \xrightarrow{\sim} \text{Hom}_\Lambda(\Gamma, X)$ be an isomorphism functorial in X for all X in $\text{mod}\,\Lambda$. Now $\Gamma \otimes_\Lambda X$ is an $\text{End}_\Lambda(X)$-module by means of the operation $h(\gamma \otimes x) = \gamma \otimes h(x)$ for all h in $\text{End}_\Lambda(X)$, γ in Γ and x in X. It is not difficult to check that this operation of $\text{End}_\Lambda(X)$ on $\Gamma \otimes_\Lambda X$ makes $\Gamma \otimes_\Lambda X$ a Γ-$\text{End}_\Lambda(X)$-bimodule. Also $\text{Hom}_\Lambda(\Gamma, X)$ is an $\text{End}_\Lambda(X)$-module by means of the operation $(hf)(\gamma) = h(f(\gamma))$ for all h in $\text{End}_\Lambda(X)$, f in $\text{Hom}_\Lambda(\Gamma, X)$ and γ in Γ. It is also not difficult to check that this operation of $\text{End}_\Lambda(X)$ on $\text{Hom}_\Lambda(\Gamma, X)$ makes $\text{Hom}_\Lambda(\Gamma, X)$ a Γ-$\text{End}_\Lambda(X)$-bimodule. Since the isomorphisms $\alpha_X : \Gamma \otimes_\Lambda X \rightarrow \text{Hom}_\Lambda(\Gamma, X)$ are functorial in X, they are Γ-$\text{End}_\Lambda(X)$-bimodule isomorphisms. In particular, letting $X = \Lambda$ we have that $\alpha_\Lambda : \Gamma \otimes_\Lambda \Lambda \rightarrow \text{Hom}_\Lambda(\Gamma, \Lambda)$ is a Γ-$\text{End}_\Lambda(\Lambda)$ bimodule isomorphism. Since $\text{End}_\Lambda(\Lambda) \simeq \Lambda^{\text{op}}$ it follows that $\alpha_\Lambda : \Gamma \rightarrow \text{Hom}_\Lambda(\Gamma, \Lambda)$ is a Γ-Λ-bimodule isomorphism.

(b) $\Rightarrow$ (a) Since Γ is a finitely generated projective left Λ-module we have that the Γ-morphisms $\beta_X : \text{Hom}_\Lambda(\Gamma, \Lambda) \otimes_\Lambda X \rightarrow \text{Hom}_\Lambda(\Gamma, X)$ given by $\beta_X(f \otimes x)(\gamma) = f(\gamma)x$ for all f in $\text{Hom}_\Lambda(\Gamma, \Lambda)$ and γ in Γ and x in X are isomorphisms functorial in X. Let $\alpha : \Gamma \rightarrow \text{Hom}_\Lambda(\Gamma, \Lambda)$ be a Γ-Λ-bimodule isomorphism. Then $\alpha \otimes X : \Gamma \otimes_\Lambda X \rightarrow \text{Hom}_\Lambda(\Gamma, \Lambda) \otimes_\Lambda X$ given by $(\alpha \otimes X)(\gamma \otimes x) = \alpha(\gamma) \otimes x$ for all γ in Γ and x in X, is a Γ-isomorphism functorial in X in $\text{mod}\,\Lambda$. Therefore the composition $\beta_X \alpha_X : \Gamma \otimes_\Lambda X \rightarrow \text{Hom}_\Lambda(\Gamma, X)$ is a Γ-isomorphism functorial in X. $\quad\square$

We now apply Proposition 4.12 to the subalgebra Λ of the skew group algebra ΛG. By definition we know that ΛG is a free left Λ-module. Therefore by Proposition 4.12, to prove that the functors $\Lambda G \otimes_\Lambda$ and $\text{Hom}_\Lambda(\Lambda G, \ )$ from $\text{mod}\,\Lambda$ to $\text{mod}\,\Gamma$ are isomorphic, it suffices to show the following.

Proposition 4.13 *Let* ϵ_1 *in* $\text{Hom}_\Lambda(\Lambda G, \Lambda)$ *be defined by* $\epsilon_1(\sum_{\sigma \in G} \lambda_\sigma \sigma) = \lambda_1$

for all $\sum_{\sigma \in G} \lambda_\sigma \sigma$ *in* ΛG. *Then the map* $f : \Lambda G \to \text{Hom}_\Lambda(\Lambda G, \Lambda)$ *given by* $f(x) = x\epsilon_1$ *for all* x *in* ΛG *is a* ΛG-Λ-*bimodule isomorphism.*

Proof By definition f is a ΛG-morphism. For each λ in Λ we have that $f((\sum_{\sigma \in G} \lambda_\sigma \sigma)(\lambda)) = f(\sum_{\sigma \in G} \lambda_\sigma \sigma(\lambda)\sigma) = \sum_{\sigma \in G} \lambda_\sigma \sigma(\lambda)\sigma\epsilon_1$. Now $(\sum \lambda_\sigma \sigma(\lambda)\sigma\epsilon_1)(\tau) = \epsilon_1(\tau \sum \lambda_\sigma \sigma(\lambda)\sigma) = \epsilon_1(\sum_{\sigma \in G} \tau(\lambda_\sigma)\tau\sigma(\lambda)\tau\sigma) = \tau(\lambda_{\tau^{-1}})\lambda$ for all τ in G. On the other hand for each λ in Λ we have that $(\sum_{\sigma \in G} \lambda_\sigma \sigma\epsilon_1)\lambda(\tau) = (\epsilon_1(\tau \sum \lambda_\sigma \sigma))\lambda = \epsilon_1(\sum \tau(\lambda_\sigma)\tau\sigma)\lambda = (\tau(\lambda_{\tau^{-1}}))\lambda$ for all τ in G. Therefore f is a right Λ-morphism and hence a bimodule morphism. We now show that f is an isomorphism.

Suppose $f(\sum \lambda_\sigma \sigma) = 0$. Then for each τ in G we have $\epsilon_1(\tau \sum \lambda_\sigma \sigma) = \tau(\lambda_{\tau^{-1}}) = 0$. So $\lambda_{\tau^{-1}} = 0$ for all τ in G which means that $\sum \lambda_\sigma \sigma = 0$, i.e. f is a monomorphism. Since ΛG is a free Λ-module with G as basis, it suffices to show that for each τ in G and λ in Λ there is some $\sum \lambda_\sigma \sigma$ in ΛG such that $(\sum \lambda_\sigma \sigma\epsilon_1)(\tau) = \lambda$ and $(\sum \lambda_\sigma \sigma\epsilon_1)(u) = 0$ for all u in $G - \{\tau\}$. But $(\tau^{-1}\lambda)\tau^{-1}$ has this desired property, so we are done. $\square$

Therefore we have proven the following.

Proposition 4.14 *Let* ΛG *be a skew group algebra. Then the functors* $\Lambda G \otimes_\Lambda$ *and* $\text{Hom}_\Lambda(\Lambda G, \ )$ *from* $\text{mod}\,\Lambda$ *to* $\text{mod}\,\Lambda G$ *are isomorphic.* $\square$

This result, which is an important tool in studying ΛG-modules, will be applied in the next chapter.

Exercises

1. Let k be a field and Γ the quiver $\circlearrowleft^\alpha$, and I the ideal in $k\Gamma$ generated by the arrow. Let $\rho = \{\alpha^2 - \alpha^3\}$. Show that $k(\Gamma)/\langle\rho\rangle \simeq (k[X]/(X^2)) \times k$, that $\langle\rho\rangle \subset I$ and that $I^n \not\subseteq \langle\rho\rangle^n$ for all n.

2. Let Λ be a basic artin R-algebra.

(a) Prove that $\mathfrak{r} = \{\lambda \in \Lambda | \lambda \text{ is nilpotent}\}$.
(b) Let Λ' be an R-subalgebra of Λ. Prove that Λ' is basic.
(c) Let e be an idempotent in Λ. Prove that $e\Lambda e$ is basic.

3. Let k be a field and Γ the quiver $\alpha \, \overset{\gamma}{\underset{\delta}{\circlearrowleft \; \circlearrowright}} \, \beta$. Let $\rho = \{\delta\gamma - \alpha^2, \alpha^3 - \alpha^2, \gamma\delta - \beta^2, \beta^3 - \beta^2, \alpha\delta - \delta\beta, \gamma\alpha - \beta\gamma\}$.

(a) Show that $\dim_k(k\Gamma/\langle\rho\rangle) = 12$.

(b) Show that the subalgebra of $k\Gamma/\langle\rho\rangle$ generated by α^2, $\gamma\alpha^2$, $\alpha^2\delta$, β^2 is isomorphic to $M_2(k)$ and use Exercise 2 to conclude that $k\Gamma/\langle\rho\rangle$ is not basic.

4. Let k be a field. In each of the cases below find the dimension of $k\Gamma$ and all indecomposable projective and all indecomposable injective representations of Γ over k up to isomorphism.

(a) $\Gamma: \cdot \rightarrow \cdot$;

(b) $\Gamma: \cdot \rightarrow \cdot \rightarrow \cdot$;

(c) $\Gamma: \cdot \rightarrow \cdot \leftarrow \cdot$;

(d) $\Gamma:$.

5. Let Γ be the quiver $\overset{1}{\cdot} \overset{\alpha}{\underset{\beta}{\rightleftarrows}} \overset{2}{\cdot}$ and let k be a field. Let $\rho = \{\alpha\beta\}$.

(a) Show that there is some t with $J^t \subset \langle\rho\rangle \subset J^2$.

(b) Find the radical of $k\Gamma/\langle\rho\rangle$.

(c) Find the indecomposable projective representations of (Γ, ρ) up to isomorphism.

(d) Find the global dimension of $k\Gamma/\langle\rho\rangle$.

6. Let $\Lambda = \left\{ \begin{pmatrix} a_{11} & 0 & 0 & 0 \\ a_{21} & a_{22} & 0 & 0 \\ a_{31} & 0 & a_{33} & 0 \\ a_{41} & a_{42} & a_{43} & a_{44} \end{pmatrix} \middle| a_{ij} \in \mathbb{C}; \mathbb{C} \text{ the complex numbers} \right\}$.

(a) Prove that Λ is a subalgebra of the 4×4 matrix algebra over $\mathbb{C}$.

(b) Find the radical $\mathfrak{r}$ of Λ.

(c) Show that Λ is basic.

(d) Find elements $\alpha_i \in \mathfrak{r} - \mathfrak{r}^2$ such that $\{\bar{\alpha}_i\}$ is a $\mathbb{C}$-basis for $\mathfrak{r}/\mathfrak{r}^2$ where $\bar{\alpha}_i$ is the coset of α_i.

(e) Find a quiver Γ and an ideal I in $\mathbb{C}\Gamma$ such that $\Lambda \simeq \mathbb{C}\Gamma/I$.

7. Let k be a field and let Γ be the quiver . For an ordered pair (i, j) of elements in k let M_{ij} be the representation given by

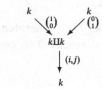

(a) Determine for which (i, j) the representation M_{ij} is indecomposable and for which (i, j) it decomposes.

(b) Prove that if M_{ij} and M_{st} are indecomposable then they are isomorphic.

Is the same true if M_{ij} and M_{st} decompose?

8. Let k be a field and Γ the quiver $\cdot \underset{\beta}{\overset{\alpha}{\leftrightarrows}} \cdot \circlearrowright^{\gamma}$. Let $\rho = \{\beta\alpha - \gamma^2, \alpha\beta\}$ and $\rho' = \{\beta\alpha - \gamma^2, \gamma^4, \alpha\beta - \alpha\gamma\beta\}$.

(a) Prove that there exists some t such that $J^t \subset \langle\rho\rangle \subset J^2$ and $J^t \subset \langle\rho'\rangle \subset J^2$.

(b) Prove that if the characteristic of k is different from 2 then $k\Gamma/\langle\rho\rangle \simeq k\Gamma/\langle\rho'\rangle$.

(c) Prove that if the characteristic of k is equal to 2 then $k\Gamma/\langle\rho\rangle \not\simeq k\Gamma/\langle\rho'\rangle$.

(d) Prove that $k\Gamma/\langle\rho, \gamma^3\rangle \simeq k\Gamma/\langle\rho', \gamma^3\rangle$ for all fields k.

9. Let k be a field, Γ the quiver $\begin{smallmatrix} & 1 & \\ \alpha\swarrow & & \searrow\beta \\ 2 & & 3 \\ & \nwarrow\gamma & \end{smallmatrix}$, M the representation

$\begin{smallmatrix} & k & \\ 1\swarrow & & \searrow 1 \\ 1 & & k \\ & \nwarrow 0 & \end{smallmatrix}$ and N the representation $\begin{smallmatrix} & k & \\ 1\swarrow & & \searrow 0 \\ 1 & & k \\ & \nwarrow 1 & \end{smallmatrix}$.

(a) Find the radical and the socle of M and N.

(b) Find the annihilator, ann M and ann N, of M and N respectively.

(c) Prove that M is a projective $(k\Gamma/(\text{ann } M))$-module and that N is an injective $(k\Gamma/(\text{ann } N))$-module.

10. Let Λ be an artin R-algebra and e a primitive idempotent.

(a) Show that $\phi: \mathrm{Hom}_\Lambda(\Lambda e, M) \to eM$ given by $\phi(f) = f(e)$ is an R-isomorphism.

(b) Interpret the result in (a) for quivers with relations and their representations.

(c) Show that $\psi: \mathrm{Hom}_\Lambda(M, D(e\Lambda)) \to D(eM)$ given by $\psi(f)(em) = f(m)(e)$ for $f \in \mathrm{Hom}_\Lambda(M, D(e\Lambda))$ and $m \in M$ is an R-isomorphism.

(d) Interpret the result in (c) for quivers with relations and their representations.

11. Let Λ be a hereditary artin algebra and Γ the valued quiver of Λ. Prove that the underlying quiver of Γ has no oriented cycles.

12. Let $\mathbb{Q}$ be the rational numbers and $\mathbb{R}$ the real numbers. Let $\Lambda = \left\{ \left(\begin{smallmatrix} a & 0 \\ b & c \end{smallmatrix}\right) | a \in \mathbb{Q}, b, c \in \mathbb{R} \right\}$.

(a) Prove that Λ is a subring of $M_2(\mathbb{R})$, the ring of 2×2 matrices over $\mathbb{R}$.

(b) Show that the proper left ideals of Λ are the ideal $\left\{ \left(\begin{smallmatrix} a & 0 \\ b & 0 \end{smallmatrix}\right) | a \in \mathbb{Q}, b \in \mathbb{R} \right\}$, the ideal $\left\{ \left(\begin{smallmatrix} 0 & 0 \\ b & c \end{smallmatrix}\right) | b, c \in \mathbb{R} \right\}$, the family I_λ indexed by $\lambda \in \mathbb{R}$, where $I_\lambda = \left\{ r\left(\begin{smallmatrix} 0 & 0 \\ 1 & \lambda \end{smallmatrix}\right) | r \in \mathbb{R} \right\}$ and the left ideal $I_\infty = \left\{ r\left(\begin{smallmatrix} 0 & 0 \\ 0 & 1 \end{smallmatrix}\right) | r \in \mathbb{R} \right\}$.

(c) Show that Λ is left artin but not right artin.

(d) Show that $\mathrm{mod}\,\Lambda$ does not have enough injectives.

13. Let $\mathbb{C}(X)$ be the field of rational functions in one variable X over $\mathbb{C}$. Then the subfield $\mathbb{C}(X^2)$ is isomorphic to $\mathbb{C}(X)$. Let M be $\mathbb{C}(X)$ as an abelian group with the natural $\mathbb{C}(X)$-module structure. Consider M as a $\mathbb{C}(X)$-$\mathbb{C}(X)$-bimodule by letting the right action be given through the isomorphism $\mathbb{C}(X) \simeq \mathbb{C}(X^2) \subset \mathbb{C}(X)$. Let $\Lambda = \left(\begin{smallmatrix} \mathbb{C}(X) & 0 \\ M & \mathbb{C}(X) \end{smallmatrix}\right)$.

(a) Find the center of Λ.

(b) Prove that $\Lambda/\mathfrak{r} \simeq \Pi_2(\mathbb{C}(X))$ as a ring but not as an algebra over the center of Λ.

(c) Find the valued quiver of Λ.

14. Let Λ be an artin algebra and $\Gamma = T_2(\Lambda)$. Prove that $\mathrm{id}_\Gamma \Gamma = \mathrm{id}_\Lambda \Lambda + 1$.

15. Let Λ be an artin algebra isomorphic to $\left(\begin{smallmatrix} T & 0 \\ {}_U M_T & U \end{smallmatrix}\right)$ where T and U are hereditary artin algebras. Prove that $\mathrm{gl.dim}\,\Lambda \leq 2$.

16. Let p be a prime number, let G be a finite p-group (i.e. the order of G is p^n for some n) and let k be a field of characteristic p.

(a) Prove that for each $g \in G$ with $g \neq 0$ the element $1 - g$ is nilpotent in kG.

(b) Let $Z(G)$ be the center of G. Prove that $\{1 - g | g \in Z(G)\}$ generates a nilpotent ideal I in kG.

(c) Prove that I is the kernel of the natural ring map from kG to $k(G/Z(G))$.

(d) Prove that kG is a local ring. (Hint: $Z(G)$ is nontrivial for each finite group G.)

17. Let k be a field and let F be a Galois extension of k with $[F:k] = n$ and Galois group G. Let G operate on F in the natural way and form the skew group ring FG. Show that $FG \simeq M_n(k)$.

18. Let k be a field, Γ the quiver $\overset{1}{\cdot} \overset{\alpha}{\underset{\beta}{\rightleftarrows}} \overset{2}{\cdot}$ and ρ the set of relations $\{\alpha\beta, \beta\alpha\}$. Let G be the cyclic group of order 2 with generator g. Let G operate on $k\Gamma$ as a group of k-automorphisms by $g(ae_1 + be_1 + c\alpha + d\beta) = (be_1 + ae_2 + d\alpha + c\beta)$ where a, b, c and d are in k.

(a) Prove that $(k\Gamma/\langle\rho\rangle)^G$, the ring of fixed points, is isomorphic to $k[X]/(X^2)$.

(b) Prove that the skew group ring $(k\Gamma/\langle\rho\rangle)G$ is isomorphic to $M_2(k[X]/(X^2))$, the ring of 2×2 matrices over $k[X]/(X^2)$.

19. Let Λ be a finite dimensional algebra over a finite field k with radical $\mathfrak{r}$. Let $U(\Lambda)$ and $U(\Lambda/\mathfrak{r})$ denote the group of units in Λ and $\Lambda/\mathfrak{r}$ respectively.

(a) Prove that $U(\Lambda/\mathfrak{r})$ is a finite group isomorphic to $\prod GL_{n_i}(k_i)$ where k_i are finite field extensions of k and n_i are some integers.

(b) Let $p: \Lambda \to \Lambda/\mathfrak{r}$ be the natural epimorphism. Prove that p induces a surjective group morphism $\bar{p}: U(\Lambda) \to U(\Lambda/\mathfrak{r})$ with kernel $\{1 + \lambda \mid \lambda \in \mathfrak{r}\}$.

(c) Prove that $U(\Lambda/\mathfrak{r})$ generates $\Lambda/\mathfrak{r}$ as a k-vector space if and only if $U(\Lambda)$ generates Λ as a k-vector space.

(d) Let $kU(\Lambda/\mathfrak{r})$ and $kU(\Lambda)$ be the group algebras of $U(\Lambda/\mathfrak{r})$ and $U(\Lambda)$ respectively. The inclusions $U(\Lambda/\mathfrak{r}) \to \Lambda/\mathfrak{r}$ and $U(\Lambda) \to \Lambda$ induce k-algebra morphisms $\phi_{\Lambda/\mathfrak{r}}: kU(\Lambda/\mathfrak{r}) \to \Lambda/\mathfrak{r}$ and $\phi_\Lambda: kU(\Lambda) \to \Lambda$. Prove that $\phi_{\Lambda/\mathfrak{r}}$ is surjective if and only if ϕ_Λ is surjective.

(Hence by (d) Λ is a quotient of a group algebra if k is different from the field with 2 elements.)

(e) Prove that if k is a field with at least three elements then $\phi_{\Lambda/\mathfrak{r}}$ is surjective.

(f) Determine when $\phi_{\Lambda/\mathfrak{r}}$ is surjective for the field k with two elements.

20. Let A be a semisimple ring, M an $A - A$-bimodule and $T(A, M)$ the tensor algebra of M over A, i.e. $T(A, M) = A \amalg M \amalg M \otimes_A M \amalg \ldots \amalg M \otimes_A M \otimes \ldots \otimes_A M \amalg \ldots$ as a twosided A-module and multiplication is induced by the tensor product. Prove that $T = T(A, M)$ is left (and right) hereditary, independent of whether $T(A, M)$ is artin or not. (Hint: First prove using the adjoint isomorphism that $P \otimes_A X$ is a left projective T-module whenever P is a $T - A$-bimodule projective as a left T-module and X is any A-module. Apply this to prove that $\overline{M} = M \amalg M \otimes_A M \amalg \ldots \amalg M \otimes_A M \otimes \ldots \otimes_A M \amalg \ldots$ is a left projective T-module. Then for each left T-module X, prove that there is an exact sequence of left T-modules

$$0 \to \overline{M} \otimes_A X \xrightarrow{g} T \otimes_A X \xrightarrow{f} X \to 0$$

where f is defined on generators by $f(t \otimes x) = tx$ for $t \in T$ and $x \in X$ and g is defined on generators by

$$g(m_1 \otimes \ldots \otimes m_i \otimes x) = m_1 \otimes \ldots \otimes m_i \otimes x - m_1 \otimes \ldots \otimes m_{i-1} \otimes m_i x.)$$

Notes

The systematic use of quivers and their representations in the representation theory of artin algebras goes back to [Ga1], where they were used in the classification of hereditary and radical squared zero algebras of finite representation type over an algebraically closed field. This point of view on modules has since then played a central role in representation theory.

Triangular matrix rings have for a long time been convenient for providing interesting examples of rings. A treatment of their homological algebra, formulated more generally for trivial extensions of abelian categories, can be found in [FoGR]. The special case of one-point extensions has been important in classification theorems (see [Rin3]).

For a further study of modules over group algebras from the point of view of the methods of the representation theory of algebras discussed in this book we refer to the texts [Ben], [CR], [Er]. For a representation theoretic treatment of skew group algebras we refer to [ReR].

IV

The transpose and the dual

In this chapter we introduce the notions of the transpose and the dual of the transpose of a module. These functors are of fundamental importance in the representation theory of artin algebras in their own right, as well as because of their close connection with almost split sequences which we discuss in the next chapter. Here we give some of the basic properties of these functors.

We give illustrations of the transpose and the dual of the transpose by showing that for certain types of artin algebras these functors are closely related to more familiar functors. These algebras include Nakayama and selfinjective algebras whose definitions we give here, in addition to hereditary algebras.

The rest of the chapter is devoted to developing a formula giving a basic relation between the lengths of the modules of morphisms between modules, which involves minimal projective presentations of modules and the dual of the transpose of modules. This formula plays an important role in several places in the book.

All rings in this chapter are artin algebras and all modules are assumed to be finitely generated.

1 The transpose

In this section we introduce the notions of the transpose and the dual of the transpose of modules and morphisms. These notions are basic to the rest of this book.

We have seen in II Proposition 4.3 that we have a duality $T = (\ )^* : \mathscr{P}(\Lambda) \to \mathscr{P}(\Lambda^{\mathrm{op}})$. Let C be in $\mathrm{mod}\,\Lambda$ and let $P_1 \xrightarrow{f} P_0 \to C \to 0$ be a minimal projective presentation. Then we have $C \simeq \mathrm{Coker}\, f$. Applying the duality T to the morphism f we get a morphism $f^*: P_0^* \to P_1^*$. Associated with f^* we have the module $\operatorname{Coker} f^*$ in $\operatorname{mod} \Lambda^{\mathrm{op}}$ which is called the **transpose** of C and denoted by $\operatorname{Tr} C$. The operation Tr does not induce a duality $\operatorname{mod} \Lambda \to \operatorname{mod} \Lambda^{\mathrm{op}}$, and in general there is not even a functor from $\operatorname{mod} \Lambda$ to $\operatorname{mod} \Lambda^{\mathrm{op}}$ sending an object C to $\operatorname{Tr} C$. But we do get a duality by replacing $\operatorname{mod} \Lambda$ with an appropriate factor category, the category modulo projectives, which we define later. Rather than showing directly that this works, we give an approach which at the same time motivates our choice of a factor category. But first we discuss the notion of factor categories in general.

By a **relation** $\mathscr{R}$ on an R-category $\mathscr{A}$ we mean R-submodules $\mathscr{R}(A, B) \subset \operatorname{Hom}(A, B)$, where $\operatorname{Hom}(A, B)$ is the R-module of morphisms from A to B, for all A and B in $\mathscr{A}$, such that under the composition map $\operatorname{Hom}(A, B) \otimes_R \operatorname{Hom}(B, C) \to \operatorname{Hom}(A, C)$, we have $\operatorname{Im}(\mathscr{R}(A, B) \otimes_R \operatorname{Hom}(B, C) \to \operatorname{Hom}(A, C)) \subset \mathscr{R}(A, C)$ and $\operatorname{Im}(\operatorname{Hom}(A, B) \otimes_R \mathscr{R}(B, C) \to \operatorname{Hom}(A, C)) \subset \mathscr{R}(A, C)$. Then $\mathscr{A}/\mathscr{R}$, the **factor category** of $\mathscr{A}$ modulo the relation $\mathscr{R}$, is defined by the following data. The objects of $\mathscr{A}/\mathscr{R}$ are the same as those of $\mathscr{A}$. The morphisms from A to B in $\mathscr{A}/\mathscr{R}$ are the elements of the factor module $\operatorname{Hom}(A, B)/\mathscr{R}(A, B)$. And the composition in $\mathscr{A}/\mathscr{R}$ is defined for A, B, C in $\mathscr{A}/\mathscr{R}$ by $(g + \mathscr{R}(B, C))(f + \mathscr{R}(A, B)) = gf + \mathscr{R}(A, C)$ for all f in $\operatorname{Hom}(A, B)$ and g in $\operatorname{Hom}(B, C)$. It is then easy to see that we have the following.

Proposition 1.1 *Let $\mathscr{A}$ be an R-category and $\mathscr{R}$ a relation on $\mathscr{A}$.*

(a) *The factor category $\mathscr{A}/\mathscr{R}$ is an R-category and $F: \mathscr{A} \to \mathscr{A}/\mathscr{R}$ given by $F(A) = A$ for all A in $\mathscr{A}$ and $F: \operatorname{Hom}(A, B) \to \operatorname{Hom}(A, B)/\mathscr{R}(A, B)$ being the canonical epimorphism, is a full and dense R-functor.*

(b) *If $G: \mathscr{A} \to \mathscr{B}$ is an R-functor between R-categories such that $G(\mathscr{R}(A, B)) = 0$ for all A and B in $\mathscr{A}$, then there is a unique functor $H: \mathscr{A}/\mathscr{R} \to \mathscr{B}$ such that $HF = G$.* □

The **morphism category** of $\mathscr{P}(\Lambda)$ is the R-category $\operatorname{Morph}(\mathscr{P}(\Lambda))$ defined by the following data. The objects of $\operatorname{Morph} \mathscr{P}(\Lambda)$ are the morphisms $f: P_1 \to P_2$ in $\mathscr{P}(\Lambda)$. The morphisms from $f: P_1 \to P_2$ to $f': P_1' \to P_2'$ are pairs (g_1, g_2) where $g_i: P_i \to P_i'$ for $i = 1, 2$ such that the diagram

$$
\begin{array}{ccc}
P_1 & \xrightarrow{\ f\ } & P_2 \\
\downarrow{\scriptstyle g_1} & & \downarrow{\scriptstyle g_2} \\
P_1' & \xrightarrow{\ f'\ } & P_2'
\end{array}
$$

commutes. Composition and addition of morphisms is componentwise. It is not hard to check that this data defines an additive R-category. Also **idempotents split**, that is, if $e: f \to f$ is an idempotent, then there is an object f' in Morph $\mathscr{P}(\Lambda)$ and morphisms $g: f \to f'$ and $h: f' \to f$ such that $e = hg$ and $gh = 1_{f'}$.

We now define the R-functor Coker: Morph $\mathscr{P}(\Lambda) \to \mathrm{mod}\,\Lambda$ by Coker$(f: P_1 \to P_2) = \mathrm{Coker}\, f$ for all $f: P_1 \to P_2$ in Morph $\mathscr{P}(\Lambda)$ and Coker$(g_1, g_2): \mathrm{Coker}\, f \to \mathrm{Coker}\, f'$ to be the unique morphism which makes the diagram

$$
\begin{array}{ccccccc}
P_1 & \xrightarrow{f} & P_2 & \longrightarrow & \mathrm{Coker}\, f & \to & 0 \\
\downarrow{\scriptstyle g_1} & & \downarrow{\scriptstyle g_2} & & \downarrow & & \\
P_1' & \xrightarrow{f'} & P_2' & \longrightarrow & \mathrm{Coker}\, f' & \to & 0
\end{array}
$$

commute.   It is straightforward to see that the R-functor Coker: Morph $\mathscr{P}(\Lambda) \to \mathrm{mod}\,\Lambda$ is dense and full but not in general faithful. Specifically, Coker$(g_1, g_2) = 0$ if and only if there is some $h: P_2 \to P_1'$ such that $f'h = g_2$. If for objects f and f' in Morph $\mathscr{P}(\Lambda)$ we define $\mathscr{R}(f, f')$ to consist of the morphisms (g_1, g_2) with the property that there is some $h: P_2 \to P_1'$ such that $f'h = g_2$, then $\mathscr{R}$ gives a relation on Morph $\mathscr{P}(\Lambda)$. Then we get the following.

Proposition 1.2 *Let $\mathscr{R}$ be the relation on Morph $\mathscr{P}(\Lambda)$ defined above. The functor Coker: Morph $\mathscr{P}(\Lambda) \to \mathrm{mod}\,\Lambda$ induces a functor G: Morph $\mathscr{P}(\Lambda)/\mathscr{R} \to \mathrm{mod}\,\Lambda$, which is an equivalence of categories.* $\square$

The duality $T: \mathscr{P}(\Lambda) \to \mathscr{P}(\Lambda^{\mathrm{op}})$ given by $P \mapsto \mathrm{Hom}_\Lambda(P, \Lambda)$ for all P in $\mathscr{P}(\Lambda)$ induces a duality $T: \mathrm{Morph}\, \mathscr{P}(\Lambda) \to \mathrm{Morph}\, \mathscr{P}(\Lambda^{\mathrm{op}})$, which sends the object $f: P_1 \to P_2$ to $f^*: P_2^* \to P_1^*$. For $f: P_1 \to P_2$ and $f': P_1' \to P_2'$ in Morph $\mathscr{P}(\Lambda)$, assume that $(g_1, g_2): f \to f'$ is in $\mathscr{R}(f, f')$, with $\mathscr{R}$ as before. There is then a morphism $h: P_2 \to P_1'$ such that $g_2 = f'h$, which gives rise to the diagram

$$
\begin{array}{ccc}
P_2'^* & \xrightarrow{f'^*} & P_1'^* \\
\downarrow{\scriptstyle g_2^*} & {\scriptstyle h^*}\nearrow & \downarrow{\scriptstyle g_1^*} \\
P_2^* & \xrightarrow{f^*} & P_1^*
\end{array}
$$

with $g_2^* = h^* f'^*$. We see that we do not necessarily have that (g_2^*, g_1^*) is in $\mathscr{R}(f'^*, f^*)$. In fact, the smallest relation $\mathscr{P}$ containing the relation $\mathscr{R}$, which we can put on Morph $\mathscr{P}(\Lambda)$ such that it is sent to itself by the duality T is generated by the following maps. For $f: P_1 \to P_2$ and $f': P_1' \to P_2'$ in Morph $\mathscr{P}(\Lambda)$ we have that $(g_1, g_2): f \to f'$ is in $\mathscr{P}(f, f')$ if there is some $h: P_2 \to P_1'$ such that either $f'h = g_2$ or $hf = g_1$. We then get the following.

Proposition 1.3 *The duality* $T: \text{Morph}\,\mathscr{P}(\Lambda) \to \text{Morph}\,\mathscr{P}(\Lambda^{op})$ *induces a duality* $\text{Tr}: \text{Morph}\,\mathscr{P}(\Lambda)/\mathscr{P} \to \text{Morph}\,\mathscr{P}(\Lambda^{op})/\mathscr{P}$ *with* $\text{Tr}: \text{Morph}\,\mathscr{P}(\Lambda^{op})/\mathscr{P} \to \text{Morph}\,\mathscr{P}(\Lambda)/\mathscr{P}$ *as inverse duality.* $\quad\square$

We now interpret these results about Morph $\mathscr{P}(\Lambda)$ and Morph $\mathscr{P}(\Lambda)/\mathscr{P}$ in terms of the category mod Λ. Since the relation $\mathscr{P}$ on Morph $\mathscr{P}(\Lambda)$ contains the relation $\mathscr{R}$ on Morph $\mathscr{P}(\Lambda)$, the image of $\mathscr{P}$ under the full and dense functor Coker: Morph $\mathscr{P}(\Lambda) \to \text{mod}\,\Lambda$ is a relation on mod Λ which we want to describe. For this the following description of the relation $\mathscr{P}$ on Morph $\mathscr{P}(\Lambda)$ is useful.

Lemma 1.4 *Let* $f: P_1 \to P_2$ *and* $f': P_1' \to P_2'$ *be in* Morph $\mathscr{P}(\Lambda)$. *Let* $\tilde{\mathscr{P}}(f, f') \subset \text{Hom}(f, f')$ *consist of the morphisms* $(g_1, g_2): f \to f'$ *with the property that there is some* $h: P_2 \to P_1'$ *such that* $f'hf = g_2 f$. *Then we have* $\mathscr{P}(f, f') = \tilde{\mathscr{P}}(f, f')$.

Proof It is clear that $\mathscr{P}(f, f') \subset \tilde{\mathscr{P}}(f, f')$. Suppose now that (g_1, g_2) is in $\tilde{\mathscr{P}}(f, f')$, and let $h: P_2 \to P_1'$ be such that $f'hf = g_2 f$. Then the diagram

$$
\begin{array}{ccc}
P_1 & \xrightarrow{\;f\;} & P_2 \\
\downarrow{\scriptstyle g_1} & & \downarrow{\scriptstyle f'h} \\
P_1' & \xrightarrow{\;f'\;} & P_2'
\end{array}
$$

commutes and $(g_1, f'h)$ is in $\mathscr{P}(f, f')$. Also $(g_1, g_2) - (g_1, f'h) = (0, g_2 - f'h)$ is in $\mathscr{P}(f, f')$, so that (g_1, g_2) is in $\mathscr{P}(f, f')$. Hence we have $\mathscr{P}(f, f') = \tilde{\mathscr{P}}(f, f')$. $\quad\square$

Suppose now that we have the commutative exact diagram

$$
\begin{array}{ccccccc}
P_1 & \xrightarrow{\;f\;} & P_2 & \xrightarrow{\;\epsilon\;} & \text{Coker}\,f & \to & 0 \\
\downarrow{\scriptstyle g_1} & & \downarrow{\scriptstyle g_2} & & \downarrow{\scriptstyle \text{Coker}(g_1, g_2)} & & \\
P_1' & \xrightarrow{\;f'\;} & P_2' & \xrightarrow{\;\epsilon'\;} & \text{Coker}\,f' & \to & 0.
\end{array}
$$

It is fairly straightforward to see that (g_1, g_2) is in $\mathscr{P}(f, f')$, i.e. there is some $h: P_2 \to P_1'$ such that $f'hf = g_2 f$ if and only if there is some $t: \text{Coker}\,f \to P_2'$ such that $\epsilon't = \text{Coker}(g_1, g_2)$. This suggests that the image of $\mathscr{P}$ in mod Λ under the functor Coker: Morph $\mathscr{P}(\Lambda) \rightarrow$ mod Λ consists of the morphisms $A \rightarrow B$ in mod Λ which can be written as a composition $A \rightarrow P \rightarrow B$ with P a projective Λ-module. That this is indeed the case follows from the above discussion and the next lemma. We say that a morphism $f: A \rightarrow B$ in mod Λ **factors through** a **projective module** if $f = hg$ with $g: A \rightarrow P$ and $h: P \rightarrow B$ with P a projective Λ-module.

Lemma 1.5 *The following are equivalent for a morphism* $f: A \rightarrow B$ *in* mod Λ.

(a) *f factors through a projective module.*
(b) *If* $g: P \rightarrow B$ *is an epimorphism with* P *projective, then there is some* $t: A \rightarrow P$ *such that* $gt = f$.
(c) *If* $g: X \rightarrow B$ *is an epimorphism, then there is some* $t: A \rightarrow X$ *such that* $gt = f$. $\qquad\qquad\square$

It is clear that the image in mod Λ of the relation $\mathscr{P}$ on Morph $\mathscr{P}(\Lambda)$ is a relation on mod Λ, which we also denote by $\mathscr{P}$, and which has the following description. For A and B in mod Λ, $\mathscr{P}(A, B)$ is the R-submodule of $\mathrm{Hom}_\Lambda(A, B)$ consisting of the morphisms $f: A \rightarrow B$ which factor through a projective module. From Lemma 1.5(b) it is clear that if $g: P \rightarrow B$ is an epimorphism with P projective then $\mathscr{P}(A, B) = \mathrm{Im}\,\mathrm{Hom}_\Lambda(A, g)$. We shall usually denote $\mathrm{Hom}_\Lambda(A, B)/\mathscr{P}(A, B)$ by $\underline{\mathrm{Hom}}_\Lambda(A, B)$ and the factor category mod $\Lambda/\mathscr{P}$ by $\underline{\mathrm{mod}}\,\Lambda$. Since Coker: Morph $\mathscr{P}(\Lambda) \rightarrow$ mod Λ is full and dense it induces an equivalence from Morph $\mathscr{P}(\Lambda)/\mathscr{P}$ to $\underline{\mathrm{mod}}\,\Lambda$ which we also denote by Coker. Identifying Morph $\mathscr{P}(\Lambda)/\mathscr{P}$ with $\underline{\mathrm{mod}}\,\Lambda$ through this equivalence we obtain that the duality Tr: Morph $\mathscr{P}(\Lambda)/\mathscr{P} \rightarrow$ Morph $\mathscr{P}(\Lambda^{\mathrm{op}})/\mathscr{P}$ induces a duality from $\underline{\mathrm{mod}}\,\Lambda$ to $\underline{\mathrm{mod}}\Lambda^{\mathrm{op}}$ which we also denote by Tr. We now collect our findings.

Proposition 1.6

(a) *The functor* Coker: Morph $\mathscr{P}(\Lambda) \rightarrow$ mod Λ *induces an equivalence* Coker: Morph $\mathscr{P}(\Lambda)/\mathscr{P} \rightarrow \underline{\mathrm{mod}}\,\Lambda$.
(b) *The compositions* $\underline{\mathrm{mod}}\,\Lambda \xrightarrow{\mathrm{Tr}} \underline{\mathrm{mod}}\,\Lambda^{\mathrm{op}} \xrightarrow{\mathrm{Tr}} \underline{\mathrm{mod}}\,\Lambda$ *and* $\underline{\mathrm{mod}}\,\Lambda^{\mathrm{op}} \xrightarrow{\mathrm{Tr}} \underline{\mathrm{mod}}\,\Lambda \xrightarrow{\mathrm{Tr}} \underline{\mathrm{mod}}\,\Lambda^{\mathrm{op}}$ *are isomorphic to the identity on* $\underline{\mathrm{mod}}\,\Lambda$ *and* $\underline{\mathrm{mod}}\,\Lambda^{\mathrm{op}}$ *respectively.* $\qquad\square$

For C in mod Λ we have a unique (up to isomorphism) decomposition $C = C_{\mathscr{P}} \coprod C'$, where $C_{\mathscr{P}}$ has no nonzero projective summands and C' is projective. Denote by $\mathrm{mod}_{\mathscr{P}}\,\Lambda$ the full subcategory of mod Λ whose objects are the C in $\operatorname{mod}\Lambda$ with $C = C_{\mathscr{P}}$, and denote by $\mathscr{P}$ the relation on $\operatorname{mod}_{\mathscr{P}}\Lambda$ induced by the relation $\mathscr{P}$ on $\operatorname{mod}\Lambda$. Denoting $\operatorname{mod}_{\mathscr{P}}\Lambda/\mathscr{P}$ by $\underline{\operatorname{mod}}_{\mathscr{P}}\Lambda$ it is easy to see that the functor $\underline{\operatorname{mod}}_{\mathscr{P}}\Lambda \to \underline{\operatorname{mod}}\Lambda$ induced by the inclusion $\operatorname{mod}_{\mathscr{P}}\Lambda \to \operatorname{mod}\Lambda$ is an equivalence of categories, since C in $\underline{\operatorname{mod}}\Lambda$ is the zero object if and only if C is projective. Hence there is also induced a duality $\operatorname{Tr}:\underline{\operatorname{mod}}_{\mathscr{P}}\Lambda \to \underline{\operatorname{mod}}_{\mathscr{P}}(\Lambda^{\operatorname{op}})$.

Even though it is on the factor categories $\underline{\operatorname{mod}}\Lambda$ and $\underline{\operatorname{mod}}_{\mathscr{P}}\Lambda$ that Tr defines a functor, it is often useful to consider the induced map between objects $\operatorname{Tr}:\operatorname{mod}\Lambda \to \operatorname{mod}(\Lambda^{\operatorname{op}})$ which we can define directly by $P_0^* \xrightarrow{f^*} P_1^* \to \operatorname{Tr} C \to 0$ being exact when $P_1 \xrightarrow{f} P_0 \to C \to 0$ is a minimal projective presentation of C. Note that if C is indecomposable nonprojective, then it is clear that $f:P_1 \to P_0$ is an indecomposable map which is not an isomorphism. Hence $f^*:P_0^* \to P_1^*$ is also an indecomposable map which is not an isomorphism, so that $\operatorname{Coker} f^* = \operatorname{Tr} C$ is clearly indecomposable. It is also not hard to see that $P_0^* \xrightarrow{f^*} P_1^* \to \operatorname{Tr} C \to 0$ is a minimal projective presentation of $\operatorname{Tr} C$. If $C = P$ is indecomposable projective, then $0 \to P \to P \to 0$ is a minimal projective presentation, but $P^* \to 0 \to 0 \to 0$ is not a minimal projective presentation of $\operatorname{Tr} P = 0$. Using these observations we obtain the following easily verified properties of the map Tr from $\operatorname{mod}\Lambda$ to $\operatorname{mod}(\Lambda^{\operatorname{op}})$.

Proposition 1.7

(a) $\operatorname{Tr}(\coprod_{i=1}^{n} A_i) \simeq \coprod_{i=1}^{n} \operatorname{Tr}(A_i)$, *where* $A_1, \ldots, A_n$ *are in* $\operatorname{mod}\Lambda$.

(b) $\operatorname{Tr} A = 0$ *if and only if* A *is projective.*

(c) $\operatorname{Tr}\operatorname{Tr} A \simeq A_{\mathscr{P}}$ *for all* A *in* $\operatorname{mod}\Lambda$.

(d) *If* A *and* B *are in* $\operatorname{mod}_{\mathscr{P}}\Lambda$, *then* $\operatorname{Tr} A \simeq \operatorname{Tr} B$ *if and only if* $A \simeq B$.

(e) $\operatorname{Tr}:\operatorname{mod}\Lambda \to \operatorname{mod}(\Lambda^{\operatorname{op}})$ *induces a bijection between the isomorphism classes of indecomposable modules in* $\operatorname{mod}_{\mathscr{P}}\Lambda$ *and the isomorphism classes of indecomposable modules in* $\operatorname{mod}_{\mathscr{P}}(\Lambda^{\operatorname{op}})$. $\square$

We now turn our attention to studying the dual of the transpose.

Let A and B be in $\operatorname{mod}\Lambda$. Then $f:A \to B$ is in $\mathscr{P}(A,B)$ if and only if $D(f):D(B) \to D(A)$ factors through an injective module, where $D:\operatorname{mod}\Lambda \to \operatorname{mod}(\Lambda^{\operatorname{op}})$ is the usual duality, i.e. there exists an injective $\Lambda^{\operatorname{op}}$-module I and $\Lambda^{\operatorname{op}}$-morphisms $g:DB \to I$ and $h:I \to DA$ with $Df = hg$. This suggests introducing the relation "modulo injectives" on $\operatorname{mod}\Lambda$, which is the dual of the relation "modulo projectives" on $\operatorname{mod}\Lambda$. Before giving the formal definition of this notion, we state the dual of Lemma 1.5.

Lemma 1.8 *The following are equivalent for a morphism $f: A \to B$ in* $\operatorname{mod} \Lambda$.

(a) *f factors through an injective module.*

(b) *If $g: A \to I$ is a monomorphism with I injective, then there is some $t: I \to B$ such that $f = tg$.*

(c) *If $g: A \to X$ is a monomorphism, then there is some $t: X \to B$ such that $f = tg$.* $\qquad\square$

For A and B in $\operatorname{mod} \Lambda$ we define $\mathscr{I}(A, B) \subset \operatorname{Hom}_\Lambda(A, B)$ to be the set of all $f: A \to B$ which factor through an injective module. Therefore, if $g: A \to I$ is a monomorphism with I injective, then

$$\mathscr{I}(A, B) = \operatorname{Im} \operatorname{Hom}_\Lambda(g, B)$$

and is therefore an R-submodule of $\operatorname{Hom}_\Lambda(A, B)$. It is also not difficult to check that the system of R-submodules $\mathscr{I}(A, B)$ defines a relation $\mathscr{I}$ on $\operatorname{mod} \Lambda$. We will often denote $\operatorname{Hom}_\Lambda(A, B)/\mathscr{I}(A, B)$ by $\overline{\operatorname{Hom}}_\Lambda(A, B)$ and $\operatorname{mod} \Lambda/\mathscr{I}$ by $\overline{\operatorname{mod}} \Lambda$. As a consequence of our discussion we have the following.

Proposition 1.9

(a) *The duality $D: \operatorname{mod} \Lambda \to \operatorname{mod} \Lambda^{\operatorname{op}}$ induces a duality $D: \underline{\operatorname{mod}} \Lambda \to \overline{\operatorname{mod}} \Lambda^{\operatorname{op}}$.*

(b) *The composition $D \operatorname{Tr}: \underline{\operatorname{mod}} \Lambda \to \overline{\operatorname{mod}} \Lambda$ is an equivalence of categories with inverse equivalence $\operatorname{Tr} D: \overline{\operatorname{mod}} \Lambda \to \underline{\operatorname{mod}} \Lambda$.* $\qquad\square$

Each C in $\operatorname{mod} \Lambda$ can be written uniquely up to isomorphism as $C = C_{\mathscr{I}} \amalg C'$ where $C_{\mathscr{I}}$ has no nonzero injective summands and C' is injective. We denote by $\operatorname{mod}_{\mathscr{I}} \Lambda$ the full subcategory of $\operatorname{mod} \Lambda$ whose objects are the C such that $C \simeq C_{\mathscr{I}}$. The relation $\mathscr{I}$ on $\operatorname{mod} \Lambda$ induces a relation on $\operatorname{mod}_{\mathscr{I}} \Lambda$ which we also denote by $\mathscr{I}$. Then letting $\operatorname{mod}_{\mathscr{I}} \Lambda/\mathscr{I} = \overline{\operatorname{mod}}_{\mathscr{I}} \Lambda$ we have that the inclusion $\operatorname{mod}_{\mathscr{I}} \Lambda \to \operatorname{mod} \Lambda$ induces an equivalence $\overline{\operatorname{mod}}_{\mathscr{I}} \Lambda \to \overline{\operatorname{mod}} \Lambda$ since C in $\overline{\operatorname{mod}} \Lambda$ is the zero object if and only if C is injective. With this terminology in mind, we list some of the basic properties of the map $D \operatorname{Tr}: \operatorname{mod} \Lambda \to \operatorname{mod} \Lambda$ which is the composition of the map $\operatorname{Tr}: \operatorname{mod} \Lambda \to \operatorname{mod}(\Lambda^{\operatorname{op}})$ with the duality $D: \operatorname{mod}(\Lambda^{\operatorname{op}}) \to \operatorname{mod} \Lambda$ and the map $\operatorname{Tr} D: \operatorname{mod} \Lambda \to \operatorname{mod} \Lambda$ which is the composition of the duality and the map $\operatorname{Tr}: \operatorname{mod}(\Lambda^{\operatorname{op}}) \to \operatorname{mod} \Lambda$. These properties are trivial consequences of Proposition 1.7.

Proposition 1.10

(a) $D \operatorname{Tr}(\coprod_{i=1}^{n} A_i) \simeq \coprod_{i=1}^{n} D \operatorname{Tr} A_i$, where $A_1, \ldots, A_n$ are in mod Λ.

(b) $D \operatorname{Tr} A = 0$ if and only if A is projective.

(c) $D \operatorname{Tr} A$ is in mod$_{\mathscr{I}} \Lambda$ for all A in mod Λ.

(d) $(\operatorname{Tr} D) D \operatorname{Tr} A \simeq A_{\mathscr{P}}$ for all A in mod Λ.

(e) If A and B are in mod$_{\mathscr{P}} \Lambda$, then $D \operatorname{Tr} A \simeq D \operatorname{Tr} B$ if and only if $A \simeq B$.

(f) $D \operatorname{Tr}: \operatorname{mod} \Lambda \to \operatorname{mod} \Lambda$ induces a bijection between the isomorphism classes of indecomposable modules in mod$_{\mathscr{P}} \Lambda$ and the isomorphism classes of indecomposable modules in mod$_{\mathscr{I}} \Lambda$ with $\operatorname{Tr} D$ as inverse. $\square$

Thus we see that associated with an indecomposable nonprojective module A is the indecomposable noninjective module $D \operatorname{Tr} A$ which, in principle, we can construct from A. Similarly, given an indecomposable noninjective module B there is associated the indecomposable nonprojective module $\operatorname{Tr} D B$, which in principle can be constructed from B. In this connection the following relation between minimal projective presentations and minimal injective copresentations of these various modules is of interest. Here a minimal injective copresentation of a module B is an exact sequence $0 \to B \xrightarrow{g_0} I_0 \xrightarrow{g_1} I_1$ with $g_0: B \to I_0$ and the induced monomorphism $g_1': \operatorname{Coker} g_0 \to I_1$ injective envelopes.

Proposition 1.11 Let $P_1 \xrightarrow{f} P_0 \to C \to 0$ be a minimal projective presentation of the indecomposable nonprojective module C. Then $0 \to D \operatorname{Tr} C \to D(P_1^*) \xrightarrow{D(f^*)} D(P_0^*) \to D(C^*) \to 0$ is exact with $0 \to D \operatorname{Tr} C \to D(P_1^*) \xrightarrow{Df^*} D(P_0^*)$ a minimal injective copresentation of the indecomposable Λ-module $D \operatorname{Tr} C$. In particular, soc $D \operatorname{Tr} C \simeq P_1 / \mathfrak{r} P_1$.

Proof Since $P_1 \xrightarrow{f} P_0 \to C \to 0$ is a minimal projective presentation of C, it is easily seen that $0 \to C^* \to P_0^* \xrightarrow{f^*} P_1^* \to \operatorname{Tr} C \to 0$ is exact and that $P_0^* \xrightarrow{f^*} P_1^* \to \operatorname{Tr} C$ is a minimal projective presentation of the indecomposable Λ^{op}-module $\operatorname{Tr} C$. Applying the functor D to this last sequence we obtain our desired result. $\square$

We also have the following dual result.

Proposition 1.12 Let $0 \to C \to I_0 \to I_1$ be a minimal injective copresentation of the indecomposable noninjective module C. Then $0 \to (DC)^* \to (DI_0)^* \to (DI_1)^* \to \operatorname{Tr} DC \to 0$ is exact with $(DI_0)^* \to (DI_1)^* \to \operatorname{Tr} DC \to 0$ a minimal projective presentation of the *indecomposable* Λ-*module* $\operatorname{Tr} DC$. *In particular,* $\operatorname{Tr} DC / \mathfrak{r} \operatorname{Tr} DC \simeq \operatorname{soc} I_1$.

□

We end this section by calculating the transpose and the dual of the transpose for some particular modules. We start out with a concrete example.

Example Let $\Lambda = k\Gamma$ where Γ is the quiver $1 \xrightarrow{\alpha} 2 \underset{\eta}{\overset{\beta}{\longrightarrow}} \begin{smallmatrix} \bullet 3 \\ \bullet 4 \end{smallmatrix}$. Then Γ^{op}

is $1 \xleftarrow{\alpha^{\mathrm{op}}} 2 \underset{\eta^{\mathrm{op}}}{\overset{\beta^{\mathrm{op}}}{\longleftarrow}} \begin{smallmatrix} \bullet 3 \\ \bullet 4 \end{smallmatrix}$. We compute $D \operatorname{Tr} S_2$ by using the category $\operatorname{Rep}(\Gamma)$.

Let P_i be the indecomposable projective representation corresponding to the vertex i. Then P_2 is $0 \longrightarrow k \begin{smallmatrix} \nearrow^1 k \\ \searrow_1 k \end{smallmatrix}$, and we have a minimal projective

presentation $0 \to P_3 \coprod P_4 \overset{(s,t)}{\to} P_2 \to S_2 \to 0$ in $\operatorname{Rep}(\Gamma)$, where s and t are not zero. This gives rise to an exact sequence $P_2^* \overset{\binom{s^*}{t^*}}{\to} P_3^* \coprod P_4^* \to$

$\operatorname{Tr} S_2 \to 0$ where s^* and t^* are not zero. Here P_2^* is $k \xleftarrow{1} k \begin{smallmatrix} \nearrow^0 \\ \searrow_0 \end{smallmatrix}$,

P_3^* is $k \xleftarrow{1} k \begin{smallmatrix} \nearrow^{1}\, k \\ \searrow_0 \end{smallmatrix}$ and P_4^* is $k \xleftarrow{1} k \begin{smallmatrix} \nearrow^0 \\ \searrow_k \end{smallmatrix}$. Since $s^*: P_2^* \to P_3^*$

and $t^*: P_2^* \to P_4^*$ are nonzero, it is easy to see that they are both

monomorphisms. Hence $\operatorname{Tr} S_2$ must be of the form $k \xleftarrow{} k \begin{smallmatrix} \nearrow^k \\ \searrow_k \end{smallmatrix}$ where

the maps are either zero or isomorphisms. If one of the maps is zero, we would get a contradiction to $\operatorname{Tr} S_2$ being indecomposable. Hence all maps are isomorphisms. Therefore after a change of basis all maps can be represented by the identity matrix. Then we get that $D \operatorname{Tr} S_2$ is given

by the representation $k \xrightarrow{1} k \begin{smallmatrix} \nearrow^{1}\, k \\ \searrow_1 k \end{smallmatrix}$, which is P_1.

Our next example shows that $\operatorname{Tr} C$ is a familiar construction for some Λ-modules C. To state this result, it is convenient to recall that $\operatorname{Ext}_\Lambda^i(C, \Lambda)$ is considered as a right Λ-module, or Λ^{op}-module, by means of the operation of Λ on Λ given by right multiplication, for all $i \geq 0$ and all C in mod Λ.

Proposition 1.13 *Let $\mathscr{C}$ be the full subcategory of* mod Λ *consisting of all Λ-modules C with* $\mathrm{pd}_\Lambda C \leq 1$. *Then the contravariant R-functors* Tr: $\mathscr{C}/\mathscr{P} \to$ mod(Λ^{op}) *and* $\mathrm{Ext}^1_\Lambda(\ ,\Lambda)$: $\mathscr{C}/\mathscr{P} \to$ mod(Λ^{op}) *are isomorphic, where $\mathscr{C}/\mathscr{P}$ is the category $\mathscr{C}$ modulo projectives.*

Proof Let $0 \to P_1 \to P_0 \to C \to 0$ be a minimal projective resolution for C in $\mathscr{C}$. Then

$$0 \to C^* \to P_0^* \to P_1^* \to \mathrm{Ext}^1_\Lambda(C,\Lambda) \to 0$$

is exact. This gives an isomorphism Tr $C \simeq \mathrm{Ext}^1_\Lambda(C,\Lambda)$ in mod(Λ^{op}) which it is not difficult to check is functorial in C. □

As an immediate consequence of this result we have the following.

Corollary 1.14 *For a hereditary artin algebra Λ the functors* Tr: $\underline{\mathrm{mod}}\ \Lambda \to$ $\underline{\mathrm{mod}}\ \Lambda^{\mathrm{op}}$ *and* $\mathrm{Ext}^1_\Lambda(\ ,\Lambda)$: $\underline{\mathrm{mod}}\ \Lambda \to \underline{\mathrm{mod}}\ \Lambda^{\mathrm{op}}$ *are isomorphic. Hence the functors D* Tr: $\underline{\mathrm{mod}}\ \Lambda \to \overline{\mathrm{mod}}\ \Lambda$ *and* $D\mathrm{Ext}^1_\Lambda(\ ,\Lambda)$: $\underline{\mathrm{mod}}\ \Lambda \to \overline{\mathrm{mod}}\ \Lambda$ *are isomorphic.* □

In this connection it is useful to make the following observation which shows that for hereditary algebras the transpose is really defined on mod Λ and not just on $\underline{\mathrm{mod}}\ \Lambda$.

Proposition 1.15 *Assume Λ is a hereditary artin algebra. Then for B and C in* mod$_{\mathscr{P}}\ \Lambda$ *we have that $\mathscr{P}(B,C) = 0$. Hence the functor* mod$_{\mathscr{P}}\ \Lambda \to$ $\underline{\mathrm{mod}}\ \Lambda$ *is an equivalence of categories.*

Proof Let $g: B \to C$ be in $\mathscr{P}(B,C)$. Then there is a commutative diagram

$$
\begin{array}{ccc}
 & P & \\
 {}^{s}\nearrow & & \searrow^{t} \\
B & \xrightarrow{\ g\ } & C
\end{array}
$$

with P a projective module. Since Λ is hereditary, we have that Im $s \subset P$ is projective. Hence Im s is a summand of B and so must be zero since B has no nonzero projective summands. Therefore $g = 0$, and hence $\mathscr{P}(B,C) = 0$. The rest of the lemma is an immediate consequence of this fact. □

As our final observation concerning the connections between the transpose and modules of projective dimension 1 we have the following.

Proposition 1.16 *Let X be a Λ-module. Then $\operatorname{pd}_\Lambda X \le 1$ if and only if $\operatorname{Hom}_\Lambda(D(\Lambda), D\operatorname{Tr} X) = 0$.*

Proof Let $P_1 \to P_0 \to \operatorname{Tr} X \to 0$ be a minimal projective presentation of $\operatorname{Tr} X$ in $\operatorname{mod}(\Lambda^{\mathrm{op}})$. Then we have the exact sequence of Λ-modules $0 \to (\operatorname{Tr} X)^* \to P_0^* \to P_1^* \to Y \to 0$ where Y has no projective indecomposable summands and $X = Y \coprod P$ with P a projective module. Since $P_0^* \to P_1^* \to Y \to 0$ is a minimal projective presentation of Y, we have that $\operatorname{pd}_\Lambda Y \le 1$ if and only if $(\operatorname{Tr} X)^* = 0$. Hence we see that $\operatorname{pd}_\Lambda X \le 1$ if and only if $\operatorname{Hom}_{\Lambda^{\mathrm{op}}}(\operatorname{Tr} X, \Lambda) = 0$. Since $\operatorname{Hom}_{\Lambda^{\mathrm{op}}}(\operatorname{Tr} X, \Lambda) = 0$ if and only if $\operatorname{Hom}_\Lambda(D(\Lambda), D\operatorname{Tr} X) = 0$, we have our desired result. $\square$

We now give an example of how the operations D and Tr can be used to show the existence of indecomposable modules of arbitrarily large length.

Example Let k be a field and $\Lambda = k[X, Y]/(X, Y)^2$. Then Λ is a local ring so that Λ is an indecomposable Λ-module, and is the only indecomposable projective Λ-module up to isomorphism. We have $\mathfrak{r}^2 = 0$ and $\mathfrak{r} \simeq S \coprod S$ where $S = \Lambda/\mathfrak{r}$ is the unique simple Λ-module up to isomorphism. If C is an indecomposable nonsimple Λ-module, we have $\mathfrak{r}C = \operatorname{soc} C$. For clearly $\mathfrak{r}C \subset \operatorname{soc} C$, and if the inclusion was proper we would have a nonzero submodule K of $\operatorname{soc} C$ such that $\operatorname{soc} C = \mathfrak{r}C \coprod K$. Then the composition $K \to C \to C/\mathfrak{r}C$ is a monomorphism, and hence there is a morphism $C/\mathfrak{r}C \to K$ such that the composition $K \to C \to K$ is the identity. Then C would be simple.

For an indecomposable Λ-module C let $t = l(C/\mathfrak{r}C)$ and $s = l(\mathfrak{r}C)$. It is easy to see we have then a minimal projective presentation $(2t - s)\Lambda \to t\Lambda \to C \to 0$. If C is simple there is a minimal projective presentation $2\Lambda \to \Lambda \to DC \to 0$ and if C is not simple there is a minimal projective presentation $(2s - t)\Lambda \to s\Lambda \to DC \to 0$. Hence for C simple we have a minimal projective presentation $\Lambda \to 2\Lambda \to \operatorname{Tr} DC \to 0$ and for C not simple we have a minimal projective presentation $s\Lambda \to (2s - t)\Lambda \to \operatorname{Tr} DC \to 0$. Hence we get by induction that $\operatorname{Tr} D^n S$, $n \ge 1$, has a minimal projective presentation $(2n - 1)\Lambda \to 2n\Lambda \to \operatorname{Tr} D^n S \to 0$. Hence there is no bound on the length of the indecomposable Λ-modules.

An artin algebra Λ is said to be of **finite representation type**, or of **finite type** for short, if there is only a finite number of indecomposable objects up to isomorphism in mod Λ. Such algebras are studied in Chapter VI. We have just seen that $k[X, Y]/(X, Y)^2$ is of infinite representation type.

2 Nakayama algebras

The interest in the functors Tr and D Tr stems from the fact that their behavior reflects important properties of the category of Λ-modules. As our first illustration of this point we show in this section that Nakayama algebras can be characterized by the property that the D Tr-orbits of simple modules consist entirely of simple modules. Nakayama algebras are of considerable interest because next to semisimple algebras they are the best understood artin algebras. Since Nakayama algebras are defined in terms of uniserial modules, we start this section with a discussion of these modules.

Let Λ be an artin algebra. A Λ-module A is called a **uniserial module** if the set of submodules is totally ordered by inclusion. We have the following useful characterizations of uniserial modules, using the notions of radical filtration and socle filtration introduced in Chapter II. The proof is left as an exercise for the reader.

Proposition 2.1 *The following are equivalent for a Λ-module A.*

(a) *A is uniserial.*

(b) *There is only one composition series for A.*

(c) *The radical filtration of A is a composition series for A.*

(d) *The socle filtration of A is a composition series for A.*

(e) *$l(A) = \mathrm{rl}(A)$.* $\square$

The following are easily verified properties of uniserial modules.

Lemma 2.2 *Suppose A is a uniserial module. Then we have the following.*

(a) *$D(A)$ is a uniserial module.*

(b) *If $0 \to A' \to A \to A'' \to 0$ is exact, then A' and A'' are uniserial modules.*

(c) *A is indecomposable with $A/\mathfrak{r}A$ and $\mathrm{soc}\,A$ simple modules.*

(d) *If $P \to A$ is a projective cover, then P is indecomposable.*

(e) *If $A \to I$ is an injective envelope, then I is indecomposable.* $\square$

An artin algebra Λ is said to be a **Nakayama algebra** if both the indecomposable projective and indecomposable injective modules are uniserial. Obviously Λ is Nakayama if and only if the indecomposable projective Λ and Λ^{op}-modules are uniserial. If k is a field then $k[X]/(X^n)$ is a Nakayama algebra for all $n > 0$.

It is worth noting that Λ having the property that every indecomposable projective module is uniserial does not necessarily imply that Λ^{op} has the same property. For example, if Λ is the path algebra $k\Gamma$ of the quiver $\underset{1}{\cdot} \rightarrow \underset{2}{\cdot} \leftarrow \underset{3}{\cdot}$, then every indecomposable projective Λ-module is uniserial, but the same is not true for Λ^{op}. Since the class of artin algebras Λ with the property that Λ is a sum of uniserial modules, i.e. every indecomposable projective module is uniserial, is of interest in itself, we point out some features of these algebras which we will need in connection with our study of Nakayama algebras.

Proposition 2.3 *The following are equivalent for an artin algebra Λ.*

(a) Λ *is a sum of uniserial modules.*
(b) $\Lambda/\mathfrak{a}$ *is a sum of uniserial modules for all ideals $\mathfrak{a}$ of Λ.*
(c) $\Lambda/\mathfrak{r}^2$ *is a sum of uniserial modules.*

Proof (a)$\Rightarrow$(b) and (b)$\Rightarrow$(c) are trivial.

(c)$\Rightarrow$(a) Let P be an indecomposable projective Λ-module. We show that $P/\mathfrak{r}^n P$ is uniserial by induction on n when $n \geq 2$.

When $n = 2$ there is nothing to prove. Suppose $n > 2$. Then $P/\mathfrak{r}^{n-1}P$ is uniserial by the induction hypothesis. If $\mathfrak{r}^{n-1}P = 0$, then $P/\mathfrak{r}^n P$ is clearly uniserial, so we can assume $\mathfrak{r}^{n-1}P \neq 0$. It follows from Proposition 2.1 that $\mathfrak{r}^i P/\mathfrak{r}^{i+1}P$ is simple for $i = 0, \ldots, n-2$. To show that $P/\mathfrak{r}^n P$ is uniserial, it is then sufficient by Proposition 2.1 to prove that $\mathfrak{r}^{n-1}P/\mathfrak{r}^n P$ is also simple. Let $Q \to \mathfrak{r}^{n-2}P$ be a projective cover. Since $\mathfrak{r}^{n-2}P/\mathfrak{r}^{n-1}P$ is simple, Q must be indecomposable and so $Q/\mathfrak{r}^2 Q$ is uniserial. But we have an epimorphism $\mathfrak{r}Q/\mathfrak{r}^2 Q \to \mathfrak{r}^{n-1}P/\mathfrak{r}^n P$ which shows that $\mathfrak{r}^{n-1}P/\mathfrak{r}^n P$ is simple.   □

Restating this result for Nakayama algebras we have the following.

Corollary 2.4 *The following are equivalent for an artin algebra Λ.*

(a) Λ *is a Nakayama algebra.*
(b) $\Lambda/\mathfrak{a}$ *is a Nakayama algebra for all ideals $\mathfrak{a}$ of Λ.*
(c) $\Lambda/\mathfrak{r}^2$ *is a Nakayama algebra.*   □

The following is a direct consequence of the definitions involved.

Lemma 2.5 *Let* Λ *be an artin algebra which is a sum of uniserial modules. Then the following are equivalent for a* Λ-*module* C.

(a) $C/\mathfrak{r}C$ *is simple.*

(b) *If* $P \to C$ *is a projective cover, then* P *is uniserial.*

(c) C *is uniserial.*

Moreover, if A *and* B *are uniserial* Λ-*modules, then* $A \simeq B$ *if and only if* $A/\mathfrak{r}A \simeq B/\mathfrak{r}B$ *and* $l(A) = l(B)$. □

The following description of the dual of the transpose for uniserial modules over Nakayama algebras is basic to our treatment of Nakayama algebras.

Proposition 2.6 *Suppose* C *is a uniserial nonprojective module of length* n *over a Nakayama algebra* Λ. *Then we have the following.*

(a) $\operatorname{Tr} C$ *and* $D \operatorname{Tr} C$ *are uniserial.*

(b) $l(C) = l(D \operatorname{Tr} C)$.

(c) *If* $P \to C$ *is a projective cover, then* $D \operatorname{Tr} C \simeq \mathfrak{r}P/\mathfrak{r}^{n+1}P$.

Proof Let $P_1 \xrightarrow{f} P \xrightarrow{g} C \to 0$ be a minimal projective presentation of C. Since C is uniserial, P is uniserial, which implies that $\operatorname{Ker} g = \mathfrak{r}^n P$ since $l(C) = n$ and is thus uniserial. Hence P_1 is uniserial. Now it is not difficult to see that $n = l(C)$ is the maximal length of a chain of nonisomorphisms $P_1 \to Q_1 \to \cdots \to Q_{n-1} \to P$ with composition f between indecomposable projective modules. Since Λ is a Nakayama algebra, the minimal projective presentation $P^* \to P_1^* \to \operatorname{Tr} C \to 0$ has the property that P^* and P_1^* are uniserial since they are indecomposable. Also the duality $T : \mathscr{P}(\Lambda) \to \mathscr{P}(\Lambda^{\mathrm{op}})$ shows that $P^* \to Q_{n-1}^* \to \cdots \to Q_1^* \to P_1^*$ is a maximal chain of nonisomorphisms with composition f^* between indecomposable projective modules. Therefore $\operatorname{Tr} C$ is uniserial and $l(\operatorname{Tr} C) = l(C)$, which gives immediately that $D \operatorname{Tr} C$ is uniserial and $l(D \operatorname{Tr} C) = l(C)$. This proves parts (a) and (b).

We now prove part (c). We know that $\operatorname{soc} D \operatorname{Tr} C \simeq P_1/\mathfrak{r}P_1$ by Proposition 1.11 and also that $P_1/\mathfrak{r}P_1 \simeq \mathfrak{r}^n P/\mathfrak{r}^{n+1}P$ since $\mathfrak{r}^n P = \operatorname{Ker} g$. Therefore $\mathfrak{r}P/\mathfrak{r}^{n+1}P$ and $D \operatorname{Tr} C$ are two uniserial modules of the same length with the same socles. Since Λ^{op} is a sum of uniserial modules and $D(D \operatorname{Tr} C)$ and $D(\mathfrak{r}P/\mathfrak{r}^{n+1}P)$ are uniserial modules of the same length and are isomorphic modulo their radicals, it follows by Lemma 2.5 that they are isomorphic. Therefore $\mathfrak{r}P/\mathfrak{r}^{n+1}P \simeq D\operatorname{Tr}C$, which finishes the proof of the proposition.   □

By duality, we have the following version of Proposition 2.6.

Proposition 2.7 *Suppose C is a uniserial noninjective module of length n over a Nakayama algebra Λ and suppose $C \to I$ is an injective envelope. Then $\operatorname{Tr}DC \simeq (\operatorname{soc}^{n+1}I)/\operatorname{soc}I$, which is a uniserial module of length n.* □

We show in Chapter VI that all indecomposable modules over a Nakayama algebra are uniserial. Hence Propositions 2.6 and 2.7 describe $D\operatorname{Tr}$ and $\operatorname{Tr}D$ for all indecomposable modules.

The rest of our discussion of Nakayama algebras uses the notion of the $D\operatorname{Tr}$-partition of the indecomposable Λ-modules which is defined as follows for arbitrary artin algebras, not just Nakayama algebras.

Let Λ be an arbitrary artin algebra. We denote by $\operatorname{ind}\Lambda$ a full subcategory of $\operatorname{mod}\Lambda$ whose objects consist of chosen representatives from isomorphism classes of indecomposable modules in $\operatorname{mod}\Lambda$. Note that the zero module is not in $\operatorname{ind}\Lambda$.

Now $D\operatorname{Tr}$ operates on $\operatorname{ind}\Lambda \cup \{0\}$ and we define $(D\operatorname{Tr})^i$ for all $i \in \mathbb{N}$ by $(D\operatorname{Tr})^0 = 1$ and $(D\operatorname{Tr})^i = D\operatorname{Tr}((D\operatorname{Tr})^{i-1})$ for $i > 0$. Further $(D\operatorname{Tr})^j$ for $-j \in \mathbb{N}$ is defined by $(D\operatorname{Tr})^j = (\operatorname{Tr}D)^{-j}$, where also $(\operatorname{Tr}D)^0 = 1$ by definition.

Let C be in $\operatorname{ind}\Lambda$. Then define the $D\operatorname{Tr}$-*orbit* of C to be the collection of indecomposable modules in $\{(D\operatorname{Tr})^iC\}_{i\in\mathbb{Z}}$. It is easily seen that the $D\operatorname{Tr}$-orbits induce a partition of the objects in $\operatorname{ind}\Lambda$ which we will refer to as the $D\operatorname{Tr}$-partition. The $D\operatorname{Tr}$-orbits of $\operatorname{ind}\Lambda$ are of three basically different types as we now describe.

Proposition 2.8 *Let $\mathcal{O}$ be a $D\operatorname{Tr}$-orbit of $\operatorname{ind}\Lambda$.*

(a) *Suppose there is a projective module P in $\mathcal{O}$. Then we have the following.*

 (i)   *$\mathcal{O}$ consists of the nonzero objects in $\{P,(D\operatorname{Tr})^{-1}P,\dots,$ $(D\operatorname{Tr})^{-i}P,\dots\}_{i\in\mathbb{N}}$*

 (ii) *$\mathcal{O}$ is finite if and only if $(D\operatorname{Tr})^{-n}P = (\operatorname{Tr}D)^nP$ is injective for some n in $\mathbb{N}$. Moreover if $(\operatorname{Tr}D)^nP$ is injective, then $\mathcal{O} = \{P,(D\operatorname{Tr})^{-1}P,\dots,(D\operatorname{Tr})^{-n}P\}$.*

(b) *Suppose $\mathcal{O}$ contains an injective module I. Then we have the following.*

 (i)   *$\mathcal{O}$ consists of the nonzero modules in $\{I,D\operatorname{Tr}I,\dots,(D\operatorname{Tr})^iI,\dots\}_{i\in\mathbb{N}}$.*

(ii) $\mathcal{O}$ is finite if and only if $(D\,\mathrm{Tr})^n I$ is projective for some $n \in \mathbb{N}$. Moreover if $(D\,\mathrm{Tr})^n I$ is projective, then $\mathcal{O} = \{I, D\,\mathrm{Tr}\,I, \ldots, (D\,\mathrm{Tr})^n I\}$.

(c) Suppose $\mathcal{O}$ contains no projective or injective modules. Then we have the following.

(i) $\mathcal{O}$ consists of $\{(D\,\mathrm{Tr})^i A\}_{i \in \mathbb{Z}}$ where A is any object in $\mathcal{O}$.

(ii) $\mathcal{O}$ is finite if and only if there is some $i > 0$ such that $(D\,\mathrm{Tr})^i A \simeq A$ for some (and hence all) A in $\mathcal{O}$. Moreover, if n is the smallest such i then $\mathcal{O} = \{A, \ldots, (D\,\mathrm{Tr})^{n-1} A\}$.

Proof (a) Since $(D\,\mathrm{Tr})^i P = 0$ for all $i > 0$, part (i) is established. We claim that if $(D\,\mathrm{Tr})^{-i} P \simeq (D\,\mathrm{Tr})^{-(i+j)} P \neq 0$ with $j > 0$, then $P \simeq (D\,\mathrm{Tr})^{-j} P = (\mathrm{Tr}\,D)^j P$ which is impossible since $j > 0$. Therefore the only way $\mathcal{O}$ can be finite is that $(D\,\mathrm{Tr})^{-(n+1)} P = 0$ for some $n \geq 0$, or equivalently, $(D\,\mathrm{Tr})^{-n} P$ is injective. It is also clear by our previous remark that if $(D\,\mathrm{Tr})^{-n} P$ is injective, then $\mathcal{O} = \{P, (D\,\mathrm{Tr})^{-1} P, \ldots, (D\,\mathrm{Tr})^{-n} P\}$.

Part (b) is the dual of (a) and (c) is easily seen to be true. □

The following is an obvious consequence of the definitions of $D\,\mathrm{Tr}$-orbits and the previous propositions.

Corollary 2.9 *Let Λ be a Nakayama algebra. If C is a uniserial module of length n, then all the modules in the $D\,\mathrm{Tr}$-orbit of C are uniserial modules of length n.* □

We can now prove our promised characterization of Nakayama algebras.

Theorem 2.10 *An artin algebra Λ is a Nakayama algebra if and only if the $D\,\mathrm{Tr}$-orbits of simple modules consist entirely of simple modules.*

Proof By Corollary 2.9 we know that if Λ is a Nakayama algebra, then the $D\,\mathrm{Tr}$-orbits of simple modules consist entirely of simple modules.

Suppose now that the $D\,\mathrm{Tr}$-orbits of simple modules consist entirely of simple modules. Let P be an indecomposable projective Λ-module. Suppose P is not simple and let $P_1 \to P \to P/\mathfrak{r}P \to 0$ be a minimal projective presentation. Then $P^* \to P_1^* \to \mathrm{Tr}(P/\mathfrak{r}P) \to 0$ is a minimal projective presentation. The fact that $D\,\mathrm{Tr}(P/\mathfrak{r}P)$, and hence $\mathrm{Tr}(P/\mathfrak{r}P)$, is simple implies that P_1^*, and hence P_1, is indecomposable. Therefore $\mathfrak{r}P/\mathfrak{r}^2 P$ is simple. This shows that $\Lambda/\mathfrak{r}^2$ is a sum of uniserial modules, which implies that Λ is a sum of uniserial modules by Proposition 2.3. Using the fact that $\operatorname{Tr} D(S)$ is simple if S is simple noninjective, we get that Λ^{op} also has the property that if T is a simple nonprojective Λ^{op}-module, then $D \operatorname{Tr} T$ is simple. Thus Λ^{op} is a sum of uniserial modules, which shows that Λ is a Nakayama algebra. $\square$

We end this preliminary discussion of Nakayama algebras by taking a more detailed look at the structure of the $D \operatorname{Tr}$-orbits of indecomposable modules containing simple Λ-modules.

Let Λ be a Nakayama algebra. We have seen in Corollary 2.9 that all the modules in a $D \operatorname{Tr}$-orbit of a uniserial module have the same length. In particular, the $D \operatorname{Tr}$-orbits $\mathcal{O}_1, \ldots, \mathcal{O}_m$ of the simple modules give a partition of the isomorphism classes of simple modules. We claim that this is the same as the block decomposition of simple modules, i.e. each $\mathcal{O}_i$ consists of the simple modules belonging to one indecomposable block of Λ. To see this it is convenient to introduce an ordering on each of the $D \operatorname{Tr}$-orbits $\mathcal{O}_i$, called the Kupisch series, which we now describe.

Let $\mathcal{O}$ be the $D \operatorname{Tr}$-orbit of some simple Λ-module, i.e. $\mathcal{O} = \mathcal{O}_i$ for some i. Since $\mathcal{O}$ consists entirely of simple Λ-modules, it is finite. Hence it follows from Proposition 1.10 that $\mathcal{O}$ contains an injective simple module if and only if it contains a projective simple module. Furthermore, if $\mathcal{O}$ contains an injective simple module S, then $\mathcal{O} = \{(D \operatorname{Tr})^0 S = S, D \operatorname{Tr} S, \ldots, (D \operatorname{Tr})^{n-1} S\}$ where the $D \operatorname{Tr}^i S$ are distinct for $i = 0, \ldots, n-1$ and $(D \operatorname{Tr})^{n-1} S$ is projective. Then $\mathcal{O}$ with this ordering is called the **Kupisch series** for the $D\operatorname{Tr}$-orbit $\mathcal{O}$. If $\mathcal{O}$ does not contain an injective module, then $\mathcal{O} = \{(D \operatorname{Tr})^0 S, D \operatorname{Tr} S, \ldots, (D \operatorname{Tr})^{n-1} S\}$, where S is an arbitrary element of $\mathcal{O}$, and n is the smallest integer such that $D \operatorname{Tr}^n S \simeq S$, and where all the $D \operatorname{Tr}^i S$ are nonisomorphic for $i = 0, \ldots, n-1$. $\mathcal{O}$ together with this ordering, which is unique up to cyclic permutations, is then called a **Kupisch series** for the $D\operatorname{Tr}$-orbit $\mathcal{O}$. For a $D \operatorname{Tr}$-orbit $\mathcal{O}$ we denote by $\widetilde{\mathcal{O}}$ the projective covers of the simple modules in $\mathcal{O}$. We call $\widetilde{\mathcal{O}}$ with the induced ordering the **Kupisch series** for $\widetilde{\mathcal{O}}$.

We have the following connection between the $D \operatorname{Tr}$-partition of the simple modules and the blocks of Λ.

Proposition 2.11 *Let Λ be a Nakayama algebra and $\{\mathcal{O}_1, \ldots, \mathcal{O}_n\}$ the $D \operatorname{Tr}$-partition of the isomorphism classes of simple Λ-modules. Let $\{\widetilde{\mathcal{O}}_1, \ldots, \widetilde{\mathcal{O}}_n\}$ be the corresponding partition of the isomorphism classes of indecomposable* projective Λ-modules, where P is in $\widetilde{\mathcal{O}}_i$ if and only if $P/\mathfrak{r}P$ is in $\mathcal{O}_i$. Then this is the block partition of the projective modules.

Proof Let $\{S_0,\ldots,S_{n-1}\}$ be a Kupisch series for $\mathcal{O}_i$, and P_j a projective cover of S_j for $j = 0, \ldots, n-1$. Since $S_{j+1} \simeq D\operatorname{Tr}S_j \simeq \mathfrak{r}P_j/\mathfrak{r}^2P_j$, and hence P_{j+1} is a projective cover for $\mathfrak{r}P_j$, for all $j = 0, \ldots, n-2$, there are nonisomorphisms $P_{j+1} \to P_j$. Hence $P_0, \ldots, P_{n-1}$ are all in the same block of Λ.

The fact that $D\operatorname{Tr}(P/\mathfrak{r}P) \simeq \mathfrak{r}P/\mathfrak{r}^2P$ for each indecomposable projective nonsimple Λ-module P also implies that if P is in $\widetilde{\mathcal{O}}_i$, then $\mathfrak{r}^jP/\mathfrak{r}^{j+1}P$ is in $\mathcal{O}_i$ for all $j = 0, \ldots, l(P)-1$. From this it follows that if P is in $\widetilde{\mathcal{O}}_i$ and P' is in $\widetilde{\mathcal{O}}_t$ with $i \neq t$, then $\operatorname{Hom}_\Lambda(P,P') = 0 = \operatorname{Hom}_\Lambda(P',P)$. This finishes the proof. $\square$

Corollary 2.12 *A Nakayama algebra Λ is an indecomposable ring if and only if all the simple Λ-modules are in the same $D\operatorname{Tr}$-orbit.* $\square$

Assume now that Λ is an indecomposable Nakayama algebra with Kupisch series $\{S_0,\ldots,S_{n-1}\}$ and $\{P_0,\ldots,P_{n-1}\}$ of simple and projective modules. Since P_{j+1} is a projective cover for $\mathfrak{r}P_j$ for all $j = 0, \ldots, n-2$, and P_0 is a projective cover of $\mathfrak{r}P_{n-1}$ if P_{n-1} is not simple, the sequence of positive integers $(a_0,\ldots,a_{n-1})$ where $a_j = l(P_j)$ for $0 \leq j \leq n-1$ has the property that $a_{j+1} \geq a_j - 1 \geq 1$ for all $j = 0,\ldots,n-2$ and $a_0 \geq a_{n-1} - 1$. Any sequence of positive integers $(a_0, \ldots, a_{n-1})$ satisfying these conditions is called an **admissible sequence**. If $\{P_0,\ldots,P_{n-1}\}$ is a Kupisch series for the indecomposable projective Λ-modules, then $(l(P_0),\ldots,l(P_{n-1}))$ is called the admissible sequence of Λ. We then have the following.

Proposition 2.13 *For any admissible sequence $(a_0,\ldots,a_{n-1})$ of positive integers there is a Nakayama algebra with this sequence as its admissible sequence.*

Proof Let $(a_0,a_1,\ldots,a_{n-1})$ be an admissible sequence of positive integers. Consider the quiver Γ

if $a_{n-1} > 1$ and

$$\underset{0}{\cdot} \xrightarrow{\alpha_0} \underset{1}{\cdot} \xrightarrow{\alpha_1} \cdots \xrightarrow{\alpha_{n-2}} \underset{n-2}{\cdot} \underset{n-1}{\cdot}$$

if $a_{n-1} = 1$. For each i there is a unique path p_i of length a_i starting at i. Then the path algebra $k\Gamma$ of Γ modulo the ideal generated by the paths p_i is a Nakayama algebra with admissible sequence $(a_0, a_1, \ldots, a_{n-1})$. The verification of this fact is left to the reader.   □

We illustrate with the following.

Example $k[X]/(X^n)$ is a Nakayama algebra with admissible sequence (n).

Example The $n \times n$ full lower triangular matrix algebra over a field is a Nakayama algebra with admissible sequence $(n, n-1, \ldots, 1)$.

The rest of this section is devoted to showing how to construct new examples of Nakayama algebras from old ones by using skew group algebras. The main result is the following.

Theorem 2.14 *Let ΛG be a skew group algebra with the order of G invertible in Λ. Then ΛG is a Nakayama algebra if and only if Λ is Nakayama.*

We begin the proof by showing that we can reduce the proof of Theorem 2.14 to proving it for Λ with the property that $\mathfrak{r}^2 = 0$ where $\mathfrak{r}$ is the radical of Λ. We know by Corollary 2.4 that an artin algebra Γ is Nakayama if and only if $\Gamma/(\operatorname{rad}\Gamma)^2$ is Nakayama. Therefore Λ is Nakayama if and only if $\Lambda/\mathfrak{r}^2$ is Nakayama and ΛG is Nakayama if and only if $\Lambda G/(\operatorname{rad}\Lambda G)^2$ is Nakayama. But by III Proposition 4.11 we have $(\operatorname{rad}\Lambda G) = \mathfrak{r}\Lambda G$ and so $(\operatorname{rad}\Lambda G)^2 = \mathfrak{r}^2\Lambda G$. Moreover we also know from Chapter III that $\Lambda G/(\mathfrak{r}\Lambda G)^2 \simeq (\Lambda/\mathfrak{r}^2)G$. Finally, the fact that the order of G is invertible in Λ implies that the order of G is invertible in $\Lambda/\mathfrak{r}^2$. Thus we have shown that to prove Theorem 2.14 it suffices to prove it under the additional hypothesis that $\mathfrak{r}^2 = 0$. This we now proceed to do. We begin with the following description of the indecomposable modules of a Nakayama algebra of radical square zero.

Lemma 2.15 *Let Λ be a Nakayama algebra with $\mathfrak{r}^2 = 0$. Then the following are equivalent for an indecomposable nonsimple Λ-module M.*

(a) *M is a projective Λ-module.*
(b) *M is an injective Λ-module.*

Proof (a) $\Rightarrow$ (b) Let M be a nonsimple indecomposable projective Λ-module. Since Λ is a Nakayama algebra, M is uniserial. Since $\mathfrak{r}^2 = 0$, it follows that $l(M) = 2$. Let $I(M)$ be an injective envelope of M. Since $\operatorname{soc} M \simeq \operatorname{soc} I(M)$ is simple, it follows that $I(M)$ is indecomposable and therefore uniserial of length at least 2. Since $\mathfrak{r}^2 = 0$, it follows that $l(I(M)) = 2$, so $M \simeq I(M)$ which means that M is injective.

(b) $\Rightarrow$ (a) This is dual of (a) $\Rightarrow$ (b). $\qquad\square$

We apply this to obtain the following homological characterization of Nakayama algebras whose square of the radical is zero.

Proposition 2.16 *Let Λ be an artin algebra with radical $\mathfrak{r}$ such that $\mathfrak{r}^2 = 0$. Then Λ is a Nakayama algebra if and only if the injective envelope $I(\Lambda)$ is a projective module.*

Proof Suppose Λ is a Nakayama algebra. Let $P \to I(P)$ be an injective envelope for an indecomposable projective Λ-module P. Since $\operatorname{soc} P$ is simple, $\operatorname{soc} I(P)$ is simple and so $I(P)$ is an indecomposable injective module. If $I(P)$ is simple then P is simple and $P \simeq I(P)$, so $I(P)$ is projective. If $I(P)$ is not simple, then by Lemma 2.15 we have that $I(P)$ is projective. So in any event, $I(P)$ is projective as well as injective, which shows that $I(\Lambda)$ is projective.

Suppose now that $I(\Lambda)$ is projective. Let P be an indecomposable projective Λ-module. We want to show that P is uniserial. If P is simple there is nothing to prove. Suppose P is not simple and let $P \to I(P)$ be an injective envelope. Since $I(P)$ is projective we know that $I(P) = Q_1 \coprod \cdots \coprod Q_t$ where the Q_i are indecomposable modules which are both projective and injective. Since $\mathfrak{r}^2 = 0$, it follows that each of the Q_i is uniserial. Since P is not simple, $P \not\subseteq \mathfrak{r}I(P)$, so there is some projection $I(P) \to Q_i$ such that $\operatorname{Im}(P \to I(P) \to Q_i)$ is not contained in $\mathfrak{r}Q_i$. Therefore the composition $P \to Q_i$ is an epimorphism and hence an isomorphism since P is indecomposable. So P is a uniserial injective module. Hence we have that the indecomposable nonsimple projective modules are uniserial injective modules. So we now have to show that if I is an indecomposable injective module, then I is uniserial.

If I is simple, there is nothing to prove. Suppose I is not simple. Let $f: P \to I$ be a projective cover for I. Let $P = P_1 \coprod \cdots \coprod P_t$ with the P_i indecomposable modules. Then $f(\operatorname{soc} P_i) \neq 0$ for some $i = 1, \ldots, t$, or else I would be simple. Hence $f|_{P_i}: P_i \to I$ is injective. Since $f: P \to I$ is a projective cover and $\operatorname{soc} I$ is simple, P_i is not simple. Therefore P_i is injective and so $f|_{P_i}: P_i \to I$ is an isomorphism. But we also know that P_i is uniserial which means that I is uniserial.   □

Combining this proposition with our previous remarks, we see that to prove Theorem 2.14 it suffices to show the following.

Proposition 2.17 *Let ΛG be an arbitrary skew group algebra. Then $I(\Lambda)$, an injective envelope of Λ, is a projective Λ-module if and only if $I(\Lambda G)$, an injective envelope of ΛG, is a projective ΛG-module.*

This proposition follows readily from some general considerations which are also of interest in their own right.

Proposition 2.18 *Let $f: \Lambda \to \Gamma$ be a morphism of artin R-algebras. Then the following are equivalent.*

(a) *Γ is a projective right Λ-module.*
(b) *Every injective Γ-module is also an injective Λ-module.*

Proof (a) $\Rightarrow$ (b) Let $0 \to A \to B \to C \to 0$ be an exact sequence of Λ-modules. Since Γ is a projective right Λ-module, we have that $0 \to \Gamma \otimes_\Lambda A \to \Gamma \otimes_\Lambda B \to \Gamma \otimes_\Lambda C \to 0$ is exact. Suppose I is an injective Γ-module. Then we have the exact commutative diagram

$$\begin{array}{ccccc}
\mathrm{Hom}_\Gamma(\Gamma \otimes_\Lambda B, I) & \to & \mathrm{Hom}_\Gamma(\Gamma \otimes_\Lambda A, I) & \to & 0 \\
\downarrow \wr & & \downarrow \wr & & \\
\mathrm{Hom}_\Lambda(B, I) & \to & \mathrm{Hom}_\Lambda(A, I)\,, & &
\end{array}$$

which means that $\mathrm{Hom}_\Lambda(B, I) \to \mathrm{Hom}_\Lambda(A, I) \to 0$ is exact. This shows that I is an injective Λ-module, since $0 \to A \to B \to C \to 0$ is an arbitrary exact sequence of Λ-modules.

(b) $\Rightarrow$ (a) Let $0 \to A \xrightarrow{f} B \xrightarrow{g} C \to 0$ be an arbitrary exact sequence of Λ-modules. Then we have the exact sequence of Γ-modules $0 \to \mathrm{Ker}(\Gamma \otimes f) \to \Gamma \otimes_\Lambda A \to \Gamma \otimes_\Lambda B$. Let I be the Γ-injective envelope of $\Gamma/\mathrm{rad}\,\Gamma$. Then I is also an injective Λ-module, so we have the exact commutative diagram

$$\begin{array}{ccccccc}
\mathrm{Hom}_\Gamma(\Gamma \otimes_\Lambda B, I) & \to & \mathrm{Hom}_\Gamma(\Gamma \otimes_\Lambda A, I) & \to & \mathrm{Hom}_\Gamma(\mathrm{Ker}(\Gamma \otimes f), I) & \to & 0 \\
\downarrow \wr & & \downarrow \wr & & & & \\
\mathrm{Hom}_\Lambda(B, I) & \to & \mathrm{Hom}_\Lambda(A, I) & \to & & & 0,
\end{array}$$

which shows that $\mathrm{Hom}_\Gamma(\mathrm{Ker}(\Gamma \otimes f), I) = 0$. This means that $\mathrm{Ker}(\Gamma \otimes f) = 0$ or that $0 \to \Gamma \otimes_\Lambda A \to \Gamma \otimes_\Lambda B$ is exact. Since this holds for all exact

Λ-sequences $0 \to A \to B \to C \to 0$, we have that Γ is a projective right Λ-module. $\qquad \square$

As an easy consequence of Proposition 2.18 we have the following which proves one direction of Proposition 2.17.

Corollary 2.19 *Let ΛG be a skew group algebra. If $I(\Lambda G)$, the ΛG-injective envelope of ΛG, is ΛG-projective, then $I(\Lambda)$, the Λ-injective envelope of Λ, is Λ-projective.*

Proof We know by III Lemma 4.5 that ΛG is both a projective right and left Λ-module by means of the inclusion $\Lambda \to \Lambda G$. Since ΛG is a projective right Λ-module, we know by Proposition 2.18 that $I(\Lambda G)$ is an injective Λ-module. Since $I(\Lambda G)$ is a projective ΛG-module and ΛG is a projective Λ-module, it follows that $I(\Lambda G)$ is also a projective Λ-module. Therefore $I(\Lambda G)$ is a Λ-module containing Λ which is both a projective and an injective Λ-module. Hence $I(\Lambda G)$ contains $I(\Lambda)$ as a summand, which means that $I(\Lambda)$ is also a projective Λ-module. $\qquad \square$

We now finish the proof of Proposition 2.17 by showing that if $I(\Lambda)$ is a projective Λ-module, then $I(\Lambda G)$ is a projective ΛG-module.

The inclusion $\Lambda \to I(\Lambda)$ induces a monomorphism of ΛG-modules $\operatorname{Hom}_\Lambda(\Lambda G, \Lambda) \to \operatorname{Hom}_\Lambda(\Lambda G, I(\Lambda))$. Since $\Lambda G \simeq \operatorname{Hom}_\Lambda(\Lambda G, \Lambda)$ as ΛG-modules, this gives a monomorphism of ΛG-modules $\Lambda G \to \operatorname{Hom}_\Lambda(\Lambda G, I(\Lambda))$. Since ΛG is a projective right ΛG-module, the fact that $I(\Lambda)$ is an injective Λ-module implies that $\operatorname{Hom}_\Lambda(\Lambda G, I(\Lambda))$ is an injective ΛG-module. This follows from the fact that for all X in $\operatorname{mod} \Lambda G$, the morphisms $h: \operatorname{Hom}_\Lambda(X, I(\Lambda)) \to \operatorname{Hom}_{\Lambda G}(X, \operatorname{Hom}_\Lambda(\Lambda G, I(\Lambda)))$ given by $h(t)(x)(z) = t(zx)$ for all t in $\operatorname{Hom}_\Lambda(X, I(\Lambda))$ and x in X and z in ΛG are isomorphisms functorial in X. But by III Proposition 4.14 we know that $\operatorname{Hom}_\Lambda(\Lambda G, I(\Lambda))$ and $\Lambda G \otimes_\Lambda I(\Lambda)$ are isomorphic ΛG-modules. Since we are assuming that $I(\Lambda)$ is a projective Λ-module, it follows that $\Lambda G \otimes_\Lambda I(\Lambda)$ is a projective ΛG-module. So ΛG is a ΛG-submodule of the ΛG-module $\operatorname{Hom}_\Lambda(\Lambda G, I(\Lambda))$ which is both a projective and injective ΛG-module. This implies that $I(\Lambda G)$ which is a summand of $\operatorname{Hom}_\Lambda(\Lambda G, I(\Lambda))$ is also a projective ΛG-module. $\qquad \square$

We end this section with the following illustration of Theorem 2.14.

Example Let $\mathbb{C}$ be the complex numbers, let $\Lambda = \mathbb{C}[X]/(X^n)$ and let G be the cyclic subgroup of $\mathbb{C}^*$, the multiplicative group of $\mathbb{C}$, generated by σ which is an nth root of unity. Define the operation of G on the $\mathbb{C}$-algebra Λ by $\sigma^i(X + (X^n)) = \sigma^i \cdot X + (X^n)$ for all $i = 1, \ldots, n$. Since n is invertible in Λ we have by Theorem 2.14 that ΛG is a Nakayama algebra since Λ is a Nakayama algebra.

3 Selfinjective algebras

This section is devoted to pointing out some special features of the module theory of selfinjective artin algebras which do not hold for arbitrary artin algebras. In particular, we show that for selfinjective algebras $D \operatorname{Tr} M$ is very closely related to $\Omega^2(M)$, the second syzygy of a module M, and that $D \operatorname{Tr} M$ and $\Omega^2(M)$ are the same for symmetric algebras, a special type of selfinjective algebras.

An artin algebra Λ is said to be **selfinjective** if it is injective as well as projective as Λ-module. Before considering the special types of self-injective algebras we are mainly interested in, we point out various characterizations and properties of selfinjective algebras.

Proposition 3.1 *The following are equivalent for an artin algebra Λ.*

(a) Λ *is selfinjective.*
(b) *A Λ-module is projective if and only if it is injective.*
(c) Λ^{op} *is selfinjective.*

Proof (a)$\Rightarrow$(b) Since every indecomposable projective Λ-module is a summand of Λ, every indecomposable projective module is injective if Λ is injective. But the numbers of isomorphism classes of indecomposable projective modules and indecomposable injective modules are the same. Hence every indecomposable injective module is projective, which establishes (b).

(b)$\Rightarrow$(a) is trivial.

(b)$\Leftrightarrow$(c) This is an immediate consequence of the duality $D: \operatorname{mod} \Lambda \to \operatorname{mod} \Lambda^{\mathrm{op}}$ and the equivalence of (a) and (b). $\square$

Our next characterization of selfinjective artin algebras is in terms of the contravariant functor $\operatorname{Hom}_\Lambda(\ , \Lambda): \operatorname{mod} \Lambda \to \operatorname{mod} \Lambda^{\mathrm{op}}$ given by $A \mapsto \operatorname{Hom}_\Lambda(A, \Lambda) = A^*$ for all A in $\operatorname{mod} \Lambda$. We have seen earlier that $\operatorname{Hom}_\Lambda(\ , \Lambda)$ induces a duality $T: \mathscr{P}(\Lambda) \to \mathscr{P}(\Lambda^{\mathrm{op}})$. Our aim now is to show that Λ is selfinjective if and only if $\text{Hom}_\Lambda(\ ,\Lambda): \text{mod}\,\Lambda \to \text{mod}\,\Lambda^{\text{op}}$ is a duality. The proof is based on the following more general considerations.

Let A be in $\text{mod}\,\Lambda$, where Λ is an artin R-algebra. Then for each X in $\text{mod}(\Lambda^{\text{op}})$ we have the morphism of R-modules $\alpha_X: X \otimes_\Lambda A \to \text{Hom}_{\Lambda^{\text{op}}}(A^*, X)$ given by $\alpha_X(x \otimes a)(f) = f(a)x$ for all x in X, for all a in A and all f in A^*. It is not difficult to check that α_X is functorial in X and that α_X is an isomorphism when A is a projective Λ-module. We now describe the kernel and cokernel of α_X for arbitrary X.

Proposition 3.2 *Let A be in $\text{mod}\,\Lambda$. Then for each X in $\text{mod}\,\Lambda^{\text{op}}$ we have an exact sequence*

$$0 \to \text{Ext}^1_{\Lambda^{\text{op}}}(\text{Tr}\,A, X) \to X \otimes_\Lambda A \overset{\alpha_X}{\to} \text{Hom}_{\Lambda^{\text{op}}}(A^*, X)$$
$$\to \text{Ext}^2_{\Lambda^{\text{op}}}(\text{Tr}\,A, X) \to 0$$

where all morphisms are functorial in X.

Proof Let $P_1 \overset{f}{\to} P_0 \overset{g}{\to} A \to 0$ be a minimal projective presentation of A. Then we have an exact sequence $0 \to A^* \overset{g^*}{\to} P_0^* \overset{f^*}{\to} P_1^* \to \text{Tr}\,A \to 0$. Since the P_i^* are projective Λ^{op}-modules for $i = 0, 1$, it is not hard to see the following for all X in $\text{mod}\,\Lambda^{\text{op}}$.

(a) $\text{Hom}_{\Lambda^{\text{op}}}(P_0^*, X) \overset{\text{Hom}_{\Lambda^{\text{op}}}(g^*, X)}{\longrightarrow} \text{Hom}_{\Lambda^{\text{op}}}(A^*, X) \to \text{Ext}^2_{\Lambda^{\text{op}}}(\text{Tr}\,A, X) \to 0$ is an exact sequence with all morphisms functorial in X.

(b) $\text{Hom}_{\Lambda^{\text{op}}}(P_1^*, X) \to \text{Ker}(\text{Hom}_{\Lambda^{\text{op}}}(g^*, X)) \to \text{Ext}^1_{\Lambda^{\text{op}}}(\text{Tr}\,A, X) \to 0$ is an exact sequence with all morphisms functorial in X.

Using these observations, it is not difficult to deduce our desired exact sequence from the commutative diagram with exact first row

$$
\begin{array}{ccccccc}
X \otimes_\Lambda P_1 & \longrightarrow & X \otimes_\Lambda P_0 & \longrightarrow & X \otimes_\Lambda A & \to 0 \\
\downarrow \wr & & \downarrow \wr & & \downarrow \alpha_X & \\
\text{Hom}_{\Lambda^{\text{op}}}(P_1^*, X) & \longrightarrow & \text{Hom}_{\Lambda^{\text{op}}}(P_0^*, X) & \overset{\text{Hom}_{\Lambda^{\text{op}}}(g^*, X)}{\longrightarrow} & \text{Hom}_{\Lambda^{\text{op}}}(A^*, X) &
\end{array}
$$

$\square$

While the exact sequences described in Proposition 3.2 are of interest in general, we are now interested only in the case $X = \Lambda$. Then $\alpha_\Lambda: A \to A^{**}$ is the usual evaluation morphism given by $\alpha_\Lambda(a)(f) = f(a)$ for all a in A and f in A^*. We recall that A is said to be **torsionless** if $\alpha_\Lambda: A \to A^{**}$ is a monomorphism and A is said to be **reflexive** if $\alpha_\Lambda: A \to A^{**}$ is an isomorphism. We have the following immediate consequence of Proposition 3.2.

Corollary 3.3 *Let A be in* $\bmod \Lambda$. *Then we have the following.*

(a) A *is torsionless if and only if* $\mathrm{Ext}^1_{\Lambda^{\mathrm{op}}}(\mathrm{Tr}\, A, \Lambda) = 0$.

(b) A *is reflexive if and only if* $\mathrm{Ext}^i_{\Lambda^{\mathrm{op}}}(\mathrm{Tr}\, A, \Lambda) = 0$ *for* $i = 1, 2$. $\square$

Combining this corollary with the fact that $\mathrm{Tr}\, A$ runs through all of $\bmod_{\mathscr{P}}(\Lambda^{\mathrm{op}})$ as A runs through all of $\bmod_{\mathscr{P}} \Lambda$, we have the following.

Proposition 3.4

(a)   *The following are equivalent for an artin algebra* Λ.

    (i) Λ *is selfinjective.*

    (ii)   *Every A in* $\bmod \Lambda$ *is torsionless.*

    (iii)   *Every A in* $\bmod \Lambda$ *is reflexive.*

(b)   *If Λ is selfinjective, then* $\mathrm{Hom}_\Lambda(\ , \Lambda) : \bmod \Lambda \to \bmod(\Lambda^{\mathrm{op}})$ *is a duality with dual inverse* $\mathrm{Hom}_{\Lambda^{\mathrm{op}}}(\ , \Lambda) : \bmod(\Lambda^{\mathrm{op}}) \to \bmod \Lambda$. $\square$

Another important property of selfinjective algebras we want to give involves the syzygy and cosyzygy functors which we now describe.

Let Λ be an arbitrary artin algebra. We define a functor $\Omega : \underline{\bmod}\, \Lambda \to \underline{\bmod}\, \Lambda$, called the **syzygy functor**, as follows. For each A in $\bmod \Lambda$ choose a fixed projective cover $P(A) \xrightarrow{h} A$ and define $\Omega(A)$ to be $\mathrm{Ker}\, h$. Suppose $f : A \to B$ is in $\bmod \Lambda$. Then there is an exact commutative diagram

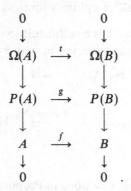

Now the morphism $t : \Omega(A) \to \Omega(B)$ we obtain in this way depends on the particular choice of g. It is not difficult to see that if we change g to $g' : P(A) \to P(B)$ we obtain a new morphism $t' : \Omega(A) \to \Omega(B)$ and that $t - t'$ is in $\mathscr{P}(\Omega A, \Omega B)$. In this way we get a morphism $\mathrm{Hom}_\Lambda(A, B) \to$

$\underline{\mathrm{Hom}}_\Lambda(\Omega(A),\Omega(B))$. Since $f \in \mathscr{P}(A,B)$ gives that $t \in \mathscr{P}(\Omega A, \Omega B)$ we obtain the morphism $\Omega: \underline{\mathrm{Hom}}_\Lambda(A,B) \to \underline{\mathrm{Hom}}_\Lambda(\Omega(A),\Omega(B))$. It is a straightforward exercise to check that this data defines a functor $\Omega: \underline{\mathrm{mod}}\,\Lambda \to \underline{\mathrm{mod}}\,\Lambda$.

Dually, we define the cosyzygy functor $\Omega^{-1}: \overline{\mathrm{mod}}\,\Lambda \to \overline{\mathrm{mod}}\,\Lambda$ as follows. For each A in $\mathrm{mod}\,\Lambda$ choose a fixed injective envelope $u: A \to I(A)$ and define $\Omega^{-1}(A)$ to be $\mathrm{Coker}\,u$. Suppose $f: A \to B$ is a morphism in $\mathrm{mod}\,\Lambda$. Then there is an exact commutative diagram

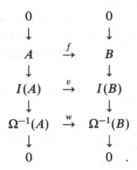

While the morphism $w: \Omega^{-1}(A) \to \Omega^{-1}(B)$ we obtain this way depends on the particular choice of v, it is not difficult to see that the image of w in $\overline{\mathrm{Hom}}_\Lambda(\Omega^{-1}(A),\Omega^{-1}(B))$ is independent of the choice of v. In this way we get a morphism $\mathrm{Hom}_\Lambda(A,B) \to \overline{\mathrm{Hom}}_\Lambda(A,B)$ whose kernel contains $\mathscr{I}(A,B)$. Thus we obtain the morphism $\Omega^{-1}: \overline{\mathrm{Hom}}_\Lambda(A,B) \to \overline{\mathrm{Hom}}_\Lambda(\Omega^{-1}(A),\Omega^{-1}(B))$. It is a straightforward exercise to check that this data defines a functor $\Omega^{-1}: \overline{\mathrm{mod}}\,\Lambda \to \overline{\mathrm{mod}}\,\Lambda$.

Although the syzygy and cosyzygy functors are important for arbitrary artin algebras, we consider them in this section only for selfinjective algebras.

Suppose Λ is a selfinjective algebra. Since the projective and injective Λ-modules coincide, we have that $\mathscr{P}(A,B) = \mathscr{I}(A,B)$ for all A and B in $\mathrm{mod}\,\Lambda$, and hence $\underline{\mathrm{mod}}\,\Lambda = \overline{\mathrm{mod}}\,\Lambda$. The duality $\mathrm{Hom}_\Lambda(\ ,\Lambda): \mathrm{mod}\,\Lambda \to \mathrm{mod}(\Lambda^{\mathrm{op}})$ induces a duality $\underline{\mathrm{Hom}}_\Lambda(\ ,\Lambda): \underline{\mathrm{mod}}\,\Lambda \to \underline{\mathrm{mod}}(\Lambda^{\mathrm{op}})$. Because Λ^{op} is also selfinjective we have the duality $\mathrm{Hom}_{\Lambda^{\mathrm{op}}}(\ ,\Lambda): \underline{\mathrm{mod}}(\Lambda^{\mathrm{op}}) \to \underline{\mathrm{mod}}\,\Lambda$ which is an inverse duality of $\mathrm{Hom}_\Lambda(\ ,\Lambda)$. It is not difficult to check that the functor $\Omega^{-1}: \underline{\mathrm{mod}}\,\Lambda \to \underline{\mathrm{mod}}\,\Lambda$ is given by $\Omega^{-1} = \mathrm{Hom}_{\Lambda^{\mathrm{op}}}(\ ,\Lambda)\Omega_{\Lambda^{\mathrm{op}}}\mathrm{Hom}_\Lambda(\ ,\Lambda)$. Straightforward calculations show the following.

Proposition 3.5 *Let Λ be a selfinjective artin algebra. The functors*

$\Omega: \underline{\mathrm{mod}}\, \Lambda \to \underline{\mathrm{mod}}\, \Lambda$ *and* $\Omega^{-1}: \underline{\mathrm{mod}}\, \Lambda \to \underline{\mathrm{mod}}\, \Lambda$ *are inverse equivalences.*
$\square$

Considering $\Omega: \underline{\mathrm{mod}}\, \Lambda \to \underline{\mathrm{mod}}\, \Lambda$ and $\Omega^{-1}: \underline{\mathrm{mod}}\, \Lambda \to \underline{\mathrm{mod}}\, \Lambda$ as maps on modules we get the following as a consequence of this proposition.

Proposition 3.6 *Let* Λ *be a selfinjective artin algebra. The map* $\Omega: \underline{\mathrm{mod}}\, \Lambda \to \underline{\mathrm{mod}}\, \Lambda$ *induces a map* $\Omega: \mathrm{mod}_{\mathscr{P}}\, \Lambda \to \mathrm{mod}_{\mathscr{P}}\, \Lambda$ *which has the following properties.*

(a) *For all A and B in* $\mathrm{mod}_{\mathscr{P}}\, \Lambda$, $A \simeq B$ *if and only if* $\Omega(A) \simeq \Omega(B)$.
(b) $\Omega(\coprod_{i=1}^{n} A_i) \simeq \coprod_{i=1}^{n} \Omega(A_i)$ *when the A_i are in* $\mathrm{mod}_{\mathscr{P}}\, \Lambda$.
(c) *For each A in* $\mathrm{mod}_{\mathscr{P}}\, \Lambda$, A *is indecomposable if and only if* $\Omega(A)$ *is indecomposable.* $\square$

It should be observed that this proposition remains valid if we substitute Ω^{-1} for Ω.

As before, assume that Λ is a selfinjective artin algebra. We define $\Omega^i: \underline{\mathrm{mod}}\, \Lambda \to \underline{\mathrm{mod}}\, \Lambda$ by induction as follows: $\Omega^0 = 1_{\underline{\mathrm{mod}}\Lambda}$ and $\Omega^{i+1} = \Omega\Omega^i$ for all $i \geq 0$. Similarly one defines Ω^{-i} for $i = 0, 1, \ldots$.

In order to explain how the functors $D\,\mathrm{Tr}$ and Ω^2 are connected we need the notion of the Nakayama automorphism of $\mathrm{mod}\,\Lambda$. The composition of dualities $\mathrm{mod}\,\Lambda \xrightarrow{\mathrm{Hom}_\Lambda(\ ,\Lambda)} \mathrm{mod}\,\Lambda^{\mathrm{op}} \xrightarrow{D} \mathrm{mod}\,\Lambda$ is an equivalence which we denote by $\mathscr{N}$ and which is called the **Nakayama automorphism**. We denote its inverse equivalence $\mathrm{Hom}_{\Lambda^{\mathrm{op}}}(\ ,\Lambda)D$ by $\mathscr{N}^{-1}$. We now have the following result.

Proposition 3.7 *Let* Λ *be a selfinjective artin algebra.*

(a) *The functors $D\,\mathrm{Tr}$, $\Omega^2 \mathscr{N}$, and $\mathscr{N}\Omega^2$ from* $\underline{\mathrm{mod}}\, \Lambda$ *to* $\underline{\mathrm{mod}}\, \Lambda$ *are isomorphic.*
(b) *The functors $\mathrm{Tr}\,D$, $\Omega^{-2}\mathscr{N}^{-1}$ and $\mathscr{N}^{-1}\Omega^{-2}$ are isomorphic.*

Proof (a) Let $P_1 \to P_0 \to A \to 0$ be a minimal projective presentation of A in $\mathrm{mod}\,\Lambda$. Then we have the exact sequence $0 \to A^* \to P_0^* \to P_1^* \to \mathrm{Tr}\,A \to 0$, which gives rise to the exact sequence $0 \to D\,\mathrm{Tr}\,A \to D(P_1^*) \to D(P_0^*) \to D(A^*) \to 0$. Because Λ is selfinjective, the $D(P_i^*)$ are projective Λ-modules and $D(P_1^*) \to D(P_0^*) \to D(A^*) \to 0$ is a minimal projective presentation of $D(A^*) = \mathscr{N}(A)$. Hence we get $D\,\mathrm{Tr}\,A \simeq \Omega^2 \mathscr{N}(A)$. We leave it to the reader to check that these isomorphisms are functorial in A and also to prove the rest of the proposition. $\square$

Of course, for those selfinjective artin algebras such that $\mathcal{N} \simeq 1_{\text{mod}\,\Lambda}$, we get that $D\operatorname{Tr} \simeq \Omega^2$ and $\operatorname{Tr}D \simeq \Omega^{-2}$ on $\underline{\text{mod}}\,\Lambda$. This brings us to the notion of a symmetric artin algebra.

An artin algebra Λ is said to be **symmetric** if $\Lambda \simeq D(\Lambda)$ as two-sided Λ-modules.

We illustrate with the following.

Example Let Λ be a local commutative selfinjective R-algebra. Then $D(\Lambda) = \operatorname{Hom}_R(\Lambda, I(R/\mathfrak{r}_R))$ is the unique indecomposable injective Λ-module and hence is isomorphic to Λ as left Λ-module. Using that Λ is commutative and the definition of the right Λ-module structure on Λ and $D(\Lambda)$, we see that we have a Λ-bimodule isomorphism between Λ and $D(\Lambda)$, so that Λ is symmetric.

Note that if Λ is a local commutative algebra with $\operatorname{soc}\Lambda$ simple, then Λ is selfinjective. For then Λ is contained in $I(\operatorname{soc}\Lambda)$, the unique indecomposable injective Λ-module up to isomorphism. Since Λ and $I(\operatorname{soc}\Lambda)$ have the same length because duality preserves length, we see that Λ is an injective Λ-module.

As a concrete example we have $\Lambda = k[X, Y]/(X^n, Y^n)$ for some $n > 0$ and k a field. Then it is easy to see that $\operatorname{soc}\Lambda$ is the simple Λ-module generated by the image of $X^{n-1}Y^{n-1}$ in Λ.

Proposition 3.8 *Suppose Λ is a symmetric artin algebra. Then we have the following.*

(a) $D \simeq \operatorname{Hom}_\Lambda(\ ,\Lambda)$.
(b) Λ *is a selfinjective algebra and* $\mathcal{N} \simeq 1_{\text{mod}\,\Lambda}$.
(c) $D\operatorname{Tr} \simeq \Omega^2$ *and* $\operatorname{Tr}D \simeq \Omega^{-2}$.

Proof (a) Suppose $g\colon \Lambda \to \operatorname{Hom}_R(\Lambda, J)$ is a two-sided Λ-isomorphism where $J = I(R/\mathfrak{r}_R)$. Then for each Λ-module X the induced morphism $\operatorname{Hom}_\Lambda(X,\Lambda) \to \operatorname{Hom}_\Lambda(X, \operatorname{Hom}_R(\Lambda, J))$ is an isomorphism which is easily seen to be a Λ^{op}-isomorphism which is functorial in X. But the usual adjointness gives an isomorphism $\operatorname{Hom}_\Lambda(X, \operatorname{Hom}_R(\Lambda, J)) \simeq \operatorname{Hom}_R(X, J) = D(X)$ which is a Λ^{op}-isomorphism functorial in X. Thus we have the Λ^{op}-isomorphism $\operatorname{Hom}_\Lambda(X,\Lambda) \to \operatorname{Hom}_R(X, J)$ which is functorial in X.

(b) and (c) follow readily from (a).   □

We now give a way of constructing from any artin algebra Λ a symmetric artin algebra of which Λ is a factor.

Proposition 3.9 *Let* Λ *be an arbitrary artin algebra. Viewing* $D(\Lambda)$ *as a two-sided* Λ-*module, the trivial extension* $\Lambda \ltimes D(\Lambda)$ *is a symmetric algebra.*

Proof It is a straightforward calculation to show that the map $t: \Lambda \ltimes D(\Lambda) \to D(\Lambda \ltimes D(\Lambda))$ given by $t(\lambda, f)(\lambda', f') = f(\lambda') + f'(\lambda)$ is a two-sided $(\Lambda \ltimes D(\Lambda))$-isomorphism.   □

We end this section by pointing out that group algebras of finite groups over fields are symmetric algebras. These are particularly important and interesting examples of symmetric algebras. We leave the proof as an exercise.

Proposition 3.10 *Let* kG *be the group algebra of the finite group* G *over the field* k. *Then the map* $t: kG \to \operatorname{Hom}_k(kG, k)$ *given by* $t(\sum_{\sigma \in G} a_\sigma \sigma)$ $(\sum_{\sigma \in G} a'_\sigma \sigma) = \sum_{\sigma \in G} a_\sigma a'_{\sigma^{-1}}$ *is a two-sided* kG-*isomorphism and so* kG *is a symmetric artin algebra.*   □

4 Defect of exact sequences

In this section we give a remarkable connection between the functors $D\operatorname{Tr}$ and $\operatorname{Tr} D$ and the structure of short exact sequences. This connection plays a fundamental role in the rest of this book. For instance, the proof of the existence of almost split sequences given in the next chapter as well as the theory of morphisms determined by modules developed in Chapter XI are based on this connection.

Associated with a short exact sequence $\delta: 0 \to A \to B \to C \to 0$ in $\operatorname{mod} \Lambda$ are the functors δ_*, the **covariant defect of the exact sequence**, and δ^*, the **contravariant defect of the exact sequence**, which are defined by the exact sequences

$$0 \to \operatorname{Hom}_\Lambda(C, \ ) \to \operatorname{Hom}_\Lambda(B, \ ) \to \operatorname{Hom}_\Lambda(A, \ ) \to \delta_* \to 0$$

and

$$0 \to \operatorname{Hom}_\Lambda(\ , A) \to \operatorname{Hom}_\Lambda(\ , B) \to \operatorname{Hom}_\Lambda(\ , C) \to \delta^* \to 0.$$

Clearly δ_* is a subfunctor of $\operatorname{Ext}^1_\Lambda(C, \ )$ and δ^* is a subfunctor of $\operatorname{Ext}^1_\Lambda(\ , A)$. Since $\operatorname{Hom}_\Lambda(A, X)$ and $\operatorname{Hom}_\Lambda(X, C)$ are finitely generated R-modules, $\delta_*(X)$ and $\delta^*(X)$ are also finitely generated R-modules for each X in $\operatorname{mod} \Lambda$. For brevity of notation as well as to avoid confusion, we denote the length of an R-module Z by $\langle Z \rangle$.

Our main objective in this section is to prove the following.

Theorem 4.1 *Let* $\delta: 0 \to A \to B \to C \to 0$ *be an exact sequence. For each Λ-module X we have* $\langle \delta_*(D \operatorname{Tr} X) \rangle = \langle \delta^*(X) \rangle$ *where δ_* and δ^* are the covariant and contravariant defects respectively of the sequence* $\delta: 0 \to A \to B \to C \to 0$.

The proof of this theorem proceeds in several steps. We begin with the following.

Proposition 4.2 *Let* $P_1 \to P_0 \to X \to 0$ *be a minimal projective presentation of X, and let Z be a Λ-module. Then there is an exact sequence*

$$0 \to \operatorname{Hom}_\Lambda(X, Z) \to \operatorname{Hom}_\Lambda(P_0, Z) \to \operatorname{Hom}_\Lambda(P_1, Z) \to \operatorname{Tr} X \otimes_\Lambda Z \to 0$$

with all morphisms functorial in Z.

Proof The exact sequence $P_0^* \to P_1^* \to \operatorname{Tr} X \to 0$ gives rise to the commutative exact diagram

$$
\begin{array}{ccccc}
P_0^* \otimes_\Lambda Z & \to & P_1^* \otimes_\Lambda Z & \to \operatorname{Tr} X \otimes_\Lambda Z \to 0 \\
\downarrow{\scriptstyle \alpha_0} & & \downarrow{\scriptstyle \alpha_1} & \\
0 \to \operatorname{Hom}_\Lambda(X, Z) \to \operatorname{Hom}_\Lambda(P_0, Z) & \to & \operatorname{Hom}_\Lambda(P_1, Z) &
\end{array}
$$

Here the morphisms $\alpha_i: P_i^* \otimes_\Lambda Z \to \operatorname{Hom}_\Lambda(P_i, Z)$ are given by $\alpha_i(f \otimes z)(x) = f(x)z$ for $f \in P_i^*, z \in Z$ and $x \in P_i$. By II Proposition 4.4 the morphisms α_i are functorial in Z and are isomorphisms since the P_i are projective. It then follows that we have our desired exact sequence and that in this exact sequence all morphisms are functorial in Z. $\qquad\square$

Before stating our next result, it is convenient to introduce the following notation. If A and B are Λ-modules, we denote the length of the R-modules $\operatorname{Hom}_\Lambda(A, B)$ by $\langle A, B \rangle$ rather than $\langle \operatorname{Hom}_\Lambda(A, B) \rangle$.

The next result is an easy consequence of Proposition 4.2.

Corollary 4.3 *Let* $P_1 \to P_0 \to X \to 0$ *be a minimal projective presentation in* $\operatorname{mod} \Lambda$. *Then for each Z in* $\operatorname{mod} \Lambda$ *we have*

$$\langle X, Z \rangle - \langle Z, D \operatorname{Tr} X \rangle = \langle P_0, Z \rangle - \langle P_1, Z \rangle.$$

Proof Let J be the R-injective envelope of $R/\operatorname{rad} R$. Since by adjointness $\operatorname{Hom}_\Lambda(Z, D \operatorname{Tr} X) \simeq \operatorname{Hom}_R(\operatorname{Tr} X \otimes_\Lambda Z, J) = D(\operatorname{Tr} X \otimes_\Lambda Z)$, we have that

$\langle \operatorname{Tr} X \otimes_\Lambda Z \rangle = \langle Z, D \operatorname{Tr} X \rangle$. Our result now follows from the exact sequence of R-modules

$$0 \to \operatorname{Hom}_\Lambda(X, Z) \to \operatorname{Hom}_\Lambda(P_0, Z) \to \operatorname{Hom}_\Lambda(P_1, Z) \to \operatorname{Tr} X \otimes_\Lambda Z \to 0$$

given in Proposition 4.2. $\square$

We now show that Theorem 4.1 is an easy consequence of Corollary 4.3.

Proof of Theorem 4.1 Let $\delta : 0 \to A \to B \to C \to 0$ be an exact sequence in mod Λ. It then follows from the definitions of δ_* and δ^* that we have the following equalities for each X in mod Λ.

$$\langle \delta^*(X) \rangle = \langle X, A \rangle - \langle X, B \rangle + \langle X, C \rangle$$
$$\langle \delta_*(D \operatorname{Tr} X) \rangle = \langle A, D \operatorname{Tr} X \rangle - \langle B, D \operatorname{Tr} X \rangle + \langle C, D \operatorname{Tr} X \rangle.$$

Subtracting the bottom row from the top row and applying Corollary 4.3 we have that

$$
\begin{aligned}
\langle \delta^*(X) \rangle - \langle \delta_*(D \operatorname{Tr} X) \rangle &= \langle P_0, A \rangle - \langle P_0, B \rangle + \langle P_0, C \rangle \\
&\quad - (\langle P_1, A \rangle - \langle P_1, B \rangle + \langle P_1, C \rangle) \\
&= \langle \delta^*(P_0) \rangle - \langle \delta^*(P_1) \rangle \\
&= 0,
\end{aligned}
$$

since the P_i are projective modules. $\square$

As an immediate consequence of Theorem 4.1 we have the following. Here we say that a morphism $g : X \to Y$ factors through $t : B \to Y$ if there is some $s : X \to B$ with $ts = g$, and $g : X \to Y$ factors through $u : X \to A$ if there is some $v : A \to Y$ with $vu = g$.

Corollary 4.4 *Let* $\delta : 0 \to A \xrightarrow{f} B \xrightarrow{g} C \to 0$ *be an exact sequence in* mod Λ. *Then for each X in* mod Λ *the following are equivalent.*

(a) *Every morphism $h : X \to C$ factors through $g : B \to C$.*
(b) *Every morphism $t : A \to D \operatorname{Tr} X$ factors through $f : A \to B$.*
(c) *For each $h : X \to C$, we have that $\operatorname{Ext}_\Lambda^1(h, A) : \operatorname{Ext}_\Lambda^1(C, A) \to \operatorname{Ext}_\Lambda^1(X, A)$ has the property that $\operatorname{Ext}_\Lambda^1(h, A)(\delta) = 0$.*
(d) *For each $t : A \to D \operatorname{Tr} X$ we have that $\operatorname{Ext}_\Lambda^1(C, t) : \operatorname{Ext}_\Lambda^1(C, A) \to \operatorname{Ext}_\Lambda^1(C, D \operatorname{Tr} X)$ has the property that $\operatorname{Ext}_\Lambda^1(C, t)(\delta) = 0$.*

Proof Since $\langle \delta^*(X) \rangle = \langle \delta_*(D \operatorname{Tr} X) \rangle$ it follows that $\delta^*(X) = 0$ if and only if $\delta_*(D \operatorname{Tr} X) = 0$. This shows that (a) and (b) are equivalent. Statement (c) is just a rewriting of (a) in terms of the functor $\operatorname{Ext}^1_\Lambda(\ , A)$ and (d) is just a rewriting of (b) in terms of the functor $\operatorname{Ext}^1_\Lambda(C, \ )$. □

As another indication of how Theorem 4.1 can be applied, we give the following result.

Proposition 4.5 *The following equalities hold for arbitrary Λ-modules.*

$$\langle \operatorname{Tor}^\Lambda_1(\operatorname{Tr} X, A) \rangle = \langle \underline{\operatorname{Hom}}_\Lambda(X, A) \rangle = \langle \operatorname{Ext}^1_\Lambda(A, D \operatorname{Tr} X) \rangle.$$

Proof Let $\delta : 0 \to \Omega^1(A) \to P \xrightarrow{f} A \to 0$ be exact with $f : P \to A$ a projective cover of A. Applying the duality D and usual adjointness isomorphisms to the exact sequence

$$0 \to \operatorname{Tor}^\Lambda_1(\operatorname{Tr} X, A) \to \operatorname{Tr} X \otimes_\Lambda \Omega^1(A) \to \operatorname{Tr} X \otimes_\Lambda P \to \operatorname{Tr} X \otimes_\Lambda A \to 0$$

we obtain the exact sequence

$$0 \to \operatorname{Hom}_\Lambda(A, D \operatorname{Tr} X) \to \operatorname{Hom}_\Lambda(P, D \operatorname{Tr} X) \to \operatorname{Hom}_\Lambda(\Omega^1(A), D \operatorname{Tr} X)$$

$$\to D(\operatorname{Tor}^\Lambda_1(\operatorname{Tr} X, A)) \to 0.$$

Thus we have that $\delta_*(D \operatorname{Tr} X) = D(\operatorname{Tor}^\Lambda_1(\operatorname{Tr} X, A))$. But we also have that $\delta_*(D \operatorname{Tr} X) = \operatorname{Ext}^1_\Lambda(A, D \operatorname{Tr} X)$ since P is projective. Hence we get $\langle \operatorname{Tor}^\Lambda_1(\operatorname{Tr} X, A) \rangle = \langle \operatorname{Ext}^1_\Lambda(A, D \operatorname{Tr} X) \rangle$. Also by Theorem 4.1 we have that $\langle \delta_*(D \operatorname{Tr} X) \rangle = \langle \delta^*(X) \rangle$. But $\delta^*(X) = \underline{\operatorname{Hom}}_\Lambda(X, A)$, so we have that $\langle \operatorname{Tor}^\Lambda_1(\operatorname{Tr} X, A) \rangle = \langle \underline{\operatorname{Hom}}_\Lambda(X, A) \rangle$, which completes the proof of the proposition. □

The above equalities are often used in computations since it is sometimes easier to compute $\underline{\operatorname{Hom}}_\Lambda(X, A)$ than either $\operatorname{Tor}^\Lambda_1(\operatorname{Tr} X, A)$ or $\operatorname{Ext}^1_\Lambda(A, D \operatorname{Tr} X)$.

As our final application of Theorem 4.1 we prove the following.

Proposition 4.6 *The equality*

$$\langle \operatorname{Ext}^1_\Lambda(X, A) \rangle = \langle \overline{\operatorname{Hom}}_\Lambda(A, D \operatorname{Tr} X) \rangle$$

holds for all Λ-modules X and A.

Proof Let $\delta: 0 \rightarrow A \rightarrow I \rightarrow \Omega^{-1}(A) \rightarrow 0$ be an exact sequence with $A \rightarrow I$ an injective envelope. Then $\delta^*(X) = \text{Ext}_\Lambda^1(X, A)$ and $\delta_*(D \text{Tr} X) = \overline{\text{Hom}}_\Lambda(A, D \text{Tr} X)$. Since $\langle \delta^*(X) \rangle = \langle \delta_*(D \text{Tr} X) \rangle$ by Theorem 4.1, we have our desired result.   □

When $\text{pd}_\Lambda X \leq 1$ we have the following simplification of the formula in Proposition 4.6.

Corollary 4.7 *Suppose* $\text{pd}_\Lambda X \leq 1$. *Then* $\langle \text{Ext}_\Lambda^1(X, A) \rangle = \langle \text{Hom}_\Lambda(A, D \text{Tr} X) \rangle$ *for all A in* $\text{mod} \Lambda$.

Proof Since $\text{pd}_\Lambda X \leq 1$, we have by Proposition 1.16 that $\text{Hom}_\Lambda(D(\Lambda), D \text{Tr} X) = 0$. Therefore $\overline{\text{Hom}}_\Lambda(A, D \text{Tr} X) = \text{Hom}_\Lambda(A, D \text{Tr} X)$ for all A, so that $\langle \text{Ext}_\Lambda^1(X, A) \rangle = \langle \text{Hom}_\Lambda(A, D \text{Tr} X) \rangle$ for all A in $\text{mod} \Lambda$ by Proposition 4.6.   □

We end this section on the defect of exact sequences with the following remark.

Let $\delta: 0 \rightarrow A \rightarrow B \rightarrow C \rightarrow 0$ be an exact sequence in $\text{mod} \Lambda$. Then the contravariant defect δ^* is a contravariant R-functor from $\text{mod} \Lambda$ to $\text{mod} R$ which vanishes on projective modules. Therefore δ^* induces a contravariant R-functor $\delta^*: \underline{\text{mod}} \Lambda \rightarrow \text{mod} R$. Hence if we define $(D\delta^*)(X)$ to be $\text{Hom}_R(\delta^*(X), J)$ where J is the R-injective envelope of $R/\text{rad} R$, we obtain the covariant R-functor $D\delta^*: \underline{\text{mod}} \Lambda \rightarrow \text{mod} R$. Also the covariant defect δ_* is a covariant functor $\text{mod} \Lambda \rightarrow \text{mod} R$ which vanishes on injective modules. Therefore δ_* induces a covariant R-functor $\delta_*: \overline{\text{mod}} \Lambda \rightarrow \text{mod} R$. Consequently the composition $\underline{\text{mod}} \Lambda \overset{D \text{Tr}}{\rightarrow} \overline{\text{mod}} \Lambda \overset{\delta_*}{\rightarrow} \text{mod} R$ which we denote by $\delta_*(D \text{Tr})$ is a covariant functor $\underline{\text{mod}} \Lambda \rightarrow \text{mod} R$.

It can be shown that the functors $D\delta^*$ and $\delta_* D \text{Tr}$ are isomorphic and hence that the R-modules $D\delta^*(X)$ and $\delta_*(D \text{Tr}(X))$ are isomorphic for all Λ-modules X. Since $\langle D\delta^*(X) \rangle = \langle \delta^*(X) \rangle$ we get the fact established in Theorem 4.1 that $\langle \delta^*(X) \rangle = \langle \delta_*(D \text{Tr} X) \rangle$ for all Λ-modules X. Despite the fact that the functors $D\delta^*$ and $\delta_* D \text{Tr}$ being isomorphic is of interest in its own right, we do not prove it since the proof involves more complicated categorical arguments than we wish to get involved with here.

Exercises

1. Find the radical and socle filtration of $\mathbb{Z}_{72}$ as a $\mathbb{Z}_{72}$-module.

2. Prove that if Λ is a Nakayama algebra and $\Lambda = \coprod_{i=1}^{s} n_i P_i$ with P_i nonisomorphic indecomposable projective modules, then the total number of nonisomorphic uniserial modules is $\sum_{i=1}^{s} l(P_i)$.

3. Let Γ be the quiver and k a field. Let M be the $k\Gamma$-module

given by the representation

$$
\begin{array}{ccc}
k & & k \\
\searrow \binom{1}{0} & & \nearrow \binom{0}{1} \\
& k \amalg k & \\
& \downarrow (1,1) & \\
& k &
\end{array}
$$

(a) Find $\operatorname{End}_{k\Gamma}(M)$.
(b) Find the socle and radical filtration of M.
(c) Find $D\operatorname{Tr} M$ and $\operatorname{Tr} DM$.

4. Let k be a field of characteristic 2 and let D_4 be the dihedral group of order 8.

(a) Find the radical and socle filtration of kD_4.
(b) Find $D\operatorname{Tr} k$, when k is the trivial kD_4-module.

5. Let k be a field, $i \in k$ and let $\Lambda_i = k\langle X, Y\rangle/I_i$ where $k\langle X, Y\rangle$ is the free k-algebra in two noncommuting variables and $I_i = \langle X^2, Y^2, XY - iYX\rangle$ is the ideal generated by X^2, Y^2 and $XY - iYX$.

(a) Determine for which i we have that Λ_i is selfinjective.
(b) Determine for which i we have that Λ_i is symmetric.

6. Let k be a field, let $\Lambda = k[X]/(X^n)$ with $n \geq 2$ and for each k-vector space V let $\langle V\rangle$ denote its length (dimension). Consider the exact sequence of Λ-modules $\delta: 0 \rightarrow k[X]/(X^{n-1}) \xrightarrow{f} k[X]/(X^n) \coprod k[X]/(X^{n-2}) \xrightarrow{g} k[X]/(X^{n-1}) \rightarrow 0$ where $f(p + (X^{n-1})) = (Xp + (X^n), p + (X^{n-2}))$ for $p + (X^{n-1}) \in k[X]/(X^{n-1})$ and $g(p + (X^n), q + (X^{n-1})) = p - Xq + (X^{n-1})$ for $p + (X^n) \in k[X]/(X^n)$ and $q + (X^{n-2}) \in k[X]/(X^{n-2})$.

(a) Prove that $\langle \delta_*(k[X]/(X^i)) \rangle = \langle \delta^*(k[X]/(X^i)) \rangle = 0$ for $i \neq n-1$.
(b) Prove that $\langle \delta_*(k[X]/(X^{n-1})) \rangle = \langle \delta^*(k[X]/(X^{n-1})) \rangle = 1$.

7. Let Λ be an artin algebra and M a module in $\text{mod}_{\mathscr{P}} \Lambda$.

(a) Prove that $\mathscr{P}(M, M)$, the group of Λ-homomorphisms from M to M which factors through a projective module, is contained in $\mathfrak{r}_{\text{End}_\Lambda(M)}$.
(b) Prove that $f \in \text{End}_\Lambda(M)$ is invertible if and only if the image $\underline{f}$ of f in $\underline{\text{End}}_\Lambda(M)$ is invertible.

8. Let Λ be an artin algebra with gl.dim $\Lambda < \infty$. Prove that there is some simple Λ-module S with $\text{Hom}_\Lambda(D\Lambda, S) = 0$.

9.

(a) Let Λ be a Nakayama algebra with admissible sequence $(5,4,3,4,3,2,2)$. Describe the simple modules with projective dimension 2.
(b) Give a description of the simple modules of projective dimension 2 for a Nakayama algebra Λ in terms of the admissible sequence. (Hint: For each maximal sub-sequence of the form $(t+i, t+i-1, \ldots, t+1, t)$, compare t with the length of the next maximal sub-sequence).

10. Let k be a field, $n \in \mathbb{N}$ and let V be an n-dimensional k-vector space. Let $T(k, V)$ be the tensor algebra of V over k, let I be the ideal generated by $\{v \otimes v | v \in V\}$ and $\Lambda = T(k, V)/I$, which is the exterior algebra of V over k. Denote by $\overline{V}^i$ the image of V^i in Λ. Prove that the following hold.

(i) $V^{n+1} \subseteq I$.
(ii) $\overline{V} = \mathfrak{r}_\Lambda$, hence Λ is a local ring.
(iii) $\dim(\overline{V}^i/\overline{V}^{i+1}) = \binom{n}{i}$ for $0 \leq i \leq n$ with $\overline{V}^0 = k$.
(iv)   For each $i < n$ there is some $v \in \overline{V}^i$ and some $\lambda \in \overline{V}$ with $\lambda v \neq 0$.
(v) $\text{soc} \Lambda$ is simple, hence Λ is selfinjective.
(vi)   Find the numbers n where Λ is symmetric.

11. Let k be a field and $S = k[X, Y]$. Prove that the S-module $(X, Y)^n/(X, Y)^{n+2}$ is an indecomposable $(k[X, Y]/(X, Y)^2)$-module for all $n \geq 0$.

12. Let Λ be an artin algebra such that $\Lambda/\mathfrak{r} \simeq \operatorname{soc}\Lambda$. Prove that Λ is selfinjective.

Notes

The transpose of a module viewed only as a module, not a functor, first appeared implicitly in the study of commutative noetherian regular local rings [Au1]. It also appeared, again implicitly, in the study of certain types of additive functors called coherent functors on the category of finitely generated modules over noetherian, not necessarily commutative, rings [Au2]. The notion given here of the transpose as a functor dates from [AuB] where it was defined and studied in the context of finitely generated modules over arbitrary noetherian rings. While the pertinence of the notion to the theory of modules over artin algebras was indicated earlier in the Ph.D. theses of Leighton, Menzin and Teter, students of Auslander, it was not until the connection of $D\operatorname{Tr}$ with almost split sequences was shown in [AuR4] that the importance of the transpose in representation theory was finally established.

Nakayama artin algebras were introduced by Nakayama in connection with his work on classical maximal orders. The Nakayama algebras are also called generalized uniserial algebras, or serial algebras. He also determined their representation theory [Nak]. The method of describing Nakayama algebras in terms of Kupisch series was given by Kupisch in [Ku1]. Further information was given in [Fu].

A systematic study of the defect functors δ^* and δ_* associated with short exact sequences was given in [Au5] where it was shown that the functors $D\delta^*(X)$ and $\delta_*(D\operatorname{Tr} X)$ of X are isomorphic. The formula given in Corollary 4.3 was developed in [AuR8] in connection with proving results given in Chapter IX. Its use in proving Theorem 4.1 is new.

Proposition 4.6 is a special case of the functorial isomorphism $D(\operatorname{Ext}^1_\Lambda(X,A)) \simeq \overline{\operatorname{Hom}}_\Lambda(A, D\operatorname{Tr} X)$, sometimes called the Auslander–Reiten formula. It is proved in [AuR4] and used there as the basis of the proof of the existence of almost split sequences (see also XI.5). Similar formulas have been proved in other contexts as an approach to establishing the existence of almost split sequences.

V

Almost split sequences

In this chapter we give an introduction to almost split sequences, a special type of short exact sequences of modules which play a central role not only in this book, but also in the representation theory of artin algebras in general. Almost split sequences are introduced via the dual notions of left almost split morphisms and right almost split morphisms. These notions give rise not only to almost split sequences but also to the equally basic concept of irreducible morphisms. In this connection the radical of the category mod Λ, which is an R-relation on mod Λ, is studied. In addition to the general theory, various examples and special types of almost split sequences are discussed, including a method for constructing almost split sequences for modules over group algebras of finite groups.

1 Almost split sequences and morphisms

Even though until now our major emphasis has been on studying modules, we have also discussed some special types of morphisms, such as projective covers, injective envelopes and right and left minimal morphisms. In this section we introduce some other special morphisms, called right and left almost split morphisms, which give rise in a natural way to the notion of almost split sequences. The section ends with a proof of the existence and uniqueness of almost split sequences.

Even though it is not quite grammatically correct, we say that a morphism $f: B \to C$ is a **split epimorphism** if $1_C: C \to C$, the identity morphism of C, factors through f. Dually, we say that a morphism $g: A \to B$ is a **split monomorphism** if 1_A factors through g. We have the following reformulation of these concepts in terms of indecomposable modules.

Lemma 1.1 *The following are equivalent for a morphism* $f: B \to C$.

(a) *The morphism* $f: B \to C$ *is a split epimorphism.*

(b) *If* $C = C_1 \coprod \cdots \coprod C_n$ *with the* C_i *indecomposable for all* $i = 1, \ldots, n$, *then the natural inclusion morphisms* $h_i: C_i \to C$ *factor through* f.

Proof It is clear that (a) implies (b). If (b) holds, let $g_i: C_i \to B$ be such that $fg_i = h_i$. Letting $p_i: C \to C_i$ be the projections associated with the decomposition $C = C_1 \coprod \cdots \coprod C_n$ for $i = 1, \ldots, n$ we get that $1_C = \sum_{i=1}^{n} fg_i p_i$, so that $f: B \to C$ is a split epimorphism. $\square$

Lemma 1.2 *The following are equivalent for a morphism* $g: A \to B$.

(a) *The morphism* $g: A \to B$ *is a split monomorphism.*

(b) *If* $A = A_1 \coprod \cdots \coprod A_n$ *with the* A_i *indecomposable for all* $i = 1, \ldots, n$, *then the associated projection morphisms* $p_i: A \to A_i$ *factor through* g.

Proof This is the dual of Lemma 1.1. $\square$

As motivation for the definitions of right almost split morphisms and left almost split morphisms we consider the following special situation.

The radicals of indecomposable projective modules and the socles of indecomposable injective modules have come up repeatedly in our past considerations. We now give a description of these important submodules in terms of morphisms.

Let P be an indecomposable projective module over an artin algebra Λ. Then the inclusion $i: \mathfrak{r}P \to P$ is not a split epimorphism since it is not an epimorphism. In addition $i: \mathfrak{r}P \to P$ has the property that any morphism $g: X \to P$ which is not a split epimorphism factors through $i: \mathfrak{r}P \to P$. For let $g: X \to P$ be a morphism which is not a split epimorphism. Then g is not an epimorphism and hence $\text{Im } g \subset \mathfrak{r}P$ since $\mathfrak{r}P$ is the unique maximal submodule of the indecomposable projective module P. Therefore $g: X \to P$ factors through $i: \mathfrak{r}P \to P$. These observations suggest the following definitions.

A morphism $f: B \to C$ is **right almost split** if (a) it is not a split epimorphism and (b) any morphism $X \to C$ which is not a split epimorphism factors through f. In view of our previous remarks, we see that $i: \mathfrak{r}P \to P$ is right almost split for all indecomposable projective modules P. Dually, a morphism $g: A \to B$ is **left almost split** if (a) it is not a split monomorphism and (b) any morphism $A \to Y$ which is not a split monomorphism factors through g. As examples of left almost split morphisms we point out that if I is an indecomposable injective module, then the epimorphism $I \rightarrow I/\operatorname{soc} I$ is left almost split. This can be seen either directly or by using the following easily proven fact.

Lemma 1.3 *A morphism $f: B \rightarrow C$ in $\operatorname{mod}\Lambda$ is right almost split if and only if $D(f): D(C) \rightarrow D(B)$, the dual of f, is left almost split.* $\square$

Let P be an indecomposable projective module. Having seen that $i: \mathfrak{r}P \rightarrow P$ is right almost split, it is natural to ask if there are other right almost split morphisms $B \rightarrow P$. If $j: C \rightarrow P$ is an inclusion which is right almost split, then condition (a) means that C is a proper submodule of P, and condition (b) means that C contains all proper submodules of P. Hence we must have $i: \mathfrak{r}P \rightarrow P$ in this case. It is clear that if $g: X \rightarrow \mathfrak{r}P$ is any morphism, then the induced morphism $X \coprod \mathfrak{r}P \rightarrow P$ is also right almost split. More generally, it is not difficult to see that a morphism $f: B \rightarrow P$ is right almost split if and only if $\operatorname{Im} f = \mathfrak{r}P$ and the induced epimorphism $B \rightarrow \operatorname{Im} f$ is a split epimorphism. We have that $i: \mathfrak{r}P \rightarrow P$ is right minimal since it is a monomorphism, and so $i: \mathfrak{r}P \rightarrow P$ is the only, up to isomorphism in $\operatorname{mod}\Lambda/P$, right almost split morphism to P which is also right minimal. We now show that this is a general fact about right almost split morphisms. But first it is convenient to make the following definition.

A morphism is said to be **minimal right almost split** if it is both right almost split and right minimal. Dually, a morphism is said to be **minimal left almost split** if it is both left almost split and left minimal. Clearly a morphism is minimal right almost split if and only if its dual is minimal left almost split. We can now prove the following.

Proposition 1.4

(a) *Let $f: B \rightarrow C$ and $f': B' \rightarrow C$ be right almost split morphisms. Then f is equivalent to f' in $\operatorname{mod}\Lambda/C$ and all morphisms equivalent to f are right almost split.*

(b) *If $f: B \rightarrow C$ is right almost split then the minimal version of f is minimal right almost split.*

(c) *If for C in $\operatorname{mod}\Lambda$ $f: B \rightarrow C$ is minimal right almost split, then f is unique up to isomorphism.*

Proof (a) Let $f: B \rightarrow C$ and $f': B' \rightarrow C$ be right almost split morphisms. Then since f' is not a split epimorphism and f is right almost split, there exists some $g: B \rightarrow B'$ with $f'g = f$. Similarly there exists some

$g': B' \to B$ with $f' = fg'$. Therefore f and f' are equivalent in mod Λ/C. The rest of (a) is trivial.

(b) This is a direct consequence of (a) and the definitions.

(c) This follows from (a) and (b) and the fact that the right minimal version of a map $f: B \to C$ is unique up to isomorphism in the corresponding equivalence class in mod Λ/C. □

We also state the dual of Proposition 1.4.

Proposition 1.5

(a) *Let $g: A \to B$ and $g': A \to B'$ be left almost split morphisms. Then g and g' are equivalent in mod $\Lambda \setminus A$ and all morphisms equivalent to g are left almost split.*

(b) *If $g: A \to B$ is left almost split then the minimal version of g, which is a minimal left almost split morphism, is unique up to isomorphism.* □

Combining these propositions with our previous remarks about indecomposable projective modules as well as indecomposable injective modules, we have the following, where (b) follows from (a) by duality.

Corollary 1.6

(a) *Let P be an indecomposable projective module. Then $i: \mathfrak{r}P \to P$ is the unique, up to isomorphism in mod Λ/P, minimal right almost split morphism.*

(b) *Let I be an indecomposable injective module. Then the natural epimorphism $t: I \to I/\operatorname{soc} I$ is the unique, up to isomorphism in mod $\Lambda \setminus I$, minimal left almost split morphism.* □

In view of these results about indecomposable projective modules and indecomposable injective modules, it is natural to ask what the situation is for decomposable projective modules, and decomposable injective modules. The following general result answers these questions.

Lemma 1.7 *Let $f: B \to C$ be a morphism.*

(a) *If f is right almost split, then C is an indecomposable module.*

(b) *If f is left almost split, then B is an indecomposable module.*

Proof (a) Suppose $C = C_1 \coprod \cdots \coprod C_n$ with the C_i indecomposable modules and $n \geq 2$. Then each of the associated inclusion morphisms

$t_i : C_i \to C$ factors through f since f is right almost split. Therefore f is a split epimorphism by Lemma 1.1, which is a contradiction.

(b) This is the dual of (a). $\square$

Having described all the right almost split morphisms $f : B \to C$ with C projective and all left almost split morphisms $g : A \to B$ with A injective, we now turn our attention to considering when right or left almost split morphisms exist for other modules. To this end, it is convenient to have the following characterization of right and left almost split morphisms in terms of indecomposable modules.

Proposition 1.8 *The following are equivalent for a morphism $f : B \to C$.*

(a) *The morphism $f : B \to C$ is right almost split.*
(b) *The morphism f is not a split epimorphism and every nonisomorphism $g : X \to C$ with X indecomposable factors through f.*

Proof (a) $\Rightarrow$ (b) This follows from the observation that if $g : X \to C$ is not an isomorphism with X indecomposable, then g is not a split epimorphism.

(b) $\Rightarrow$ (a) Since we are assuming that f is not a split epimorphism, we only have to show that if $g : X \to C$ is not a split epimorphism, then g factors through f. Write X as $X_1 \amalg \cdots \amalg X_n$ with the X_i indecomposable modules. Then none of the morphisms $g|_{X_i} : X_i \to C$ is a split epimorphism because g is not a split epimorphism. Hence each $g|_{X_i}$ factors through f which means that $g : X \to C$ factors through f. $\square$

We now state the dual of Proposition 1.8.

Proposition 1.9 *The following are equivalent for a morphism $g : A \to B$.*

(a) *The morphism g is left almost split.*
(b) *The morphism $g : A \to B$ is not a split monomorphism and every non-isomorphism $A \to Y$ with Y indecomposable factors through g.* $\square$

We now give examples which show that the existence of almost split morphisms is not restricted to indecomposable projective or injective modules.

Example Let p be a prime element in a principal ideal domain R and let Λ be the local commutative artin ring $R/(p^n)$ with $n \geq 2$. Using the structure theorem for finitely generated modules over principal ideal domains, one sees that $R/(p), R/(p^2), \cdots, R/(p^n)$ are all the indecomposable Λ-modules. So Λ is a Nakayama algebra. Since we have natural inclusion morphisms $R/(p) \to R/(p^2) \to \cdots \to R/(p^n)$, it follows that $R/(p^n)$ is the unique indecomposable injective Λ-module as well as the unique indecomposable projective Λ-module. We also have the natural epimorphisms $R/(p^n) \to R/(p^{n-1}) \to \cdots \to R/(p)$.

It is not difficult to check that we have the nonsplit exact sequences

$$0 \to R/(p) \xrightarrow{g_1} R/(p^2) \xrightarrow{f_1} R/(p) \to 0,$$
$$0 \to R/(p^2) \xrightarrow{g_2} (R/(p)) \amalg (R/(p^3)) \xrightarrow{f_2} R/(p^2) \to 0,$$
$$\vdots \qquad\qquad \vdots \qquad\qquad \vdots$$
$$0 \to R/(p^{n-1}) \xrightarrow{g_{n-1}} (R/(p^{n-2})) \amalg (R/(p^n)) \xrightarrow{f_{n-1}} R/(p^{n-1}) \to 0,$$

where the $f_i : (R/(p^{i-1})) \amalg (R/(p^{i+1})) \to R/(p^i)$ are induced by the inclusions $R/(p^{i-1}) \to R/(p^i)$ and the epimorphisms $R/(p^{i+1}) \to R/(p^i)$ for $i = 1, \ldots, n-1$. The morphisms $g_i : R/(p^i) \to (R/(p^{i-1})) \amalg (R/(p^{i+1}))$ are induced by the same inclusions $R/(p^i) \to R/(p^{i+1})$ but for the epimorphisms $R/(p^i) \to R/(p^{i-1})$ for $i = 1, \ldots, n-1$ we switch sign.

Since the $R/(p^i)$ for $i = 1, \ldots, n$ are indecomposable modules, it follows that all the g_i are left minimal and all the f_i are right minimal morphisms. Also using the description of the indecomposable Λ-modules given above, it is a straightforward calculation using Propositions 1.8 and 1.9 to show that the f_i are right almost split morphisms and all the g_i are left almost split morphisms.

These examples suggest not only that for an arbitrary artin algebra each indecomposable module C has a minimal right almost split morphism $M \to C$ and a minimal left almost split morphism $C \to N$, but the ones for nonprojective and noninjective modules match up nicely in the form of short exact sequences. The rest of this section is devoted to proving that this is indeed the case. To this end it is useful to have the following variation of Lemma 1.7 and Proposition 1.8.

Proposition 1.10 *The following are equivalent for a morphism $f : B \to C$.*

(a) *f is right almost split.*

(b) *The morphism f is not a split epimorphism, the module C is indecomposable and if X is an indecomposable module not isomorphic to C, then every morphism $g : X \to C$ factors through f.*

Proof (a) $\Rightarrow$ (b) This is an easy consequence of the previous results.

(b) $\Rightarrow$ (a) Since we are assuming that f is not a split epimorphism, it suffices to show by Proposition 1.8 that all nonisomorphisms $Y \to C$ with Y indecomposable factor through f. Since we are assuming that this is the case if Y is not isomorphic to C, it suffices to show that every nonisomorphism $g : C \to C$ factors through f. Since $g : C \to C$ is not an isomorphism we know that $A = \operatorname{Im} g$ is a proper submodule of C. Let $A = A_1 \coprod \cdots \coprod A_n$ with the A_i indecomposable modules. The fact that each A_i is a proper submodule of C implies that none of the A_i are isomorphic to C. Therefore each of the inclusion morphisms $A_i \to C$ factors through f, which means that the inclusion $A \to C$ factors through f. Therefore $g : C \to C$, which is the composition $C \to \operatorname{Im} g \to C$, factors through f. $\qquad\square$

The dual of this result is the following.

Proposition 1.11 *The following are equivalent for a morphism* $g : A \to B$.

(a) *The morphism* $g : A \to B$ *is left almost split.*
(b) *The morphism* g *is not a split monomorphism, the module* A *is indecomposable and if* Y *is an indecomposable module not isomorphic to* A, *then every morphism* $h : A \to Y$ *factors through* g. $\qquad\square$

We now apply these results to investigating right almost split morphisms $B \to C$ when C is not projective and left almost split morphisms $A \to B$ when A is not injective.

Proposition 1.12 *Suppose* $f : B \to C$ *is a minimal right almost split morphism with* C *not a projective module. Then we have the following.*

(a) *The module* C *is indecomposable and* f *is an epimorphism.*
(b) *The exact sequence* $0 \to \operatorname{Ker} f \xrightarrow{g} B \xrightarrow{f} C \to 0$ *has the following properties.*
   (i) $\operatorname{Ker} f \simeq D \operatorname{Tr} C$.
   (ii) g *is a minimal left almost split morphism.*

Proof (a) Since f is right almost split, we know by Lemma 1.7 that C is indecomposable. Let $h : P \to C$ be a projective cover of C. Then h is not a split epimorphism since C is not projective. Hence h factors through $f : B \to C$, which implies that f is an epimorphism.

(b) We first show that $\operatorname{Ker} f$ is indecomposable. Suppose $\operatorname{Ker} f =$

$A_1 \coprod \cdots \coprod A_n$ with the A_i indecomposable modules. Since $g: \mathrm{Ker}\, f \to B$ is not a split monomorphism there is some A_i such that the projection $\mathrm{Ker}\, f \to A_i$ induced by the given decomposition does not factor through g. Therefore the commutative pushout diagram

$$
\begin{array}{ccccccccc}
0 & \to & \mathrm{Ker}\, f & \overset{g}{\to} & B & \overset{f}{\to} & C & \to & 0 \\
 & & \downarrow u & & \downarrow v & & \| & & \\
0 & \to & A_i & \overset{s}{\to} & A_i \times^{\mathrm{Ker}\, f} B & \overset{t}{\to} & C & \to & 0
\end{array}
$$

has the property that t is not a split epimorphism. Also if $h: X \to C$ is not a split epimorphism, then it factors through t since it factors through f. Thus t is right almost split. But it is also right minimal since A_i is indecomposable. Hence t is also minimal right almost split, which implies that $v: B \to A_i \times^{\mathrm{Ker}\, f} B$ is an isomorphism. Therefore $u: \mathrm{Ker}\, f \to A_i$ is an isomorphism and hence $\mathrm{Ker}\, f$ is indecomposable.

Since C is indecomposable and f is not a split epimorphism it follows that g is left minimal.

We now prove that g is left almost split and that $\mathrm{Ker}\, f \simeq D\,\mathrm{Tr}\, C$ simultaneously. Let Y be any indecomposable module not isomorphic to $D\,\mathrm{Tr}\, C$ and $h: \mathrm{Ker}\, f \to Y$ any morphism. Two cases occur. If Y is injective, then Y is not isomorphic to $\mathrm{Ker}\, f$ since g is not a split monomorphism. Suppose now that Y is not injective. Because $Y \not\simeq D\,\mathrm{Tr}\, C$ we have that $\mathrm{Tr}\, DY \not\simeq C$. But since f is a right almost split morphism, all morphisms $\mathrm{Tr}\, DY \to C$ factor through f. Therefore by IV Corollary 4.4 every morphism $h: \mathrm{Ker}\, f \to Y$ factors through g. It follows that $\mathrm{Ker}\, f \not\simeq Y$ since g is not a split monomorphism. But then we have $\mathrm{Ker}\, f \simeq D\,\mathrm{Tr}\, C$ since $\mathrm{Ker}\, f$ is indecomposable. We have now also proved that any morphism $h: \mathrm{Ker}\, f \to Y$ with Y indecomposable and $Y \not\simeq \mathrm{Ker}\, f$ factors through g and hence g is left almost split by Proposition 1.11. This finishes the proof of the proposition. $\square$

We now give the dual of Proposition 1.12.

Proposition 1.13 *Suppose* $g: A \to B$ *is a minimal left almost split morphism with A not injective. Then we have the following.*

(a) *The module A is indecomposable and g is a monomorphism.*

(b) *The exact sequence* $0 \to A \overset{g}{\to} B \overset{f}{\to} \mathrm{Coker}\, g \to 0$ *has the following properties.*

    (i) $\mathrm{Coker}\, g \simeq \mathrm{Tr}\, DA.$

(ii) *f is a minimal right almost split morphism.*   □

Propositions 1.12 and 1.13 suggest the following definition.

An exact sequence $0 \to A \xrightarrow{g} B \xrightarrow{f} C \to 0$ is called an **almost split sequence** if g is left almost split and f is right almost split. It is clear that an exact sequence $0 \to A \xrightarrow{g} B \xrightarrow{f} C \to 0$ is almost split if and only if $0 \to D(C) \xrightarrow{D(f)} D(B) \xrightarrow{D(g)} D(A) \to 0$ is almost split. Summarizing some of our previous results we have the following.

Proposition 1.14 *The following are equivalent for an exact sequence* $0 \to A \xrightarrow{g} B \xrightarrow{f} C \to 0.$

(a) *The sequence is an almost split sequence.*
(b) *The morphism f is minimal right almost split.*
(c) *The morphism g is minimal left almost split.*
(d) *The module A is indecomposable and f is right almost split.*
(e) *The module C is indecomposable and g is left almost split.*
(f) *The module C is isomorphic to* $\operatorname{Tr} DA$ *and g is left almost split.*
(g) *The module A is isomorphic to* $D \operatorname{Tr} C$ *and f is right almost split.*

Proof (a) ⇒ (b) Since g is left almost split, the module A is indecomposable by Lemma 1.7. Hence f is right minimal, and consequently it is minimal right almost split.

(b) ⇒ (c), (c) ⇒ (d) and (d) ⇒ (e) are immediate consequences of Propositions 1.12, 1.13 and Lemma 1.7 together with Proposition 1.12 respectively.

(e) ⇒ (f) Since C is indecomposable, g is left minimal and therefore g is minimal left almost split. Hence $C \simeq \operatorname{Tr} DA$ by Proposition 1.13.

(f) ⇒ (g) Since g is left almost split, A is an indecomposable module. Therefore $\operatorname{Tr} DA \simeq C$ is indecomposable. Hence g is minimal left almost split. So by Proposition 1.13 we have that f is right almost split with $A \simeq D \operatorname{Tr} C$.

(g) ⇒ (a) Since f is right almost split, we know that C is indecomposable. Therefore $A \simeq D \operatorname{Tr} C$ is indecomposable and so f is also right minimal. Hence f is a minimal right almost split morphism, so by Proposition 1.12 we have that g is a minimal left almost split morphism. Hence the sequence $0 \to A \xrightarrow{g} B \xrightarrow{f} C \to 0$ is almost split.   □

We are now ready to prove the following existence theorem for almost split sequences.

Theorem 1.15

(a) *If C is an indecomposable nonprojective module, then there is an almost split sequence* $0 \to A \xrightarrow{g} B \xrightarrow{f} C \to 0$.

(b) *If A is an indecomposable noninjective module, then there is an almost split sequence* $0 \to A \xrightarrow{g} B \xrightarrow{f} C \to 0$.

Proof (a) In view of Proposition 1.14 it suffices to show that if C is an indecomposable nonprojective module, then there is an exact sequence $0 \to D\operatorname{Tr} C \xrightarrow{g} E \xrightarrow{f} C \to 0$ with f right almost split. Since $D\operatorname{Tr} C$ is not injective we know there is some nonsplit exact sequence

$$0 \to D\operatorname{Tr} C \xrightarrow{h} B \xrightarrow{j} V \to 0.$$

If every morphism $C \to V$ factors through j, then every morphism $D\operatorname{Tr} C \to D\operatorname{Tr} C$ factors through h by IV Corollary 4.4. Then h would be a split monomorphism, which contradicts the hypothesis that the sequence does not split. Letting $\Gamma = \operatorname{End}_\Lambda(C)^{\operatorname{op}}$ we get the exact sequence of Γ-modules

$$\operatorname{Hom}_\Lambda(C, B) \xrightarrow{\operatorname{Hom}_\Lambda(C, j)} \operatorname{Hom}_\Lambda(C, V) \to \operatorname{Coker} \operatorname{Hom}_\Lambda(C, j) \to 0$$

with $\operatorname{Coker} \operatorname{Hom}_\Lambda(C, j) \neq 0$. Hence there is a morphism $t: C \to V$ such that its image in $\operatorname{Coker} \operatorname{Hom}_\Lambda(C, j)$ generates a simple Γ-module. Thus we have the pullback diagram

$$
\begin{array}{ccccccccc}
0 & \to & D\operatorname{Tr} C & \xrightarrow{g} & E & \xrightarrow{f} & C & \to & 0 \\
  &   & \| & & \downarrow & & \downarrow t & & \\
0 & \to & D\operatorname{Tr} C & \xrightarrow{h} & B & \xrightarrow{j} & V & \to & 0.
\end{array}
$$

We claim that the exact sequence $0 \to D\operatorname{Tr} C \xrightarrow{g} E \xrightarrow{f} C \to 0$ has our desired property that f is right almost split. First of all, it does not split. For if it did, t would factor through j, which would imply that the image of t in $\operatorname{Coker} \operatorname{Hom}_\Lambda(C, j)$ is zero, a contradiction. Suppose now that $s: X \to C$ is not a split epimorphism. We want to show that s factors through f or, what is the same thing, that the exact sequence $0 \to D\operatorname{Tr} C \to E \times_C X \to X \to 0$ in the pullback diagram

$$
\begin{array}{ccccccccc}
0 & \to & D\operatorname{Tr} C & \xrightarrow{h} & E \times_C X & \xrightarrow{v} & X & \to & 0 \\
  &   & \| & & \downarrow & & \downarrow s & & \\
0 & \to & D\operatorname{Tr} C & \xrightarrow{g} & E & \xrightarrow{f} & C & \to & 0
\end{array}
$$

splits. To do this it suffices by IV Corollary 4.4 to show that every morphism $w: C \to X$ factors through v.

Now the composition $sw: C \to C$ is not an isomorphism since s is not a split epimorphism. Hence the composition $tsw: C \to V$ is in $(\mathrm{rad}\, \Gamma)\Gamma t$. Since the image of Γt in $\mathrm{Coker}\,\mathrm{Hom}_\Lambda(C, j)$ is a simple module, tsw goes to zero in $\mathrm{Coker}\,\mathrm{Hom}_\Lambda(C, j)$. In other words tsw factors through j. This means, by the basic properties of pullbacks, that sw factors through f. But again by the basic properties of pullbacks, this means that $w: C \to X$ factors through v. This shows that f is right almost split, finishing the proof of part (a) of the theorem.

(b) This is dual to (a).   □

Having proven the existence of almost split sequences, we now explain in what sense they are unique.

Theorem 1.16 *The following are equivalent for two almost split sequences* $0 \to A \xrightarrow{g} B \xrightarrow{f} C \to 0$ *and* $0 \to A' \xrightarrow{g'} B' \xrightarrow{f'} C' \to 0$.

(a) $C \simeq C'$.

(b) $A \simeq A'$.

(c) *The sequences are isomorphic in the sense that there is a commutative diagram*

$$
\begin{array}{ccccccccc}
0 & \to & A & \xrightarrow{g} & B & \xrightarrow{f} & C & \to & 0 \\
 & & \wr\wr & & \wr\wr & & \wr\wr & & \\
0 & \to & A' & \xrightarrow{g'} & B' & \xrightarrow{f'} & C' & \to & 0
\end{array}
$$

with the vertical morphisms isomorphisms.

Proof (a) ⇔ (b) This follows from the fact that $A \simeq D\,\mathrm{Tr}\, C$ and $A' \simeq D\,\mathrm{Tr}\, C'$ established in Proposition 1.14.

(a) ⇔ (c) Clearly (c) implies (a). Suppose there is an isomorphism $h: C \to C'$. Since f is minimal right almost split, the composition $hf: B \to C'$ is also minimal right almost split. Therefore $hf: B \to C'$ and $f': B' \to C'$ are two minimal right almost split morphisms, so they are isomorphic in $\mathrm{mod}\,\Lambda/C'$ by Proposition 1.4. This shows (a) ⇔ (c).   □

Note that the exact sequences described for $R/(p^n)$ on page 141 are almost split sequences.

Combining Corollary 1.6 with Theorem 1.15 we obtain the following.

Corollary 1.17 *We have the following for an artin algebra* Λ.

(a) *For each indecomposable Λ-module C there is a unique, up to isomorphism in* $\operatorname{mod}\Lambda/C$, *minimal right almost split morphism* $f:B \to C$.

(b) *For each indecomposable Λ-module A there is a unique, up to isomorphism in* $\operatorname{mod}\Lambda \setminus A$, *minimal left almost split morphism* $g:A \to E$. $\square$

2 Interpretation and examples

This section is devoted to giving some basic interpretations of almost split sequences as generators of the socle of $\operatorname{Ext}^1_\Lambda(C, D\operatorname{Tr} C)$ viewed either as an $\operatorname{End}_\Lambda(C)^{\mathrm{op}}$-module or as an $\operatorname{End}_\Lambda(D\operatorname{Tr} C)$-module. From this it follows that any nonsplit exact sequence $0 \to D\operatorname{Tr} C \to B \to C \to 0$ is almost split if C is indecomposable with $\underline{\operatorname{End}}_\Lambda(C)$ a division ring. Using this fact we construct almost split sequences where the left or right hand term is simple.

We begin with the following result.

Proposition 2.1 *Let C be an indecomposable nonprojective Λ-module. Then* $\operatorname{Ext}^1_\Lambda(C, D\operatorname{Tr} C)$ *has a simple socle both as an* $\operatorname{End}_\Lambda(C)^{\mathrm{op}}$-module *and as an* $\operatorname{End}_\Lambda(D\operatorname{Tr} C)$-module. *These socles coincide and each nonzero element of this socle is an almost split sequence.*

Proof Let $\delta: 0 \to D\operatorname{Tr} C \overset{f}{\to} B \overset{g}{\to} C \to 0$ be an almost split sequence and let $\epsilon: 0 \to D\operatorname{Tr} C \overset{f'}{\to} B' \overset{g'}{\to} C \to 0$ be a nonzero element in the $\operatorname{End}_\Lambda(C)^{\mathrm{op}}$-socle of $\operatorname{Ext}^1_\Lambda(C, D\operatorname{Tr} C)$. Since any nonisomorphism $h:C \to C$ factors through g, it follows that δ is annihilated by $\operatorname{rad}\operatorname{End}_\Lambda(C)^{\mathrm{op}}$ and is hence in the socle. Using that δ is almost split, we have an exact commutative diagram

$$
\begin{array}{ccccccccc}
\delta:0 & \longrightarrow & D\operatorname{Tr} C & \overset{f}{\longrightarrow} & B & \overset{g}{\longrightarrow} & C & \longrightarrow & 0 \\
 & & \| & & \downarrow{\scriptstyle s} & & \downarrow{\scriptstyle t} & & \\
\epsilon:0 & \longrightarrow & D\operatorname{Tr} C & \overset{f'}{\longrightarrow} & B' & \overset{g'}{\longrightarrow} & C & \longrightarrow & 0
\end{array}
$$

and consequently $\operatorname{Ext}^1_\Lambda(t, D\operatorname{Tr} C)(\epsilon) = \delta$. Since $t \in \operatorname{End}_\Lambda(C)^{\mathrm{op}}$ does not annihilate ϵ, t is not in $\operatorname{rad}\operatorname{End}_\Lambda(C)^{\mathrm{op}}$ and is therefore an isomorphism since $\operatorname{End}_\Lambda(C)^{\mathrm{op}}$ is a local ring. Hence we have that $\epsilon = \operatorname{Ext}^1_\Lambda(t^{-1}, D\operatorname{Tr} C)(\delta)$. This shows that ϵ is in the $\operatorname{End}_\Lambda(C)^{\mathrm{op}}$-submodule of $\operatorname{Ext}^1_\Lambda(C, D\operatorname{Tr} C)$ generated by δ. Therefore the $\operatorname{End}_\Lambda(C)^{\mathrm{op}}$-socle is simple and generated by δ.

That the $\operatorname{End}_\Lambda(D\operatorname{Tr} C)$-submodule of $\operatorname{Ext}^1_\Lambda(C, D\operatorname{Tr} C)$ generated by δ

is simple and equal to the End($D \operatorname{Tr} C$)-socle follows by duality. Since in each of the two socles the nonzero elements are the almost split sequences, the socles must coincide. □

We now proceed to give a series of characterizations of almost split sequences.

Proposition 2.2 *Let C be a nonprojective indecomposable Λ-module and $\delta: 0 \to D \operatorname{Tr} C \xrightarrow{f} B \xrightarrow{g} C \to 0$ a nonsplit exact sequence. Then the following are equivalent.*

(a) *δ is an almost split sequence.*

(b) *Each nonisomorphism $h: C \to C$ factors through $g: B \to C$.*

(c) *$\operatorname{Im}(\operatorname{Hom}_\Lambda(C, g)) = \operatorname{rad}(\operatorname{End}_\Lambda(C)^{\mathrm{op}})$.*

(d) *$\delta^*(C)$ is a simple $\operatorname{End}_\Lambda(C)^{\mathrm{op}}$-module, where δ^* denotes the contravariant defect of δ.*

(e) *$\delta^*(X) = 0$ for each indecomposable module X which is not isomorphic to C.*

(f) *δ generates the socle of $\operatorname{Ext}_\Lambda^1(C, D \operatorname{Tr} C)$ as an $\operatorname{End}_\Lambda(C)^{\mathrm{op}}$-module.*

(b′) *Each nonisomorphism $h: D \operatorname{Tr} C \to D \operatorname{Tr} C$ factors through $f: D \operatorname{Tr} C \to B$.*

(c′) *$\operatorname{Im}(\operatorname{Hom}_\Lambda(f, D \operatorname{Tr} C)) = \operatorname{rad}(\operatorname{End}_\Lambda(D \operatorname{Tr} C))$.*

(d′) *$\delta_*(D \operatorname{Tr} C)$ is a simple $\operatorname{End}_\Lambda(D \operatorname{Tr} C)$-module, where δ_* is the covariant defect.*

(e′) *$\delta_*(X) = 0$ for each indecomposable module X which is not isomorphic to $D \operatorname{Tr} C$.*

(f′) *δ generates the socle of $\operatorname{Ext}_\Lambda^1(C, D \operatorname{Tr} C)$ as an $\operatorname{End}_\Lambda(D \operatorname{Tr} C)$-module.*

Proof We prove that (a), (b), (c), (d), (e) and (f) are equivalent. That (a), (b′), (c′), (d′), (e′) and (f′) are equivalent follows by duality.

Since C is an indecomposable module, $\operatorname{End}_\Lambda(C)$ is local and therefore (b), (c) and (d) are clearly equivalent. By Proposition 2.1 we have that (a) and (f) are equivalent. But $\delta^*(C)$ is isomorphic to the $\operatorname{End}_\Lambda(C)^{\mathrm{op}}$-submodule of $\operatorname{Ext}_\Lambda^1(C, D \operatorname{Tr} C)$ generated by δ. Hence $\delta^*(C)$ is a simple $\operatorname{End}_\Lambda(C)^{\mathrm{op}}$-module if and only if δ is an almost split sequence by Proposition 2.1. Therefore (a), (b), (c), (d) and (f) are equivalent.

Since C is an indecomposable module and δ is a nonsplit exact sequence it follows by Proposition 1.10 that (e) is equivalent to g being right almost split. But then δ is an almost split sequence by Proposition 1.14 since $\operatorname{Ker} g$ is isomorphic to $D \operatorname{Tr} C$. Hence also (e) and (a) are equivalent. This concludes the proof of the proposition. □

As our first application of Proposition 2.2 we show that an exact sequence $0 \to A \to B \to C \to 0$ is determined up to isomorphism by the modules A, B and C if it is almost split. Also split exact sequences have this property. A further discussion of sequences with this property which we call **rigid exact sequences** will be given later in Chapter XI.

Proposition 2.3 *Let* $\delta : 0 \to A \xrightarrow{f} B \xrightarrow{g} C \to 0$ *and* $\epsilon : 0 \to A \xrightarrow{f'} B \xrightarrow{g'} C \to 0$ *be exact sequences. If* δ *is almost split or split then* δ *and* ϵ *are isomorphic sequences.*

Proof We only give the proof when δ is almost split, and leave the rest as an exercise. Consider $\delta^*(C)$ and $\epsilon^*(C)$ as $\operatorname{End}_\Lambda(C)^{\mathrm{op}}$-modules. Letting $l(X)$ in this proof denote the length of X as an $\operatorname{End}_\Lambda(C)^{\mathrm{op}}$-module we obtain $l(\delta^*(C)) = l(\operatorname{Hom}_\Lambda(C, A)) - l(\operatorname{Hom}_\Lambda(C, B)) + l(\operatorname{Hom}_\Lambda(C, C)) = l(\epsilon^*(C))$. By Proposition 2.2 we have $l(\delta^*(C)) = 1$ since δ is an almost split sequence. Therefore $\epsilon^*(C)$ is also a simple $\operatorname{End}_\Lambda(C)^{\mathrm{op}}$-module and since $A \simeq D \operatorname{Tr} C$, Proposition 2.2 gives that ϵ is also almost split. □

From the definition of an almost split sequence it appears that one needs to know all modules in order to determine whether a given short exact sequence is almost split or not. However, Proposition 2.2 gives a powerful way of determining whether an exact sequence $0 \to A \to B \to C \to 0$ with C indecomposable and $A \simeq D \operatorname{Tr} C$ is almost split or not. In some cases, as we see in the next result, this should in principle be easy.

Corollary 2.4

(a) *Let* C *be an indecomposable module with* $\underline{\operatorname{End}}_\Lambda(C)$ *a division ring. Then the following are equivalent for a short exact sequence* $\delta : 0 \to D \operatorname{Tr} C \to B \to C \to 0$.

    (i) δ *is almost split.*

    (ii) δ *does not split.*

    (iii) $B \not\simeq C \coprod D \operatorname{Tr} C$.

(b) *Let* A *be an indecomposable module with* $\overline{\operatorname{End}}_\Lambda(A)$ *a division ring. Then the following are equivalent for a short exact sequence* $\delta : 0 \to A \to B \to \operatorname{Tr} D A \to 0$.

    (i) δ *is almost split.*

    (ii) δ *does not split.*

    (iii) $B \not\simeq A \coprod \operatorname{Tr} DA.$

Proof (a) Let C be an indecomposable module with $\underline{\operatorname{End}}_\Lambda(C)$ a division ring and let $\delta : 0 \to D\operatorname{Tr}C \xrightarrow{f} B \xrightarrow{g} C \to 0$ be an exact sequence. Since g is an epimorphism, $\operatorname{Im}\operatorname{Hom}_\Lambda(C,g)$ contains the ideal of all morphisms $f : C \to C$ which factor through a projective module. Since C is indecomposable and $\underline{\operatorname{End}}_\Lambda(C)$ is a division ring, we then have that $\delta^*(C)$ is a simple $\operatorname{End}_\Lambda(C)^{\mathrm{op}}$-module if and only if δ is not split. Hence by Proposition 2.2 we have that δ is almost split if and only if δ is not split, so that (i) and (ii) are equivalent. The equivalence of (ii) and (iii) follows directly from Proposition 2.3.

    (b) This follows by duality.   □

    We illustrate with the following.

Example Let $\Lambda = k\Gamma$ where k is a field and Γ is the quiver

Denote as usual by S_i the simple module corresponding to the vertex i and by P_i the corresponding indecomposable projective module, for $i = 1, 2, 3, 4$. In Chapter IV we showed that $D\operatorname{Tr}S_2 = P_1$. Viewing the modules as representations we have the exact sequence

$$0 \to \begin{matrix} k \\ \downarrow{\scriptstyle 1} \\ k \\ {}_{1}\swarrow \ \searrow{}_{1} \\ k \quad\ \ k \end{matrix} \ \xrightarrow{g}\ \begin{matrix} k \\ \downarrow{\scriptstyle\binom{1}{1}} \\ k\amalg k \\ {}_{(1\,0)}\swarrow \ \searrow{}_{(0\,1)} \\ k \quad\ \ k \end{matrix} \ \xrightarrow{f}\ \begin{matrix} 0 \\ \downarrow \\ k \\ {}_{0}\swarrow \ \searrow{}_{0} \\ 0 \quad\ \ 0 \end{matrix} \to 0$$

where $g = (1, \binom{1}{1}, 1, 1)$ and $f = (0, (1, -1), 0, 0)$, according to the numbering in the quiver. This sequence does not split since the middle term is indecomposable. Hence it is almost split.

    We now illustrate how to use this result to construct almost split sequences where the right hand term is simple nonprojective or the left hand term is simple noninjective. The answers are particularly easy when the right hand term is simple injective or the left hand term is simple projective, and also when the algebra is symmetric.

Proposition 2.5 *Let Λ be an artin algebra and $P_1 \xrightarrow{f} P_0 \to S \to 0$ a minimal projective presentation of a simple nonprojective module S.*

(a) *There is an almost split sequence $0 \to D\operatorname{Tr} S \to (Df^*)^{-1}(\operatorname{soc}(DP_0^*)) \to S \to 0$.*

(b) *If S is also injective, there is an almost split sequence $0 \to D\operatorname{Tr} S \to DP_1^* \to S \to 0$.*

(c) *If Λ is a symmetric algebra, there is an almost split sequence $0 \to \Omega^2 S \to f^{-1}(S) \to S \to 0$.*

Proof (a) Let S be a simple nonprojective Λ-module and let $P_1 \xrightarrow{f} P_0 \to S \to 0$ be a minimal projective presentation. Apply $\operatorname{Hom}_\Lambda(\ ,\Lambda)$ to obtain the minimal projective Λ^{op}-presentation $P_0^* \xrightarrow{f^*} P_1^* \to \operatorname{Tr} S \to 0$ of $\operatorname{Tr} S$. Then apply the duality to obtain the minimal injective copresentation $0 \to D\operatorname{Tr} S \to DP_1^* \xrightarrow{Df^*} DP_0^*$ of $D\operatorname{Tr} S$. Since this resolution is minimal, the sequence $0 \to D\operatorname{Tr} S \to (Df^*)^{-1}\operatorname{soc}(DP_0^*) \to \operatorname{soc}(DP_0^*) \to 0$ does not split. However $\operatorname{soc}(DP_0^*) \simeq S$, hence we get a nonsplit exact sequence $0 \to D\operatorname{Tr} S \to (Df^*)^{-1}\operatorname{soc}(DP_0^*) \to S \to 0$ which is almost split by Corollary 2.4.

(b) When S is simple injective, then $S \simeq DP_0^*$, so that $\operatorname{soc}(DP_0^*) \simeq DP_0^*$ and hence $(Df^*)^{-1}(\operatorname{soc}(DP_0^*)) = DP_1^*$. Then we use part (a) to complete (b).

(c) When Λ is symmetric, the exact sequence $0 \to D\operatorname{Tr} S \to DP_1^* \xrightarrow{Df^*} DP_0^*$ is isomorphic to the exact sequence $0 \to \Omega^2 S \to P_1 \xrightarrow{f} P_0$. Hence the sequence in (a) becomes the sequence $0 \to \Omega^2 S \to f^{-1}(S) \to S \to 0$. □

By duality we also have the following result.

Proposition 2.6 *Let Λ be an artin algebra and $0 \to S \to I_0 \xrightarrow{g} I_1$ a minimal injective copresentation of a simple noninjective Λ-module S.*

(a) *There is an almost split sequence $0 \to S \to (DI_1)^*/(Dg)^*(\mathfrak{r}((DI_0)^*)) \to \operatorname{Tr} DS \to 0$.*

(b) *If S is also projective, there is an almost split sequence $0 \to S \to (DI_1)^* \to \operatorname{Tr} DS \to 0$.*

(c) *If Λ is a symmetric algebra, there is an almost split sequence $0 \to$*
    $S \to I_1/g(\mathfrak{r}I_0) \to \Omega^{-2}S \to 0$.   □

We illustrate with the following.

Example Let Γ be the quiver

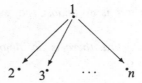

Then the path algebra $k\Gamma$ is finite dimensional and the simple module S_1
corresponding to the vertex 1 is injective nonprojective. It is easy to see
that we have an almost split sequence $0 \to D\,\mathrm{Tr}\,S_1 \to I_2 \coprod \cdots \coprod I_n \to$
$S_1 \to 0$, where I_i denotes the injective envelope of the simple module
corresponding to the vertex i.

When Λ is a selfinjective artin algebra and $0 \to A \to B \to C \to 0$ is
an exact sequence, we have an exact commutative diagram

$$
\begin{array}{ccccccccc}
 & & 0 & & 0 & & 0 & & \\
 & & \downarrow & & \downarrow & & \downarrow & & \\
\epsilon: & 0 \longrightarrow & \Omega A & \longrightarrow & K & \longrightarrow & \Omega C & \longrightarrow & 0 \\
 & & \downarrow & & \downarrow & & \downarrow & & \\
\eta: & 0 \longrightarrow & P(A) & \longrightarrow & P(A) \amalg P(C) & \longrightarrow & P(C) & \longrightarrow & 0 \\
 & & \downarrow & & \downarrow & & \downarrow & & \\
\delta: & 0 \longrightarrow & A & \longrightarrow & B & \longrightarrow & C & \longrightarrow & 0 \\
 & & \downarrow & & \downarrow & & \downarrow & & \\
 & & 0 & & 0 & & 0 & & \\
\end{array}
$$

where $P(A) \to A$ and $P(C) \to C$ are projective covers. We shall see in
Chapter X that ϵ is almost split if and only if δ is almost split. Hence
starting with an almost split sequence $0 \to A \to B \to C \to 0$ where C is
simple, we can use this procedure to construct many other almost split
sequences.

3 Almost split sequences with projective or injective middle terms

For an indecomposable nonprojective module C we have in addition to the important projective cover $P \to C$ also the minimal right almost split morphism $B \to C$. These morphisms are both unique up to isomorphism, and the main object of this section is to investigate when they coincide. We shall see that knowing that a projective cover is in addition right almost split gives additional interesting information on C.

If $0 \to A \overset{g}{\to} P \overset{f}{\to} C \to 0$ is an almost split sequence with P projective, then $f: P \to C$ is clearly a projective cover since it is right minimal. Hence asking when projective covers and minimal right almost split morphisms coincide is the same thing as asking when almost split sequences have projective middle terms.

We have seen in Section 2 that when S is a simple module which is projective and not injective, there is an almost split sequence $0 \to S \to P \overset{f}{\to} \mathrm{Tr}\, DS \to 0$ where P is projective. The next result provides an example with projective middle term when the left hand side is not simple projective.

Denote by $k[X_1, \ldots, X_n]$ the polynomial ring over a field k in the indeterminates $X_1, \ldots, X_n$. Let $\Lambda = k[X_1, \ldots, X_n]/(X_1, \ldots, X_n)^2$ where $(X_1, \ldots, X_n)$ is the ideal generated by $X_1, \ldots, X_n$. Then Λ is a commutative local ring of k-dimension $n + 1$ with the unique maximal ideal $\mathfrak{m} = (x_1, \ldots, x_n)$ where $x_i = X_i + (X_1, \ldots, X_n)^2$. Also $S = \Lambda/\mathfrak{m}$ is the unique simple Λ-module and S has dimension 1 over k.

Proposition 3.1 *Suppose* $\Lambda = k[X_1, \ldots, X_n]/(X_1, \ldots, X_n)^2$ $(n \geq 1)$ *and* $\Lambda/\mathfrak{m}$ *is the simple* Λ-module. *Let* $f: \Lambda/\mathfrak{m} \to n\Lambda$ *be the* Λ-*morphism such that* $f(1) = (x_1, x_2, \ldots, x_n)$ *in* $n\Lambda$. *Then the exact sequence* $0 \to \Lambda/\mathfrak{m} \overset{f}{\to} n\Lambda \overset{g}{\to} \mathrm{Coker}\, f \to 0$ *is an almost split sequence and* $n\Lambda \overset{g}{\to} \mathrm{Coker}\, f = \mathrm{Tr}\, D(\Lambda/\mathfrak{m})$ *is a projective cover.*

Proof Let $n\Lambda \overset{u}{\to} \Lambda \overset{v}{\to} \Lambda/\mathfrak{m} \to 0$ be the minimal projective presentation of $\Lambda/\mathfrak{m}$ where $v: \Lambda \to \Lambda/\mathfrak{m}$ is the canonical epimorphism and $u((\lambda_1, \ldots, \lambda_n)) = \sum_{i=1}^{n} \lambda_i x_i$ for all $(\lambda_1, \ldots, \lambda_n)$ in $n\Lambda$. Then we have the exact sequence $0 \to (\Lambda/\mathfrak{m})^* \overset{v^*}{\to} \Lambda^* \overset{u^*}{\to} (n\Lambda)^* \to \mathrm{Tr}(\Lambda/\mathfrak{m}) \to 0$. Using the usual identification of Λ with Λ^*, we see that $\mathrm{Im}\, v^* = \mathfrak{m}$, so that $\mathrm{Coker}\, v^*$ is identified with $\Lambda/\mathfrak{m}$. Identifying $n\Lambda$ with $(n\Lambda)^*$ through dual basis we obtain the exact sequence $0 \to \Lambda/\mathfrak{m} \overset{f}{\to} n\Lambda \overset{g}{\to} \mathrm{Tr}\, D(\Lambda/\mathfrak{m}) \to 0$ with g a projective cover. Hence this sequence is almost split by Corollary 2.4 since it is not split and $\mathrm{End}_\Lambda(\Lambda/\mathfrak{m}) = k$, a division ring. $\square$

Before proving the main result of this section we point out some basic properties of almost split sequences which are generally useful as well as needed for the proof of our main theorem.

Lemma 3.2 *Let* $0 \to A \xrightarrow{f} B \xrightarrow{g} C \to 0$ *be an almost split sequence.*

(a) *If X is a proper submodule of C, then the exact sequence $0 \to A \to g^{-1}(X) \to X \to 0$ splits.*

(b) *If Y is a nonzero submodule of A, then the exact sequence $0 \to A/Y \to B/f(Y) \to C \to 0$ splits.*

(c) *If $\mathfrak{a}$ is a two-sided ideal in Λ such that $\mathfrak{a}A \neq 0$, then $0 \to A/\mathfrak{a}A \to B/\mathfrak{a}B \to C/\mathfrak{a}C \to 0$ is split exact.*

(d) *If C is not simple, then $0 \to \operatorname{soc} A \to \operatorname{soc} B \to \operatorname{soc} C \to 0$ is exact.*

Proof (a) and (b) follow immediately from the definition of almost split sequences and are left as exercises.

(c) Since $\mathfrak{a}A \neq 0$, it follows from (b) that $0 \to A/\mathfrak{a}A \to B/f(\mathfrak{a}A) \to C \to 0$ is a split exact sequence. Tensoring with $\Lambda/\mathfrak{a}$ we obtain the split exact sequence $0 \to A/\mathfrak{a}A \to B/\mathfrak{a}B \to C/\mathfrak{a}C \to 0$.

(d) For a Λ-module X we know that $\operatorname{Hom}_\Lambda(\Lambda/\mathfrak{r}, X) = \operatorname{soc} X$ under the identification $f \mapsto f(1)$ for all f in $\operatorname{Hom}_\Lambda(\Lambda/\mathfrak{r}, X)$. Since C is not simple, C is not a summand of $\Lambda/\mathfrak{r}$. Hence by the definition of almost split sequences we have that $0 \to \operatorname{Hom}_\Lambda(\Lambda/\mathfrak{r}, A) \to \operatorname{Hom}_\Lambda(\Lambda/\mathfrak{r}, B) \to \operatorname{Hom}_\Lambda(\Lambda/\mathfrak{r}, C) \to 0$ is exact, giving our result. $\square$

We now describe the almost split sequences $0 \to A \to B \to C \to 0$ with B either a projective or an injective module.

Theorem 3.3 *Let* $0 \to A \xrightarrow{f} B \xrightarrow{g} C \to 0$ *be an almost split sequence.*

(a) *The following are equivalent.*

   (i)   *The module B is injective.*

   (ii)   *The morphism f is an injective envelope.*

   (iii) *C is a nonprojective simple module which is not a composition factor of $\mathfrak{r}P/\operatorname{soc} P$ for any projective module P.*

(b) *The following are equivalent.*

   (i)   *The module B is projective.*

   (ii)   *The morphism g is a projective cover.*

   (iii) *A is a noninjective simple module which is not a composition factor of $\mathfrak{r}I/\operatorname{soc} I$ for any injective module I.*

Proof We only prove (a) since (b) then follows by duality.

The equivalence of the first two statements is easy and left to the reader. To establish the equivalence of (i) and (iii), assume first that B is an injective module. Let S be a simple submodule of B. Since B is injective, there is a summand B' of B which is an injective envelope of S. If $g(S) \subset C$ is not zero, then $g|_{B'} : B' \to C$ would be a monomorphism and hence an isomorphism since C is indecomposable. Therefore $g : B \to C$ would split, which is a contradiction. We can then conclude $g(\text{soc } B) = 0$. But this implies that C is simple, because otherwise $0 \to \text{soc } A \to \text{soc } B \to \text{soc } C \to 0$ is exact by Lemma 3.2 (d). Thus C is a nonprojective simple module.

We next show that C is not a composition factor of $\mathfrak{r}P / \text{soc } P$ for any projective module P. Suppose that P is a projective module such that C is a composition factor of $\mathfrak{r}P / \text{soc } P$. Let $Q \to C$ be a projective cover for C. Then there is a nonzero morphism $u : Q \to \mathfrak{r}P / \text{soc } P$ since C is a composition factor of $\mathfrak{r}P / \text{soc } P$. Because Q is projective, u factors through some $t : Q \to \mathfrak{r}P$. Then $X = \text{Im } t$ is a submodule of $\mathfrak{r}P$ which is an indecomposable nonsimple module such that $X/\mathfrak{r}X \simeq C$. Thus there is an epimorphism $h : X \to C$ which is not a split epimorphism. Since $g : B \to C$ is right almost split, there is a morphism $h' : X \to B$ such that $gh' = h$. Because B is injective, there is some $s : P \to B$ such that $s|_X = h'$. Then we have $h'(X) \subset \mathfrak{r}B$ since $X \subset \mathfrak{r}P$. Hence the composition gh' is zero since C is simple, which means that $h : X \to C$ is zero, and this is a contradiction.

We now prove that (iii) implies (i), completing the proof of (a). In order to carry out this proof it is convenient to have the following facts. Let P be an indecomposable projective Λ-module. Then we have the exact sequence of Λ^{op}-modules $0 \to \text{Hom}_\Lambda(P, \mathfrak{r}) \to \text{Hom}_\Lambda(P, \Lambda) \to \text{Hom}_\Lambda(P, \Lambda/\mathfrak{r}) \to 0$. Since $\text{Hom}_\Lambda(P, \Lambda)$ is an indecomposable projective Λ^{op}-module and $\text{Hom}_\Lambda(P, \Lambda/\mathfrak{r})$ is a nonzero semisimple Λ^{op}-module, it follows that $\text{Hom}_\Lambda(P, \Lambda/\mathfrak{r})$ is simple and $\text{Hom}_\Lambda(P, \mathfrak{r}) = \text{Hom}_\Lambda(P, \Lambda)\mathfrak{r}$. Suppose P is not simple and $P/\mathfrak{r}P = C$ has the property that C is not a composition factor of $\mathfrak{r}/ \text{soc } \mathfrak{r}$. Then $\text{Hom}_\Lambda(P, \text{soc } \mathfrak{r}) = \text{Hom}_\Lambda(P, \mathfrak{r})$ and so the monomorphism $\text{Hom}_\Lambda(C, \Lambda) \to \text{Hom}_\Lambda(P, \Lambda)$ of Λ^{op}-modules induced by the epimorphism $P \to C$ has $\text{Hom}_\Lambda(P, \mathfrak{r})$ as its image. Hence $0 \to \text{Hom}_\Lambda(C, \Lambda) \to \text{Hom}_\Lambda(P, \Lambda) \to \text{Hom}_\Lambda(P, \Lambda/\mathfrak{r}) \to 0$ is exact with $\text{Hom}_\Lambda(P, \Lambda/\mathfrak{r})$ simple.

We now return to showing that (iii) implies (i) in (a). Suppose C is a nonprojective simple Λ-module which is not a composition factor of

$r/\mathrm{soc}\,r$. Let $P_1 \xrightarrow{h} P_0 \to C \to 0$ be a minimal projective presentation of C. Then we have the exact sequence

$$0 \to C^* \to P_0^* \xrightarrow{h^*} P_1^* \to \mathrm{Tr}\,C \to 0.$$

By our previous discussion $\mathrm{Im}\,h^*$ is the simple Λ^{op}-module P_0^*/rP_0^*. Therefore applying the duality, we get the exact sequence

$$\delta: 0 \to D\,\mathrm{Tr}\,C \xrightarrow{f'} D(P_1^*) \xrightarrow{g'} C \to 0$$

since $C \simeq D(P_0^*/rP_0^*)$. This sequence does not split since f' is an injective envelope. Since $\mathrm{End}_\Lambda C$ is a division ring, it follows by Corollary 2.4 that δ is an almost split sequence. Now $D(P_1^*)$ is injective and since δ is isomorphic to the original almost split sequence, it follows that B is injective. □

As an immediate consequence of this theorem we have the following.

Corollary 3.4 *Let S be a simple Λ-module.*

(a) *If S is not projective and not a composition factor of $rP/\mathrm{soc}\,P$ for any projective Λ-module, then S is a quotient of an injective module.*

(b) *If S is not injective and not a composition factor of $rI/\mathrm{soc}\,I$ for any injective Λ-module I, then S is a submodule of a projective module.* □

While we have given examples of artin algebras which have almost split sequences with projective or injective middle term, it is worth noting that there are artin algebras which do not have such almost split sequences. Suppose Λ is a local artin algebra with $r^2 \neq 0$ (for example, $\Lambda = k[X_1, \ldots, X_n]/(X_1, \ldots, X_n)^t$ with $t \geq 3$ and $n \geq 1$). Then Λ/r, the only simple Λ-module, is a composition factor of $r/\mathrm{soc}\,r$ since this is a nonzero module. Therefore, by Theorem 3.3, there is no almost split sequence with an injective middle term. Dually, there is no almost split sequence with a projective middle term.

By contrast, artin algebras Λ with $r^2 = 0$ which are not semisimple always have almost split sequences whose middle terms are either projective or injective. This follows directly from the following.

Proposition 3.5 *Let Λ be an artin algebra with $r^2 = 0$ and let $0 \to A \xrightarrow{f} B \xrightarrow{g} C \to 0$ be an almost split sequence.*

(a) *The module B is injective if and only if C is a nonprojective simple module. Moreover, if B is injective, then $f: A \to B$ is an injective envelope.*

(b) *The module B is projective if and only if A is a noninjective simple module. Moreover, if B is projective, then $g: B \to C$ is a projective cover.*

Proof (a) Let Λ be an artin algebra with $\mathfrak{r}^2 = 0$. We then have that $\mathfrak{r}/\operatorname{soc}\mathfrak{r} = 0$, so no simple module is a composition factor of $\mathfrak{r}/\operatorname{soc}\mathfrak{r}$. Hence (a) follows from Theorem 3.3.

Statement (b) follows from (a) by duality. □

When $0 \to S \to P \to \operatorname{Tr} DS \to 0$ is an almost split sequence with P projective we can deduce some extra information about the module $\operatorname{Tr} DS$, in addition to the information on generators and relations given by having a minimal projective presentation.

A nonprojective Λ-module C is said to be **almost projective** if whenever $g: P \to C$ is a projective cover, the induced morphism $g^{-1}(X) \to X$ is a split epimorphism for all proper submodules X of C. We have the following.

Proposition 3.6 *Let $g: P \to C$ be a projective cover which is minimal right almost split. Then we have the following.*

(a) *If A is indecomposable with $\operatorname{Ext}^1_\Lambda(C, A) \neq 0$, then A is simple and isomorphic to $D \operatorname{Tr} C$.*

(b) *C is almost projective.*

Proof (a) If A is indecomposable with $\operatorname{Ext}^1_\Lambda(C, A) \neq 0$, we have a nonsplit exact sequence $0 \to A \to E \to C \to 0$, and hence an exact commutative diagram

$$
\begin{array}{ccccccccc}
0 & \to & A & \to & E & \xrightarrow{f} & C & \to & 0 \\
  &   & \downarrow u & & \downarrow s & & \parallel & & \\
0 & \to & D \operatorname{Tr} C & \to & P & \xrightarrow{g} & C & \to & 0 \\
  &   & \downarrow v & & \downarrow t & & \parallel & & \\
0 & \to & A & \to & E & \xrightarrow{f} & C & \to & 0.
\end{array}
$$

Since A is indecomposable, $f: E \to C$ is right minimal. Hence $ts: E \to E$

is an isomorphism, and consequently $u: A \to D \operatorname{Tr} C$ is an isomorphism. In addition we know from Theorem 3.3 that $D \operatorname{Tr} C$ is simple.

(b) This is a direct consequence of the definitions.   □

4 Almost split sequences for group algebras

While the proof of the existence of almost split sequences for arbitrary artin algebras given in Section 1 is fairly constructive, it is sometimes possible to give other ways of constructing almost split sequences which are particularly well suited to special types of artin algebras. This section is devoted to illustrating this point in the case of group algebras. We give an application to integral representation rings.

Unless stated to the contrary, we assume throughout this section that k is a field of characteristic $p > 0$ and G is a finite group whose order is divisible by p. Then kG is not a semisimple k-algebra, so the theory of almost split sequences developed so far applies to mod kG. In particular the trivial kG-module k has an almost split sequence $0 \to A \to E \xrightarrow{f} k \to 0$. Then for each indecomposable module X we obtain the exact sequence of kG-modules $0 \to X \otimes_k A \to X \otimes_k E \xrightarrow{X \otimes_k f} X \to 0$. Our main aim is to show that the epimorphisms $X \otimes_k f : X \otimes_k E \to X$ are either split epimorphisms or right almost split morphisms and to determine for which X they are right almost split morphisms.

We know that $A \simeq D \operatorname{Tr}(k)$ in the almost split sequence $0 \to A \to E \to k \to 0$. Since kG is a symmetric algebra we know that $D \operatorname{Tr}(k) \simeq \Omega^2(k)$, the second syzygy of k, by IV Proposition 3.8. We now show that $X \otimes_k \Omega^2(k) \simeq \Omega^2(X) \coprod Q$ for some projective kG-module Q.

Lemma 4.1 *Let*

$$\cdots \to P_i \xrightarrow{d_i} P_{i-1} \to \cdots \to P_1 \xrightarrow{d_1} P_0 \xrightarrow{d_0} k \to 0$$

be a minimal projective resolution of k. Then for each kG-module X we have the following.

(a) *The exact sequence*

$$\cdots \to X \otimes_k P_i \to \cdots \to X \otimes_k P_1 \to X \otimes_k P_0 \to X \to 0$$

is a (not necessarily minimal) projective resolution of X.

(b) $X \otimes_k \Omega^i(k) \xrightarrow{\sim} \Omega^i(X) \coprod Q_i$ *with Q_i a projective module for all $i > 0$.*

Proof By III Proposition 3.1 the $X \otimes_k P_i$ are projective kG-modules since the P_i are projective kG-modules. Parts (a) and (b) follow trivially from this fact. □

Thus we see that if X is an indecomposable nonprojective module, the exact sequence $0 \to X \otimes_k \Omega^2(k) \to X \otimes_k E \overset{X \otimes_k f}{\to} X \to 0$ is isomorphic to $0 \to \Omega^2(X) \coprod Q \to X \otimes_k E \overset{X \otimes_k f}{\to} X \to 0$ with Q projective. Since kG is a symmetric algebra, Q is also injective. Hence the exact sequence $0 \to X \otimes_k \Omega^2(k) \to X \otimes_k E \overset{X \otimes_k f}{\to} X \to 0$ can be written as the sum of exact sequences $0 \to \Omega^2(X) \coprod Q \overset{g \coprod 1}{\to} (X \otimes_k E)_0 \coprod Q \overset{((X \otimes_k f)_0, 0)}{\to} X \to 0$. Consequently $X \otimes_k f$ is a split epimorphism or a right almost split morphism according to whether $(X \otimes_k f)_0$ is a split epimorphism or a right almost split morphism. Also $X \otimes_k f$ is right almost split if and only if $0 \to \Omega^2(X) \to (X \otimes_k E)_0 \overset{(X \otimes_k f)_0}{\to} X \to 0$ is an almost split sequence. We use these observations and notation freely throughout this section.

Our proof that $X \otimes_k f : X \otimes_k E \to X$ is either a split epimorphism or a right almost split morphism is based on the following comparison with the morphism $\mathrm{Hom}_{kG}(X, X \otimes_k f) : \mathrm{Hom}_{kG}(X, X \otimes_k E) \to \mathrm{Hom}_{kG}(X, X)$.

Lemma 4.2 *Let* $0 \to \Omega^2(k) \to E \overset{f}{\to} k \to 0$ *be an almost split sequence. Then the following are equivalent for an indecomposable kG-module X.*

(a) $\mathrm{rad}\,\mathrm{End}_{kG}(X)$ *is contained in* $\mathrm{Im}\,\mathrm{Hom}_{kG}(X, X \otimes_k f)$.

(b) $X \otimes_k f : X \otimes_k E \to X$ *is either a split epimorphism or a right almost split morphism.*

Proof (a) $\Rightarrow$ (b) Since $\mathrm{End}_{kG}(X)$ is a local ring, $\mathrm{Im}\,\mathrm{Hom}_{kG}(X, X \otimes_k f) \supset \mathrm{rad}\,\mathrm{End}_{kG}(X)$ implies that $\mathrm{Im}\,\mathrm{Hom}_{kG}(X, X \otimes_k f)$ is $\mathrm{End}_{kG}(X)$ or $\mathrm{rad}\,\mathrm{End}_{kG}(X)$. If $\mathrm{Im}\,\mathrm{Hom}_{kG}(X, X \otimes_k f) = \mathrm{End}_{kG}(X)$, then $X \otimes_k f$ is a split epimorphism. Suppose now that $\mathrm{Im}\,\mathrm{Hom}_{kG}(X, X \otimes_k f) = \mathrm{rad}\,\mathrm{End}_{kG}(X)$. Then the exact sequence

$$(*) \quad 0 \to \Omega^2(X) \to (X \otimes_k E)_0 \overset{(X \otimes_k f)_0}{\to} X \to 0$$

also has the property that $\mathrm{Im}\,\mathrm{Hom}_{kG}(X, (X \otimes_k f)_0) = \mathrm{rad}\,\mathrm{End}_{kG}(X)$. This means that an endomorphism $X \to X$ factors through $(X \otimes_k f)_0$ if and only if it is not an isomorphism. Then it follows from Proposition 2.2 that the exact seqeunce $(*)$ is almost split since it is not split, $\Omega^2(X) \simeq D \,\mathrm{Tr}\, X$ and every endomorphism $X \to X$ which is not an automorphism factors through $(X \otimes_k f)_0$. Hence $(X \otimes_k f)_0$ is right almost split and therefore $X \otimes_k f$ is right almost split.

(b) $\Rightarrow$ (a) This is trivial. $\qquad\square$

In view of this lemma it is of interest to have a description of $\operatorname{Im} \operatorname{Hom}_{kG}(X, X \otimes_k f)$. Since our description is in terms of traces of linear transformations, we first give some notation and recall some basic facts we will need concerning traces of linear transformations.

Let $h: V \to V$ be a k-linear transformation. We denote by $tr(h)$ the **trace** of h which is the sum of the diagonal elements in any matrix representing h with respect to some basis of V. Recall that $tr(h) = 0$ if h is nilpotent and $tr(h) = \dim_k V \cdot 1$ if h is the identity.

Proposition 4.3 *Let* $0 \to \Omega^2(k) \to E \xrightarrow{f} k \to 0$ *be an almost split sequence and let* X *be a* kG-*module. Then the following are equivalent for an element* h *in* $\operatorname{End}_{kG}(X)$.

(a) $h \in \operatorname{Im}(\operatorname{Hom}_{kG}(X, X \otimes_k E) \xrightarrow{\operatorname{Hom}_{kG}(X,X\otimes_k f)} \operatorname{End}_{kG}(X))$.
(b) $tr(hg) = 0$ *for all* g *in* $\operatorname{End}_{kG}(X)$.

Before giving the proof of this result, we will deduce some consequences.

Corollary 4.4 *Let* $0 \to \Omega^2(k) \to E \to k \to 0$ *be an almost split sequence. For each* kG-*module* X *we have the following.*

(a) $\operatorname{Im}(\operatorname{Hom}_{kG}(X, X \otimes_k E) \to \operatorname{End}_{kG}(X))$ *contains* $\operatorname{rad} \operatorname{End}_{kG}(X)$.

(b) $X \otimes_k E \xrightarrow{X \otimes_k f} X$ *is a split epimorphism if and only if* $tr(h) = 0$ *for all* h *in* $\operatorname{End}_{kG}(X)$.

Proof (a) Suppose h is in $\operatorname{rad} \operatorname{End}_{kG}(X)$. Then hg is in $\operatorname{rad} \operatorname{End}_{kG}(X)$ for all g in $\operatorname{End}_{kG}(X)$. Therefore hg is nilpotent for all g in $\operatorname{End}_{kG}(X)$ and so $tr(hg) = 0$ for all g in $\operatorname{End}_{kG}(X)$. Hence h is in $\operatorname{Im}(\operatorname{Hom}_{kG}(X, X \otimes_k E) \to \operatorname{End}_{kG}(X))$ by Proposition 4.3.

(b) This is a trivial consequence of Proposition 4.3. $\qquad\square$

Combining Lemma 4.2 and Corollary 4.4 we have the following.

Theorem 4.5 *Let* $0 \to \Omega^2(k) \to E \xrightarrow{f} k \to 0$ *be an almost split sequence and* X *an indecomposable* kG-*module.*

(a) $X \otimes_k f : X \otimes_k E \to X$ *is a split epimorphism if and only if* $tr(h) = 0$
for all h in $\text{End}_{kG}(X)$. *Otherwise* $X \otimes_k f : X \otimes_k E \to X$ *is right almost split.*

(b) *If p does not divide* $\dim_k X$, *then* $X \otimes_k f : X \otimes_k E \to X$ *is right almost split.*

(c) *If k is algebraically closed, then* $X \otimes_k f : X \otimes_k E \to X$ *is right almost split if and only if p does not divide* $\dim_k X$.

Proof (a) This is an easy consequence of Lemma 4.2 and Corollary 4.4.

(b) If p does not divide $\dim_k X$, then $tr(1_X) \neq 0$ in k. So $X \otimes_k f : X \otimes_k E \to X$ is right almost split by (a).

(c) Suppose k is algebraically closed. Then the elements of the local ring $\text{End}_{kG}(X)$ can be written as $v1 + h$ with v in k, 1 the identity on X and h a nilpotent element. Here we use that when Λ is a local k-algebra, then $\Lambda/\mathfrak{r}$ is a division k-algebra and hence isomorphic to k, so that the composition $k \to \Lambda \to \Lambda/\mathfrak{r}$ is an isomorphism. We then get $tr(v1 + h) = v(tr(1))$. Hence it follows that $tr(v1 + h) \neq 0$ if and only if $v \neq 0$ and p does not divide $\dim_k X$. Therefore, by part (a), we have that $X \otimes_k f : X \otimes_k E \to X$ is right almost split if and only if p does not divide $\dim_k X$. $\square$

The following is an immediate consequence of this theorem.

Corollary 4.6 *Let X be an indecomposable projective kG-module. Then we have the following.*

(a) $tr(h) = 0$ *for all h in* $\text{End}_{kG}(X)$.

(b) *p divides* $\dim_k X$. $\square$

Theorem 4.5 can be formulated in terms of almost split sequences as follows.

Theorem 4.7 *Let $0 \to \Omega^2(k) \to E \xrightarrow{f} k \to 0$ be an almost split sequence and assume that X is an indecomposable nonprojective kG-module. Then the exact sequence $0 \to \Omega^2(X) \to (X \otimes_k E)_0 \xrightarrow{(X \otimes_k f)_0} X \to 0$ has the following properties.*

(a) *It is either split or almost split.*

(b) *It is split if and only if $tr(h) = 0$ for all h in* $\text{End}_{kG}(X)$.

(c) *It is almost split if p does not divide* $\dim_k X$.

(d) *Suppose k is algebraically closed. Then the sequence is almost split if and only if p does not divide* $\dim_k X$. $\square$

We now finish the proof of Theorem 4.5 by giving a proof of Proposition 4.3. Our proof of Proposition 4.3 is based on the following generalities.

Lemma 4.8 *For B and C in* $\mathrm{mod}\,kG$ *we have the following.*

(a) *The map of vector spaces* $\eta : D(B) \to \mathrm{Hom}_k(B \otimes_k C, C)$ *given by* $\eta(h)(b \otimes c) = h(b)c$ *for all h in* $D(B) = \mathrm{Hom}_k(B, k)$ *and b in B and c in C is a kG-module morphism.*

(b) *The vector space isomorphism* $\delta : B \otimes_k C \to \mathrm{Hom}_k(D(B), C)$ *given by* $\delta(b \otimes c)(h) = h(b)c$ *for all h in* $D(B)$*, for all b in B and c in C is a kG-isomorphism functorial in B and C.*

Proof This is left to the reader to check. □

As a consequence of this lemma we obtain the following result.

Proposition 4.9 *Let A, B and C be in* $\mathrm{mod}\,kG$*. Then we have the following.*

$$\beta : \mathrm{Hom}_{kG}(A, B \otimes_k C) \to \mathrm{Hom}_{kG}(D(B) \otimes_k A, C)$$

given by $\beta(f)(h \otimes a) = \eta(h)(f(a))$ *for all f in* $\mathrm{Hom}_{kG}(A, B \otimes_k C)$*, for all a in A and h in* $D(B)$*, where* $\eta(h)(b \otimes c) = h(b)c$ *for all b in B and c in C, is an isomorphism functorial in A, B and C.*

Proof By Lemma 4.8 we know that $\delta : B \otimes_k C \to \mathrm{Hom}_k(D(B), C)$ given by $\delta(b \otimes c)(h) = h(b)c$ for all b in B and c in C as well as all h in $D(B)$ is a kG-isomorphism functorial in B and C. This induces the isomorphism

$$\mathrm{Hom}_{kG}(A, \delta) : \mathrm{Hom}_{kG}(A, B \otimes_k C) \to \mathrm{Hom}_{kG}(A, \mathrm{Hom}_k(D(B), C))$$

functorial in A, B and C. By III Proposition 3.4, we have the canonical isomorphism $\alpha : \mathrm{Hom}_{kG}(A, \mathrm{Hom}_k(D(B), C)) \to \mathrm{Hom}_{kG}(D(B) \otimes_k A, C)$ which is also functorial in A, B and C. It is not difficult to check that the composition $\alpha \, \mathrm{Hom}_{kG}(A, \delta)$ is our morphism $\beta : \mathrm{Hom}_{kG}(A, B \otimes_k C) \to \mathrm{Hom}_{kG}(D(B) \otimes_k A, C)$. Therefore β is an isomorphism functorial in A, B and C. □

It is easily checked that the usual vector space isomorphism $\epsilon : D(A) \otimes_k A \to \mathrm{End}_k(A)$, given by $\epsilon(h \otimes a)(x) = h(x)a$ for all h in $D(A)$, for all a and x in A, is a kG-isomorphism whose inverse can be described as follows. Let $\{a_1, \ldots, a_d\}$ be a k-basis for A with dual basis $\{h_1, \ldots, h_d\}$. Then it is easily seen that the map $\mu : \mathrm{End}_k(A) \to D(A) \otimes_k A$ given by $\mu(g) = \sum_{i=1}^d (h_i \otimes g(a_i))$ for all g in $\mathrm{End}_k(A)$ is the inverse of

ϵ and therefore a kG-isomorphism. It is also not difficult to see that $\sum_{i=1}^{d} h_i(g(a_i)) = tr(g)$ for all g in $End_k(A)$.

We now use Proposition 4.9 with $A = B$ and $C = k$, the trivial kG-module. Then $\beta: Hom_{kG}(A, A) \to Hom_{kG}(D(A) \otimes_k A, k)$ given by $\beta(f)(h \otimes a) = h(f(a))$ for all h in $D(A)$ and a in A is an isomorphism. From these remarks it follows that the isomorphism $v: End_{kG}(A) \to Hom_{kG}(End_k(A), k)$ which is the composition

$$Hom_{kG}(A, A) \xrightarrow{\beta} Hom_{kG}(D(A) \otimes_k A, k) \xrightarrow{Hom_{kG}(\mu, k)} Hom_{kG}(End_k(A), k)$$

is given by $v(f)(g) = tr(fg)$ for all f in $End_{kG}(A)$ and g in $End_k(A)$. We are particularly interested in the following property of the isomorphism v.

Lemma 4.10 *Let A be in* $mod\, kG$.

(a) *An element h in $End_{kG}(A)$ has the property that $v(h): End_k(A) \to k$ is a split kG-epimorphism if and only if there is some $g \in End_{kG}(A)$ such that $tr(hg) \neq 0$.*

(b) *The trivial module k is a kG-summand of $End_k(A)$ if and only if there is some h in $End_{kG}(A)$ such that $tr(h) \neq 0$.*

Proof (a) Obviously $v(h): End_k(A) \to k$ is a split kG-epimorphism if and only if there is a kG-morphism $t: k \to End_k(A)$ such that $v(f)t \neq 0$. The isomorphism $Hom_{kG}(k, End_k(A)) \to End_k(A)^G = End_{kG}(A)$ given by $t \mapsto t(1)$ shows that there is a kG-morphism $t: k \to End_k(A)$ such that $v(h)t \neq 0$ if and only if there is some g in $End_{kG}(A)$ such that $v(h)(g) \neq 0$. Using the fact that $v(h)(g) = tr(hg)$, we have our desired result.

(b) This is a trivial consequence of (a) and the fact that $v: End_{kG}(A) \to Hom_{kG}(End_k(A), k)$ is an isomorphism. $\square$

We are now ready to prove Proposition 4.3. Suppose that k, the trivial kG-module, is not projective (i.e. kG is not semisimple) and let $0 \to \Omega^2(k) \to E \xrightarrow{f} k \to 0$ be an almost split sequence ending in k. Let X be a kG-module. Then we have the following commutative exact diagram.

$$
\begin{array}{ccc}
Hom_{kG}(X, X \otimes_k E) & \xrightarrow{Hom_{kG}(X, X \otimes_k f)} & Hom_{kG}(X, X) \\
\downarrow \wr & & \downarrow \wr \\
Hom_{kG}(D(X) \otimes_k X, E) & \xrightarrow{Hom_{kG}(D(X) \otimes_k X, f)} & Hom_{kG}(D(X) \otimes_k X, k) \\
\downarrow \wr & & \downarrow \wr \\
Hom_{kG}(Hom_k(X, X), E) & \xrightarrow{Hom_{kG}(Hom_k(X, X), f)} & Hom_{kG}(Hom_k(X, X), k)
\end{array}
$$

Since $f: E \to k$ is right almost split, we have that $\mathrm{Im}(\mathrm{Hom}_{kG}(\mathrm{Hom}_k(X,X), E) \to \mathrm{Hom}_{kG}(\mathrm{Hom}_k(X,X), k))$ consists of the morphisms $g: \mathrm{Hom}_k(X,X) \to k$ which are not split epimorphisms. Then by Lemma 4.10 (a) we have that

$$\mathrm{Im}(\mathrm{Hom}_{kG}(X, X \otimes_k E) \overset{\mathrm{Hom}_{kG}(X, X \otimes f)}{\to} \mathrm{Hom}_{kG}(X,X))$$

consists of the h in $\mathrm{Hom}_{kG}(X,X)$ such that $tr(hg) = 0$ for all g in $\mathrm{Hom}_{kG}(X,X)$. Therefore the proof of Proposition 4.3, and hence the proof of Theorem 4.5, is now complete. □

The rest of this section is devoted to showing how the results about almost split sequences over group algebras can be used to give a description of the integral representation ring of kG where k is a field of characteristic $p > 0$ and $G \simeq \mathbb{Z}/p\mathbb{Z}$. But first we define what is meant by the integral representation ring of a group algebra kG.

For each A in $\mathrm{mod}\, kG$ we denote the isomorphism class of A in $\mathrm{mod}\, kG$ by $[A]$. Then the collection of all $[A]$ with A in $\mathrm{mod}\, kG$ is a set. Let $\mathrm{Rep}(kG)$ be the free abelian group with basis elements $[A]$ modulo the subgroup generated by $[A] + [B] - [A \coprod B]$ for all A and B in $\mathrm{mod}\, kG$. We denote the image of $[A]$ in $\mathrm{Rep}(kG)$ also by $[A]$. By the Krull–Schmidt theorem we know that $\mathrm{Rep}(kG)$ is a free abelian group with basis elements $[A]$ where A runs through all indecomposable kG-modules. We now make $\mathrm{Rep}(kG)$ a commutative ring by defining $[A] \cdot [B] = [A \otimes_k B]$. The fact that this multiplication is well defined is a consequence of the isomorphism $A \otimes_k (B \coprod C) \simeq (A \otimes_k B) \coprod (A \otimes_k C)$. The unit element is $[k]$ where k is the trivial kG-module. It is commutative because $A \otimes_k B \simeq B \otimes_k A$. It is left to the reader to finish the proof that this multiplication makes $\mathrm{Rep}(kG)$ a commutative ring which is a free $\mathbb{Z}$-algebra.

Suppose now that k is a field of characteristic $p > 0$ and G is a finite cyclic group of order p. Then $kG \simeq k[X]/(X^p - 1)$ where X is an indeterminate. Since char $k = p$, we have that $(X^p - 1) = (X - 1)^p$ so $kG \simeq k[X]/(X - 1)^p$ which is a commutative Nakayama local ring. Let $A_i = k[X]/(X - 1)^i$ for $i = 1, \ldots, p$. Then $\{A_1, A_2, \ldots, A_p\}$ is a complete set of nonisomorphic indecomposable modules and $\dim_k A_i = i$. We have $A_1 \simeq k$, the trivial module, and $A_p \simeq kG$, the unique indecomposable projective kG-module. We also know all the almost split sequences. They are

$$0 \to k \to A_2 \to k \to 0,$$
$$0 \to A_i \to A_{i+1} \coprod A_{i-1} \to A_i \to 0$$

for $p > i > 1$.

Since p does not divide $\dim_k A_i$ for $1 \leq i < p$, we have by Theorem 4.7 that $0 \rightarrow A_i \rightarrow A_2 \otimes_k A_i \rightarrow A_i \rightarrow 0$ is an almost split sequence for $1 \leq i < p$. Therefore we have that $A_2 \otimes_k A_i \simeq A_{i+1} \coprod A_{i-1}$ when $p > i \geq 2$. We also know that $A_2 \otimes_k A_p \simeq 2A_p$ since it is a free kG-module of dimension $2p$, and finally $A_2 \otimes_k A_1 \simeq A_2$. So we know all the products $[A_2] \cdot [A_j]$ for $1 \leq j \leq p$.

Now we have the surjective $\mathbb{Z}$-algebra morphism $\xi : \mathbb{Z}[X_2, \ldots, X_p] \rightarrow$ Rep(kG) given by $\xi(X_i) = [A_i]$ for $i = 2, \ldots, p$. Of course we have $\xi(1) = [A_1]$. By our calculations above we know that the ideal $\mathfrak{a}$ generated by the elements $X_2^2 - X_3 - 1, X_2 X_3 - X_4 - X_2, \ldots, X_2 X_{p-1} - X_p - X_{p-2}$ and $X_2 X_p - 2X_p$ is in Ker ξ. We now show that $\mathfrak{a} = $ Ker ξ.

We begin by showing that $1, x_2, \ldots, x_p$ generate $\mathbb{Z}[X_2, \ldots, X_p]/\mathfrak{a}$ as an abelian group where $x_i = X_i + \mathfrak{a}$. To do this, it suffices to show that every monomial is a linear combination of the x_u and 1. But this will obviously be the case if we show that all monomials $x_i x_j$ are linear combinations of the x_u and 1. It is clear that all monomials $x_2 x_j$ are linear combinations of x_u and 1. Suppose t is an integer satisfying $2 < t \leq p$ and that we have shown that all monomials $x_i x_j$ are linear combinations of 1 and the x_u when $2 \leq i < t$ and j satisfies $2 \leq j \leq p$. We now want to show that all the monomials of the form $x_t x_j$ with $2 \leq j \leq p$ are linear combinations of the x_u and 1.

Now we have that $x_2 x_{t-1} = x_t + x_{t-2}$. So we have $x_2 x_{t-1} x_j - x_{t-2} x_j = x_t x_j$. But by the inductive hypothesis $x_{t-1} x_j$ and $x_{t-2} x_j$ are linear combinations of the x_u and 1. Therefore $x_2(x_{t-1} x_j)$ is also a linear combination of the x_u and 1. Hence the $x_t x_j$ are a linear combination of the x_u and 1 for all $2 \leq j \leq p$. This finishes the proof that the x_u and 1 generate the abelian group $\mathbb{Z}[X_2, \ldots, X_p]/\mathfrak{a}$.

We are now ready to give our description of the $\mathbb{Z}$-algebra Rep(kG).

Proposition 4.11 *The $\mathbb{Z}$-algebra surjection $\xi : \mathbb{Z}[X_2, \ldots, X_p] \rightarrow$ Rep(kG) given by $\xi(X_i) = [A_i]$ induces an isomorphism $\mathbb{Z}[X_2, \ldots, X_p]/\mathfrak{a} \xrightarrow{\sim} $ Rep(kG) where $\mathfrak{a}$ is the ideal $(X_2^2 - X_3 - 1, X_2 X_3 - X_4 - X_2, \ldots, X_2 X_{p-1} - X_p - X_{p-2}, X_2 X_p - 2X_p)$.*

Proof We have already seen that the x_u and 1 generate the abelian group $\mathbb{Z}[X_2, \ldots, X_p]/\mathfrak{a}$. Since the $[A_0], [A_1], \ldots, [A_p]$ are linearly independent in Rep(kG), it follows that the x_u together with 1 are $\mathbb{Z}$-linearly independent in $\mathbb{Z}[X_2, \ldots, X_p]/\mathfrak{a}$. Thus the set $\{1, x_2, \ldots, x_p\}$ is a basis of the abelian group $\mathbb{Z}[X_2, \ldots, X_p]/\mathfrak{a}$. Therefore the induced map $\mathbb{Z}[X_2, \ldots, X_p]/\mathfrak{a} \rightarrow$ Rep(kG) is an isomorphism of $\mathbb{Z}$-algebras. $\square$

5 Irreducible morphisms

Let $f: A \to B$ be a minimal left almost split morphism and $g: B \to C$ a minimal right almost split morphism. Associated with a decomposition $B = \coprod B_i$, are the morphisms $p_i f: A \to B_i$ and $g j_i: B_i \to C$, where p_i is the ith projection and j_i is the ith embedding. We show that these morphisms are not arbitrary, but are irreducible in a sense that we now define. This section is devoted to a study of irreducible morphisms and their basic properties.

A morphism $g: B \to C$ in mod Λ is called **irreducible** if g is neither a split monomorphism nor a split epimorphism, and if $g = ts$ for some $s: B \to X$ and $t: X \to C$, then s is a split monomorphism or t is a split epimorphism. It is easily seen that a morphism $g: B \to C$ is irreducible if and only if $D(g): D(C) \to D(B)$ is irreducible.

Before giving the connection with almost split sequences, we make some simple but important observations about irreducible morphisms.

Lemma 5.1

(a) *If $g: B \to C$ is an irreducible morphism in mod Λ, then g is either a monomorphism or an epimorphism.*

(b) *If $g: B \to C$ is an irreducible monomorphism, then B is a summand of all proper submodules of C containing B.*

(c) *If $g: B \to C$ is an irreducible epimorphism, then C is a summand of B/I for all submodules I of B such that $0 \neq I \subset \operatorname{Ker} g$.*

Proof (a) Let $g: B \to C$ be an irreducible morphism. Consider the factorization $B \xrightarrow{s} B/\operatorname{Ker} g \xrightarrow{t} C$ of g. Then either s is a split monomorphism, or t must be a split epimorphism. But s being a split monomorphism implies that g is a monomorphism and t being a split epimorphism implies that g is an epimorphism. This completes the proof of (a).

The rest of the proof is left as an exercise. $\square$

Another basic property of irreducible morphisms is that they are minimal.

Lemma 5.2 *If $g: B \to C$ is an irreducible morphism, then g is both a right minimal and a left minimal morphism.*

Proof We only prove that g is a right minimal morphism. Assume $h: B \to B$ is such that $g = gh$. Since g is not a split epimorphism, h is a split monomorphism. But then h is an isomorphism since B is of finite length, and hence $g: B \to C$ is right minimal. □

We now give the precise connection between irreducible morphisms and right and left almost split morphisms.

Theorem 5.3

(a) *Let C be an indecomposable module. Then a morphism $g: B \to C$ is irreducible if and only if there exists some morphism $g': B' \to C$ such that the induced morphism $(g, g'): B \coprod B' \to C$ is a minimal right almost split morphism.*

(b) *Let A be an indecomposable module. Then a morphism $g: A \to B$ is irreducible if and only if there exists some morphism $g': A \to B'$ such that the induced morphism $\left(\begin{smallmatrix} g \\ g' \end{smallmatrix} \right): A \to B \coprod B'$ is a minimal left almost split morphism.*

Proof We only prove (a) since (b) follows from (a) by duality. Assume first that $g: B \to C$ is irreducible and let $h: E \to C$ be the minimal right almost split morphism which exists by Corollary 1.17. Since g is not a split epimorphism, g factors through h, i.e. $g = hf$ for some $f: B \to E$. Since h is not a split epimorphism, f is a split monomorphism, and therefore $E \simeq B \coprod B'$ where $B' = \operatorname{Coker} f$. Considering this isomorphism as an identification we have that $h': B' \to C$, where $h' = h|_{B'}$, is such that $(g, h'): B \coprod B' \to C$ is a minimal right almost split morphism.

Next assume that $h: E \to C$ is a minimal right almost split morphism. Let $E = B \coprod B'$ and let $g: B \to C$, with $g = h|_B$. Assume that $g = st$ for some morphisms $t: B \to X$ and $s: X \to C$ with s not a split epimorphism. Then, since h is a right almost split morphism, there exists $\left(\begin{smallmatrix} u \\ v \end{smallmatrix} \right): X \to B \coprod B'$ with $s = (g, g') \left(\begin{smallmatrix} u \\ v \end{smallmatrix} \right)$ where $g' = h|_{B'}$. We then obtain the following commutative diagram

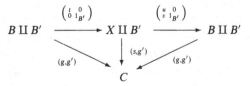

Since $h = (g, g')$ is right minimal, $\left(\begin{smallmatrix} ut & 0 \\ vt & 1_{B'} \end{smallmatrix} \right)$ is an isomorphism. Hence $ut: B \to B$ is an isomorphism, showing that t is a split monomorphism. This shows that g is an irreducible morphism. □

Because in an almost split sequence $0 \to A \overset{f}{\to} B \overset{g}{\to} C \to 0$ we have that f is minimal left almost split and g is minimal right almost split, Theorem 5.3 also gives a connection between irreducible morphisms and almost split sequences. We then get the following examples of irreducible morphisms by reinterpreting some results from Section 3.

Example Assume that $r^2 = 0$ and let S be a simple noninjective module and $C = \operatorname{Tr} DS$. Then we know from Section 3 that in the almost split sequence $0 \to S \to B \overset{g}{\to} C \to 0$ the morphism $g: B \to C$ is a projective cover. It is now not hard to show that a morphism $h: X \to C$ which is not a split epimorphism is irreducible if and only if the induced morphism $X/rX \to C/rC$ is a monomorphism.

An important consequence of Theorem 5.3 is that we can get information about minimal left almost split morphisms from knowing the minimal right almost split morphisms and conversely. In particular this can be used to study the structure of almost split sequences. We illustrate this point by giving some information about the middle term in an almost split sequence.

Proposition 5.4 *An indecomposable module X is not a summand of the middle term of any almost split sequence if and only if X is a simple module which is not a composition factor of $r/\operatorname{soc} r$.*

Proof Assume first that X is not a summand of the middle term of any almost split sequence, and let $g: B \to X$ be minimal right almost split. If B' is an indecomposable summand of B, then $g' = g|_{B'}: B' \to X$ is irreducible by Theorem 5.3, and hence again by Theorem 5.3 there is a minimal left almost split morphism $h: B' \to X \coprod Y$ for some Y. If B' is not injective, there is an almost split sequence $0 \to B' \to X \coprod Y \to \operatorname{Coker} h \to 0$ contradicting the assumption on X. Hence B' is injective and therefore also B is injective. If X is not projective, we know by Theorem 3.3 that X must be a simple module which is not a composition factor of $r/\operatorname{soc} r$. If X is projective, then $B = rX$, which implies that B must be 0 since B is injective. Hence X is simple projective and consequently not a composition factor of $r/\operatorname{soc} r$ in this case either.

Assume conversely that X is a simple module which is not a composition factor of $r/\operatorname{soc} r$. If X is projective there is no irreducible morphism to X, hence X is not a summand of the middle term of any almost split sequence. If X is not projective, then in the almost split sequence

$0 \to Z \to Y \to X \to 0$ we have that Y is injective by Theorem 3.3. Hence if $h: A \to X$ is irreducible with A an indecomposable module, then A is a summand of Y by Theorem 5.3, and consequently A is injective. This shows that X is not a summand of the middle term of any almost split sequence. □

Using Proposition 5.4 one sees immediately that for an artin algebra Λ with $\mathfrak{r}^2 = 0$, the middle term of any almost split sequence does not contain a semisimple summand.

Our next application is to give the structure of the almost split sequences whose middle terms have a nonzero projective injective summand. This result is of a similar flavor to the structure of the almost split sequences with projective or injective middle term. In addition to illustrating how to use Theorem 5.3 this result will also be applied later.

Proposition 5.5

(a) *Let $\delta: 0 \to A \to B \to C \to 0$ be an almost split sequence. If B has an indecomposable projective injective summand P, then $l(P) \geq 2$ and δ is isomorphic to the sequence*

$$\epsilon: 0 \to \mathfrak{r}P \xrightarrow{\binom{-i}{p}} P \amalg \mathfrak{r}P/\operatorname{soc} P \xrightarrow{(q,j)} P/\operatorname{soc} P \to 0,$$

where $i: \mathfrak{r}P \to P$ and $j: \mathfrak{r}P/\operatorname{soc} P \to P/\operatorname{soc} P$ are the natural inclusion morphisms and $p: \mathfrak{r}P \to \mathfrak{r}P/\operatorname{soc} P$ and $q: P \to P/\operatorname{soc} P$ are the natural quotient morphisms.

(b) *If P is indecomposable projective injective with $l(P) \geq 2$, there is some almost split sequence $\delta: 0 \to A \to B \to C \to 0$ such that P is a summand of B.*

Proof (a) Assume $\delta: 0 \to A \to B \to C \to 0$ is an almost split sequence with $B = P \amalg B'$ and P an indecomposable projective injective module. It follows from Proposition 5.4 that $l(P) \geq 2$. We have a minimal right almost split morphism $i: \mathfrak{r}P \to P$ and a minimal left almost split morphism $q: P \to P/\operatorname{soc} P$ and $\mathfrak{r}P$ and $P/\operatorname{soc} P$ are indecomposable. It then follows from Theorem 5.3 that $A \simeq \mathfrak{r}P$ and $C \simeq P/\operatorname{soc} P$ and that δ must be of the form $0 \to \mathfrak{r}P \xrightarrow{\binom{i}{f}} P \amalg X \xrightarrow{(q,g)} P/\operatorname{soc} P \to 0$ with $f: \mathfrak{r}P \to X$ and $g: X \to P/\operatorname{soc} P$. Then we have $l(X) = l(\mathfrak{r}P) + l(P/\operatorname{soc} P) - l(P) = l(P) - 2 = l(\mathfrak{r}P) - 1 = l(P/\operatorname{soc} P) - 1$, so that f is an epimorphism and g is a monomorphism. Since $P/\operatorname{soc} P$ has a unique maximal submodule $\mathfrak{r}P/\operatorname{soc}P$, we must have $g(X) = \mathfrak{r}P/\operatorname{soc}P$, so g can be chosen to be the natural inclusion $j:\mathfrak{r}P/\operatorname{soc}P \to P/\operatorname{soc}P$. Hence $(q,j):P\coprod \mathfrak{r}P/\operatorname{soc}P \to P/\operatorname{soc}P$ is right almost split. Since ϵ is an exact sequence and $\mathfrak{r}P$ is indecomposable, it must be almost split.

(b) If P is indecomposable projective injective with $l(P) \geq 2$, we have an irreducible morphism $P \to P/\operatorname{soc}P$. Hence P is a summand of B in the almost split sequence $0 \to A \to B \to P/\operatorname{soc}P \to 0$. $\square$

In addition to the connection with almost split sequences, irreducible morphisms have other interesting properties and give rise to new classes of indecomposable modules.

Proposition 5.6 *Let* $\delta:0 \to A \xrightarrow{f} B \xrightarrow{g} C \to 0$ *be an exact sequence.*

(a) *f is an irreducible morphism if and only if δ is not a split exact sequence and for any morphism $h:X \to C$ there is either a morphism $t:X \to B$ with $h = gt$ or a morphism $s:B \to X$ with $g = hs$.*
(b) *If f is an irreducible morphism, then $\operatorname{End}_\Lambda(C)$ is a local ring, and hence C is an indecomposable module.*

Proof (a) Assume first that f is an irreducible morphism. Then δ does not split. Let $h:X \to C$ be an arbitrary morphism. Consider the following exact commutative diagram.

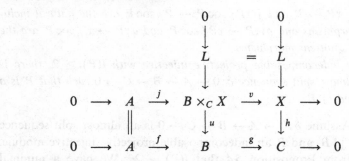

Since f is irreducible, either j is a split monomorphism or u is a split epimorphism. In the first case there exists a morphism $t:X \to B$ such that $h = gt$ and in the second case there exists a morphism $s:B \to X$ such that $g = hs$.

Assume now conversely that δ does not split and that each morphism $h:X \to C$ satisfies the property that there exists either a morphism $t:X \to B$ with $h = gt$ or a morphism $s:B \to X$ with $hs = g$. Since δ does not split, f is not a split monomorphism. Assume that $f = vu$

with $u: A \to Y$ and $v: Y \to B$. Since f is a monomorphism, u is a monomorphism. Consider the induced exact commutative diagram

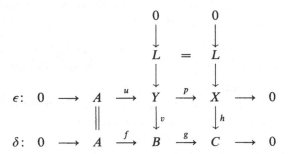

with $X = \operatorname{Coker} u$. From our assumption there is either a morphism $t: X \to B$ such that $gt = h$ or a morphism $s: B \to X$ with $g = hs$. In the first case the sequence ϵ splits and hence u is a split monomorphism. In the second case, since h is then an epimorphism, v is also an epimorphism. Then from the exact commutative diagram

$$
\begin{array}{ccccccccc}
\mu: & 0 & \longrightarrow & L & \longrightarrow & Y & \overset{v}{\longrightarrow} & B & \longrightarrow & 0 \\
 & & & \| & & \downarrow{\scriptstyle p} & & \downarrow{\scriptstyle g} & & \\
\nu: & 0 & \longrightarrow & L & \longrightarrow & X & \overset{h}{\longrightarrow} & C & \longrightarrow & 0
\end{array}
$$

one sees that μ is a split exact sequence. Hence v is a split epimorphism, and consequently f is an irreducible morphism.

(b) Assume that f is an irreducible monomorphism. We first observe that since g is an epimorphism, g does not factor through any nonisomorphism $h: C \to C$. From (a) it then follows that $\operatorname{Im} \operatorname{Hom}_\Lambda(C, g)$ is the set of nonisomorphisms in $\operatorname{End}_\Lambda(C)$. Obviously $\operatorname{Im} \operatorname{Hom}_\Lambda(C, g)$ is a right $\operatorname{End}_\Lambda(C)$-ideal of $\operatorname{End}_\Lambda(C)$. Now let s be in $\operatorname{Im} \operatorname{Hom}_\Lambda(C, g)$. Since s is not an isomorphism, it is not a monomorphism. Therefore ts is not a monomorphism for each t in $\operatorname{End}_\Lambda(C)$ which means that ts is in $\operatorname{Im} \operatorname{Hom}_\Lambda(C, g)$ for each t in $\operatorname{End}_\Lambda(C)$. Hence $\operatorname{Im} \operatorname{Hom}_\Lambda(C, g)$, the set of nonisomorphisms in $\operatorname{End}_\Lambda(C)$, is an ideal in $\operatorname{End}_\Lambda(C)$, which shows that $\operatorname{End}_\Lambda(C)$ is a local ring. $\qquad\square$

We also state the dual result.

Proposition 5.7 *Let* $\delta: 0 \to A \overset{f}{\to} B \overset{g}{\to} C \to 0$ *be an exact sequence.*

(a) *g is an irreducible morphism if and only if δ is not a split exact se-*

quence and for any morphism $h: A \rightarrow X$ there is either a morphism
$t: B \rightarrow X$ such that $h = tf$ or a morphism $s: X \rightarrow B$ with $f = sh$.
(b) If g is an irreducible morphism, then $\text{End}_\Lambda(A)$ is a local ring, and
hence A is an indecomposable module. $\square$

Clearly any indecomposable nonprojective module C is of the form
$\text{Coker} f$ where $f: A \rightarrow B$ is an irreducible monomorphism, as is seen
by considering the almost split sequence $0 \rightarrow A \xrightarrow{f} B \xrightarrow{g} C \rightarrow 0$. If
however we require B to be indecomposable, then Proposition 5.6 imposes
restrictions on the modules occurring as $\text{Coker} f$.

Corollary 5.8

(a) *If $f: A \rightarrow B$ is an irreducible monomorphism with B indecomposable,
then each irreducible morphism $h: X \rightarrow \text{Coker} f$ is an epimorphism.*
(b) *If $g: B \rightarrow C$ is an irreducible epimorphism with B indecomposable,
then each irreducible morphism $h: \text{Ker} g \rightarrow Y$ is a monomorphism.*

Proof (a) Let $f: A \rightarrow B$ be an irreducible monomorphism with B indecomposable. Assume that $h: X \rightarrow \text{Coker} f$ is an irreducible monomorphism. By Proposition 5.6 there is then some $t: X \rightarrow B$ with $gt = h$
where $g: B \rightarrow \text{Coker} f$ is the natural morphism. Since g is not a split
epimorphism, t must be a split monomorphism, and hence an isomor-
phism since B is assumed to be indecomposable. This contradicts the
fact that g is a proper epimorphism. Hence each irreducible morphism
$h: X \rightarrow \text{Coker} f$ is an epimorphism.
 (b) This follows by duality. $\square$

We now know that if C is an indecomposable nonprojective module
such that there is an irreducible monomorphism $h: X \rightarrow C$, then C
is not of the form $\text{Coker} f$ where $f: A \rightarrow B$ is irreducible and B is
indecomposable. This is for example the case for $C = P / \text{soc} P$ when P
is indecomposable projective injective of length at least 3, since we then
have an irreducible monomorphism $\mathfrak{r}P / \text{soc} P \rightarrow P / \text{soc} P$.

We now characterize almost split sequences in terms of irreducible
morphisms.

Proposition 5.9 *Let $\delta: 0 \rightarrow A \xrightarrow{f} B \xrightarrow{g} C \rightarrow 0$ be an exact sequence. Then
δ is an almost split sequence if and only if f and g are both irreducible.*

Proof It follows from Theorem 5.3 that if δ is an almost split sequence, then f and g are both irreducible.

Assume conversely that both f and g are irreducible. We then know by Propositions 5.6 and 5.7 that A and C are both indecomposable and noninjective and nonprojective respectively. Let $\epsilon: 0 \to D\operatorname{Tr} C \xrightarrow{f'} B' \xrightarrow{g'} C \to 0$ be the almost split sequence with right hand term C. We then get the following commuting diagram

$$
\begin{array}{ccccccccc}
\delta: & 0 & \longrightarrow & A & \xrightarrow{f} & B & \xrightarrow{g} & C & \longrightarrow & 0 \\
& & & \downarrow{\scriptstyle s} & & \downarrow{\scriptstyle t} & & \| & & \\
\epsilon: & 0 & \longrightarrow & D\operatorname{Tr} C & \xrightarrow{f'} & B' & \xrightarrow{g'} & C & \longrightarrow & 0
\end{array}
$$

Since g is irreducible and g' is not a split epimorphism, t is a split monomorphism, so there exists $h: B' \to B$ with $ht = 1_B$. Hence we get $f = htf = hf's$. Since f' is a minimal left almost split morphism, hf' is not a split epimorphism. Therefore s is a split monomorphism. But then s is an isomorphism since $D\operatorname{Tr} C$ is indecomposable. It follows that t is also an isomorphism proving that δ is isomorphic to ϵ. $\qquad\square$

6 The middle term of an almost split sequence

Whereas the end terms A and C of an almost split sequence $0 \to A \to B \to C \to 0$ are indecomposable, the middle term B will usually decompose. The number of indecomposable summands in a sum decomposition of B is an important invariant of C, which we denote by $\alpha(C)$, and it measures in a sense the complexity of morphisms to C. For any positive integer t there is an indecomposable module over some artin algebra Λ with $\alpha(C) = t$. For example let $C = \operatorname{Tr} DS$ where S is the unique simple $(k[X_1, \ldots, X_t]/(X_1, \ldots, X_t)^2)$-module, as studied in Section 3. However, for a given algebra there are strong limitations on which numbers can occur. We here give two preliminary results, one which shows that the number 1 always occurs for an arbitrary nonsemisimple algebra Λ, and one which shows that there is a bound on the $\alpha(C)$ for a given algebra Λ.

For the first main result in this section we shall be particularly interested in indecomposable nonprojective modules C having a minimal projective presentation $P_1 \xrightarrow{h} P_0 \to C \to 0$ with P_0 and P_1 indecomposable. It is easy to construct examples of such modules. For instance, if

P is indecomposable projective and B is a submodule of P with $B/\mathfrak{r}B$ simple, then $C = P/B$ has this property. Our first aim is to show that $\alpha(C) = 1$ if in addition to P_0 and P_1 being indecomposable, we also have $\operatorname{Im} h \not\subset \mathfrak{r}^2 P_0$. To see this the following result is useful.

Proposition 6.1 *Let* $h: P_1 \to P_0$ *be a nonzero morphism between inde-composable projective modules which is not an epimorphism, and let* $C = \operatorname{Coker} h$. *Let* $0 \to A \xrightarrow{f} \coprod_{i \in I} B_i \xrightarrow{g} C \to 0$ *be an almost split sequence where each* B_i *is indecomposable. Then we have the following.*

(a) *There is a unique* j *such that* B_j *has maximal length.*

(b) *The induced irreducible morphisms* $g_j: B_j \to C$ *and* $f_j: A \to B_j$ *are epimorphisms and monomorphisms respectively.*

(c) *The induced irreducible morphisms* $\coprod_{i \neq j} B_i \to C$ *and* $A \to \coprod_{i \neq j} B_i$ *are monomorphisms and epimorphisms respectively.*

Proof Let B_j be a summand of maximal length in the decomposition $B = \coprod_{i \in I} B_i$. Since P_0 is an indecomposable projective module, C has a unique maximal submodule. Therefore not all g_i can be monomor-phisms. Hence g_j must be an epimorphism since B_j has maximal length. $D \operatorname{Tr} C$ has a simple socle since it is contained in $D(P_1^*)$, which is an indecomposable injective module because P_1 is an indecomposable pro-jective module. This guarantees that not all the f_i can be epimorphisms. Again, since B_j is an indecomposable summand of B of maximal length, f_j has to be a monomorphism. Writing $B' = \coprod_{i \neq j} B_i$ and computing lengths we get $l(C) + l(A) = l(B_j) + l(B')$. But $l(B_j) > \max\{l(C), l(A)\}$ since g_j is an epimorphism and f_j is a monomorphism. Therefore we get $l(B') < \min\{l(C), l(A)\}$, which implies that the induced morphism $A \to B'$ is an epimorphism and the induced morphism $B' \to C$ is a monomorphism. This proves our claims. $\qquad\square$

The next result describes a class of indecomposable modules C such that there are no irreducible monomorphisms to C.

Lemma 6.2 *Let* M *be an indecomposable module and* N *a nonzero sub-module of* M *which is a summand of all maximal submodules* X *of* M *containing* N. *Then we have the following.*

(a) M/N *is indecomposable.*

(b) *There are no irreducible monomorphisms to* M/N *from any module.*

Proof (a) This is left as an exercise.

(b) Consider the exact sequence $0 \to N \overset{i}{\to} M \overset{p}{\to} M/N \to 0$, and assume $h: C \to M/N$ is an irreducible monomorphism. Then it follows from our assumptions that the induced exact sequence $0 \to N \overset{i}{\to} p^{-1}(\operatorname{Im} h) \to \operatorname{Im} h \to 0$ splits. Hence we get that $h = pg$ for some $g: C \to M$. Since M is indecomposable and $N \neq 0$, it follows that p is not a split epimorphism. Since $h: C \to M/N$ is irreducible, $g: C \to M$ must be a split monomorphism and consequently an isomorphism. This contradicts the fact that h is a proper monomorphism. □

We have seen in Section 5 that if $f: A \to B$ is an irreducible monomorphism and B is an indecomposable module, then any irreducible morphism to $\operatorname{Coker} f$ is an epimorphism. This result can be considered as a special case of Lemma 6.2 since by Lemma 5.1 we have that $f(A)$ is a summand in every maximal submodule of B containing $f(A)$. Here we shall apply Lemma 6.2 in another situation, as a step towards proving the existence of an indecomposable module C with $\alpha(C) = 1$.

Proposition 6.3 *Let* $P_1 \overset{h}{\to} P_0 \to C \to 0$ *be a minimal projective presentation for a nonzero module* C *with* P_0 *and* P_1 *indecomposable projective modules and* $\operatorname{Im} h \not\subset r^2 P_0$. *Then the almost split sequence* $\delta: 0 \to A \to B \to C \to 0$ *has the property that* B *is indecomposable.*

Proof Clearly $M = P_0/r \operatorname{Im} h$ is indecomposable with a unique maximal submodule $rM = rP_0/r \operatorname{Im} h$. Further, since $r \operatorname{Im} h \subset r^2 P_0$, we get that $r^2 M = r^2 P_0/r \operatorname{Im} h$. Let $N = \operatorname{Im} h/r \operatorname{Im} h$ which is a simple submodule of $rP_0/r \operatorname{Im} h$. However, since $\operatorname{Im} h \not\subset r^2 P_0$ we have that $N \not\subset r^2 M$. Therefore N is a summand of rM. Hence M and $C = M/N$ are indecomposable with N a summand of all maximal submodules of M containing N. It then follows from Lemma 6.2 that all irreducible morphisms $g: X \to C$ are epimorphisms. From Proposition 6.1 we have that the middle term B in the almost split sequence $\delta: 0 \to A \to B \to C \to 0$ has a decomposition $B = B' \coprod B''$ with B' indecomposable and such that if $B'' \neq 0$ the induced morphism $g'': B'' \to C$ is an irreducible monomorphism. But we have just seen this cannot happen. Hence we get $B = B'$ and B is therefore indecomposable. □

As a consequence of Proposition 6.3 we get the following.

Theorem 6.4 *For each non-semisimple artin algebra Λ there exists an almost split sequence $\delta : 0 \to A \to B \to C \to 0$ with B indecomposable.*

Proof Since Λ is not semisimple, there exists an indecomposable projective module P_0 with $\mathfrak{r}P_0 \neq 0$. Let $P' \to \mathfrak{r}P_0$ be a projective cover and let P_1 be an indecomposable summand of P' with induced morphism $h : P_1 \to P_0$. Then we get that $\operatorname{Im} h \not\subseteq \mathfrak{r}^2 P_0$, so for $C = \operatorname{Coker} h$ we have $\alpha(C) = 1$ by Proposition 6.3. This completes the proof of the theorem. $\square$

We shall in Chapter X give methods for computing new almost split sequences from old ones. In particular this will provide us with more examples of indecomposable modules C with $\alpha(C) = 1$ since the invariant α is often preserved by these computations. Also in Chapter XI we discuss other ways of obtaining almost split sequences with indecomposable middle term.

For a given artin algebra Λ we denote by $\alpha(\Lambda)$ the numerical invariant which is the supremum of the $\alpha(C)$ for C indecomposable and not projective. It is obvious that each $\alpha(C)$ is finite and our next aim is to show that $\alpha(\Lambda)$ is also finite. For this we need to compare the lengths of the modules C and $D \operatorname{Tr} C$, and of the modules A and B when there is an irreducible morphism between them. These preliminary results are also useful in other contexts, and will be applied in the next chapter.

Lemma 6.5 *For an artin algebra Λ let n be an integer such that $l(P) \leq n$ for each indecomposable projective left and projective right Λ-module P. Then we have $l(D \operatorname{Tr} C) \leq l(C)n^2$ for all nonzero modules C in $\operatorname{mod} \Lambda$ or $\operatorname{mod} \Lambda^{\mathrm{op}}$.*

Proof We only prove the claim for C in $\operatorname{mod} \Lambda$ since the rest follows by duality. Let C be a nonzero Λ-module and let $P_1 \to P_0 \to C \to 0$ be a minimal projective presentation of C. For each projective module Q in $\operatorname{mod} \Lambda$ or $\operatorname{mod} \Lambda^{\mathrm{op}}$, we have that $l(Q/\mathfrak{r}Q)$ gives the number of summands in a decomposition of Q into indecomposable modules. Since $l(P_0/\mathfrak{r}P_0) \leq l(C)$, we then get $l(P_0) \leq l(P_0/\mathfrak{r}P_0)n \leq l(C)n$. Therefore we get $l(D \operatorname{Tr} C) = l(\operatorname{Tr} C) \leq l(P_1^*) \leq l(P_1^*/\mathfrak{r}P_1^*)n = l(P_1/\mathfrak{r}P_1)n \leq l(P_0)n \leq l(C)n^2$, which completes the proof of the claim. $\square$

Using Lemma 6.5 we obtain the following.

Proposition 6.6 *Let Λ be an artin algebra, n an integer such that $l(P) \leq n$*

for each indecomposable projective left and projective right Λ-module P and $f: A \to B$ an irreducible morphism.

(a) *If A is indecomposable, then $l(B) \leq l(A)(n^2 + 1)$.*

(b) *If B is indecomposable, then $l(A)/(n^2 + 1) \leq l(B)$.*

Proof We only prove (a) since (b) then follows from (a) by duality. If A is injective, f is an epimorphism and there is nothing to prove. If A is not injective, consider the almost split sequence $0 \to A \to E \to \operatorname{Tr} DA \to 0$. By Theorem 5.3 B is a summand of E, so $l(B) \leq l(E) = l(A) + l(\operatorname{Tr} DA)$. Using Lemma 6.5 we get $l(A) + l(\operatorname{Tr} DA) \leq l(A) + n^2 l(A) = l(A)(1 + n^2)$. $\square$

We now use these results to get a bound for $\alpha(\Lambda)$.

Theorem 6.7 *If n is the maximal length of the indecomposable projective modules over Λ and Λ^{op}, then $\alpha(\Lambda) \leq (n^2 + 1)^2$.*

Proof Let $0 \to A \to B \to C \to 0$ be an almost split sequence and let $B = \coprod_{i=1}^{m} B_i$ be a decomposition of B into a sum of indecomposable modules. By Proposition 6.6 we have $l(A)(1/(n^2+1)) \leq l(B_i)$ for each i and $l(B) \leq l(A)(n^2+1)$, and hence $l(A)(m/(n^2+1)) \leq l(B) \leq l(A)(n^2+1)$. From this we conclude that $m \leq (n^2 + 1)^2$, and consequently $\alpha(\Lambda) \leq (n^2 + 1)^2$. $\square$

By induction we obtain the following easy consequence.

Corollary 6.8 *Let Λ be an artin algebra, n the maximal length of the indecomposable projective modules over Λ and Λ^{op} and A an indecomposable Λ-module. Then the number of indecomposable modules B (up to isomorphism) with a chain of irreducible morphisms of length m between indecomposable modules ending (starting) in B and starting (ending) in A is bounded by $(n^2 + 1)^{2m}$.* $\square$

To illustrate the point that the invariant $\alpha(\Lambda)$ is a measure of the complexity of the modules over Λ, we show that if $\alpha(\Lambda) = 1$, then Λ is a Nakayama algebra of Loewy length 2. First observe that since there is some almost split sequence, Λ is not semisimple. For each nonprojective indecomposable Λ-module B we then have that there is no irreducible monomorphism $f: A \to B$ and for each indecomposable noninjective Λ-module B there is no irreducible epimorphism $g: B \to C$. Hence in any almost split sequence $0 \to A \to B \to C \to 0$ we have that B is both projective and injective and therefore A and C are simple. But then each indecomposable projective and each indecomposable injective Λ-module is either simple or of length 2. This shows that Λ is a Nakayama algebra of Loewy length 2.

7 The radical

Closely connected with the study of irreducible morphisms is the radical of mod Λ which is an R-relation in mod Λ when Λ is an artin R-algebra, as studied in Chapter IV. This section is devoted to introducing the radical, developing the basic properties of this R-relation in mod Λ and giving the connections with irreducible morphisms between indecomposable modules over an artin R-algebra Λ.

Let Λ be an artin R-algebra and let A and B be in mod Λ. Define $\mathrm{rad}_\Lambda(A, B)$, **the radical** of $\mathrm{Hom}_\Lambda(A, B)$ by $\mathrm{rad}_\Lambda(A, B) = \{f \in \mathrm{Hom}_\Lambda(A, B)|$ hfg is not an isomorphism for any $g: X \to A$ and $h: B \to X$ with X in ind $\Lambda\}$. Using the notion of an R-relation in an R-category we have the following basic result about the radical in mod Λ.

Proposition 7.1 *Let Λ be an artin R-algebra. Then the radical* rad_Λ *is an R-relation in* mod Λ.

Proof We first prove that $\mathrm{rad}_\Lambda(A, B)$ is a subgroup of $\mathrm{Hom}_\Lambda(A, B)$ for each pair of modules A and B in mod Λ. Let f_1 and f_2 be in $\mathrm{rad}_\Lambda(A, B)$ and let $g: X \to A$ and $h: B \to X$ be Λ-morphisms with X indecomposable in mod Λ. Since f_1 and f_2 are in $\mathrm{rad}_\Lambda(A, B)$, hf_1g and hf_2g are not isomorphisms. Since X is indecomposable the set of nonisomorphisms in $\mathrm{Hom}_\Lambda(X, X)$ forms a subgroup and therefore $hf_1g - hf_2g = h(f_1 - f_2)g$ is not an isomorphism. Since X, g and h were arbitrary it follows that $f_1 - f_2$ is in $\mathrm{rad}_\Lambda(A, B)$.

In order to complete the proof we also have to prove that if $f \in \mathrm{rad}_\Lambda(A, B)$ and $f' \in \mathrm{Hom}_\Lambda(B, C)$ then $f'f \in \mathrm{rad}_\Lambda(A, C)$, and dually if $f \in \mathrm{rad}_\Lambda(A, B)$ and $f'' \in \mathrm{Hom}_\Lambda(E, A)$ then $ff'' \in \mathrm{rad}_\Lambda(E, B)$. We prove the first part of this. Let $f \in \mathrm{rad}_\Lambda(A, B)$ and $f' \in \mathrm{Hom}_\Lambda(B, C)$ and let X be an indecomposable module in mod Λ and $g: X \to A$ and $h: C \to X$ morphisms in mod Λ. Then $h(f'f)g = (hf')fg$ is not an isomorphism since f is in $\mathrm{rad}_\Lambda(A, B)$. Hence $f'f$ is in $\mathrm{rad}_\Lambda(A, C)$. This shows that the radical is a relation on mod Λ.

It is not hard to see that $\mathrm{rad}_\Lambda(A, B)$ is not only a subgroup of

$\text{Hom}_\Lambda(A, B)$ but an R-submodule, which then shows that rad_Λ is an R-relation on $\text{mod}\,\Lambda$. $\qquad\qquad\qquad\qquad\qquad\qquad\qquad\qquad\qquad\qquad\square$

As for any R-relation on $\text{mod}\,\Lambda$ we define the powers of the radical inductively. Hence for A and B in $\text{mod}\,\Lambda$ and a natural number n we have that $\text{rad}_\Lambda^n(A, B) = \{f \in \text{Hom}_\Lambda(A, B)|\ \text{there exist } X \text{ in } \text{mod}\,\Lambda, \text{ morphisms } g \in \text{rad}_\Lambda(A, X) \text{ and } h \in \text{rad}_\Lambda^{n-1}(X, B) \text{ with } f = hg\}$. Finally we define $\text{rad}_\Lambda^\infty(A, B) = \bigcap_{n\in\mathbb{N}} \text{rad}_\Lambda^n(A, B)$. Before going on to give the connection with irreducible morphisms, some comments of a general nature may be helpful.

Any R-relation on an R-category behaves nicely with respect to decomposition. If A and B are in $\text{mod}\,\Lambda$ and $A \simeq \coprod_{i=1}^n A_i$ and $B \simeq \coprod_{j=1}^m B_j$ are decompositions of A and B into sums of modules A_i and B_j, with $\alpha_i: A_i \to A$ and $\beta_j: B \to B_j$ the induced inclusions and projections, we have that a morphism $f: A \to B$ is in $\text{rad}_\Lambda^n(A, B)$ if and only if $\beta_j f \alpha_i$ is in $\text{rad}_\Lambda^n(A_i, B_j)$ for all $i = 1, \ldots, n$ and $j = 1, \ldots, m$.

We also want to make the following observation.

Lemma 7.2 *For A and B in $\text{mod}\,\Lambda$ with Λ an artin R-algebra there exists an $n \in \mathbb{N}$ such that $\text{rad}_\Lambda^n(A, B) = \text{rad}_\Lambda^\infty(A, B)$.*

Proof For A and B in $\text{mod}\,\Lambda$ we have that $\text{Hom}_\Lambda(A, B)$ is a finitely generated R-module, hence of finite length. Therefore the descending chain $\text{Hom}_\Lambda(A, B) \supset \text{rad}_\Lambda(A, B) \supset \cdots \supset \text{rad}_\Lambda^n(A, B) \supset \cdots$ becomes stable, and hence there exists some $n \in \mathbb{N}$ such that $\text{rad}_\Lambda^\infty(A, B) = \bigcap_{m\in\mathbb{N}} \text{rad}_\Lambda^m(A, B) = \text{rad}_\Lambda^n(A, B)$. $\qquad\qquad\qquad\qquad\qquad\qquad\qquad\qquad\qquad\qquad\square$

We now give the desired connection between irreducible morphisms and the radical.

Proposition 7.3 *Let $f: A \to B$ be a morphism between indecomposable modules A and B in $\text{mod}\,\Lambda$. Then f is irreducible if and only if $f \in \text{rad}_\Lambda(A, B) - \text{rad}_\Lambda^2(A, B)$.*

Proof We first observe that if M is an indecomposable module and X an arbitrary module, the radical $\text{rad}_\Lambda(M, X)$ is the set of morphisms from M to X which are not split monomorphisms and that the radical $\text{rad}_\Lambda(X, M)$ is the set of morphisms from X to M which are not split epimorphisms. Our claim then follows directly from the definitions. $\qquad\square$

We next give a description of the morphisms in $\operatorname{rad}_\Lambda^n(A, B)$ for $n \geq 2$ by using irreducible morphisms.

Proposition 7.4 *Let A and B be indecomposable modules in* $\operatorname{mod}\Lambda$ *and let* $f \in \operatorname{rad}_\Lambda^n(A, B)$ *with $n \geq 2$. Then we have the following.*

(a) (i) *There exist a natural number $s \geq 1$, indecomposable modules $B_1, \ldots, B_s$, morphisms $f_i \in \operatorname{rad}_\Lambda(A, B_i)$ and morphisms $g_i \colon B_i \to B$ with each g_i a sum of compositions of $n-1$ irreducible morphisms between indecomposable modules such that $f = \sum_{i=1}^s g_i f_i$.*

    (ii) *If $f \in \operatorname{rad}_\Lambda^n(A, B) - \operatorname{rad}_\Lambda^{n+1}(A, B)$ then at least one of the f_i in (i) is irreducible and $f = u + v$ where u is not zero and is a sum of compositions of n irreducible morphisms between indecomposable modules and $v \in \operatorname{rad}_\Lambda^{n+1}(A, B)$.*

(b) (i) *There exist a natural number $t \geq 1$, indecomposable modules $A_1, \ldots, A_t$, morphisms $f_i \colon A \to A_i$ and $g_i \in \operatorname{rad}_\Lambda(A_i, B)$ with each f_i a sum of compositions of $n-1$ irreducible morphisms between indecomposable modules such that $f = \sum_{i=1}^t g_i f_i$.*

    (ii) *If $f \in \operatorname{rad}_\Lambda^n(A, B) - \operatorname{rad}_\Lambda^{n+1}(A, B)$ then at least one of the g_i in (i) is irreducible.*

Proof We prove statement (a) by induction on n. Statement (b) then follows by duality.

For $n = 2$ let $g \colon M \to B$ be a minimal right almost split morphism. Then there are some s and indecomposable modules $B_1, \ldots, B_s$ such that $M \simeq \coprod_{i=1}^s B_i$. Let $g_i \colon B_i \to B$ be the induced morphisms. Since $f \in \operatorname{rad}_\Lambda^2(A, B)$, there is a morphism $f' \colon A \to M$ such that $f = g f'$. Let $f_i \colon A \to B_i$ be the morphism induced by f' and the decomposition $M \simeq \coprod_{i=1}^s B_i$. Then we get that $f_i \in \operatorname{rad}_\Lambda(A, B_i)$, the morphisms g_i are irreducible and $f = \sum_{i=1}^s g_i f_i$. Moreover if $f \notin \operatorname{rad}_\Lambda^3(A, B)$ then not all f_i are in $\operatorname{rad}_\Lambda^2(A, B_i)$. Hence for at least one $i \in \{1, \ldots, s\}$ we have that f_i is irreducible. This establishes the claim for $n = 2$.

Assume now that $f \in \operatorname{rad}_\Lambda^n(A, B)$ where $n \geq 3$. Then there are some X in $\operatorname{mod}\Lambda$ and morphisms $g \colon A \to X$ and $h \colon X \to B$ with $g \in \operatorname{rad}_\Lambda(A, X)$ and $h \in \operatorname{rad}_\Lambda^{n-1}(X, B)$ and $f = hg$. Let $X = \coprod_{i=1}^t X_i$ be a decomposition of X into a sum of indecomposable modules, and let $g_i \colon A \to X_i$ and $h_i \colon X_i \to B$ be the induced morphisms. Then we have that $f = \sum_{i=1}^t h_i g_i$. Now each h_i is in $\operatorname{rad}_\Lambda^{n-1}(X_i, B)$, so since $n - 1 \geq 2$ there are by induction for each $i = 1, \ldots, t$ indecomposable modules C_{ij} for $j = 1, \ldots, s_i$, morphisms $h_{ij} \in \operatorname{rad}_\Lambda(X_i, C_{ij})$ and $h'_{ij} \colon C_{ij} \to B$, where each h'_{ij} is a sum of compositions of $n-2$ irreducible morphisms between indecomposable modules, such that $h_i = \sum_{j=1}^{s_i} h'_{ij}h_{ij}$. Since $h_{ij}g_i : A \to C_{ij}$ is in $\mathrm{rad}^2_\Lambda(A, C_{ij})$ we have by induction that $h_{ij}g_i = \sum_{p=1}^{q_{ij}} u'_{ijp}u_{ijp}$ where $u_{ijp} \in \mathrm{rad}_\Lambda(A, E_{ijp})$ and $u'_{ijp} : E_{ijp} \to C_{ij}$ are irreducible morphisms and each E_{ijp} is indecomposable. We then have $f = \sum_{i=1}^{t} h_i g_i = \sum_{i=1}^{t} (\sum_{j=1}^{s_i} h'_{ij}h_{ij})g_i = \sum_{i=1}^{t} \sum_{j=1}^{s_i} h'_{ij}(h_{ij}g_i) = \sum_{i=1}^{t} \sum_{j=1}^{s_i} h'_{ij} \sum_{p=1}^{q_{ij}} u'_{ijp}u_{ijp}$. Since each $h'_{ij}u'_{ijp}$ is a sum of compositions of $n-1$ irreducible morphisms and each u_{ijp} is in $\mathrm{rad}_\Lambda(A, E_{ijp})$, this completes the proof of (a)(i).

To prove (a)(ii) observe that when $f \notin \mathrm{rad}^{n+1}_\Lambda(A, B)$, then not all u_{ijp} can be in $\mathrm{rad}^2_\Lambda(A, E_{ijp})$ since each $h'_{ij}u'_{ijp}$ is in $\mathrm{rad}^n_\Lambda(A, B)$. This shows that at least one of the u_{ijp} is irreducible. $\qquad \square$

We next consider the associated subfunctors of the representable functors $\mathrm{Hom}_\Lambda(\ , B)$ and $\mathrm{Hom}_\Lambda(B, \ )$ for B in $\mathrm{mod}\,\Lambda$ induced by the radical of $\mathrm{mod}\,\Lambda$. Define for each B in $\mathrm{mod}\,\Lambda$ and $n \in \mathbb{N}$ the subfunctors $\mathrm{rad}^n_\Lambda(\ , B)$ of $\mathrm{Hom}_\Lambda(\ , B)$ and $\mathrm{rad}^n_\Lambda(B, \ )$ of $\mathrm{Hom}_\Lambda(B, \ )$ by $\mathrm{rad}^n_\Lambda(\ , B)(A) = \mathrm{rad}^n_\Lambda(A, B)$ and $\mathrm{rad}^n_\Lambda(B, \ )(A) = \mathrm{rad}^n_\Lambda(B, A)$ respectively on modules. If $h : A \to A'$ is a morphism in $\mathrm{mod}\,\Lambda$, then the morphism $\mathrm{Hom}_\Lambda(h, B) : \mathrm{Hom}_\Lambda(A', B) \to \mathrm{Hom}_\Lambda(A, B)$ induces a morphism $\mathrm{rad}^n_\Lambda(A', B) \to \mathrm{rad}^n_\Lambda(A, B)$. For an R-functor $F : \mathrm{mod}\,\Lambda \to \mathrm{mod}\,R$ we denote by $\mathrm{Supp}\,F$ **the support** of the functor F, which is the full subcategory of $\mathrm{ind}\,\Lambda$ whose objects are the X in $\mathrm{ind}\,\Lambda$ with $F(X) \neq 0$. We have the following useful consequence of Proposition 7.4.

Proposition 7.5 *Let A and B be indecomposable modules in $\mathrm{mod}\,\Lambda$ and let f be a nonzero element in $\mathrm{rad}_\Lambda(A, B)$.*

(a) *If $\mathrm{rad}^n_\Lambda(\ , B) = 0$ for some $n \in \mathbb{N}$, then f is a sum of compositions of irreducible morphisms between indecomposable modules.*

(b) *If $\mathrm{rad}^n_\Lambda(A, \ ) = 0$ for some $n \in \mathbb{N}$, then f is a sum of compositions of irreducible morphisms between indecomposable modules.*

Proof (a) Let B be an indecomposable Λ-module with $\mathrm{rad}^n_\Lambda(\ , B) = 0$ and let f be a nonzero morphism in $\mathrm{rad}_\Lambda(A, B)$ where A is indecomposable. Since $\mathrm{rad}^n_\Lambda(A, B) = 0$, there is some $t < n$ such that $f \in \mathrm{rad}^t_\Lambda(A, B) - \mathrm{rad}^{t+1}_\Lambda(A, B)$. Then by Proposition 7.4(a)(ii) we have that $f = u + v$ where u is the sum of compositions of irreducible morphisms between indecomposable modules and $v \in \mathrm{rad}^{t+1}_\Lambda(A, B)$. If $v \neq 0$ choose $s > t$ such that $v \in \mathrm{rad}^s_\Lambda(A, B) - \mathrm{rad}^{s+1}_\Lambda(A, B)$, and apply Proposition 7.4 again. Since $\mathrm{rad}^n_\Lambda(A, B) = 0$ we see that f is a sum of compositions of irreducible morphisms between indecomposable modules.

(b) This follows from (a) by duality. $\qquad\square$

From the subfunctors $\text{rad}_\Lambda^n(\ ,B)$ of $\text{Hom}_\Lambda(\ ,B)$ we may form the quotient functors $\text{Hom}_\Lambda(\ ,B)/\text{rad}_\Lambda^n(\ ,B)$ as follows. For a module X $(\text{Hom}_\Lambda(\ ,B)/\text{rad}_\Lambda^n(\ ,B))(X)$ is the R-module $\text{Hom}_\Lambda(X,B)/\text{rad}_\Lambda^n(X,B)$. Since for a Λ-morphism $f\colon X \to Y$ the morphism $\text{Hom}_\Lambda(f,B)$ takes the R-submodule $\text{rad}_\Lambda^n(Y,B)$ of $\text{Hom}_\Lambda(Y,B)$ into the R-submodule $\text{rad}_\Lambda^n(X,B)$ of $\text{Hom}_\Lambda(X,B)$, there is induced a unique map from $\text{Hom}_\Lambda(Y,B)/\text{rad}_\Lambda^n(Y,B)$ to $\text{Hom}_\Lambda(X,B)/\text{rad}_\Lambda^n(X,B)$. This is the value of the quotient functor on f. One defines the covariant quotient functor $\text{Hom}_\Lambda(B,\ )/\text{rad}_\Lambda^n(B,\ )$ dually.

We have the following easily verified result about these quotient functors.

Lemma 7.6 *Let B be in* $\text{mod}\,\Lambda$. *Then we have the following.*

(a) *The supports of* $\text{Hom}_\Lambda(\ ,B)/\text{rad}_\Lambda^n(\ ,B)$ *and* $\text{Hom}_\Lambda(B,\ )/\text{rad}_\Lambda^n(B,\ )$ *are finite for each* $n \in \mathbb{N}$.

(b) *The support of* $\text{Hom}_\Lambda(\ ,B)$ *is finite if and only if there exists some* $n \in \mathbb{N}$ *with* $\text{rad}_\Lambda^n(\ ,B) = 0$.

(c) *The support of* $\text{Hom}_\Lambda(B,\ )$ *is finite if and only if there exists some* $n \in \mathbb{N}$ *with* $\text{rad}_\Lambda^n(B,\ ) = 0$.

Proof (a) Since we have seen that $\text{rad}_\Lambda^n(A,B_1 \coprod B_2) \simeq \text{rad}_\Lambda^n(A,B_1) \coprod \text{rad}_\Lambda^n(A,B_2)$ it is clearly enough to prove the statement for B indecomposable. If A is indecomposable and $\text{rad}_\Lambda^i(A,B) \neq \text{rad}_\Lambda^{i+1}(A,B)$ there is by Proposition 7.4 a composition of i irreducible morphisms from A to B. Since there is only a finite number of indecomposable modules X with an irreducible morphism to a given indecomposable module Y, there is only a finite number of such modules A up to isomorphism for each i. It follows that the support of $\text{Hom}_\Lambda(\ ,B)/\text{rad}_\Lambda^n(\ ,B)$ is finite for each $n \in \mathbb{N}$. By duality we also get that the support of $\text{Hom}_\Lambda(B,\ )/\text{rad}_\Lambda^n(B,\ )$ is finite for each $n \in \mathbb{N}$.

In order to prove statement (b) first observe that if $\text{rad}_\Lambda^n(\ ,B) = 0$ for some $n \in \mathbb{N}$ then by (a) the support of $\text{Hom}_\Lambda(\ ,B)$ is finite. Assume conversely that the support of $\text{Hom}_\Lambda(\ ,B)$ is finite. Then there is a bound n on the Loewy length of all $\text{End}_\Lambda(A)$ with A in $\text{Supp}\,\text{Hom}_\Lambda(\ ,B)$. Hence the composition of tn nonisomorphisms within $\text{Supp}\,\text{Hom}_\Lambda(\ ,B)$ is zero where t is the number of modules in $\text{Supp}\,\text{Hom}_\Lambda(\ ,B)$. This then shows that $\text{rad}_\Lambda^{tn}(\ ,B) = 0$.

Statement (c) follows from (b) by duality. $\qquad\square$

We have the following direct consequence of Lemma 7.6.

Theorem 7.7 *The following are equivalent for an artin algebra* Λ.

(a) Λ *is of finite representation type.*
(b) *The support of* $\operatorname{Hom}_\Lambda(\ ,B)$ *is finite for each* B *in* $\operatorname{mod}\Lambda$.
(c) *For each* B *in* $\operatorname{mod}\Lambda$ *there is some* $n \in \mathbb{N}$ *with* $\operatorname{rad}_\Lambda^n(\ ,B) = 0$.
(d) *The support of* $\operatorname{Hom}_\Lambda(B,\ )$ *is finite for each* B *in* $\operatorname{mod}\Lambda$.
(e) *For each* B *in* $\operatorname{mod}\Lambda$ *there is some* n *with* $\operatorname{rad}_\Lambda^n(B,\ ) = 0$.

Proof It follows from the definition of finite representation type that (a) implies (d). To prove that (d) implies (a) consider the functor $\operatorname{Hom}_\Lambda(\Lambda,\ )$. Part (d) then states that $\operatorname{Hom}_\Lambda(\Lambda,\ )$ has finite support, but this is clearly the same as saying that $\operatorname{ind}\Lambda$ is finite. That (d) and (e) are equivalent follows from Lemma 7.6. The rest follows by duality. $\square$

We also have the following consequence for finite representation type.

Theorem 7.8 *Let* Λ *be of finite representation type and let* $f \in \operatorname{rad}_\Lambda(A,B)$ *with* A *and* B *indecomposable modules in* $\operatorname{mod}\Lambda$. *Then* f *is a sum of compositions of irreducible morphisms between indecomposable modules in* $\operatorname{mod}\Lambda$.

Proof By Theorem 7.7 we have that $\operatorname{Supp}\operatorname{Hom}_\Lambda(\ ,B)$ is finite and hence $\operatorname{rad}_\Lambda^n(\ ,B) = 0$ for some $n \in \mathbb{N}$. Therefore it follows from Proposition 7.5 that f is a sum of compositions of irreducible morphisms between indecomposable modules. $\square$

We saw in Lemma 7.2 that for each pair of modules A and B in $\operatorname{mod}\Lambda$ there is an $n \in \mathbb{N}$ such that $\operatorname{rad}_\Lambda^n(A,B) = \operatorname{rad}_\Lambda^\infty(A,B)$. We want to show that when the support of the functor $\operatorname{Hom}_\Lambda(\ ,B)$ is infinite, then the chain of subfunctors $\cdots \subset \operatorname{rad}_\Lambda^n(\ ,B) \subset \operatorname{rad}_\Lambda^{n-1}(\ ,B) \subset \cdots \subset \operatorname{Hom}_\Lambda(\ ,B)$ is a proper descending chain of subfunctors. This is a direct consequence of the following.

Proposition 7.9 *If* B *is in* $\operatorname{mod}\Lambda$ *with the support of* $\operatorname{Hom}_\Lambda(\ ,B)$ *infinite, then*

$$\operatorname{rad}_\Lambda^n(\ ,B)/\operatorname{rad}_\Lambda^{n+1}(\ ,B) \neq 0$$

for all $n \in \mathbb{N}$.

The proof of this is based on results about minimal morphisms in Chapter I and the following result.

Lemma 7.10 *Let B be in* $\operatorname{mod}\Lambda$. *Then for each $n \geq 1$ there exist a Λ-module C_n in* $\operatorname{mod}\Lambda$ *and a morphism $f_n: C_n \to B$ such that $f_n = f_{n-1}h_n$ with $h_n \in \operatorname{rad}_\Lambda(C_n, C_{n-1})$ and such that $\operatorname{Im}\operatorname{Hom}_\Lambda(X, f_n) = \operatorname{rad}_\Lambda^n(X, B)$ for all X in* $\operatorname{mod}\Lambda$, *where $C_0 = B$ and $f_0 = 1_B$.*

Proof The proof goes by induction based upon the existence of right almost split morphisms.

Let $C_0 = B$ and $f_0 = 1_B$. For $n = 1$ decompose B as a sum of indecomposable modules $B = \coprod_{i=1}^t B_i$ and let $h_i: B_i' \to B_i$ be minimal right almost split. Then let $C_1 = \coprod_{i=1}^t B_i'$ and let $f_1 = h_1 \coprod h_2 \coprod \cdots \coprod h_t$. Then clearly $\operatorname{Im}\operatorname{Hom}_\Lambda(X, f_1) = \operatorname{rad}_\Lambda(X, B)$ for all Λ-modules X. For the inductive step assume we have proved the statement for n for all modules in $\operatorname{mod}\Lambda$ and want to prove the claim for $n + 1$. So assume $f_n: C_n \to B$ is such that $\operatorname{Im}\operatorname{Hom}_\Lambda(X, f_n) = \operatorname{rad}_\Lambda^n(X, B)$ for all X in $\operatorname{mod}\Lambda$ and let $h: Y \to C_n$ be such that $\operatorname{Im}\operatorname{Hom}_\Lambda(X, h) = \operatorname{rad}_\Lambda(X, C_n)$ for all X in $\operatorname{mod}\Lambda$. Let $C_{n+1} = Y$ and $f_{n+1} = f_n h$. Then we clearly have $\operatorname{Im}\operatorname{Hom}_\Lambda(X, f_{n+1}) \subset \operatorname{rad}_\Lambda^{n+1}(X, B)$ for all X in $\operatorname{mod}\Lambda$.

For the converse inclusion let X be arbitrary in $\operatorname{mod}\Lambda$ and let $\alpha \in \operatorname{rad}_\Lambda^{n+1}(X, B)$. Then there is some Y in $\operatorname{mod}\Lambda$, some $\alpha' \in \operatorname{rad}_\Lambda(X, Y)$ and $\alpha'' \in \operatorname{rad}_\Lambda^n(Y, B)$ with $\alpha = \alpha''\alpha'$. Hence there is by induction hypothesis some $\beta: Y \to C_n$ with $\alpha'' = f_n\beta$. Since $\alpha' \in \operatorname{rad}_\Lambda(X, Y)$, we have $\beta\alpha' \in \operatorname{rad}_\Lambda(X, C_n)$. Hence there is some $\beta': X \to C_{n+1}$ with $h\beta' = \beta\alpha'$. But then $\alpha = \alpha''\alpha' = f_n\beta\alpha' = f_n h\beta' = f_{n+1}\beta'$, which shows that $\operatorname{rad}_\Lambda^{n+1}(X, B) \subset \operatorname{Im}\operatorname{Hom}_\Lambda(X, f_{n+1})$. This completes the proof of the lemma. $\qquad\square$

We are now ready to prove Proposition 7.9. Assume B is in $\operatorname{mod}\Lambda$ with $\operatorname{Supp}\operatorname{Hom}_\Lambda(\ , B)$ infinite. Since the support of $\operatorname{Hom}_\Lambda(\ , B)/\operatorname{rad}_\Lambda^n(\ , B)$ is finite for all n, we have that $\operatorname{rad}_\Lambda^n(\ , B) \neq 0$ for all n. Hence the modules C_n and the morphisms $f_n: C_n \to B$ from Lemma 7.10 such that $\operatorname{Im}\operatorname{Hom}_\Lambda(X, f_n) = \operatorname{rad}_\Lambda^n(X, B)$ for all X in $\operatorname{mod}\Lambda$ are nonzero and we have $f_n \in \operatorname{rad}_\Lambda^n(C_n, B)$. Then according to I Theorem 2.2 there exists a decomposition of each C_n as a sum $C_n' \coprod C_n''$ such that $f_n|_{C_n''} = 0$ and $f_n|_{C_n'}$ is right minimal. If we write $f_n' = f_n|_{C_n'}$, we claim that the image $\bar{f}_n' \in \operatorname{rad}_\Lambda^n(C_n', B)/\operatorname{rad}_\Lambda^{n+1}(C_n', B)$ of $f_n' \in \operatorname{rad}_\Lambda^n(C_n', B)$ is nonzero. Otherwise we have $f_n' \in \operatorname{rad}_\Lambda^{n+1}(C_n', B) = \operatorname{Im}\operatorname{Hom}_\Lambda(C_{n+1}, f_{n+1})$. Hence there is some $g: C_n' \to C_{n+1}$ such that $f_n' = f_{n+1}g$. However $f_{n+1} = f_n h$ with

$h \in \mathrm{rad}_\Lambda(C_{n+1}, C_n)$ and therefore $f'_n = f_{n+1}g = f_n hg$. Using the projection p of C_n onto C'_n according to the decomposition $C_n = C'_n \coprod C''_n$ we get that $f'_n = f'_n phg$. But since f'_n was right minimal, phg is an isomorphism. This contradicts that $h \in \mathrm{rad}_\Lambda(C_{n+1}, C_n)$. So our claim follows and the proof of the proposition is complete. $\qquad\square$

As an immediate consequence of Proposition 7.9 we have the following result.

Corollary 7.11 *For a Λ-module B in mod Λ the following are equivalent.*

(a) *The support of* $\mathrm{Hom}_\Lambda(\ ,B)$ *is infinite.*

(b) $\mathrm{rad}_\Lambda^n(\ ,B)/\mathrm{rad}_\Lambda^{n+1}(\ ,B) \neq 0$ *for all $n \in \mathbb{N}$.*

(c) *The support of* $\mathrm{Hom}_\Lambda(\ ,B)/\mathrm{rad}_\Lambda^\infty(\ ,B)$ *is infinite.* $\qquad\square$

Exercises

1. Let Λ be an artin algebra such that every indecomposable Λ-module has a simple socle.

(a) Show that each indecomposable projective Λ-module is uniserial.

(b) Show that the length of $C/\mathrm{r}C$ is less than or equal to 2 for each indecomposable Λ-module C.

(c) Show that $C/\mathrm{soc}\,C$ is either uniserial or the sum of two uniserial modules for each indecomposable Λ-module C.

(d) Show that the number of indecomposable summands of the middle term of any almost split sequence is at most two.

(e) Give an example of an algebra Λ satisfying these properties without being Nakayama.

2. Let Λ be an artin algebra and let I be an indecomposable injective nonprojective Λ-module.

(a) Show that the following are equivalent.

    (i)   There is an irreducible epimorphism $f: P \to I$ with P indecomposable projective noninjective.

    (ii) $D\,\mathrm{Tr}\,I$ is simple and the middle term P of the almost split sequence $0 \to D\,\mathrm{Tr}\,I \to P \to I \to 0$ is indecomposable projective with $l(\mathrm{soc}\,P) = 2$.

    (iii) $I \simeq P/S_1$ where P is an indecomposable projective module, S_1 is simple, $\mathrm{r}P = S_1 \coprod X$ with $X \neq 0$ and $\mathrm{Ext}_\Lambda^1(\Lambda/\mathrm{r}, S_1)$ is a simple Λ-module.

(b) Suppose I satisfies the equivalent conditions of (a) and let P and S_1 be as above with $\operatorname{soc} P = S_1 \coprod T$. Show that the middle term M of the almost split sequence $0 \to P \to M \to \operatorname{Tr} DP \to 0$ is isomorphic to $I \coprod P/T$ and that $\operatorname{Tr} DP \simeq P/\operatorname{soc} P$.

3. Let Λ be an artin algebra and S a simple projective Λ-module of injective dimension 1.

(a) Show that $\operatorname{Hom}_\Lambda(\operatorname{Tr} DS, \Lambda) = 0$ and that $\operatorname{pd} \operatorname{Tr} DS = 1$.
(b) Show that $\operatorname{End}_\Lambda(\operatorname{Tr} DS)$ is a division ring.

4. Let P be an indecomposable projective noninjective nonsimple module and let $0 \to P \to B \to \operatorname{Tr} DP \to 0$ be an almost split sequence. Show that B is indecomposable if and only if $\mathfrak{r} P$ is indecomposable and P is not a summand of $\mathfrak{r}$.

5. Let Λ be an artin algebra and $\mathfrak{a}$ an ideal in Λ. For each Λ-module M consider the quotient module $M/\mathfrak{a}M$ and the submodule $_\mathfrak{a}M = \{m \in M | \mathfrak{a}m = 0\}$ of M.

(a) Prove that $(\Lambda/\mathfrak{a}) \otimes_\Lambda M \simeq M/\mathfrak{a}M$ and that $\operatorname{Hom}_\Lambda(\Lambda/\mathfrak{a}, M) \simeq {}_\mathfrak{a}M$ for all M in $\operatorname{mod} \Lambda$.
(b) Let C be an indecomposable Λ-module with $\mathfrak{a}C = 0$ and C not a summand of $\Lambda/\mathfrak{a}$ and let $0 \to A \xrightarrow{f} B \xrightarrow{g} C \to 0$ be an almost split sequence.
  (i) Show that the induced sequence $0 \to \operatorname{Hom}_\Lambda(\Lambda/\mathfrak{a}, A) \to \operatorname{Hom}_\Lambda(\Lambda/\mathfrak{a}, B) \to \operatorname{Hom}_\Lambda(\Lambda/\mathfrak{a}, C) \to 0$ with $\operatorname{Hom}_\Lambda(\Lambda/\mathfrak{a}, C) \simeq C$ is exact.
  (ii) Show that $\operatorname{Hom}_\Lambda(\Lambda/\mathfrak{a}, g)$ is right almost split in $\operatorname{mod}(\Lambda/\mathfrak{a})$.
  (iii) Let $P_1 \to P_0 \to C \to 0$ be a minimal projective presentation of C as a Λ-module. Prove that $P_1/\mathfrak{a}P_1 \to P_0/\mathfrak{a}P_0 \to C \to 0$ is isomorphic to a sequence $Q_0 \coprod Q_1 \xrightarrow{(0,f)} P_0/\mathfrak{a}P_0 \to C \to 0$ with $Q_1 \xrightarrow{f} P_0/\mathfrak{a}P_0 \to C \to 0$ a minimal projective presentation of C as a $(\Lambda/\mathfrak{a})$-module and Q_0 a projective $(\Lambda/\mathfrak{a})$-module.
  (iv) Prove that the induced sequence $0 \to \operatorname{Hom}_\Lambda(\Lambda/\mathfrak{a}, A) \to \operatorname{Hom}_\Lambda(\Lambda/\mathfrak{a}, B) \to C \to 0$ is isomorphic to a sequence $0 \to D\operatorname{Tr}_{\Lambda/\mathfrak{a}} C \coprod DQ_0^* \xrightarrow{(\alpha \coprod 1)} B' \coprod DQ_0^* \xrightarrow{(\beta, 0)} C \to 0$ where $0 \to D\operatorname{Tr}_{\Lambda/\mathfrak{a}} C \to B' \to C \to 0$ is almost split in $\operatorname{mod} \Lambda/\mathfrak{a}$ and DQ_0^* is an injective $(\Lambda/\mathfrak{a})$-module (Here $*$ is used for $\operatorname{Hom}_{\Lambda/\mathfrak{a}}(\ , \Lambda/\mathfrak{a})$.)

(v)   Prove that $D \operatorname{Tr}_{\Lambda/\mathfrak{a}} C$ is a submodule of $D \operatorname{Tr}_{\Lambda} C$.

(c)   Let A be an indecomposable Λ-module with $\mathfrak{a}A = 0$, A not an injective $(\Lambda/\mathfrak{a})$-module and let $0 \to A \to B \to C \to 0$ be an almost split sequence.

   (i)   Prove that $0 \to A \to B/\mathfrak{a}B \to C/\mathfrak{a}C \to 0$ is exact and isomorphic to a sequence $0 \to A \overset{\binom{\alpha}{0}}{\to} B' \coprod Q \overset{(\beta \coprod 1_Q)}{\to} C' \coprod Q \to 0$ with $0 \to A \overset{\alpha}{\to} B' \overset{\beta}{\to} C' \to 0$ an almost split sequence in $\operatorname{mod} \Lambda/\mathfrak{a}$ and Q a projective $(\Lambda/\mathfrak{a})$-module.

   (ii)   Prove that $\operatorname{Tr} D_{\Lambda/\mathfrak{a}}A$ is a quotient of $\operatorname{Tr} D_{\Lambda}A$.

(d)   Let $\mathfrak{a}$ be an ideal in Λ such that $\Lambda/\mathfrak{a}$ is a projective right Λ-module. Let C be as in (b). Prove then that Q_0 in (b)(iii) is zero or equivalently that $D \operatorname{Tr}_{\Lambda/\mathfrak{a}} C \simeq \operatorname{Hom}_{\Lambda}(\Lambda/\mathfrak{a}, D \operatorname{Tr}_{\Lambda} C)$.

(e)   Let $\mathfrak{a}$ be an ideal in Λ such that $\Lambda/\mathfrak{a}$ is a projective left Λ-module and let A be as in (c). Prove then that Q in (c)(ii) is zero or equivalently that $\operatorname{Tr} D_{\Lambda/\mathfrak{a}}A \simeq (\Lambda/\mathfrak{a}) \otimes_{\Lambda} A$.

(f)   Assume Λ is a selfinjective algebra and $\mathfrak{a}$ is an ideal with $\mathfrak{a} \subset \operatorname{soc} \Lambda$. Prove that for each indecomposable $(\Lambda/\mathfrak{a})$-module C with C not projective we have $D \operatorname{Tr}_{\Lambda/\mathfrak{a}} C \simeq D \operatorname{Tr}_{\Lambda} C$ and for each indecomposable noninjective $(\Lambda/\mathfrak{a})$-module A we have that $\operatorname{Tr} D_{\Lambda/\mathfrak{a}}A \simeq \operatorname{Tr} D_{\Lambda}A$.

6. Let $0 \to A \overset{f}{\to} B \overset{g}{\to} C \to 0$ be an almost split sequence over an artin algebra Λ and let $P_1 \overset{h}{\to} P_0 \to C \to 0$ be a minimal projective presentation of C. Show that the following statements are equivalent.

(i) A is simple.

(ii)   The morphism $B \overset{g}{\to} C$ is an essential epimorphism.

(iii) $0 \to P_1/\mathfrak{r}P_1 \to P_0/h(\mathfrak{r}P_1) \to C \to 0$ is an almost split sequence.

7. Let A and B be indecomposable modules over an artin algebra Λ and X an arbitrary module.

(a)   Prove that $\operatorname{rad}_{\Lambda}(A, A) = \mathfrak{r}_{\operatorname{End}(A)}$.

(b)   Prove that $\operatorname{rad}_{\Lambda}(A, B) = \begin{cases} \mathfrak{r}_{\operatorname{End}(A)^{\operatorname{op}}} \operatorname{Hom}_{\Lambda}(A, B) & \text{if } A \simeq B, \\ \operatorname{Hom}_{\Lambda}(A, B) & \text{if } A \not\simeq B. \end{cases}$

(c)   Prove that $\operatorname{rad}_{\Lambda}(A, X) = \{f \in \operatorname{Hom}_{\Lambda}(A, X) | f$ is not a split monomorphism$\}$ and that $\operatorname{rad}_{\Lambda}(X, B) = \{f \in \operatorname{Hom}_{\Lambda}(X, B) | f$ is not a split epimorphism$\}$.

8. Let Λ be an artin algebra and let A be in mod Λ.

(a) Prove that if A is indecomposable, then $\mathrm{rad}_\Lambda(A, A) = \mathrm{rad}_\Lambda^2(A, A)$, where rad_Λ denotes the radical in mod Λ.

(b) Decide for which A in mod Λ we have $\mathrm{rad}_\Lambda(A, A) \neq \mathrm{rad}_\Lambda^2(A, A)$.

9. Let Λ be an artin algebra such that $D\,\mathrm{Tr}\,C \simeq \Omega^2 C$ for all C in mod Λ.

(a) Prove that Λ has no simple injective nonprojective modules. (Hint: Use information on almost split sequences whose right hand term is simple injective.)

(b) Prove that all simple Λ-modules are torsionless.

(c) Prove that $\mathrm{soc}\,P \simeq P/\mathfrak{r}P$ for each projective Λ-module P.

(d) Prove that Λ is selfinjective.

10. Let M be a module over an artin algebra Λ.

(a) Show that for each Λ-module X, the abelian group $\mathrm{Ext}_\Lambda^1(M, X)$ has a natural structure as an $\mathrm{End}_\Lambda(M)^{\mathrm{op}}$-module with the property that the two-sided ideal $\mathcal{P}(M, M)$ of $\mathrm{End}_\Lambda(M)^{\mathrm{op}}$ annihilates $\mathrm{Ext}_\Lambda^1(M, X)$ and hence each $\mathrm{Ext}_\Lambda^1(M, X)$ is an $\underline{\mathrm{End}}_\Lambda(M)^{\mathrm{op}}$-module.

(b) Show that the functor $F: \mathrm{mod}\,\Lambda \to \mathrm{mod}\,\underline{\mathrm{End}}_\Lambda(M)^{\mathrm{op}}$ given by $F(X) = \mathrm{Ext}_\Lambda^1(M, X)$ for all X in mod Λ and $F(f)$ for all f in $\mathrm{Hom}_\Lambda(X, Y)$ is $\mathrm{Ext}_\Lambda^1(M, f): \mathrm{Ext}_\Lambda^1(M, X) \to \mathrm{Ext}_\Lambda^1(M, Y)$ for all X and Y in mod Λ has the following properties.

   (i) If $\mathrm{Ext}_\Lambda^1(M, \Lambda) = 0$, then there is an X in mod Λ such that $F(X)$ is a generator for $\underline{\mathrm{End}}_\Lambda(M)^{\mathrm{op}}$, i.e. $\underline{\mathrm{End}}_\Lambda(M)^{\mathrm{op}}$ is in add $F(X)$.

   (ii) If $\mathrm{Ext}_\Lambda^1(M, M \amalg \Lambda) = 0$, then F is dense, i.e. given any $\underline{\mathrm{End}}_\Lambda(M)^{\mathrm{op}}$-module Z there is an X in mod Λ such that $F(X) \simeq Z$.

   (iii) Show that if $\mathrm{Ext}_\Lambda^1(M, M \amalg \Lambda) = 0$, then the functor $F: \mathrm{mod}\,\Lambda \to \mathrm{mod}\,\underline{\mathrm{End}}_\Lambda(M)^{\mathrm{op}}$ induces a surjective map from the isomorphism classes of indecomposable Λ-modules to the isomorphism classes of indecomposable $\underline{\mathrm{End}}_\Lambda(M)^{\mathrm{op}}$-modules. Hence $\underline{\mathrm{End}}_\Lambda(M)^{\mathrm{op}}$ is of finite representation type if Λ is of finite representation type.

Notes

The existence of almost split sequences, also called Auslander–Reiten sequences, was first observed around 1971 for artin algebras of finite representation type in connection with describing the projective presentations of simple modules over Auslander algebras, a class of algebras we discuss in Chapter VI. Two very different approaches to proving that right and left almost split morphisms (also called sink maps and source maps [Rin3]) existed for arbitrary artin algebras were then pursued. One approach was showing that for arbitrary artin algebras Λ the simple functors from mod Λ to abelian groups are finitely presented [AuR2]. The other approach, which gave the existence of almost split sequences, not just almost split morphisms, was based on the guess that the ends of an almost split sequence $0 \to A \to B \to C$ were related by $A \simeq D\operatorname{Tr} C$ and a careful homological algebra study of $\operatorname{Ext}^1_\Lambda(C, D\operatorname{Tr} C)$ including its description as $D\overline{\operatorname{End}}_\Lambda(C)$ [AuR4]. There then followed a series of papers [AuR5] [AuR6] [AuR7] where most of the notions and results in this chapter were established. In particular, it was here that irreducible morphisms were first discussed, including a connection with radical series of functors. The bimodule $\operatorname{Irr}(A, B)$ of irreducible morphisms was investigated by Bautista and Ringel (see Chapter VII). Recently the theory of irreducible maps has been further developed through the notion of the degree of an irreducible map [Liu1].

Although almost split sequences now play a fundamental role in the representation theory of artin algebras, it took several years after their introduction before their significance began to be appreciated. It was essentially the notion of an Auslander–Reiten-quiver, a device for studying all left and right almost split morphisms simultaneously (see Chapter VII for details), that made the difference. For example much of the early work on hereditary algebras and selfinjective algebras of finite representation type was concerned with describing their Auslander–Reiten-quivers, or equivalently, their almost split sequences (see [Rin2], [Rie]). And this interest persists to this day.

Existence theorems for almost split sequences have been proved also in certain subcategories of mod Λ for an artin algebra Λ [AuS2], and in contexts other than artin algebras (see [AuR11]). It is worth noting that the notion of almost split sequences has proven to be useful in such diverse fields as modular group representations, the theory of orders, algebraic singularity theory and model theory of modules. The construction we give for almost split sequences for group algebras is taken from [AuC].

The fact that there is always an almost split sequence with indecomposable middle term was proved in [AuR5] for finite representation type and in [M2] in general. The construction given here is taken from [ButR]. For an algebra Λ of finite representation type one has that $\alpha(\Lambda) \leq 4$ [BauB]. This result has been further generalized in [Liu3] [Kra2].

VI
Finite representation type

The artin algebras of finite representation type are in some sense the simplest kinds of artin algebras, and a lot of effort has been put into understanding and classifying various classes of algebras of finite representation type. Often these algebras serve as a test case and inspiration for what might be true more generally. For example existence of almost split sequences was first proved in this context.

This chapter is devoted to studying algebras of finite representation type. We start by giving a criterion for finite type in terms of irreducible morphisms, which we apply to describe all indecomposable modules over Nakayama algebras. Using the special features of group algebras developed in Chapter III, we describe which group algebras over fields are of finite representation type. A criterion for finite representation type is also given in terms of generators and relations for the Grothendieck group of artin algebras. The chapter ends with a discussion of the endomorphism algebra of a Λ-module M containing all indecomposable modules as a summand when Λ is of finite representation type. These algebras are called Auslander algebras.

1 A criterion for finite representation type

In this section we use almost split morphisms to give a criterion for an artin algebra Λ to be of finite representation type. Using this criterion we show that Λ is of finite type if there is a bound on the length of the indecomposable Λ-modules.

Denote as before by ind Λ a fixed full subcategory of mod Λ whose objects consist of a complete set of nonisomorphic indecomposable Λ-modules. We first define an equivalence relation on the objects of ind Λ. Two modules A and B in ind Λ are said to be related by an irreducible morphism if there exists an irreducible morphism $f : A \to B$. We call an equivalence class under the equivalence relation generated by this relation a **component** of ind Λ. Then A and B are in the same component if and only if there exist a natural number n, indecomposable modules X_i, for $i = 1, \cdots, n$, and for each i either an irreducible morphism $f_i : X_i \to X_{i+1}$ or an irreducible morphism $g_i : X_{i+1} \to X_i$ with $X_1 = A$ and $X_n = B$. We prove that if an indecomposable artin algebra Λ has a component $\mathscr{C}$ where all the modules are bounded in length by some number t, then Λ is of finite representation type and ind Λ consists of a single component.

If A and B are indecomposable modules and $\operatorname{rad}^n_\Lambda(A, B) \neq \operatorname{rad}^{n+1}_\Lambda(A, B)$, there is by V Proposition 7.4 a chain of n irreducible morphisms between indecomposable modules from A to B. Hence we get the following.

Lemma 1.1 *Let A be in* ind Λ. *If B is in* $\operatorname{Supp}(\operatorname{Hom}(\ , A) / \operatorname{rad}^\infty_\Lambda(\ , A))$ *or in* $\operatorname{Supp}(\operatorname{Hom}(A, \ ) / \operatorname{rad}^\infty_\Lambda(A, \ ))$, *then A and B are in the same component of* ind Λ. $\quad\square$

When Λ is of finite representation type, there is, as noted in V Section 7, a bound on the lengths of chains of nonisomorphisms with nonzero compositions between indecomposable modules. The following useful technical lemma shows that there is a similar result under the more general assumption that the indecomposable modules involved have bounded length.

Lemma 1.2 *Let $n \in \mathbb{N}$ and for each $i \in \mathbb{Z}$, let A_i be an indecomposable module with $l(A_i) \leq n$, and let $f_i : A_i \to A_{i+1}$ be nonisomorphisms. Then $l(\operatorname{Im}(f_{i+2^m-2} \cdots f_i)) \leq \max\{n - m, 0\}$ for each $m \in \mathbb{N}$.*

Proof The proof of this goes by induction on m. If $m = 1$, consider the morphism $f_i : A_i \to A_{i+1}$. Since f_i is not an isomorphism, then either f_i is not a monomorphism or f_i is not an epimorphism. In the first case $l(\operatorname{Im} f_i) \leq l(A_i) - 1$ and in the second case $l(\operatorname{Im} f_i) \leq l(A_{i+1}) - 1$. Hence we have $l(\operatorname{Im} f_i) \leq n - 1 \leq \max\{n - 1, 0\}$.

Assume now that the lemma is proved for m and consider

$$f_{i+2^{m+1}-2} \cdots f_{i+2^m} f_{i+2^m-1} f_{i+2^m-2} \cdots f_i : A_i \to A_{i+2^{m+1}-1}.$$

Writing $f = f_{i+2^m-1}$, $g = f_{i+2^m-2} \cdots f_i$ and $h = f_{(i+2^m)+2^m-2} \cdots f_{i+2^m}$ we have the sequence of morphisms

$$A_i \xrightarrow{g} A_{i+2^m-1} \xrightarrow{f} A_{i+2^m} \xrightarrow{h} A_{i+2^{m+1}-1}.$$

Assume that $l(\operatorname{Im}(hfg)) \not\leq \max\{n-m-1, 0\}$, that is $l(\operatorname{Im}(hfg)) \geq n-m > 0$.

We will prove that this implies that $f = f_{i+2^m-1}$ is an isomorphism, which is a contradiction.

By the induction assumption we have $l(\mathrm{Im}\, g) \leq \max\{n - m, 0\}$ and $l(\mathrm{Im}\, h) \leq \max\{n - m, 0\}$. Further we obviously have

$$
\begin{aligned}
l(\mathrm{Im}(hfg)) &\leq \min\{l(\mathrm{Im}\, g), l(\mathrm{Im}(fg)), l(\mathrm{Im}(hf)), l(\mathrm{Im}\, h)\} \\
&\leq \max\{n - m, 0\}.
\end{aligned}
$$

Hence we get

$$
\begin{aligned}
l(\mathrm{Im}(hfg)) &= l(\mathrm{Im}(hf)) = l(\mathrm{Im}\, h) = l(\mathrm{Im}(fg)) \\
&= l(\mathrm{Im}\, g) = n - m > 0.
\end{aligned}
$$

Now $h|_{\mathrm{Im}(fg)} : \mathrm{Im}(fg) \to \mathrm{Im}(h)$ is an isomorphism, since $\mathrm{Im}(hfg) \subset \mathrm{Im}\, h$ and the modules have the same length. Hence we have $A_{i+2^m} \simeq \mathrm{Im}(fg) \coprod \mathrm{Ker}\, h$. Since by assumption A_{i+2^m} is indecomposable and $l(\mathrm{Im}(fg)) = n - m > 0$, we must have $\mathrm{Ker}\, h = 0$. Therefore we get $\mathrm{Im}(fg) = A_{i+2^m}$, so that f is an epimorphism. Next consider $hf|_{\mathrm{Im}\, g} : \mathrm{Im}\, g \to \mathrm{Im}(hf)$. Again since $\mathrm{Im}(hfg) \subset \mathrm{Im}(hf)$ and the modules have the same length, we have that $hf|_{\mathrm{Im}\, g}$ is an isomorphism. Therefore we get $A_{i+2^m-1} = \mathrm{Im}\, g \coprod \mathrm{Ker}(hf)$. Since $l(\mathrm{Im}\, g) = n-m > 0$ and A_{i+2^m-1} is indecomposable, it follows that $\mathrm{Ker}(hf) = 0$ and hence f is a monomorphism. Hence we get that $f = f_{i+2^m-1}$ is an isomorphism, contradicting the hypothesis. We can now conclude that $l(\mathrm{Im}(hfg)) \leq \max\{n-(m+1), 0\}$ and this completes the induction proof. $\qquad\square$

As an immediate consequence of this lemma we get the following.

Corollary 1.3 *If $f_i : A_i \to A_{i+1}$ are nonisomorphisms between indecomposable modules A_i for $i = 1, \ldots, 2^n - 1$ and $l(A_i) \leq n$ for all i, then $f_{2^n-1} \cdots f_1 = 0$.* $\qquad\square$

We can now prove the main result of this section.

Theorem 1.4 *Let Λ be an indecomposable artin algebra and $\mathscr{C}$ a component of $\mathrm{ind}\, \Lambda$ such that the length of the objects in $\mathscr{C}$ is bounded. Then Λ is of finite representation type and $\mathscr{C} = \mathrm{ind}\, \Lambda$.*

Proof Let n be a positive integer such that $l(C) \leq n$ for all C in $\mathscr{C}$, and let A be in $\mathscr{C}$. If for some B in $\mathrm{ind}\, \Lambda$ we have $\mathrm{rad}_\Lambda^{2^n}(A, B) \neq 0$, there is by V Proposition 7.4 a chain of 2^n nonisomorphisms between indecomposable modules with nonzero composition from A to B. This is a contradiction to Corollary 1.3, so we have $\text{rad}_\Lambda^{2^n}(A, \ ) = 0$, and similarly $\text{rad}_\Lambda^{2^n}(\ , A) = 0$. Then it follows by V Lemma 7.6 and Lemma 1.1 that $\text{Supp}(A, \ )$ and $\text{Supp}(\ , A)$ are finite and contained in $\mathscr{C}$. Hence we have $\text{Hom}_\Lambda(A, B) = 0 = \text{Hom}_\Lambda(B, A)$ if B is in $\text{ind}\,\Lambda$ and not in $\mathscr{C}$. In particular there is some indecomposable projective module P with $\text{Hom}_\Lambda(P, A) \neq 0$, and hence P is in $\mathscr{C}$. Let $P_1, \ldots, P_n$ be the projective modules in $\text{ind}\,\Lambda \cap \mathscr{C}$. Then $\text{Hom}_\Lambda(P_i, Q) = \text{Hom}_\Lambda(Q, P_i) = 0$ for all i and all projective modules Q in $\text{ind}\,\Lambda\text{-}\mathscr{C}$. Since Λ is indecomposable, it follows that all indecomposable projective Λ-modules are in $\mathscr{C}$ (see II Section 5). Hence $\text{Supp}(Q, \ )$ is finite for each indecomposable projective module Q, and consequently $\text{Supp}(\Lambda, \ )$ is finite. Since $\text{Supp}(\Lambda, \ )$ contains all indecomposable modules and all modules in $\text{Supp}(\Lambda, \ )$ are in $\mathscr{C}$, we have that Λ is of finite representation type and $\mathscr{C} = \text{ind}\,\Lambda$.   □

The following result, confirming the first Brauer–Thrall conjecture, is now a direct consequence.

Corollary 1.5 *An artin algebra Λ is of finite representation type if and only if there is a bound on the lengths of the indecomposable Λ-modules.*   □

For an artin algebra Λ the easiest types of indecomposable modules are the simple, projective and injective ones. Starting with these modules we can try to construct new indecomposable modules. If after finding a finite set of indecomposable modules we are convinced that we have found all, it may not be easy to prove that this is actually the case. However, Theorem 1.4 provides a method for giving such a proof when we have a finite set of indecomposable modules which is a candidate for being all. This involves computing almost split sequences and minimal right and left almost split morphisms for the modules in our finite set. Here it is important that, as we have seen in V Section 2, there are criteria for deciding whether an exact sequence is almost split without knowing all the indecomposable modules. In addition the information on almost split sequences with simple end terms given in V Section 2 is useful. Note also that if in carrying out this procedure we encounter indecomposable modules which were not in our original list, we modify our list accordingly.

We illustrate this method on two concrete examples. Another illustration is given in the next section.

Example Let $\Lambda = k\Gamma$, where k is a field and Γ is the quiver $\underset{1}{\cdot} \leftarrow \underset{2}{\cdot} \rightarrow \underset{3}{\cdot}$.
We have for the three vertices 1, 2, 3 the simple modules S_1, S_2 and
S_3, the indecomposable projective modules $P_1 = S_1, P_2, P_3 = S_3$ and
the indecomposable injective modules $I_1, I_2 = S_2$ and I_3. Computing
$D\operatorname{Tr}$ for the nonprojective modules in these lists one gets $D\operatorname{Tr} I_1 \simeq S_3$,
$D\operatorname{Tr} I_3 \simeq S_1$, $D\operatorname{Tr} I_2 \simeq P_2$. We have exact sequences

$$0 \to S_3 \to P_2 \to I_1 \to 0,$$

$$0 \to S_1 \to P_2 \to I_3 \to 0,$$

and

$$0 \to P_2 \to I_1 \coprod I_3 \to S_2 \to 0.$$

They are all not split and the left hand term is obtained by applying
$D\operatorname{Tr}$ to the right hand term. Further we see that $\operatorname{End}_\Lambda(S_1)$, $\operatorname{End}_\Lambda(S_2)$ and
$\operatorname{End}_\Lambda(S_3)$ are all isomorphic to the field k. Then by V Corollary 2.4 the
above sequences are almost split. Considering in addition the minimal
right almost split morphism $S_1 \coprod S_3 \to P_2$ and the minimal left almost
split morphisms $I_1 \to S_2$ and $I_3 \to S_2$ we see that our six modules
constitute a component $\mathscr{C}$ of $\operatorname{ind} \Lambda$. Since Λ is clearly an indecomposable
algebra, we have $\mathscr{C} = \operatorname{ind} \Lambda$ by Theorem 1.4.

Example Let $T = k[X]/(X^2)$ and let $S = k[X]/(X)$ be the simple T-
module. Let $i: S \to T$ be a monomorphism, $p: T \to S$ an epimorphism
and $f: T \to T$ a morphism with $\operatorname{Im} f = S$. For the triangular ma-
trix algebra $\Lambda = \left(\begin{smallmatrix} T & 0 \\ T & T \end{smallmatrix}\right)$ consider the set of indecomposable Λ-modules
$\{(T, T, 1_T), (0, T, 0), (T, 0, 0), (0, S, 0), (S, S, 1_S), (S, T, i), (T, S, p), (S, 0, 0),$
$(T, T, f)\}$. By first computing $D\operatorname{Tr}$ for the indecomposable nonprojective modules in the set, it is not hard to see that the following are
almost split sequences.

$$
\begin{aligned}
0 &\to (T, T, f) \to & (S, T, i) \amalg (T, 0, 0) & \to (S, 0, 0) \to 0, \\
0 &\to (0, S, 0) \to & (T, S, p) \amalg (0, T, 0) & \to (T, T, f) \to 0, \\
0 &\to (0, T, 0) \to & (T, T, f) & \to (T, 0, 0) \to 0, \\
0 &\to (S, T, i) \to & (0, S, 0) \amalg (S, 0, 0) \amalg (T, T, 1_T) & \to (T, S, p) \to 0, \\
0 &\to (S, S, 1_S) \to & (S, T, i) & \to (0, S, 0) \to 0, \\
0 &\to (S, 0, 0) \to & (T, S, p) & \to (S, S, 1_S) \to 0, \\
0 &\to (T, S, p) \to & (S, S, 1_S) \amalg (T, T, f) & \to (S, T, i) \to 0.
\end{aligned}
$$

The indecomposable projective Λ-modules are $(T, T, 1_T)$ and $(0, T, 0)$
and we have minimal right almost split morphisms $(S, T, i) \to (T, T, 1_T)$

and $(0, S, 0) \rightarrow (0, T, 0)$. The indecomposable injective Λ-modules are $(T, T, 1_T)$ and $(T, 0, 0)$, and we have minimal left almost split morphisms $(T, T, 1_T) \rightarrow (T, S, p)$ and $(T, 0, 0) \rightarrow (S, 0, 0)$.

Since Λ is an indecomposable algebra, and we have a finite set of indecomposable modules closed under irreducible morphisms, we conclude that there are no other indecomposable Λ-modules.

The second Brauer–Thrall conjecture says that for an artin algebra Λ of infinite representation type over an infinite field there is an infinite number of positive integers n such that there is an infinite number of indecomposable modules of length n. This result has been proven for Λ an algebra over an algebraically closed field, but it is beyond the scope of this book to give a proof of this result. However, the following partial result follows from the material we have developed.

Proposition 1.6 *Let Λ be an artin algebra, χ an infinite cardinal and assume there are χ nonisomorphic indecomposable modules of length n. Then there exists an infinite number of integers $m > n$ such that there are at least χ indecomposable modules of length m.*

Proof Let I be a set of cardinality χ and $\{A_i | i \in I\}$ a set of nonisomorphic indecomposable Λ-modules of length n. Since there is only a finite number of indecomposable projective modules and χ is an infinite cardinal, there are a subset J of I of cardinality χ and an indecomposable projective Λ-module P such that $\text{Hom}_\Lambda(P, A_i) \neq 0$ for all $i \in J$. It follows by V Lemma 7.6 that $\text{Hom}_\Lambda(P, A_i)/\text{rad}_\Lambda^{2^n}(P, A_i) \neq 0$ for only a finite number of the modules A_i. Hence there is a subset K of I of cardinality χ such that $\text{rad}_\Lambda^{2^n}(P, A_i) \neq 0$ for all $i \in K$. Choose a morphism $f_i \neq 0$ in $\text{rad}_\Lambda^{2^n}(P, A_i)$ for $i \in K$. By V Proposition 7.4 we have for each i that $f_i = \sum_{l=1}^{t_i} g_{il2^n} \cdots g_{il1}$, where each $g_{ilj} : X_{ilj} \rightarrow X_{il(j+1)}$ is a morphism between indecomposable modules and $X_{il1} = P$ and $X_{il2^n} = A_i$ for all l with $1 \leq l \leq t_i$, and g_{ilj} with $2 \leq j \leq n$ can be chosen to be irreducible. Since all the f_i are nonzero, there is for each $i \in K$ some l with $1 \leq l \leq t_i$ such that the composition $g_{il2^n} \cdots g_{il2}$ of $2^n - 1$ nonisomorphisms is nonzero. Then it follows by Corollary 1.3 that there is some j such that writing $B_i = X_{ilj}$ we have $l(B_i) > n$. Since all $g_{ilj} : X_{ilj} \rightarrow X_{il(j+1)}$ are irreducible for $2 \leq j \leq 2^n$ and X_{il2^n} is equal to A_i, which has length n, it follows from V Proposition 6.6 that the X_{ilj} with $2 \leq j \leq 2^n$ are bounded in length, say by n_0. By V Corollary 6.8 there is only a finite number of nonisomorphic indecomposable modules which can be reached by at most 2^n irreducible morphisms starting at a given B_i. This shows that an indecomposable module X can be isomorphic to some B_i for only a finite number of i. Hence we have χ nonisomorphic indecomposable modules B_i with $n < l(B_i) \leq n_0$, and consequently there is some m with $n < m \leq n_0$ such that χ of them have length m. $\qquad\square$

2 Nakayama algebras

In this section we show how to use Theorem 1.4 to obtain the structure of the indecomposable modules over a Nakayama algebra. Using this structure we investigate $\operatorname{Hom}_\Lambda(A, B)$ as a module over $\operatorname{End}_\Lambda(A)^{\mathrm{op}}$ and over $\operatorname{End}_\Lambda(B)$ where A and B are indecomposable modules over a Nakayama algebra Λ.

We have the following main result in this section.

Theorem 2.1 *We have the following for a Nakayama algebra Λ.*

(a) *Every module in* ind Λ *is uniserial, and hence a factor of an indecomposable projective module.*

(b) Λ *is of finite representation type.*

Proof Assume that Λ is an indecomposable Nakayama algebra and let $\mathscr{C}$ be the set of uniserial modules in ind Λ. We want to show that $\mathscr{C}$ is a component of ind Λ and hence $\mathscr{C}$ is ind Λ by using Theorem 1.4. We know from IV Lemma 2.5 that a uniserial module C over a Nakayama algebra is uniquely determined up to isomorphism by its top $C/\mathfrak{r}C = S$ and its length $l(C) = t$. We denote this module by $S^{(t)}$. Let $\{S_0, S_1, \cdots, S_{n-1}\}$ be a Kupisch series for Λ. We then have for each i and t, where the lower index is calculated modulo n, a natural inclusion $f_{i+1}^{(t)} : S_{i+1}^{(t)} \to S_i^{(t+1)}$ and a natural epimorphism $p_i^{(t+1)} : S_i^{(t+1)} \to S_i^{(t)}$. For each i and t such that $S_i^{(t)}$ is not projective consider the sequence

$$\delta : 0 \to S_{i+1}^{(t)} \xrightarrow{\begin{pmatrix} f_{i+1}^{(t)} \\ p_{i+1}^{(t)} \end{pmatrix}} S_i^{(t+1)} \amalg S_{i+1}^{(t-1)} \xrightarrow{\left(-p_i^{(t+1)}, f_{i+1}^{(t-1)}\right)} S_i^{(t)} \to 0.$$

It is not hard to see that this is an exact sequence by considering lengths. Further, IV Proposition 2.6 gives $S_{i+1}^{(t)} \simeq D \operatorname{Tr} S_i^{(t)}$, and every nonisomorphism from $S_i^{(t)}$ to itself factors through $f_{i+1}^{(t-1)}$, hence through $\left(-p_i^{(t+1)}, f_{i+1}^{(t-1)}\right)$. Therefore δ is an almost split sequence by V Proposition 2.2. We now prove that $\mathscr{C}$ is a component of ind Λ.

Let $f:A \to B$ be an irreducible morphism between indecomposable Λ-modules with A or B uniserial. We want to prove that then both are uniserial. By duality it is enough to consider the case that B is uniserial. We have two cases to consider. If B is projective, then $A \simeq \mathfrak{r}B$ and is uniserial. If B is not projective, then $B \simeq S_i^{(t)}$ for some i and t and from the above discussion A is either isomorphic to $S_i^{(t+1)}$ or $S_{i+1}^{(t-1)}$. In both cases A is uniserial. This finishes the proof that $\mathscr{C}$ is a component as well as the proof of the theorem. □

The above theorem and its proof give explicit information about the indecomposable modules and the almost split sequences for Nakayama algebras. We now interpret some of the concepts and results we have discussed so far for this class of algebras.

By Theorem 2.1 a Nakayama algebra Λ is of finite representation type. From the structure of the almost split sequences we see that for a nonprojective indecomposable Λ-module C we have $\alpha(C) = 1$ if C is simple and $\alpha(C) = 2$ if C is not simple. This shows that $\alpha(\Lambda) = 1$ if and only if Λ is not semisimple and each nonprojective indecomposable module is simple, which is the case if and only if Λ has Loewy length 2. In view of the last comment in V Section 6 we have now proved that for an arbitrary artin algebra Λ, we have $\alpha(\Lambda) = 1$ if and only if Λ is a Nakayama algebra of Loewy length 2. If Λ is a Nakayama algebra of Loewy length greater than 2, we have $\alpha(\Lambda) = 2$.

When A and B are modules over an artin algebra Λ, then $\mathrm{Hom}_\Lambda(A, B)$ is in a natural way an $\mathrm{End}_\Lambda(A)^{\mathrm{op}}$-module and an $\mathrm{End}_\Lambda(B)$-module. In general it is, nevertheless, hard to describe the structure of the module $\mathrm{Hom}_\Lambda(A, B)$. It is, however, possible for Nakayama algebras to use our description of the modules to get some information on $\mathrm{Hom}_\Lambda(A, B)$ when A and B are indecomposable.

We start with the following preliminary observation.

Lemma 2.2 *Assume that Λ is a Nakayama algebra.*

(a) *Let $f:A \to C$ and $h:A \to E$ be epimorphisms between indecomposable modules. Given a morphism $g:E \to C$ there is a morphism $\tilde{g}:A \to A$ such that $gh = f\tilde{g}$, and g is an epimorphism if and only if $\tilde{g}$ is an isomorphism.*

(b) *Let $f:C \to B$ and $g:E \to B$ be monomorphisms between indecomposable modules. Given a morphism $h:C \to E$ there is a morphism*

$\tilde{h}: B \to B$ *such that* $\tilde{h}f = gh$, *and* h *is a monomorphism if and only if* $\tilde{h}$ *is an isomorphism.*

Proof (a) If n denotes the length of the uniserial module A, then A is a uniserial module of maximal length over the Nakayama algebra $\Lambda/\mathfrak{r}^n$. Since A is a factor of an indecomposable projective module, A is then a projective $(\Lambda/\mathfrak{r}^n)$-module and C and E are $(\Lambda/\mathfrak{r}^n)$-modules, so we have that $f: A \to C$ and $h: A \to E$ are projective covers over $\Lambda/\mathfrak{r}^n$. The claim follows by using this fact.

(b) This follows from (a) by duality. □

The next result gives explicit information about the $\mathrm{End}_\Lambda(A)^{\mathrm{op}}$-submodules and $\mathrm{End}_\Lambda(B)$-submodules of $\mathrm{Hom}_\Lambda(A, B)$.

Proposition 2.3 *Let A and B be indecomposable modules over a Nakayama algebra Λ, and let $\Gamma = \mathrm{End}_\Lambda(A)^{\mathrm{op}}$ and $\Sigma = \mathrm{End}_\Lambda(B)$.*

(a) *For f and g in $\mathrm{Hom}_\Lambda(A, B)$ the following are equivalent.*

    (i) $\Gamma g \subset \Gamma f$.

    (ii) $\mathrm{Im}\, g \subset \mathrm{Im}\, f$.

    (iii) $\Sigma g \subset \Sigma f$.

(b) *For $f \in \mathrm{Hom}_\Lambda(A, B)$ we have $\Gamma f = \Sigma f = \{g \in \mathrm{Hom}_\Lambda(A, B) \mid \mathrm{Im}\, g \subset \mathrm{Im}\, f\}$.*

(c) *Sending Γf to $\mathrm{Im}\, f$ gives a one to one inclusion preserving correspondence between the Γ-submodules of $\mathrm{Hom}_\Lambda(A, B)$ (which is the same as the Σ-submodules of $\mathrm{Hom}_\Lambda(A, B)$) and the Λ-submodules X of B with the property that $X/\mathfrak{r}X \simeq A/\mathfrak{r}A$ and $l(X) \le l(A)$.*

Proof (a) If $\Gamma g \subset \Gamma f$ we have $g = fs$ for some $s \in \mathrm{End}_\Lambda(A)$ and hence $\mathrm{Im}\, g \subset \mathrm{Im}\, f$. If $\mathrm{Im}\, g \subset \mathrm{Im}\, f$ there is by Lemma 2.2 a morphism $h: A \to A$ with $g = fh$, so that $\Gamma g \subset \Gamma f$. The second part follows similarly.

(b) This is a direct consequence of (a).

(c) Since B is a uniserial Λ-module, part (a) shows that all Γ-submodules and all Σ-submodules of $\mathrm{Hom}_\Lambda(A, B)$ are of the form $\Gamma f = \Sigma f$ for some $f \in \mathrm{Hom}_\Lambda(A, B)$. It is clear that if $X = \mathrm{Im}\, f$ for some $f \in \mathrm{Hom}_\Lambda(A, B)$, then $X/\mathfrak{r}X \simeq A/\mathfrak{r}A$ and $l(X) \le l(A)$. If conversely X is a submodule of B with this property, then we have an epimorphism $f: A \to X$ since the indecomposable Λ-modules are uniserial modules determined by their top and length. Then f induces a morphism $f: A \to B$ with $\mathrm{Im}\, f = X$. □

We state explicitly the following direct consequence.

Corollary 2.4 *Let A and B be indecomposable modules over a Nakayama algebra Λ. Then we have the following.*

(a) $\operatorname{Hom}_\Lambda(A, B)$ *is uniserial both as an* $\operatorname{End}_\Lambda(A)^{\mathrm{op}}$-*module and as an* $\operatorname{End}_\Lambda(B)$-*module, and its length is the number of times $A/\mathfrak{r}A$ occurs as a composition factor of* $\operatorname{soc}^n B$ *where $n = l(A)$.*

(b) $\operatorname{End}_\Lambda(A)$ *is a Nakayama algebra.*

Proof (a) For a submodule X of the uniserial module B we have $l(X) \le l(A) = n$ if and only if $X \subset \operatorname{soc}^n B$. Further, it is easy to see that the submodules X of $\operatorname{soc}^n B$ with $X/\mathfrak{r}X \simeq A/\mathfrak{r}A$ are in one to one correspondence with the simple composition factors isomorphic to $A/\mathfrak{r}A$ in the composition series for $\operatorname{soc}^n B$. Our claim now follows from Proposition 2.3.

(b) This follows directly from (a). □

3 Group algebras of finite representation type

In the previous section we illustrated how to use the criterion for an artin algebra to be of finite representation type by showing that Nakayama algebras are of finite representation type. This section is devoted to showing how a different technique for proving finite representation type can be used. In particular, we prove that if G is a finite group and k a field of characteristic $p > 0$ dividing the order of G, then kG is of finite representation type if and only if the Sylow p-subgroups of G are cyclic.

Our results in this section are all based on the following observation.

Lemma 3.1 *Let Λ be an R-subalgebra of the artin R-algebra Γ.*

(a) *Suppose Λ is a two-sided summand of Γ, i.e. $\Gamma = \Lambda \coprod C$ as a two-sided Λ-module. Then Λ is of finite representation type if Γ is of finite representation type.*

(b) *Suppose X is a Γ-summand of $\Gamma \otimes_\Lambda X$ for all X in $\operatorname{mod}\Gamma$. Then Γ is of finite representation type if Λ is of finite representation type.*

Proof (a) Let Γ be of finite representation type, and let Y be an indecomposable Λ-module. Then $\Gamma \otimes_\Lambda Y = (\Lambda \otimes_\Lambda Y) \coprod (C \otimes_\Lambda Y)$ as Λ-modules, so Y is a Λ-summand of $\Gamma \otimes_\Lambda Y$ viewed as a Λ-module.

Suppose $\{A_1, \ldots, A_t\}$ is a complete set of nonisomorphic indecomposable Γ-modules. Then $\Gamma \otimes_\Lambda Y = \coprod_{i=1}^t n_i A_i$ and so Y is a summand of A_i viewed as a Λ-module for some $i = 1, \ldots, t$. Therefore the nonisomorphic indecomposable Λ-summands of all the A_i give a complete set of nonisomorphic indecomposable Λ-modules. Therefore Λ is of finite representation type.

(b) Let $\{B_1, \ldots, B_t\}$ be a complete set of nonisomorphic indecomposable Λ-modules. Suppose X is an indecomposable Γ-module. Then viewing X as a Λ-module, we have that $X \simeq \coprod_{j=1}^t n_j B_j$ and so $\Gamma \otimes_\Lambda X \simeq \coprod_{j=1}^t (n_j \Gamma \otimes_\Lambda B_j)$. Since X is a Γ-summand of $\Gamma \otimes_\Lambda X$, we have that X is a Γ-summand of $\Gamma \otimes_\Lambda B_i$ for some i. Therefore the nonisomorphic indecomposable Γ-summands of all the $\Gamma \otimes_\Lambda B_i$ give a complete set of nonisomorphic indecomposable Γ-modules. Since this set is obviously finite, Γ is of finite representation type. $\qquad\square$

As our first application of this lemma we prove the following.

Proposition 3.2 *Let ΛG be a skew group algebra with G a finite group whose order is invertible in Λ. Then ΛG is of finite representation type if and only if Λ is of finite representation type.*

Proof We have already shown in III Lemma 4.5 (b) that Λ is a two-sided summand of ΛG. Therefore by Lemma 3.1, if ΛG is of finite representation type then Λ is of finite representation type. We also showed in III Lemma 4.8 that if the order of G is invertible in Λ, then X is a ΛG-summand of $\Lambda G \otimes_\Lambda X$ for all ΛG-modules X. Hence by Lemma 3.1 we have that ΛG is of finite representation type if Λ is of finite representation type. $\qquad\square$

Throughout the rest of this section we assume that k is a field of characteristic $p > 0$ and all groups are finite groups. Our main aim now is to prove the following.

Theorem 3.3 *Suppose G is a finite group whose order is divisible by p. Then kG is of finite representation type if and only if every Sylow p-subgroup of G is a cyclic group.*

Our proof of this theorem goes in two steps. Let P be a Sylow p-subgroup of G. We first show that kG is of finite representation type if and only if kP is of finite representation type. We then finish the proof by showing that if H is a p-group, then kH is of finite representation type if and only if H is cyclic.

Lemma 3.4 *Let H be a subgroup of G. Then the k-subspace $k[G - H]$ of kG with basis $G - H$ is a two-sided kH-submodule of kG such that $kG = kH \coprod k[G - H]$ as a two-sided kH-module.*

Proof This follows from the fact that $H(G-H) = G-H = (G-H)H$. $\square$

As an immediate consequence of this lemma and Lemma 3.1(a) we have the following.

Proposition 3.5 *Suppose kG is of finite representation type. Then kH is of finite representation type for all subgroups H of G.* $\square$

We investigate next which subgroups H of G have the property that X is a kG-summand of $kG \otimes_{kH} X$ for all X in $\mathrm{mod}\,kG$. To this end it is convenient to have the following general result.

Proposition 3.6 *Let Λ be an R-subalgebra of the R-algebra Γ. Then the following are equivalent for a Γ-module X.*

(a) X *is a Γ-summand of $\Gamma \otimes_\Lambda X$.*

(b) *If $0 \to A \to B \to C \to 0$ is an exact sequence of Γ-modules which splits as an exact sequence of Λ-modules, then $\mathrm{Hom}_\Gamma(X,B) \to \mathrm{Hom}_\Gamma(X,C) \to 0$ is exact.*

(c) *If $0 \to A \to B \to X \to 0$ is an exact sequence of Γ-modules which splits as an exact sequence of Λ-modules, then it splits as an exact sequence of Γ-modules.*

(d) *The epimorphism $m: \Gamma \otimes_\Lambda X \to X$ of Γ-modules given by the multiplication map splits as a Γ-epimorphism.*

Proof (a) $\Rightarrow$ (b) Suppose $0 \to A \to B \to C \to 0$ is an exact sequence of Γ-modules which splits as an exact sequence of Λ-modules. Using that the bimodule ${}_\Gamma\Gamma_\Lambda$ gives rise to the pair of adjoint functors $\Gamma \otimes_\Lambda : \mathrm{mod}\,\Lambda \to \mathrm{mod}\,\Gamma$ and $\mathrm{Hom}_\Gamma(\Gamma, \ ): \mathrm{mod}\,\Gamma \to \mathrm{mod}\,\Lambda$ we get the exact commutative diagram

$$0 \to \mathrm{Hom}_\Gamma(\Gamma \otimes_\Lambda X, A) \to \mathrm{Hom}_\Gamma(\Gamma \otimes_\Lambda X, B) \to \mathrm{Hom}_\Gamma(\Gamma \otimes_\Lambda X, C)$$
$$\wr\wr \qquad\qquad \wr\wr \qquad\qquad \wr\wr$$
$$0 \to \quad \mathrm{Hom}_\Lambda(X, A) \quad \to \quad \mathrm{Hom}_\Lambda(X, B) \quad \to \quad \mathrm{Hom}_\Lambda(X, C) \quad \to 0.$$

Hence it follows that $\mathrm{Hom}_\Gamma(\Gamma \otimes_\Lambda X, B) \to \mathrm{Hom}_\Gamma(\Gamma \otimes_\Lambda X, C) \to 0$ is exact. Since X is a Γ-summand of $\Gamma \otimes_\Lambda X$, it follows that $\text{Hom}_\Gamma(X, B) \to$ $\text{Hom}_\Gamma(X, C) \to 0$ is exact because the functor $\text{Hom}_\Gamma(\ , Y)$ commutes with sums.

(b) $\Rightarrow$ (c) Since the exact sequence of Γ-modules $0 \to A \to B \xrightarrow{g} X \to 0$ splits as an exact sequence of Λ-modules, we have that $\text{Hom}_\Gamma(X, B) \to$ $\text{Hom}_\Gamma(X, X) \to 0$ is exact. Therefore there is a Γ-morphism $f : X \to B$ such that $gf : X \to X$ is the identity. This means that $0 \to A \to B \to X \to 0$ splits as an exact sequence of Γ-modules.

(c) $\Rightarrow$ (d) The Λ-morphism $f : X \to \Gamma \otimes_\Lambda X$ given by $f(x) = 1 \otimes x$ for all x in X has the property that $mf = 1_X$. Therefore $\Gamma \otimes_\Lambda X \to X$ is a Γ-epimorphism which splits as a Λ-epimorphism. Hence it splits as a Γ-epimorphism.

(d) $\Rightarrow$ (a) This is trivial. $\qquad\qquad\qquad\qquad\qquad\qquad\qquad\qquad\square$

Suppose Λ is an R-subalgebra of the R-algebra Γ. A Γ-module X is said to be **relatively projective** over Λ if it satisfies the equivalent conditions of Proposition 3.6.

Our aim now is to establish the following.

Proposition 3.7 *The following are equivalent for a subgroup H of G.*

(a) *The trivial kG-module k is relatively projective over kH.*

(b) *$[G : H]$, the index of H in G, is not divisible by p.*

(c) *H contains a Sylow p-subgroup of G.*

(d) *Every kG-module is relatively projective over kH.*

Proof (a) $\Rightarrow$ (b) We first observe the following. Let $\{1 = \sigma_1, \sigma_2, \ldots, \sigma_t\}$ be a set of left coset representatives of H in G. Then G is the disjoint union $H \cup \sigma_2 H \cup \cdots \cup \sigma_t H$ and so $kG = kH \coprod k[\sigma_2 H] \coprod \cdots \coprod k[\sigma_t H]$ as a right kH-module, where each $k[\sigma_i H]$ is the k-subspace of kG with basis the elements in $\sigma_i H$. But for each i we have $k[\sigma_i H] \simeq kH$ as a right kH-module. Therefore $kG \otimes_{kH} X \simeq tX$ as a k-vector space, and so $\dim_k(kG \otimes_{kH} X) = t \dim_k X$. We will use this fact shortly.

Let $k[G/H]$ be the k-vector space with basis the left cosets of H in G. Then the operation of G on G/H given by $\tau(\sigma H) = \tau\sigma H$ gives $k[G/H]$ a left kG-module structure. Unless stated to the contrary, this is the only way we will consider $k[G/H]$ as a kG-module.

Consider now the map $f : kG \otimes_{kH} k \to k[G/H]$ given by $f(\sum_{\sigma \in G} t_\sigma \sigma \otimes b) = b(\sum_{\sigma \in G} t_\sigma \sigma H)$. It is not difficult to see that this is an epimorphism of kG-modules. By our previous remark $\dim_k(kG \otimes_{kH} k) = [G : H] \dim_k k = [G : H]$ which is clearly the same as

$\dim_k k[G/H]$. Therefore $f: kG \otimes_{kH} k \to k[G/H]$ is an isomorphism. Moreover, if we define $g: k[G/H] \to k$ by $g(\sum_{\sigma H \in G/H} t_{\sigma H} \sigma H) = \sum_{\sigma H \in G/H} t_{\sigma H}$ and $m: kG \otimes_{kH} k \to k$ is the multiplication map, then g is a morphism of kG-modules such that the diagram

$$
\begin{array}{ccc}
kG \otimes_{kH} k & \xrightarrow{f} & k[G/H] \\
\downarrow m & & \downarrow g \\
k & = & k
\end{array}
$$

commutes. Therefore m is a split epimorphism if and only if g is a split epimorphism. Since $k^G = k$, g is a split epimorphism if and only if there is some x in $k[G/H]^G$ such that $g(x) = 1$. But $\sum_{\sigma H \in G/H} t_{\sigma H} \sigma H$ is in $k[G/H]^G$ if and only if all the $t_{\sigma H}$ are the same. So $g(k[G/H])^G$ consists of the elements $[G:H]t$ for all t in k. Hence g is a split epimorphism if and only if p does not divide $[G:H]$.

(b) $\Leftrightarrow$ (c) This is trivial.

(b) $\Rightarrow$ (d) This proof requires some preliminary considerations of a fairly general nature.

Let H be an arbitrary subgroup of G and let $1 = \sigma_1, \ldots, \sigma_t$ be left coset representatives of H in G. Suppose X and Y are kG-modules and $f: X \to Y$ is a kH-module morphism. Define $\tilde{f}: X \to Y$ by $\tilde{f}(x) = \sum_{i=1}^t \sigma_i f(\sigma_i^{-1} x)$ for all x in X. Suppose we change coset representatives to $\sigma_1 h_1, \ldots, \sigma_t h_t$ with the h_i in H. Then we have $\sum_{i=1}^t \sigma_i h_i f((\sigma_i h_i)^{-1} x) = \sum_{i=1}^t \sigma_i h_i f(h_i^{-1} \sigma_i^{-1} x) = \sum_{i=1}^t \sigma_i h_i h_i^{-1} f(\sigma_i^{-1} x) = \sum_{i=1}^t \sigma_i f(\sigma_i^{-1} x)$ for all x in X. Thus $\tilde{f}$ is independent of the particular coset representatives used to define it.

We now show that $\tilde{f}$ is a kG-morphism. For let σ^{-1} be in G. Then $\{\sigma^{-1}\sigma_1, \ldots, \sigma^{-1}\sigma_t\}$ is also a set of left coset representatives of H in G, and we have $\tilde{f}(x) = \sum_{i=1}^t \sigma^{-1}\sigma_i f(\sigma_i^{-1}\sigma x) = \sum_{i=1}^t \sigma_i f(\sigma_i^{-1} x)$. Hence we get $\tilde{f}(\sigma x) = \sum_{i=1}^t \sigma_i f(\sigma_i^{-1}(\sigma x)) = \sigma \sum_{i=1}^t \sigma^{-1}\sigma_i f(\sigma_i^{-1}\sigma x) = \sigma \tilde{f}(x)$ for all x in X, which shows that $\tilde{f}$ is a kG-morphism.

We now return to proving (b) $\Rightarrow$ (d). Suppose $t = [G:H]$ and $1/t \in k$. Let X be a kG-module. Define $f: X \to kG \otimes_{kH} X$ to be the kH-morphism defined by $f(x) = 1/t(1 \otimes x)$. Then we get $\tilde{f}(x) = 1/t \sum_{i=1}^t \sigma_i f(\sigma_i^{-1}(x)) = 1/t \sum_{i=1}^t \sigma_i(1 \otimes \sigma_i^{-1}(x)) = 1/t \sum_{i=1}^t (\sigma_i \otimes \sigma_i^{-1}(x))$ for all x in X. Therefore we have $m\tilde{f}(x) = 1/t \sum_{i=1}^t \sigma_i \sigma_i^{-1}(x) = x$, so that $m\tilde{f} = 1_X$ and hence $m: \Gamma \otimes_\Lambda X \to X$ is a split Γ-epimorphism.

(d) $\Rightarrow$ (a) This is trivial. $\qquad \square$

Assume that kP is of finite representation type for each Sylow p-

subgroup P of G. Then every kG-module is relatively projective over kH by Proposition 3.7, and hence kG is of finite representation type by Lemma 3.1. Since kP is of finite representation type if kG is, by Proposition 3.5, it only remains to prove that if G is a p-group, then kG is of finite representation type if and only if G is cyclic, in order to finish the proof of Theorem 3.3.

Proposition 3.8 *Suppose G is a cyclic p-group of order p^n. Then the k-algebras kG and $k[X]/(X^{p^n})$ are isomorphic, so kG is of finite representation type.*

Proof Let σ be a generator of the cyclic group G. Define $f : k[Y] \to kG$ by $f(\sum_{i=0}^{m} a_i Y^i) = \sum_{i=0}^{m} a_i \sigma^i$. It is not difficult to see that f is a surjective k-algebra map with $\operatorname{Ker} f = (Y^{p^n} - 1)$. Since k is of characteristic $p > 0$ we know that $(Y^{p^n} - 1) = (Y - 1)^{p^n}$. Now $k[Y - 1] = k[Y]$ so we have that $k[Y]/((Y - 1)^{p^n}) = k[Y - 1]/((Y - 1)^{p^n})$. Letting $X = Y - 1$, we have that $kG \simeq k[X]/(X)^{p^n}$. Since $k[X]/(X)^{p^n}$ is a Nakayama algebra it follows that kG is of finite representation type. $\qquad\square$

In order to show that if G is a noncyclic p-group, then kG is of infinite representation type, we will first consider the quotient group of G by its commutator subgroup $[G, G]$. For this we need the following result about p-groups.

Lemma 3.9 *Let G be a p-group of order p^n and let $[G, G]$ be the commutator subgroup of G. If $G/[G, G]$ is cyclic, then $G \simeq \mathbb{Z}/p^n\mathbb{Z}$.*

Proof We use induction on n. If $n = 1$, then $G \simeq \mathbb{Z}/p\mathbb{Z}$ and we are done. Suppose the order of G is p^{n+1}. Since G is a p-group, we know that the center Z of G is not $\{1\}$. Therefore there is a subgroup H of Z of order p which is of course a normal subgroup of G. Let $f : G \to G/H$ be the canonical surjective homomorphism of groups. Then f induces a surjection $[G, G] \to [G/H, G/H]$ of the commutator subgroup of G to the commutator subgroup of G/H. Thus f induces a surjection $h : G/[G, G] \to (G/H)/[G/H, G/H]$ which implies that $(G/H)/[G/H, G/H]$ is cyclic. Since the order of G/H is p^n, we know by the induction hypothesis that G/H is cyclic.

Let σ in G be such that $f(\sigma)$ generates G/H. We claim that $\langle \sigma \rangle \supset H$. If not, then we have $\langle \sigma \rangle \cap H = \{1\}$. Since H is contained in the center of G, then the subgroup J of G generated by $\langle \sigma \rangle$ and H is the product $\langle \sigma \rangle \times H$

which is abelian and not cyclic. But $J = G$ since the order of J is $p^n p$. Hence $G/[G, G] = G$ is not cyclic, which is a contradiction. Therefore we get $H \subset \langle \sigma \rangle$ which means that $G = \langle \sigma \rangle$, i.e. G is cyclic. $\square$

We are now ready to finish the proof of Theorem 3.3. Suppose now that G is a noncyclic p-group. Then $G/[G, G]$ is not cyclic by Lemma 3.4 and hence $G/[G, G] \simeq (\mathbb{Z}/p^{n_1}\mathbb{Z}) \times \cdots \times (\mathbb{Z}/p^{n_t}\mathbb{Z})$ with $t \geq 2$. Therefore there is a surjection $G \to (\mathbb{Z}/p\mathbb{Z}) \times (\mathbb{Z}/p\mathbb{Z})$. This induces a surjection $kG \to k[(\mathbb{Z}/p\mathbb{Z}) \times (\mathbb{Z}/p\mathbb{Z})]$ of k-algebras. Now we have the k-algebra isomorphisms $k[(\mathbb{Z}/p\mathbb{Z}) \times (\mathbb{Z}/p\mathbb{Z})] \simeq k[X, Y]/(X^p - 1, Y^p - 1) \simeq k[X, Y]/((X - 1)^p, (Y - 1)^p) \simeq k[S, T]/(S^p, T^p)$ where $S = (X - 1)$ and $T = (Y - 1)$. Since $(S^p, T^p) \subset (S, T)^2$, we have the surjective k-algebra morphism $k[S, T]/(S^p, T^p) \to k[S, T]/(S, T)^2$. So we have a surjection of k-algebras $kG \to k[S, T]/(S, T)^2$ which shows that kG is of infinite representation type since $k[S, T]/(S, T)^2$ is of infinite representation type. The fact that $k[S, T]/(S, T)^2$ is of infinite representation type was shown in IV Section 1. This finishes the proof of Theorem 3.3. $\square$

4 Grothendieck groups

In I Section 1 we discussed the Grothendieck group $K_0(\text{f.l.}\,\Lambda)$ of finite length modules over a ring Λ. In this section we show how almost split sequences can be used in the study of the Grothendieck group when Λ is an artin algebra and hence f.l. $\Lambda = \text{mod}\,\Lambda$.

Denote by $K_0(\text{mod}\,\Lambda, 0)$ the free abelian group with $\text{ind}\,\Lambda$ as basis. For each A in $\text{ind}\,\Lambda$ let $[A]$ denote the corresponding basis element of $K_0(\text{mod}\,\Lambda, 0)$ and for $M \simeq \coprod_{i \in I} n_i A_i$ with $n_i \in \mathbb{N}$ and $A_i \in \text{ind}\,\Lambda$, let $[M]$ denote $\sum_{i \in I} n_i [A_i]$ in $K_0(\text{mod}\,\Lambda, 0)$. For each short exact sequence $\delta : 0 \to A \to B \to C \to 0$ in $\text{mod}\,\Lambda$ consider the element $[A] - [B] + [C]$ in $K_0(\text{mod}\,\Lambda, 0)$ which we shall denote by $[\delta]$. Let H be the subgroup of $K_0(\text{mod}\,\Lambda, 0)$ generated by the elements $[\delta]$ where δ runs through all short exact sequences of $\text{mod}\,\Lambda$. The main result in this section is that H is generated by the elements $[\delta]$ where δ runs through the almost split sequences if and only if Λ is of finite representation type.

Recall from IV Section 4 that for an artin R-algebra Λ we denote by $\langle A, B \rangle$ the R-length of $\text{Hom}_\Lambda(A, B)$ for each pair of Λ-modules A and B. Now $\langle \ , \ \rangle : \text{ind}\,\Lambda \times \text{ind}\,\Lambda \to \mathbb{Z}$ determines a bilinear form from $K_0(\text{mod}\,\Lambda, 0) \times K_0(\text{mod}\,\Lambda, 0)$ to $\mathbb{Z}$ which we also denote by $\langle \ , \ \rangle$. With each C in $\text{ind}\,\Lambda$ let $\delta_{[C]}$ in $K_0(\text{mod}\,\Lambda, 0)$ denote the following elements.

(i) If C is not projective $\delta_{[C]} = [\delta_C]$ where δ_C is the almost split sequence $0 \to A \to B \to C \to 0$.

(ii) If C is projective then $\delta_{[C]} = [C] - [\mathfrak{r}C]$.

We have the following elementary result where l_C denotes the length of $\mathrm{End}_\Lambda(C)/\mathrm{rad}(\mathrm{End}_\Lambda(C))$ as an R-module.

Proposition 4.1

(a) *For each X in* $\mathrm{ind}\,\Lambda$ *we have* $\langle [X], \delta_{[C]} \rangle = 0$ *if $X \not\simeq C$, and* $\langle [C], \delta_{[C]} \rangle = l_C$.

(b) *For each $x \in K_0(\mathrm{mod}\,\Lambda, 0)$ we have* $x = \sum_{C \in \mathrm{ind}\,\Lambda}(\langle x, \delta_{[C]} \rangle / l_C)[C]$.

(c) $\{\delta_{[C]}\}_{C \in \mathrm{ind}\,\Lambda}$ *is linearly independent in* $K_0(\mathrm{mod}\,\Lambda, 0)$.

Proof (a) Let C be a nonprojective module in $\mathrm{ind}\,\Lambda$ and let $\delta_C : 0 \to A \to B \to C \to 0$ be an almost split sequence. Then the contravariant defect δ_C^* of δ_C satisfies $\delta_C^*(X) = 0$ for each X in $\mathrm{ind}\,\Lambda$ with $X \not\simeq C$ and $\delta_C^*(C)$ is a simple $\mathrm{End}_\Lambda(C)^{\mathrm{op}}$-module according to V Proposition 2.2. Letting as usual $\langle \delta_C^*(X) \rangle$ denote the length of $\delta_C^*(X)$ as an R-module we obtain $\langle [X], \delta_{[C]} \rangle = \langle \delta_C^*(X) \rangle = 0$ if X is in $\mathrm{ind}\,\Lambda$ and $X \not\simeq C$ and $\langle [C], \delta_{[C]} \rangle = \langle \delta_C^*(C) \rangle = l_C$. Hence (a) holds if C is nonprojective.

Now let C be projective. Then for all X in $\mathrm{ind}\,\Lambda$ with $X \not\simeq C$ we have $\mathrm{Hom}_\Lambda(X, C) = \mathrm{Hom}_\Lambda(X, \mathfrak{r}C)$ and so $\langle [X], \delta_{[C]} \rangle = 0$. Further $\mathrm{Hom}_\Lambda(C, \mathfrak{r}C) = \mathrm{rad}(\mathrm{End}_\Lambda C)$ and therefore $\langle [C], \delta_{[C]} \rangle = l_C$.

Parts (b) and (c) follow directly from (a). $\qquad\square$

As a direct consequence of this result we obtain the following criterion for two modules A and B to be isomorphic.

Theorem 4.2

(a) *The following are equivalent for x and y in $K_0(\mathrm{mod}\,\Lambda, 0)$.*

  (i) $x = y$.

  (ii) $\langle x, [C] \rangle = \langle y, [C] \rangle$ *for all C in* $\mathrm{ind}\,\Lambda$.

  (iii) $\langle [C], x \rangle = \langle [C], y \rangle$ *for all C in* $\mathrm{ind}\,\Lambda$.

(b) *The following are equivalent for A and B in* $\mathrm{mod}\,\Lambda$.

  (i) $A \simeq B$.

  (ii) $\langle A, C \rangle = \langle B, C \rangle$ *for all C in* $\mathrm{ind}\,\Lambda$.

  (iii) $\langle C, A \rangle = \langle C, B \rangle$ *for all C in* $\mathrm{ind}\,\Lambda$.

Proof (a) If $x = y$ then we obviously have $\langle x, [C] \rangle = \langle y, [C] \rangle$. Conversely, if $\langle x, [C] \rangle = \langle y, [C] \rangle$ for all C in ind Λ, then $\langle x, \delta_{[C]} \rangle = \langle y, \delta_{[C]} \rangle$ for all C in ind Λ. Hence we get $\langle x - y, \delta_{[C]} \rangle = 0$ for all C in ind Λ. But Proposition 4.1(b) gives that $x - y = \sum_{C \in \text{ind } \Lambda} (\langle x - y, \delta_{[C]} \rangle / l_C)[C]$ which is then 0. The second equivalence follows by duality.

(b) This is a direct consequence of (a). □

Using $K_0(\text{mod } \Lambda, 0)$ and $\delta_{[C]}$ we have the following characterization of finite representation type.

Theorem 4.3 *The following are equivalent for an artin algebra Λ.*

(a) Λ *is of finite representation type.*
(b) $\{\delta_{[C]}\}_{C \in \text{ind } \Lambda}$ *generates* $K_0(\text{mod } \Lambda, 0)$.
(c) $\{\delta_{[C]}\}_{C \in \text{ind } \Lambda}$ *is a basis for* $K_0(\text{mod } \Lambda, 0)$.

Proof Since $\{\delta_{[C]}\}_{C \in \text{ind } \Lambda}$ is linearly independent according to Proposition 4.1(c), we have that (b) and (c) are equivalent.

We next prove that (a) implies (b). So assume Λ is of finite representation type. Then $K_0(\text{mod } \Lambda, 0)$ is a finitely generated free abelian group with basis $\{[C]\}_{C \in \text{ind } \Lambda}$. For x in $K_0(\text{mod } \Lambda, 0)$ we have that $y = \sum_{C \in \text{ind } \Lambda} (\langle [C], x \rangle / l_C) \delta_{[C]}$ is an element of $K_0(\text{mod } \Lambda, 0)$ since l_C divides $\langle [C], [X] \rangle$ for all X in ind Λ. But then

$$\langle [C], y \rangle = (\langle [C], x \rangle / l_C) \langle [C], \delta_{[C]} \rangle = \langle [C], x \rangle$$

for all C in ind Λ. Hence we get $x = y$ by Theorem 4.2. This shows that $\{\delta_{[C]}\}_{C \in \text{ind } \Lambda}$ generates $K_0(\text{mod } \Lambda, 0)$.

Conversely, assume that $\{\delta_{[C]}\}_{C \in \text{ind } \Lambda}$ generates $K_0(\text{mod } \Lambda, 0)$. In particular $[D\Lambda] = \sum_{C \in \text{ind } \Lambda} a_C \delta_{[C]}$ with $a_C \neq 0$ for only finitely many C in ind Λ. Since $\langle X, D\Lambda \rangle$ is nonzero for all X in ind Λ and $\langle X, D\Lambda \rangle = a_X l_X$ one has that ind Λ is finite. □

As a consequence of this we prove our desired result on the relations for the Grothendieck group. As usual we let $\mathscr{P}(\Lambda)$ denote the full subcategory of mod Λ of projective modules.

Theorem 4.4 *For an artin algebra Λ the following are equivalent, where $\phi : K_0(\text{mod } \Lambda, 0) \to K_0(\text{mod } \Lambda)$ is the natural map.*

(a) Λ *is of finite representation type.*
(b) $\{\delta_{[C]}\}_{C \in \text{ind } \Lambda - \mathscr{P}(\Lambda)}$ *generates* $\text{Ker } \phi$.

(c) $\{\delta_{[C]}\}_{C\in\text{ind}\,\Lambda-\mathscr{P}(\Lambda)}$ *is a basis for* $\text{Ker}\,\phi$.

Proof Parts (b) and (c) are equivalent since $\{\delta_{[C]}\}_{C\in\text{ind}\,\Lambda-\mathscr{P}(\Lambda)}$ is a linearly independent set by Proposition 4.1(c).

Assume first that Λ is of finite representation type. Recall from I Theorem 1.7 that $K_0(\text{mod}\,\Lambda)$ is a free abelian group with basis $\{\phi([S])\}_{S\in\mathscr{L}}$, where $\mathscr{L}$ denotes a set of representatives from each of the isomorphism classes of simple Λ-modules, and that $\phi[C] = \sum_{S\in\mathscr{L}} m_S\phi[S]$ where C has a composition series with m_S composition factors isomorphic to S for S in $\mathscr{L}$. Therefore ϕ maps the subgroup K of $K_0(\text{mod}\,\Lambda, 0)$ generated by $\{\delta_{[C]}\}_{C\in\mathscr{P}(\Lambda)}$ isomorphically onto $K_0(\text{mod}\,\Lambda)$. By Theorem 4.3 it follows that $\{\delta_{[C]}\}_{C\in\text{ind}\,\Lambda-\mathscr{P}(\Lambda)}$ forms a basis for a complement of K in $K_0(\text{mod}\,\Lambda, 0)$ which is obviously contained in $\text{Ker}\,\phi$. Hence $\{\delta_{[C]}\}_{C\in\text{ind}\,\Lambda-\mathscr{P}(\Lambda)}$ generates $\text{Ker}\,\phi$.

Conversely, assume $\{\delta_{[C]}\}_{C\in\text{ind}\,\Lambda-\mathscr{P}(\Lambda)}$ generates $\text{Ker}\,\phi$. Since $\text{Ker}\,\phi$ is a complement to K, where K is as above, then $\{\delta_{[C]}\}_{C\in\text{ind}\,\Lambda}$ generates $K_0(\text{mod}\,\Lambda, 0)$. But then Λ is of finite representation type by Theorem 4.3.

$\square$

5 Auslander algebras

In studying artin algebras of finite representation type, it has proven useful to consider another class of artin algebras called Auslander algebras. An artin algebra Γ is an **Auslander algebra** if it satisfies the following conditions: (a) gl.dim $\Gamma \leq 2$ and (b) if $0 \to \Gamma \to I_0 \to I_1 \to I_2 \to 0$ is a minimal injective resolution of Γ, then I_0 and I_1 are projective Γ-modules. This section is mainly devoted to showing how to construct Auslander algebras from artin algebras of finite type and the other way around. These constructions give an inverse bijection between the Morita equivalence classes of artin algebras of finite representation type and Morita equivalence classes of Auslander algebras. As will become apparent, the module theories of artin algebras of finite representation type and their associated Auslander algebras are intimately related and it is this module theoretic as well as ring theoretic relationship which has proven to be of use in studying algebras of finite representation type.

Let Λ be an artin algebra. A Λ-module M is said to be an **additive generator** for Λ if add $M = \text{mod}\,\Lambda$. Clearly a module M is an additive generator for Λ if and only if every indecomposable Λ-module is isomorphic to a summand of M. From this it follows that Λ is of finite representation type if and only if Λ has an additive generator. In particular, let $\{M_1, \ldots, M_t\}$ be a complete set of representatives of the isomorphism classes of indecomposable modules for an artin algebra Λ of finite representation type. Then $M = \coprod_{i=1}^{t} M_i$ is an additive generator for Λ and a Λ-module M' is an additive generator for Λ if and only if M is isomorphic to a summand of M'.

Suppose now that M is an additive generator for an artin algebra Λ of finite representation type and let $\Gamma_M = \operatorname{End}_\Lambda(M)^{\mathrm{op}}$. It then follows from II Proposition 2.1(c) that the functor $\operatorname{Hom}_\Lambda(M, \ ): \operatorname{mod}\Lambda \to \operatorname{mod}\Gamma_M$ induces an equivalence between $\operatorname{mod}\Lambda$ and the full subcategory $\mathscr{P}(\Gamma_M)$ of $\operatorname{mod}\Gamma_M$ consisting of the projective Γ_M-modules. Hence if M' is another additive generator for Λ, then the categories of projective Γ_M- and $\Gamma_{M'}$-modules are equivalent, which means that the algebras Γ_M and $\Gamma_{M'}$ are Morita equivalent. Thus associated with the artin algebra Λ of finite representation type are the unique, up to Morita equivalence, algebras Γ_M with M an additive generator for Λ. Our main aim in this section is to show that these algebras Γ_M are exactly the Auslander algebras. These results form the basis for using Auslander algebras to study artin algebras of finite representation type.

For an artin algebra Λ and a module M in $\operatorname{mod}\Lambda$ we will throughout this section denote by Γ_M the artin algebra $\operatorname{End}_\Lambda(M)^{\mathrm{op}}$. We begin by pointing out some crucial homological facts concerning the algebras Γ_M when Λ is an artin algebra of finite representation type and M is an additive generator for $\operatorname{mod}\Lambda$. First we give a description of the projective resolutions of Γ_M-modules.

Lemma 5.1 *Let Λ be an artin algebra of finite representation type, M an additive generator and let X be in $\operatorname{mod}\Gamma_M$. Then we have the following.*

(a) *Suppose $P_1 \xrightarrow{h} P_0 \to X \to 0$ is a projective Γ_M-presentation for X. Then there is an exact sequence $0 \to A_2 \xrightarrow{f} A_1 \xrightarrow{g} A_0$ of Λ-modules such that the induced exact sequence of projective Γ_M-modules $0 \to \operatorname{Hom}_\Lambda(M, A_2) \xrightarrow{\operatorname{Hom}_\Lambda(M,f)} \operatorname{Hom}_\Lambda(M, A_1) \xrightarrow{\operatorname{Hom}_\Lambda(M,g)} \operatorname{Hom}_\Lambda(M, A_0)$ gives a Γ_M-projective resolution of X with the morphism $\operatorname{Hom}_\Lambda(M, g)$ isomorphic to the morphism h.*

(b) *$\operatorname{pd}_{\Gamma_M} X \leq 2$.*

Proof (a) Since $\operatorname{Hom}_\Lambda(M, \ ): \operatorname{mod}\Lambda \to \operatorname{mod}\Gamma_M$ induces an equivalence between $\operatorname{mod}\Lambda$ and $\mathscr{P}(\Gamma_M)$, there is a morphism $A_1 \xrightarrow{g} A_0$ in $\operatorname{mod}\Lambda$ such that the induced morphism $\text{Hom}_\Lambda(M,g)$ is isomorphic to h. Then the exact sequence of Λ-modules $0 \to \text{Ker}\, g \to A_1 \overset{g}{\to} A_0$ has our desired properties.

(b) This follows directly from (a). □

Applying Lemma 5.1 we obtain the following.

Proposition 5.2 *Let Λ be an artin algebra of finite representation type and M an additive generator for* $\text{mod}\,\Lambda$.

(a) *If Λ is semisimple, then Γ_M is semisimple and Morita equivalent to Λ.*
(b) *If Λ is not semisimple, then* $\text{gl.dim}\,\Gamma_M = 2$.

Proof (a) If Λ is semisimple, then M is a projective generator, and hence Γ_M is Morita equivalent to Λ and consequently also semisimple.

(b) Assume Λ is not semisimple. Then there is some simple Λ-module S such that the projective cover $f : P \to S$ is not an isomorphism. Then $\text{Hom}(M, f) : \text{Hom}_\Lambda(M, P) \to \text{Hom}_\Lambda(M, S)$ is a nonzero morphism between indecomposable projective Γ_M-modules which is not a monomorphism. From this it follows that $\text{gl.dim}\,\Gamma_M \geq 2$, and consequently $\text{gl.dim}\,\Gamma_M = 2$ by Lemma 5.1. □

Throughout the rest of this discussion we will be mainly concerned with nonsemisimple artin algebras of finite representation type. It is convenient to make the following definition.

Let Σ be an artin algebra. The **dominant dimension** of a Σ-module A, which we denote by $\text{dom.dim}_\Sigma A$, is the maximum integer t (or ∞) having the property that if $0 \to A \to I_0 \to I_1 \to \cdots \to I_t \to \cdots$ is a minimal injective resolution of A, then I_j is projective for all $j < t$ (or ∞).

Our aim now is to show that if Γ_M is obtained from a nonsemisimple artin algebra Λ of finite representation type, then $\text{dom.dim}_\Gamma \Gamma_M = 2$. To this end we point out the following.

Lemma 5.3 *Let Λ be an artin algebra of finite representation type and M an additive generator for* $\text{mod}\,\Lambda$.

(a) *If I is an injective Λ-module, then $\text{Hom}_\Lambda(M, I)$ is an injective Γ_M-module.*
(b) *Let $0 \to A \to I_0 \to I_1$ be a minimal injective copresentation in $\text{mod}\,\Lambda$ of the Λ-module A. Then $0 \to \text{Hom}_\Lambda(M, A) \to \text{Hom}_\Lambda(M, I_0) \to$*

$\mathrm{Hom}_\Lambda(M, I_1)$ *is a minimal injective* Γ_M*-copresentation for the projective* Γ_M*-module* $\mathrm{Hom}_\Lambda(M, A)$.

(c) *A* Γ_M*-module is a projective injective module if and only if it is isomorphic to* $\mathrm{Hom}_\Lambda(M, I)$ *for some injective* Λ*-module* I.

(d) *The functor* $\mathrm{Hom}_\Lambda(M,\ ): \mathrm{mod}\,\Lambda \to \mathrm{mod}\,\Gamma_M$ *induces an equivalence between the category* $\mathscr{I}(\Lambda)$ *of the injective* Λ*-modules and the category of projective injective* Γ_M*-modules.*

Proof (a) Let I be an injective Λ-module. We now show that $\mathrm{Hom}_\Lambda(M, I)$ is an injective Γ_M-module by showing that $\mathrm{Ext}^i_{\Gamma_M}(X, \mathrm{Hom}_\Lambda(M, I)) = 0$ for all $i > 0$ and X in $\mathrm{mod}\,\Gamma_M$. We have already seen in Lemma 5.1 that if X is in $\mathrm{mod}\,\Gamma_M$, then there is an exact sequence $0 \to A \to B \to C$ of Λ-modules such that we have a projective Γ_M-resolution $0 \to \mathrm{Hom}_\Lambda(M, A) \to \mathrm{Hom}_\Lambda(M, B) \to \mathrm{Hom}_\Lambda(M, C) \to X \to 0$ of X. This gives rise to the sequence

(*) $\quad \mathrm{Hom}_{\Gamma_M}(\mathrm{Hom}_\Lambda(M, C), \mathrm{Hom}_\Lambda(M, I)) \to \mathrm{Hom}_{\Gamma_M}(\mathrm{Hom}_\Lambda(M, B), \mathrm{Hom}_\Lambda(M, I)) \to$
$\quad\quad \mathrm{Hom}_{\Gamma_M}(\mathrm{Hom}_\Lambda(M, A), \mathrm{Hom}_\Lambda(M, I)) \to 0$

which is isomorphic to the sequence

(**) $\quad\quad\quad \mathrm{Hom}_\Lambda(C, I) \to \mathrm{Hom}_\Lambda(B, I) \to \mathrm{Hom}_\Lambda(A, I) \to 0$

since $\mathrm{Hom}_\Lambda(M,\ ): \mathrm{mod}\,\Lambda \to \mathscr{P}(\Gamma)$ is an equivalence of categories. Since I is an injective Λ-module, the sequence (**) is exact. This means that the sequence (*) is exact, so that $\mathrm{Ext}^i_{\Gamma_M}(X, \mathrm{Hom}_\Lambda(M, I)) = 0$ for $i > 0$. Hence $\mathrm{Hom}_\Lambda(M, I)$ is an injective Γ_M-module.

(b) If $0 \to A \to I_0 \to I_1$ is a minimal injective copresentation of the Λ-module A, it follows from (a) that $0 \to \mathrm{Hom}_\Lambda(M, A) \to \mathrm{Hom}_\Lambda(M, I_0) \to \mathrm{Hom}_\Lambda(M, I_1)$ is an injective Γ_M-copresentation of $\mathrm{Hom}_\Lambda(M, A)$. That it is also minimal follows easily from the facts that $0 \to A \to I_0 \to I_1$ is a minimal injective Λ-copresentation and that $\mathrm{Hom}_\Lambda(M,\ ): \mathrm{mod}\,\Lambda \to \mathscr{P}(\Gamma_M)$ is an equivalence of categories.

(c) By part (a) we know that the projective Γ_M-modules $\mathrm{Hom}_\Lambda(M, I)$ with I an injective Λ-module are also injective. Suppose now that P is a projective injective Γ_M-module. Since P is projective we have that $P \simeq \mathrm{Hom}_\Lambda(M, A)$ for some Λ-module A. Let $A \to I$ be a Λ-injective envelope of A. Since $\mathrm{Hom}_\Lambda(M, A)$ is injective, the monomorphism $\mathrm{Hom}_\Lambda(M, A) \to \mathrm{Hom}_\Lambda(M, I)$ of Γ_M-modules splits. But this means that the monomorphism $A \to I$ splits. Hence the monomorphism $A \to I$ is an isomorphism since it is an essential split monomorphism. Thus we get $P \simeq \mathrm{Hom}_\Lambda(M, I)$, our desired result.

(d) This is a trivial consequence of (c). $\qquad\qquad\square$

Continuing our discussion we obtain the following.

Proposition 5.4 *Let Λ be a nonsemisimple artin algebra of finite representation type and M an additive generator for* $\mod \Lambda$. *Then we have the following.*

(a) $\mathrm{gl.dim}\,\Gamma_M = 2 = \mathrm{dom.dim}\,\Gamma_M$, *so that Γ_M is an Auslander algebra.*

(b) *Let Q be a projective injective Γ_M-module such that $\mathrm{add}\,Q$ is the category of all projective injective Γ_M-modules. Then $\mathrm{End}_{\Gamma_M}(Q)^{\mathrm{op}}$ is Morita equivalent to Λ.*

Proof (a) We have already seen that $\mathrm{gl.dim}\,\Gamma_M = 2$. Let $0 \to M \to I_0 \to I_1$ be part of a minimal injective Λ-resolution of M. Then by Lemma 5.3 we have that $0 \to \mathrm{Hom}_\Lambda(M, M) \to \mathrm{Hom}_\Lambda(M, I_0) \overset{\alpha}{\to} \mathrm{Hom}_\Lambda(M, I_1)$ is part of a minimal injective Γ_M-resolution of Γ_M. Hence we get $\mathrm{dom.dim}_\Gamma \Gamma_M \geq 2$. Because $\mathrm{gl.dim}\,\Gamma_M = 2$, we have that $\mathrm{Coker}\,\alpha$ is injective. Since it cannot be projective because of the minimality of the injective resolution of Γ_M, we conclude that $\mathrm{dom.dim}_\Gamma \Gamma_M = 2$.

(b) By Lemma 5.3 we have that $\mathrm{add}\,\mathrm{Hom}_\Lambda(M, D(\Lambda))$ is the category of all projective injective Γ_M-modules. Therefore $\mathrm{add}\,Q = \mathrm{add}\,\mathrm{Hom}_\Lambda(M, D(\Lambda))$ which means that $\mathrm{End}_{\Gamma_M}(Q)$ and $\mathrm{End}_{\Gamma_M}(\mathrm{Hom}_\Lambda(M, D\Lambda))$ are Morita equivalent. But $\mathrm{End}_{\Gamma_M}(\mathrm{Hom}_\Lambda(M, D\Lambda)) \simeq \mathrm{End}_\Lambda(D(\Lambda)) \simeq \Lambda^{\mathrm{op}}$. Therefore $\mathrm{End}_{\Gamma_M}(Q)^{\mathrm{op}}$ and Λ are Morita equivalent. $\qquad\square$

In view of Proposition 5.4 it is natural to ask if an artin algebra Σ satisfying $\mathrm{gl.dim}\,\Sigma = 2 = \mathrm{dom.dim}_\Sigma \Sigma$ is necessarily isomorphic to some Γ_M obtained from a nonsemisimple artin algebra Λ of finite representation type with M an additive generator for Λ. Our aim is to show that this is indeed the case. Our proof depends on the following general result.

Lemma 5.5 *Let Σ be an arbitrary artin algebra.*

(a) *Suppose A is a Σ-module with $\mathrm{pd}_\Sigma A = n < \infty$. Then we have $\mathrm{Ext}^n_\Sigma(A, \Sigma) \neq 0$.*

(b) *Suppose $\mathrm{gl.dim}\,\Sigma = n < \infty$. Then we have the following.*

(i) $\mathrm{id}_\Sigma \Sigma = \mathrm{gl.dim}\,\Sigma$.

(ii) *Let $0 \to \Sigma \to I_0 \to I_1 \to \cdots \to I_n \to 0$ be a minimal injective resolution of Σ. If I is an indecomposable injective Σ-module, then I is isomorphic to a summand of I_j for some j.*

Proof (a) Let $0 \to P_n \overset{f_n}{\to} P_{n-1} \to \cdots \to P_1 \to P_0 \overset{f_0}{\to} A \to 0$ be a minimal projective resolution of a Σ-module A with $\mathrm{pd}_\Sigma A = n$. Suppose $\mathrm{Ext}^n_\Sigma(A, \Sigma) = 0$. Then $\mathrm{Ext}^n_\Sigma(A, P) = 0$ for all projective Σ-modules P. In particular we have $\mathrm{Ext}^n_\Sigma(A, P_n) = 0$ which implies that $\mathrm{Hom}(f_n, P_n)\colon \mathrm{Hom}_\Sigma(P_{n-1}, P_n) \to \mathrm{Hom}_\Sigma(P_n, P_n)$ is an epimorphism. Thus there is a morphism $g\colon P_{n-1} \to P_n$ such that the composition $gf_n\colon P_n \to P_n$ is the identity. This means f_n is a split monomorphism, which contradicts the hypothesis that $\mathrm{pd}_\Sigma A = n$.

(b)(i) Since $\mathrm{gl.dim}\,\Sigma = n < \infty$, we know that $\mathrm{id}_\Sigma \Sigma \le n$. We also know that there is a simple Σ-module S such that $\mathrm{pd}_\Sigma S = n$. Hence we get $\mathrm{Ext}^n_\Sigma(S, \Sigma) \ne 0$ by part (a), and consequently $\mathrm{id}_\Sigma \Sigma = n$.

(ii) Let I be an indecomposable Σ-module. Then I is the injective envelope of a simple Σ-module S. Since $\mathrm{pd}_\Sigma S = m \le n$, we know by part (a) that $\mathrm{Ext}^m_\Sigma(S, \Sigma) \ne 0$. From this it follows that $\mathrm{Hom}_\Sigma(S, I_m) \ne 0$. Therefore there is an indecomposable injective summand I' of I_m such that $\mathrm{Hom}_\Sigma(S, I') \ne 0$. Hence I' is an injective envelope of S which means that $I \simeq I'$. $\square$

We are now ready to give a positive answer to our previous question.

Proposition 5.6 *Let Σ be an artin algebra satisfying $\mathrm{gl.dim}\,\Sigma = 2 = \mathrm{dom.dim}_\Sigma \Sigma$. Let Q be a Σ-module such that $\mathrm{add}\,Q$ is the category of projective injective Σ-modules. Then we have the following.*

(a) $\Lambda = \mathrm{End}_\Sigma(Q)^{\mathrm{op}}$ *is of finite representation type.*

(b) Σ *is isomorphic to* $\mathrm{End}_\Lambda(M)^{\mathrm{op}}$, *where M is an additive generator for* Λ.

Proof (a) Since $\mathrm{gl.dim}\,\Sigma = 2$, we have by Lemma 5.5 that $\mathrm{id}_\Sigma \Sigma = 2$. Moreover, since $\mathrm{dom.dim}\,\Sigma = 2$ we know that the minimal injective Σ-resolution $0 \to \Sigma \to I_0 \to I_1 \to I_2 \to 0$ of Σ has the property that I_0 and I_1 are projective modules. Now by Lemma 5.5 we know that if I is an indecomposable injective module, then I is a summand of $I_0 \coprod I_1 \coprod I_2$. If I is projective, then I must be isomorphic to a summand of $I_0 \coprod I_1$ since it is not a summand of I_2. So $\mathrm{add}(I_0 \coprod I_1)$ is the category of projective injective Σ-modules. If I is not projective, then I is isomorphic to a summand of I_2. From this it follows that if $P_1 \to P_0 \to I \to 0$ is a minimal projective presentation of I, then P_0 and P_1 are in $\mathrm{add}(I_0 \coprod I_1)$. Therefore, if $\mathscr{C}$ is the subcategory of $\mathrm{mod}\,\Sigma$ consisting of the modules whose projective presentations consist of projective injective modules, then $\mathscr{C} \supset \mathscr{I}(\Lambda)$.

On the other hand, suppose X is in $\mathscr{C}$. Then there is an exact sequence $P_1 \xrightarrow{f} P_0 \to X \to 0$ with P_1 and P_0 projective injective modules. Since $\mathrm{id}_\Sigma \operatorname{Ker} f \leq 2$ and P_1 and P_0 are injective, it follows that X is an injective Σ-module, and so $\mathscr{C} = \mathscr{I}(\Lambda)$. Hence $\mathscr{C}$ has only a finite number of nonisomorphic indecomposable Σ-modules.

Suppose now that $\operatorname{add} Q$ is the category of projective injective Σ-modules. Then $\operatorname{add} Q = \operatorname{add}(I_0 \coprod I_1)$, so $\mathscr{I}(\Lambda)$ consists of the X in $\operatorname{mod} \Lambda$ such that there is a projective presentation $Q_1 \to Q_0 \to X \to 0$ with Q_1 and Q_0 in $\operatorname{add} Q$. Therefore letting $\Lambda = \operatorname{End}_\Sigma(Q)^{\mathrm{op}}$, we know by II Proposition 2.5 that the functor $\operatorname{Hom}_\Sigma(Q, \ ): \operatorname{mod} \Sigma \to \operatorname{mod} \Lambda$ induces an equivalence $\operatorname{Hom}_\Sigma(Q, \ ): \mathscr{I}(\Sigma) \to \operatorname{mod} \Lambda$. Therefore Λ is of finite representation type and the Λ-module $\operatorname{Hom}_\Sigma(Q, D(\Sigma)) = M$ has the property that $\operatorname{add} M = \operatorname{mod} \Lambda$. Moreover, the equivalence of categories $\operatorname{Hom}_\Sigma(Q, \ ): \mathscr{I}(\Sigma) \to \operatorname{mod} \Lambda$ gives an isomorphism $\operatorname{End}_\Lambda(M) \simeq \operatorname{End}_\Sigma(D(\Sigma))$. Since $\operatorname{End}_\Sigma(D(\Sigma)) \simeq \Sigma^{\mathrm{op}}$, it follows that $\Sigma \simeq \operatorname{End}_\Lambda(M)^{\mathrm{op}} = \Gamma_M$, where Λ is of finite representation type and M is an additive generator for Λ. $\square$

Summarizing our results so far we have the following.

Theorem 5.7 *For each artin algebra Λ of finite representation type choose an additive generator $M(\Lambda)$ and for each Auslander algebra Γ choose a projective injective module $Q(\Gamma)$ such that $\operatorname{add} Q(\Gamma)$ is precisely the category of projective injective Γ-modules. Then the maps $\Lambda \mapsto \operatorname{End}_\Lambda(M(\Lambda))^{\mathrm{op}}$ and $\Gamma \mapsto \operatorname{End}_\Gamma(Q(\Gamma))^{\mathrm{op}}$ induce inverse bijections between the Morita equivalence classes of nonsemisimple artin algebras of finite representation type and the Morita equivalence classes of nonsemisimple Auslander algebras.* $\square$

For an Auslander algebra Γ_M the module theory is closely connected with the module theory for the triangular matrix algebra $T_2(\Lambda)$. In particular we have the following relationship.

Proposition 5.8 *Let Λ be an artin algebra of finite representation type and M an additive generator for $\operatorname{mod} \Lambda$. Then the Auslander algebra Γ_M is of finite representation type if and only if $T_2(\Lambda)$ is of finite representation type.*

Proof Denote as in Chapter IV by $\operatorname{Morph} \mathscr{P}(\Gamma_M)$ the category of morphisms $f: P_1 \to P_2$ in $\mathscr{P}(\Gamma_M)$ and by $\mathscr{R}$ the relations on $\operatorname{Morph} \mathscr{P}(\Gamma_M)$ given by, for $f: P_1 \to P_2$ and $f': P_1' \to P_2'$, $\mathscr{R}(f, f')$ being the pairs (g_1, g_2) with $g_1: P_1 \to P_1'$ and $g_2: P_2 \to P_2'$, such that there is some $h: P_2 \to P_1'$ with $f'h = g_2$. Then by IV Proposition 1.2 the functor $\operatorname{Coker}: \operatorname{Morph} \mathscr{P}(\Gamma_M)/\mathscr{R} \to \operatorname{mod} \Gamma_M$ defined by sending $f: P_1 \to P_2$ to $\operatorname{Coker} f$, induces an equivalence of categories $G: \operatorname{Morph} \mathscr{P}(\Gamma_M) \to \operatorname{mod} \Gamma_M$. Further we have an equivalence $\operatorname{Hom}_\Lambda(M, \ ): \operatorname{mod} \Lambda \to \mathscr{P}(\Gamma_M)$, which clearly induces an equivalence $\operatorname{Hom}_\Lambda(M, \ ): \operatorname{mod} T_2(\Lambda) \to \operatorname{Morph} \mathscr{P}(\Gamma_M)$.

The objects of the form $1_P: P \to P$ and $P \to 0$ go to zero via the functor $\operatorname{Coker}: \operatorname{Morph} \mathscr{P}(\Gamma_M) \to \operatorname{mod} \Gamma_M$. Denote by $\mathscr{C}$ the full additive subcategory of $\operatorname{Morph} \mathscr{P}(\Gamma_M)$ containing all indecomposable objects in $\operatorname{Morph} \mathscr{P}(\Gamma_M)$ not isomorphic to an object of the form $1_P: P \to P$ or $P \to 0$ with P indecomposable in $\mathscr{P}(\Gamma_M)$. Let $f: P_1 \to P_2$ be an indecomposable object X in $\mathscr{C}$. Since $f: P_1 \to P_2$ is not a split epimorphism, it follows directly from the definitions that 1_X is not in $\mathscr{R}(f, f)$. Hence we have $\mathscr{R}(f, f) \subset \operatorname{rad} \operatorname{End}_{\mathscr{C}}(X)$, so that $\operatorname{End}_{\Gamma_M}(GX)$ is a local ring and consequently GX is indecomposable. Hence there is induced a one to one correspondence between the indecomposable objects in $\mathscr{C}$ and the indecomposable Γ_M-modules. Since there is only a finite number of indecomposable modules P in $\mathscr{P}(\Gamma_M)$, there is only a finite number of indecomposable objects in $\operatorname{Morph} \mathscr{P}(\Gamma_M)$ of the form $1_P: P \to P$ or $P \to 0$. Hence Γ is of finite representation type if and only if $\operatorname{Morph} \mathscr{P}(\Gamma_M)$ has only a finite number of indecomposable objects, which is the case if and only if $T_2(\Lambda)$ is of finite representation type. $\square$

The functor $\operatorname{Hom}_\Lambda(M, \ ): \operatorname{mod} T_2(\Lambda) \to \operatorname{Morph} \mathscr{P}(\Gamma_M)$ associates with a morphism $f: A \to B$ in $\operatorname{mod} \Lambda$ a projective presentation $\operatorname{Hom}_\Lambda(M, A) \xrightarrow{\operatorname{Hom}(M, f)} \operatorname{Hom}_\Lambda(M, B) \to \operatorname{Coker} \operatorname{Hom}(M, f) \to 0$ of $\operatorname{Coker} \operatorname{Hom}(M, f)$ in $\operatorname{mod} \Gamma_M$. We show that this way we get a connection between right almost split morphisms in $\operatorname{mod} \Lambda$ and projective presentations of simple Γ_M-modules.

Our discussion is based on the following general observation. Let Σ be an arbitrary artin algebra. A morphism $f: P \to P''$ in $\mathscr{P}(\Sigma)$ is said to be right almost split in $\mathscr{P}(\Sigma)$ if the morphism f is not a split epimorphism and any $g: P' \to P''$ in $\mathscr{P}(\Sigma)$ which is not a split epimorphism factors through f.

Lemma 5.9 *Let* Σ *be an arbitrary artin algebra. Then the following are equivalent for a morphism* $f : P \to Q$ *in* $\mathscr{P}(\Sigma)$.

(a) *The morphism* $f : P \to Q$ *is right almost split in* $\mathscr{P}(\Sigma)$.
(b) Q *is an indecomposable module and* $\operatorname{Im} f = \mathfrak{r}Q$.

Proof (a) $\Rightarrow$ (b) Let X be a maximal submodule of Q with projective cover $p : P_0 \to X$. Then $ip : P_0 \to Q$ is not a split epimorphism and hence factors through f. Therefore $X \subset \operatorname{Im} f$. Thus all maximal submodules of Q are equal to $\operatorname{Im} f$ and hence Q is indecomposable and $\mathfrak{r}Q = \operatorname{Im} f$.
    (b) $\Rightarrow$ (a) This is left as an exercise. $\square$

Applying these general observations we obtain the following.

Lemma 5.10 *Let* Γ_M *be an Auslander algebra for a nonsemisimple artin algebra* Λ *of finite representation type. Then the following are equivalent for a morphism* $f : B \to C$ *in* $\operatorname{mod} \Lambda$.

(a) *The morphism* $f : B \to C$ *is right almost split in* $\operatorname{mod} \Lambda$.
(b) *The* Γ_M*-morphism* $\operatorname{Hom}(M, f) : \operatorname{Hom}_\Lambda(M, B) \to \operatorname{Hom}_\Lambda(M, C)$ *of projective* Γ_M*-modules is right almost split in* $\mathscr{P}(\Gamma_M)$.
(c) $\operatorname{Hom}_\Lambda(M, C)$ *is an indecomposable projective* Γ_M*-module and* $\operatorname{Im} \operatorname{Hom}(M, f) = \mathfrak{r}\operatorname{Hom}_\Lambda(M, C)$.

Proof This follows easily from Lemma 5.9 using the equivalence of categories $\operatorname{Hom}_\Lambda(M, \ ) : \operatorname{mod} \Lambda \to \mathscr{P}(\Gamma_M)$. $\square$

These results give us the following connections between minimal projective resolutions in Auslander algebras and minimal right almost split morphisms and almost split sequences for algebras of finite representation type.

Proposition 5.11 *Let* Γ_M *be an Auslander algebra for a nonsemisimple artin algebra* Λ *of finite representation type. Let* S *be a simple* Γ_M*-module and let* C *be the unique, up to isomorphism,* Λ*-module such that* $\operatorname{Hom}_\Lambda(M, C)$ *is a projective cover for* S.

(a) *The following are equivalent.*

   (i) $\operatorname{pd}_{\Gamma_M} S = 0$.
   (ii) $\operatorname{Hom}_\Lambda(M, C) = S$.
   (iii) C *is a simple projective* Λ*-module.*

(b) *The following are equivalent.*

    (i) $\mathrm{pd}_{\Gamma_M} S = 1$.

    (ii) *C is a nonsimple projective Λ-module.*

    (iii) $0 \to \mathrm{Hom}_\Lambda(M, \mathfrak{r}C) \to \mathrm{Hom}_\Lambda(M, C) \to S \to 0$ *is a minimal projective resolution of S.*

(c) $\mathrm{pd}_{\Gamma_M} S = 2$ *if and only if C is not projective.*

(d) *If C is not projective and $0 \to A \to B \to C \to 0$ is an almost split sequence, then $0 \to \mathrm{Hom}_\Lambda(M, A) \to \mathrm{Hom}_\Lambda(M, B) \to \mathrm{Hom}_\Lambda(M, C) \to S \to 0$ is a minimal projective resolution of S.*

Proof (a) Clearly S is a projective Γ_M-module if and only if $S \simeq \mathrm{Hom}_\Lambda(M, C)$, or equivalently, $\mathrm{Hom}_\Lambda(M, C)$ is a simple projective Γ_M-module. But $\mathrm{Hom}_\Lambda(M, C)$ is a simple Γ_M-module if and only if any nonzero morphism $P \to \mathrm{Hom}_\Lambda(M, C)$ with P a projective Γ_M-module is an epimorphism and hence a split epimorphism. This is equivalent to C having the property that any nonzero morphism $B \xrightarrow{g} C$ in $\mathrm{mod}\,\Lambda$ is a split epimorphism. It is easy to see that the only indecomposable C in $\mathrm{mod}\,\Lambda$ with this property are the simple projective Λ-modules.

(b), (c) and (d) Suppose $C = P$ is a nonsimple indecomposable projective module. Then we know that the inclusion $\mathfrak{r}P \to P$ is minimal right almost split. Therefore by Lemma 5.10 we have that the inclusion $\mathrm{Hom}_\Lambda(M, \mathfrak{r}P) \to \mathrm{Hom}_\Lambda(M, P)$ is right almost split in $\mathscr{P}(\Gamma_M)$ so its image is $\mathfrak{r}\mathrm{Hom}_\Lambda(M, P)$. Therefore we have the minimal projective Γ_M-resolution $0 \to \mathrm{Hom}_\Lambda(M, \mathfrak{r}P) \to \mathrm{Hom}_\Lambda(M, P) \to S \to 0$, and so $\mathrm{pd}_{\Gamma_M} S = 1$.

Assume now that C is not projective, and consider the almost split sequence $0 \to A \xrightarrow{f} B \xrightarrow{g} C \to 0$. Then we have the exact sequence of projective Γ_M-modules $0 \to \mathrm{Hom}_\Lambda(M, A) \xrightarrow{\mathrm{Hom}(M,f)} \mathrm{Hom}_\Lambda(M, B) \xrightarrow{\mathrm{Hom}(M,g)} \mathrm{Hom}_\Lambda(M, C)$. Then $\mathrm{Hom}(M, g)$ is a right almost split morphism in $\mathscr{P}(\Gamma_M)$, so we get $\mathrm{Im}\,\mathrm{Hom}(M, g) = \mathfrak{r}\mathrm{Hom}_\Lambda(M, C)$ and hence $\mathrm{Coker}\,\mathrm{Hom}(M, g) \simeq S$. Consequently we obtain the Γ_M-projective resolution

(*) $0 \to \mathrm{Hom}_\Lambda(M, A) \to \mathrm{Hom}_\Lambda(M, B) \to \mathrm{Hom}_\Lambda(M, C) \to S \to 0$

which is minimal since $\mathrm{Hom}_\Lambda(M, A)$ and $\mathrm{Hom}_\Lambda(M, C)$ are indecomposable projective Γ_M-modules. Hence we get $\mathrm{pd}_{\Gamma_M} S = 2$ when C is not projective. Therefore the proof of the proposition is finished.   □

In III Section 1 we defined the associated quiver for a finite dimensional algebra over an algebraically closed field k, and more generally a quiver with valuation for any artin algebra. Using the previous discussion we give a description within mod Λ of the quiver with valuation associated with an Auslander algebra Γ_M of an artin algebra Λ of finite representation type.

The vertices of the quiver of Γ_M are by definition in one to one correspondence with the isomorphism classes of simple Γ_M-modules, hence with a complete set of isomorphism classes of indecomposable Λ-modules. For an indecomposable Λ-module X denote by $[X]$ the associated vertex in the quiver, and by S_X the corresponding simple Γ_M-module. There is then an arrow from $[X]$ to $[Y]$ if and only if $\operatorname{Ext}^1_{\Gamma_M}(S_X, S_Y) \neq 0$, that is if and only if in the minimal projective presentation $P \to \operatorname{Hom}_\Lambda(M, X) \to S_X \to 0$ we have that $\operatorname{Hom}_\Lambda(M, Y)$ is a summand of P. By Proposition 5.11 this is the case if and only if there is an irreducible morphism $Y \to X$.

For the associated valuation (a, b) of the arrow from $[X]$ to $[Y]$, we know from III Section 1 that b is the multiplicity of $\operatorname{Hom}_\Lambda(M, Y)$ as a summand of P. Hence b is in our case the multiplicity of Y in E when $E \to X$ is minimal right almost split. In the next chapter we show that a is the multiplicity of X in F when $Y \to F$ is minimal left almost split.

We illustrate with the following concrete example.

Example Let Λ be the Nakayama algebra with admissible sequence $(3, 3, 3)$. Then the quiver of Γ_M looks as follows where the dotted lines are identified.

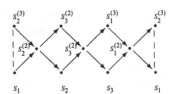

Exercises

1. Let Λ be an artin algebra of finite representation type.

(a) Prove that $\Lambda/\mathfrak{a}$ is of finite representation type for each ideal $\mathfrak{a}$ in Λ.

(b) Give an example of an artin R-algebra Λ of finite representation type where all artin R-subalgebras are also of finite representation type and give an example where this is not true.

2 Let Λ be as in V Exercise 2. Prove that Λ is of finite representation type.

3 Let Λ be an artin algebra. Assume that for each indecomposable injective Λ-module I there exists an $i \in \mathbb{N}$ with $(D \operatorname{Tr})^i I$ projective and for each indecomposable summand C of $I / \operatorname{soc} I$ there is an $i \in \mathbb{N}$ with $(D \operatorname{Tr})^i C$ projective.

(a) Prove that for each indecomposable Λ-module A in a component containing a projective module there is an $i \in \mathbb{N}$ such that $(D \operatorname{Tr})^i A$ is projective.

(b) Prove that for each indecomposable Λ-module A in a component containing an injective module there is an $i \in \mathbb{N}$ such that $(\operatorname{Tr} D)^i A$ is injective.

(c) Prove that Λ is of finite representation type.

4 Let Λ be an artin algebra such that each indecomposable projective left Λ-module is uniserial. Let M be a sum of uniserial modules.

(a) Prove that any indecomposable submodule of M is uniserial. (Hint: Prove that if $N \subset M$ and the radical length $rl(M)$ is n, then N contains a projective Λ/r^n-summand.)

(b) Prove that if Λ is a Nakayama algebra then all indecomposable Λ-modules are uniserial.

(This exercise gives an alternative proof of the fact that all indecomposable modules over a Nakayama algebra are uniserial.)

5 Let Λ be an artin R-algebra and denote by $\langle X \rangle$ the length of a finite length R-module. Let M and N be in $\operatorname{mod}\Lambda$ and assume $\langle \operatorname{Hom}_\Lambda(M, X) \rangle = \langle \operatorname{Hom}_\Lambda(N, X) \rangle$ for all modules X in $\operatorname{mod}\Lambda$. Show the following without using almost split sequences.

(a) Prove that if $M \neq 0$ then $N \neq 0$ and $\operatorname{Hom}_\Lambda(M, N) \neq 0$.

(b) Prove that if $M \neq 0$ there is an exact sequence $0 \to K \to nM \to N$ such that $0 \to \operatorname{Hom}_\Lambda(M, K) \to \operatorname{Hom}_\Lambda(M, nM) \to \operatorname{Hom}_\Lambda(M, N) \to 0$ is exact.

(c) Prove that for the sequence $0 \to K \to nM \to N$ in (b) $0 \to \operatorname{Hom}_\Lambda(N, K) \to \operatorname{Hom}_\Lambda(N, nM) \to \operatorname{Hom}_\Lambda(N, N) \to 0$ is also exact, and conclude that N is a summand of nM.

(d) Now prove by induction on the number of indecomposable summands in N that $M \simeq N$.

(This gives an alternative proof of Theorem 4.2(b).)

6. Let Λ be an artin algebra which is a sum of uniserial modules. Prove that dom.dim $\Lambda \geq 1$ if and only if Λ is a Nakayama algebra.

7. Show that an artin algebra Λ is of finite representation type if and only if for each Λ-module M the artin algebra $\Gamma_M = \text{End}_\Lambda M$ has the following property. There are a finite number of indecomposable Γ_M-modules $U_1, \ldots, U_t$ such that (a) $\coprod_{i=1}^{t} U_i$ is in $\text{add}\,\Omega^2(Y)$ for some Y in $\text{mod}\,\Gamma_M$ and (b) for each Γ_M-module X we have that $\Omega^2(X)$ is in $\text{add}(\coprod_{i=1}^{t} U_i)$.

Notes

The study of artin algebras of finite representation type splits naturally into three different but connected parts; one is determining which artin algebras have finite representation type, another is describing the indecomposable modules, up to isomorphism, for artin algebras of finite representation type and the last is finding the module theoretic consequences of an artin algebra having finite representation type. Prior to 1970 there were few classes of nonsemisimple artin algebras of finite representation type for which this program had been carried out. Amongst these were the Nakayama algebras and the modular group algebras, i.e. the group algebras of finite groups over fields of characteristic $p \neq 0$ dividing the order of the group. The complete story for Nakayama algebras was given in [Nak]. The fact that a modular group algebra is of finite representation type if and only if the p-Sylow-subgroups of the group are cyclic where p is the characteristic of the field was proven in [Hi]. The determination of the indecomposable modules over group algebras of finite type, at least over an algebraically closed field, was given independently in [Ku2] and [J].

In this same period Brauer and Thrall posed two problems about finite representation type for arbitrary finite dimensional algebras which were known as the first and second Brauer–Thrall conjectures. In this connection it is perhaps of some historical interest that Brauer insisted in private conversation that these were just problems suggested by what was known for group algebras, not conjectures. Conjectures or not, their solution occupied a great deal of attention in the beginning of the contemporary study of the representation theory of artin algebras. The first Brauer–Thrall conjecture was proven for finite dimensional algebras in [Roi] and subsequently for artin rings in [Au4]. The proof presented here is a modification of the proof in [Ya] based on Corollary 1.3, the Harada–Sai Lemma, which is valid in any abelian category with the same proof [HarS]. The second Brauer–Thrall conjecture says that for a finite dimensional algebra over an infinite field, there is an infinite number of dimensions which have an infinite number of nonisomorphic indecomposable modules when the algebra is of infinite representation type. That it suffices to show only that there is some dimension with an infinite number of nonisomorphic indecomposable modules was proven in [Sm]. The conjecture was established in [Bau3] for the case that k is an algebraically closed field. This proof, which is the culmination of many people's work, in particular the fundamental work in [BauGRS], together with [Bo0][Bo3], is much too involved to be presented here. Somewhat simplified proofs can be found in [BretT] and [Fi].

This chapter is devoted primarily to giving various module theoretic descriptions of when artin algebras are of finite representation type, not with the problem of which algebras are of finite representation type. The fact that there is a connection between an algebra being of finite representation type and the structure of the relations defining the Grothendieck group of the artin algebras was observed first in [Bu] where it was shown that the almost split sequences generate the relations for the Grothendieck group when the artin algebra is of finite representation type. The converse was proven in [Au8].

The criterion given in Theorem 1.4 for an indecomposable artin algebra to be of finite representation type is essentially the criterion given in [Au6]. The criterion has been extended and applied to other situations such as Cohen–Macaulay modules over isolated singularities [AuR10], [So]. Based on the theory of coverings and lists of algebras of infinite type in [Bo2], [HapV] there is now a nonroutine procedure for deciding whether an algebra is of finite type, provided the algebra is given in an appropriate form, as a path algebra modulo relations which are either zero relations or commutativity relations.

The algebras known now as Auslander algebras were first introduced in [Au3] where their connections with artin algebras of finite representation type were studied. The homological properties of Auslander algebras have served as an inspiration for parts of the theory of coverings [BoG] and hammocks [Bren1]. The analogue of Auslander algebras has also been established for orders [AuRo].

The fact that two modules M and N over an artin algebra are isomorphic if $\langle X, M \rangle = \langle X, N \rangle$ for all indecomposable Λ-modules X was proven in [Au7]. Exercise 5 is modelled on an entirely different proof which is valid in much more general settings than finitely generated modules over an artin algebra is given in [Bo4]. For instance, this result is true for coherent sheaves over projective varieties defined over fields.

VII

The Auslander–Reiten-quiver

In this chapter we introduce a quiver called the Auslander–Reiten quiver, or for short the AR-quiver, of any artin algebra Λ. The definition is motivated by the interpretation of the ordinary quiver of an Auslander algebra in terms of the associated algebra of finite representation type given at the end of the previous chapter.

We start by giving the construction of the AR-quiver and a criterion which can be read off directly from the AR-quiver ensuring that the composition of some irreducible morphisms in mod Λ is not zero. The AR-quiver often decomposes into a union of infinite components and the possible structures of such components and other full subquivers of the AR-quiver are studied. Here combinatorial results play a crucial role and these combinatorial results will also be applied in the next chapter dealing with hereditary artin algebras.

1 The Auslander–Reiten-quiver

In this section we introduce the Auslander–Reiten-quiver of an artin algebra and give some of its basic properties. We illustrate with several examples, and give the connection between the Auslander–Reiten-quiver of an algebra of finite representation type and the ordinary quiver of its Auslander algebra.

In VI Section 1 we introduced for an artin algebra Λ an equivalence relation on ind Λ. The equivalence relation is generated by M being related to N if there is an irreducible morphism from M to N or from N to M. On the other hand we have seen in VI Section 5 that the quiver of the Auslander algebra Γ of an artin algebra Λ of finite representation type has vertices in one to one correspondence with ind Λ. Denoting the vertex corresponding to a module M by $[M]$, there is an arrow

$[M] \rightarrow [N]$ between two vertices if and only if there is an irreducible morphism $M \rightarrow N$. If (a,b) is the valuation of the arrow $[M] \rightarrow [N]$, then a is the number of copies of M in a sum decomposition of E into indecomposable modules, where $E \rightarrow N$ is minimal right almost split.

Motivated by these observations we define for any artin algebra Λ an associated valued quiver Γ_Λ as follows. The vertices of Γ_Λ are in one to one correspondence with the objects of ind Λ, and are denoted by $[M]$ for M in ind Λ. There is an arrow $[M] \rightarrow [N]$ if and only if there is an irreducible morphism $M \rightarrow N$. The arrow has valuation (a,b) if there is a minimal right almost split morphism $aM \coprod X \rightarrow N$ where M is not a summand of X, and a minimal left almost split morphism $M \rightarrow bN \coprod Y$ where N is not a summand of Y. The vertices corresponding to projective modules are called **projective vertices** and those corresponding to injective modules are called **injective vertices**. Then $D\operatorname{Tr}$ induces a map from the nonprojective vertices to the noninjective vertices. This map is called the **translation** of Γ_Λ, and is denoted by $D\operatorname{Tr}$ or by τ. The valued quiver Γ_Λ together with the translation τ is called the **Auslander–Reiten quiver** of Λ, or the **AR-quiver** of Λ for short.

We illustrate the concept of an AR-quiver with some concrete examples.

Example Let Δ be the quiver $\overset{1}{\cdot} \overset{2}{\longrightarrow} \overset{3}{\cdot}$ and let k be a field. The path algebra $k\Delta$ is a Nakayama algebra with admissible sequence $(3,2,1)$. We know from VI Theorem 2.1 what the indecomposable $k\Delta$-modules look like and how they are tied together to form almost split sequences. There are six indecomposable modules: the three projective modules P_3, P_2 and P_1 of length 1, 2 and 3 respectively, the two simple nonprojective modules $P_1/\mathfrak{r}P_1$, $P_2/\mathfrak{r}P_2$, and the sixth module is $P_1/\mathfrak{r}^2 P_1$.

The almost split sequences are

$$0 \rightarrow P_2 \rightarrow P_1 \amalg P_2/\mathfrak{r}P_2 \rightarrow P_1/\mathfrak{r}^2 P_1 \rightarrow 0,$$
$$0 \rightarrow P_2/\mathfrak{r}P_2 \rightarrow P_1/\mathfrak{r}^2 P_1 \rightarrow P_1/\mathfrak{r}P_1 \rightarrow 0 \text{ and}$$
$$0 \rightarrow P_3 \rightarrow P_2 \rightarrow P_2/\mathfrak{r}P_2 \rightarrow 0.$$

Hence we get that the AR-quiver $\Gamma_{k\Delta}$ is

where the broken arrows indicate the translation.

Example For an indecomposable Nakayama algebra Λ we know the structure of the indecomposable modules and the almost split sequences, so we can easily construct the AR-quiver.

Note that the AR-quiver for these algebras depends only on their admissible sequences. For example the AR-quivers of the algebra $\mathbb{Z}/8\mathbb{Z}$ and the algebra $k[X]/(X^3)$ where k denotes a field are the following.

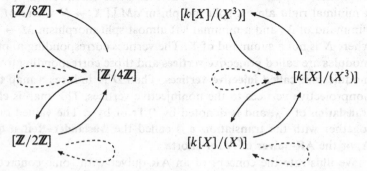

Example Let Λ be the path algebra of the quiver $\underset{1}{\cdot} \leftarrow \underset{2}{\cdot} \rightarrow \underset{3}{\cdot}$ over a field k. As computed in VI Section 1 the indecomposable Λ-modules are the simple modules S_1, S_2, S_3 corresponding to the vertices 1, 2 and 3, the projective cover P_2 of S_2 and the injective envelopes I_1 of S_1 and I_3 of S_3. The almost split sequences are of the form

$$
\begin{array}{ccccccccc}
0 & \to & S_3 & \to & P_2 & \to & I_1 & \to & 0 \\
0 & \to & P_2 & \to & I_1 \amalg I_3 & \to & S_2 & \to & 0 \\
0 & \to & S_1 & \to & P_2 & \to & I_3 & \to & 0.
\end{array}
$$

Hence the AR-quiver is

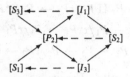

Example Let $\Lambda = \left(\begin{smallmatrix} T & 0 \\ T & T \end{smallmatrix} \right)$ with $T = k[X]/(X^2)$ where k is a field. Using the description of the indecomposable Λ-modules and the almost split sequences given at the end of VI Section 1, we get that the AR-quiver Γ_Λ is the following quiver where we again indicate the translation by a broken arrow.

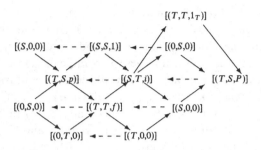

Here the modules $(0, T, 0)$ and $(T, T, 1_T)$ are projective and $(T, 0, 0)$ and $(T, T, 1_T)$ are injective. The full subquiver of Γ_Λ obtained by removing these vertices lies on a Moebius band.

There is another interesting structure on the AR-quiver Γ_Λ of an artin algebra Λ, given by a partially defined map σ on the valued arrows of Γ_Λ called the **semitranslation** of Γ_Λ. Let $\alpha \colon [M] \to [N]$ be an arrow in Γ_Λ, and assume that N is not projective. Since we have an irreducible morphism $M \to N$, there is then by V Proposition 1.12 and V Theorem 5.3 an irreducible morphism $D \operatorname{Tr} N \to M$. Then $\sigma(\alpha)$ is defined to be the corresponding arrow $[D \operatorname{Tr} N] \to [M]$.

Clearly σ is one to one on the arrows ending in nonprojective vertices, and the image under σ consists of the arrows starting in noninjective vertices. Hence σ^{-1} is defined on the arrows starting in noninjective vertices. Note that since there is never an irreducible morphism from an indecomposable injective module to an indecomposable projective module, we have for any arrow α that either $\sigma(\alpha)$ or $\sigma^{-1}(\alpha)$ is defined.

If (a, b) is the valuation of the arrow α and (c, d) the valuation of $\sigma(\alpha)$, we have an almost split sequence $0 \to D \operatorname{Tr} N \to E \to N \to 0$ with $E \simeq aM \coprod X$ where M is not a summand of X and $E \simeq dM \coprod Y$ where M is not a summand of Y. Hence it follows that $a = d$. In order to show $b = c$ we shall need a different interpretation of the valuation of an arrow in the AR-quiver.

We have in V Section 7 given an interpretation of the existence of irreducible morphisms in terms of the radical of the module category. We now exploit this connection further. For A and B in ind Λ, we shall see that we can view the abelian group $\operatorname{Irr}(A, B) = \operatorname{rad}_\Lambda(A, B) / \operatorname{rad}_\Lambda^2(A, B)$ as a T_B-T_A^{op}-bimodule in a natural way, where T_X denotes the division algebra $\operatorname{End}_\Lambda(X) / \mathfrak{r}_{\operatorname{End}_\Lambda(X)}$ for each X in ind Λ.

If A and B are indecomposable Λ-modules, then for each natural number n we have that $\operatorname{rad}_\Lambda^n(A, B)$ is an $\operatorname{End}_\Lambda(B)$-$\operatorname{End}_\Lambda(A)^{\mathrm{op}}$-subbimodule of the $\operatorname{End}_\Lambda(B)$-$\operatorname{End}_\Lambda(A)^{\mathrm{op}}$-bimodule $\operatorname{Hom}_\Lambda(A,B)$. Further, for each Λ-module M, the radical of $\operatorname{End}_\Lambda(M)$ as a ring and the radical of $\operatorname{Hom}_\Lambda(M,M)$ in the category $\operatorname{mod}\Lambda$ coincide, and therefore we could have used the same notation for these two radicals. However, when we are taking powers of these radicals it would lead to confusion.

We have for each n that $\mathfrak{r}_{\operatorname{End}_\Lambda(B)}\operatorname{rad}^n_\Lambda(A,B) \subset \operatorname{rad}^{n+1}_\Lambda(A,B)$ and $\operatorname{rad}^n_\Lambda(A,B)\mathfrak{r}_{\operatorname{End}_\Lambda(A)} \subset \operatorname{rad}^{n+1}_\Lambda(A,B)$. Hence $\operatorname{rad}^n_\Lambda(A,B)/\operatorname{rad}^{n+1}_\Lambda(A,B)$ becomes a T_B-T_A^{op}-bimodule. In particular, when A and B are indecomposable modules, the interesting group $\operatorname{rad}_\Lambda(A,B)/\operatorname{rad}^2_\Lambda(A,B) = \operatorname{Irr}(A,B)$ becomes a T_B-T_A^{op}-vector space.

In order to investigate the T_B-T_A^{op}-bimodule $\operatorname{Irr}(A,B)$ more closely, it is convenient to introduce the following notation. For $f_i \in \operatorname{Hom}_\Lambda(A,B)$, $i = 1,\ldots,n$, we write $nA \xrightarrow{(f_i)} B$ to denote the morphism given by $(f_i)(a_i) = \sum_{i=1}^n f_i(a_i)$ for $(a_i) \in nA$ and $A \xrightarrow{(f_i)^{\mathrm{tr}}} nB$ to denote the morphism given by $(f_i)^{\mathrm{tr}}(a) = (b_j)$ with $b_j = f_j(a)$ for $j = 1,\ldots,n$. Our aim is now to prove that if the arrow $[A] \to [B]$ has valuation (a,b) then a is the dimension of $\operatorname{Irr}(A,B)$ as a T_A^{op}-vector space and b is the dimension of $\operatorname{Irr}(A,B)$ as a T_B-vector space.

We shall need the following.

Lemma 1.1 *Let A and B be indecomposable nonisomorphic Λ-modules and let $f_1, f_2, \ldots, f_n$ be Λ-morphisms from A to B and let $\overline{f}_i$ denote the coset in $\operatorname{rad}_\Lambda(A,B)/\operatorname{rad}^2_\Lambda(A,B)$ of the element f_i.*

(a) *If the induced Λ-morphism $nA \xrightarrow{(f_i)} B$ is irreducible, then $\{\overline{f}_1, \overline{f}_2, \ldots, \overline{f}_n\}$ is a linearly independent set of elements of the T_A^{op}-vector space $\operatorname{rad}_\Lambda(A,B)/\operatorname{rad}^2_\Lambda(A,B)$.*

(b) *If the induced Λ-morphism $A \xrightarrow{(f_i)^{\mathrm{tr}}} nB$ is irreducible, then $\{\overline{f}_1, \overline{f}_2, \ldots, \overline{f}_n\}$ is a linearly independent set of elements of the T_B-vector space $\operatorname{rad}_\Lambda(A,B)/\operatorname{rad}^2_\Lambda(A,B)$.*

Proof We prove (a), and then (b) follows by duality. Assume that the induced Λ-morphism $nA \xrightarrow{(f_i)} B$ is irreducible, and that $a_1\overline{f}_1 + a_2\overline{f}_2 + \cdots + a_n\overline{f}_n = 0$ with each a_i in T_A^{op}. Now lift the elements a_i in $\operatorname{End}_\Lambda(A)^{\mathrm{op}}/\mathfrak{r}_{\operatorname{End}_\Lambda(A)^{\mathrm{op}}}$ to morphisms α_i from A to A. Then the relation above states that the morphism $f_1\alpha_1 + f_2\alpha_2 + \cdots + f_n\alpha_n$ from A to B is in $\operatorname{rad}^2_\Lambda(A,B)$. However, if there is some i such that a_i is nonzero, then the corresponding α_i is an isomorphism, and therefore the morphism $\alpha = (\alpha_i)^{\mathrm{tr}} : A \to nA$ induced by the α_i is a split monomorphism.

Since $nA \xrightarrow{(f_i)} B$ is irreducible there exists by V Theorem 5.3 a morphism $g : A' \to B$ such that $nA \coprod A' \xrightarrow{((f_i),g)} B$ is a minimal right almost split morphism. But this implies that there also exists a morphism $h : A'' \to B$ where $A'' \simeq A' \coprod \operatorname{Coker} \alpha$, such that $A \coprod A'' \xrightarrow{(f_1\alpha_1 + f_2\alpha_2 + \cdots + f_n\alpha_n,\, h)} B$ is a minimal right almost split morphism. Therefore by V Theorem 5.3, $f_1\alpha_1 + f_2\alpha_2 + \cdots + f_n\alpha_n : A \to B$ is irreducible and this contradicts the fact that $f_1\alpha_1 + f_2\alpha_2 + \cdots + f_n\alpha_n$ is in $\operatorname{rad}^2_\Lambda(A, B)$. Therefore $\{\bar{f}_1, \bar{f}_2, \ldots, \bar{f}_n\}$ is a linearly independent set of elements in the T_A^{op}-vector space $\operatorname{rad}_\Lambda(A, B)/\operatorname{rad}^2_\Lambda(A, B)$. $\square$

We next establish the converse of this result. Let $g : X \to B$ be a minimal right almost split morphism ending in B. Decompose X as $mA \coprod X'$ where X' has no indecomposable summand isomorphic to A. Let $\bar{g}_i$ for $i = 1, 2, \ldots, m$ be the elements of the T_A^{op}-vector space $\operatorname{rad}_\Lambda(A, B)/\operatorname{rad}^2_\Lambda(A, B)$ corresponding to the coordinates of g using this decomposition of X. Dually, let $h : A \to Y$ be a minimal left almost split morphism starting in A. Decompose Y as $nB \coprod Y'$ where Y' has no indecomposable summand isomorphic to B. Let $\bar{h}_i$ for $i = 1, 2, \ldots, n$ be the elements of the T_B-vector space $\operatorname{rad}_\Lambda(A, B)/\operatorname{rad}^2_\Lambda(A, B)$ corresponding to the coordinates of h with respect to this decomposition of Y.

Lemma 1.2 *Let the notation be as above. Then we have the following.*

(a) *The set $\{\bar{g}_1, \bar{g}_2, \ldots, \bar{g}_m\}$ generates the T_A^{op}-vector space $\operatorname{rad}_\Lambda(A, B)/\operatorname{rad}^2_\Lambda(A, B)$.*

(b) *The set $\{\bar{h}_1, \bar{h}_2, \ldots, \bar{h}_n\}$ generates the T_B-vector space $\operatorname{rad}_\Lambda(A, B)/\operatorname{rad}^2_\Lambda(A, B)$.*

Proof We only prove (a) since (b) follows from (a) by duality. Let $\bar{f}$ be any element of the T_A^{op}-vector space $\operatorname{rad}_\Lambda(A, B)/\operatorname{rad}^2_\Lambda(A, B)$, where f is a morphism from A to B. Then since g is right almost split and f is not an isomorphism we have that $f = gh$ for some $h : A \to X$. Considering the fixed decomposition of X as $mA \coprod X'$ as described above, we get that $f = (g|_{mA})ph + (g|_{X'})qh$, where p is the projection onto the submodule mA of X and q is the projection onto the submodule X' of X according to the decomposition of X. Since X' has no summands isomorphic to A, the map qh is in $\operatorname{rad}_\Lambda(A, X')$ and therefore $(g|_{X'})qh$ is in $\operatorname{rad}^2_\Lambda(A, B)$. Hence we get $\bar{f} = \overline{(g|_{mA})ph}$. Using the decomposition $mA = A \coprod A \coprod \cdots \coprod A$ the map ph corresponds to a set of elements $\{\alpha_i\}_{i=1}^m \subset \operatorname{End}_\Lambda(A)^{\mathrm{op}}$ such that $\bar{f} = \sum \overline{\alpha_i g_i}$. This completes the proof of part (a). $\square$

Combining the last two lemmas, we now get the following connection between the valuation of the quiver Γ_Λ and the dimensions of $\mathrm{Irr}(A, B)$ as a T_B-vector space and as a T_A^{op}-vector space.

Proposition 1.3 *Let A and B be indecomposable Λ-modules, and assume there is an irreducible morphism from A to B. Then we have the following.*

(a) *The multiplicity of A as a summand of M when there is a minimal right almost split morphism $f: M \to B$ is equal to the dimension of $\mathrm{Irr}(A, B)$ as a T_A^{op}-vector space.*

(b) *The multiplicity of B as a summand of N when there is a minimal left almost split morphism $g: A \to N$ is equal to the dimension of $\mathrm{Irr}(A, B)$ as a T_B-vector space.* □

Let B be a nonprojective indecomposable module and α an arrow from $[A]$ to $[B]$ in Γ_Λ. In order to prove that the valuation of the arrow $\sigma(\alpha)$ from $[D\,\mathrm{Tr}\,B]$ to $[A]$ in Γ_Λ is (b, a) if the valuation of the arrow α in Γ_Λ is (a, b) we also need the following result.

Lemma 1.4 *Let Λ be an artin R-algebra.*

(a) *For each nonprojective indecomposable Λ-module B we have that $D\,\mathrm{Tr}$ induces an R-isomorphism between T_B and $T_{D\,\mathrm{Tr}\,B}$.*

(b) *For each noninjective indecomposable Λ-module A we have that $\mathrm{Tr}\,D$ induces an R-isomorphism between T_A and $T_{\mathrm{Tr}\,DA}$.*

(c) *If A and B are indecomposable modules with B not a projective module, then the dimension of $\mathrm{Irr}(A, B)$ as a T_B-vector space is the same as the dimension of $\mathrm{Irr}(D\,\mathrm{Tr}\,B, A)$ as a $T_{D\,\mathrm{Tr}\,B}^{\mathrm{op}}$-vector space.*

Proof We know from IV Proposition 1.9(b) that $D\,\mathrm{Tr}: \underline{\mathrm{mod}}\Lambda \to \overline{\mathrm{mod}}\Lambda$ is an equivalence of categories. Therefore $D\,\mathrm{Tr}$ induces a ring isomorphism $D\,\mathrm{Tr}: \underline{\mathrm{End}}_\Lambda(B) \to \overline{\mathrm{End}}_\Lambda(D\,\mathrm{Tr}\,B)$. Since B is an indecomposable module we have that $\mathrm{End}_\Lambda(B)$ is local and since B is nonprojective the ideal $\mathscr{P}(B, B)$, consisting of the morphisms from B to B factoring through a projective module is contained in $\mathfrak{r}_{\mathrm{End}_\Lambda(B)}$. Similarly, since $D\,\mathrm{Tr}\,B$ is a noninjective indecomposable module, the ideal $\mathscr{I}(D\,\mathrm{Tr}\,B, D\,\mathrm{Tr}\,B)$, consisting of the morphisms from $D\,\mathrm{Tr}\,B$ to $D\,\mathrm{Tr}\,B$ factoring through an injective module, is contained in $\mathfrak{r}_{\mathrm{End}_\Lambda(D\,\mathrm{Tr}\,B)}$. Hence $D\,\mathrm{Tr}$ induces an isomorphism from T_B to $T_{D\,\mathrm{Tr}\,B}$. Further, this isomorphism is an R-isomorphism.

Statement (b) is the dual of statement (a). In order to prove (c), let $l_R(X)$ as usual denote the length of the module X over R and let for a division algebra S $\dim_S V$ denote the dimension of V as a vector space over S. We have from Proposition 1.3 that

$$\dim_{T_A^{\mathrm{op}}}(\mathrm{Irr}(A,B)) = \dim_{T_A}(\mathrm{Irr}(D\operatorname{Tr} B, A)).$$

Therefore we have

$$
\begin{aligned}
l_R(\mathrm{Irr}(A,B)) &= \dim_{T_B}(\mathrm{Irr}(A,B))l_R(T_B) \\
&= l_R(T_A^{\mathrm{op}})\dim_{T_A^{\mathrm{op}}}(\mathrm{Irr}(A,B)) \\
&= l_R(T_A)\dim_{T_A}(\mathrm{Irr}(D\operatorname{Tr} B, A)) \\
&= l_R(\mathrm{Irr}(D\operatorname{Tr} B, A)) \\
&= l_R(T_{D\operatorname{Tr} B}^{\mathrm{op}})\dim_{T_{D\operatorname{Tr} B}^{\mathrm{op}}}(\mathrm{Irr}(D\operatorname{Tr} B, A)).
\end{aligned}
$$

This shows that $\dim_{T_B}(\mathrm{Irr}(A,B)) = \dim_{T_{D\operatorname{Tr} B}^{\mathrm{op}}}(\mathrm{Irr}(D\operatorname{Tr} B, A))$. $\qquad\square$

As an immediate consequence of this result we have the following.

Proposition 1.5 *If α is an arrow in Γ_Λ from $[B]$ to $[C]$ with valuation (a,b) and C is nonprojective then (b,a) is the valuation of the arrow $\sigma(\alpha)$ from $[D\operatorname{Tr} C]$ to $[B]$ in Γ_Λ.* $\qquad\square$

Before giving the connection between the AR-quiver of Λ and the ordinary valued quiver of the Auslander algebra associated with Λ when Λ is an artin algebra of finite representation type, we need the concept of the transpose of a valued quiver. Let Δ be a valued quiver. Then the **transposed** valued quiver Δ^{tr} has the same underlying quiver as Δ, and if an arrow α has valuation (a, b) in Δ, it has valuation (b, a) in Δ^{tr}.

Theorem 1.6 *Let Λ be an artin algebra of finite representation type and let Γ be the associated Auslander algebra. Then the ordinary quiver of Γ is the opposite of the transpose of the Auslander–Reiten-quiver of Λ.*

Proof As in VI Section 5 let M be in $\operatorname{mod} \Lambda$ such that $\operatorname{add} M = \operatorname{mod} \Lambda$. For each indecomposable module B in $\operatorname{mod} \Lambda$ let S_B be the associated simple Γ-module where $\Gamma = \operatorname{End}_\Lambda(M)^{\mathrm{op}}$ is the Auslander algebra associated with Λ. Let $[S_B]$ denote the vertex in the quiver of Γ associated with the simple module S_B. For two indecomposable modules B and A in $\operatorname{mod} \Lambda$ we have already observed in VI Section 5 that $\operatorname{Ext}^1_\Gamma(S_B, S_A) \neq 0$ if and only if there is an irreducible morphism from A to B in $\operatorname{mod} \Lambda$. Hence the underlying nonvalued quiver of the AR-quiver is the same as the opposite of the underlying nonvalued quiver of Γ. Recall that if $[S_A] \overset{(a,b)}{\leftarrow} [S_B]$ is an arrow with valuation in the quiver of Γ, then by definition we have $b = \dim_{\text{End}_\Gamma(S_A)} \text{Ext}^1_\Gamma(S_B, S_A)$ and $a = \dim_{\text{End}_\Gamma(S_B)^{\text{op}}} \text{Ext}^1_\Gamma(S_B, S_A)$. The corresponding valued arrow in the AR-quiver of Λ is $[A] \overset{(a',b')}{\rightarrow} [B]$, where we have seen in VI Section 5 that $b = a'$. We have $a' = \dim_{T^{\text{op}}_A} \text{Irr}(A, B)$ and $b' = \dim_{T_B} \text{Irr}(A, B)$. Note that $T_A \simeq \text{End}_\Gamma(S_A)$ for the following reason: Since $\text{Hom}_\Lambda(M, \ ): \text{mod}\,\Lambda \rightarrow \mathscr{P}(\Gamma)$ is an equivalence of categories, $\text{End}_\Lambda(A)$ and $\text{End}_\Gamma(\text{Hom}_\Lambda(M, A))$ are isomorphic. $P = \text{Hom}_\Lambda(M, A)$ is an indecomposable projective Γ-module with $P/\mathfrak{r}_\Gamma P \simeq S_A$. Since for any $f: P \rightarrow P$ we have $f(\mathfrak{r}P) \subset \mathfrak{r}P$, there is induced a natural ring homomorphism $\text{End}_\Gamma(P) \rightarrow \text{End}_\Gamma(S_A)$, which is surjective since P is projective. Hence $\text{End}_\Gamma(S_A)$ is the unique division algebra which is a factor of the local algebra $\text{End}_\Gamma(P)$, and consequently we have $\text{End}_\Gamma(S_A) \simeq T_A$ as R-algebras. We then have $l_R(\text{Ext}^1_\Gamma(S_B, S_A)) = bl_R(\text{End}_\Gamma S_A) = al_R(\text{End}_\Gamma S_B)$ and $l_R(\text{Irr}(A, B)) = a'(l_R(\text{End}_\Gamma S_A)) = b'(l_R(\text{End}_\Gamma S_B))$. Since $b = a'$, it follows that $a = b'$. This shows that the AR-quiver of Λ is the opposite of the transpose of the valued quiver of Γ. $\quad\square$

We remark that it is often possible to reconstruct an artin algebra of finite representation type from its AR-quiver. This can be accomplished through the following construction. For let Γ_Λ be the AR-quiver of a basic algebra Λ over a field k, and assume that all valuations are trivial. For each nonprojective vertex x in Γ_Λ we have a relation m_x on the quiver Γ_Λ, called the **mesh relation**, and defined by $m_x = \sum_{\{\alpha \in (\Gamma_\Lambda)_1 | e(\alpha) = x\}} \alpha\sigma(\alpha)$ where σ is the semitranslation. It can be proved that if k is an algebraically closed field of characteristic different from 2, then the path algebra of Γ_Λ over k modulo the mesh relations is isomorphic to the opposite of an Auslander algebra for Λ.

2 Shape of Auslander–Reiten-quivers and finite type

We have seen in Chapter VI that if for an indecomposable artin algebra Λ there is a finite component with respect to irreducible morphisms, then Λ is of finite representation type. In this section we illustrate how the structure of the AR-quiver together with the semitranslation helps in applying this criterion. We end the section with some information on the structure of the AR-quiver.

We start by formulating our criterion for finite representation type in terms of AR-quivers.

Theorem 2.1 *The following are equivalent for an indecomposable artin algebra Λ.*

(a) Λ *is of finite representation type.*

(b) Γ_Λ *is a finite connected quiver.*

(c) Γ_Λ *has a finite component.* □

We now proceed to give examples of how to compute components of AR-quivers without knowing all almost split sequences to start with. In our examples the components turn out to be finite, showing that the algebras are of finite representation type.

The computations are based on the following principles. Assume that we have computed a full subquiver of the AR-quiver and assume that there is a vertex $[A]$ such that each immediate predecessor $[B]$ of $[A]$ in the AR-quiver is in the full subquiver, along with the part corresponding to the almost split sequence with B on the left if B is not injective, and corresponding to the minimal left almost split morphism for B if B is injective. If we have an irreducible morphism $A \to C$ where C is indecomposable nonprojective, there is an irreducible morphism $D\operatorname{Tr} C \to A$, and by assumption the part corresponding to the almost split sequence with left hand term $D\operatorname{Tr} C$ is already in our full subquiver. In particular the valued arrow $[A] \overset{(a,b)}{\to} [C]$, corresponding to the valued arrow $[D\operatorname{Tr} C] \overset{(b,a)}{\to} [A]$, is there. We know that there is an irreducible morphism $A \to P$ with P indecomposable projective if and only if A is a summand of $\mathfrak{r}P$, which we can check from knowing the structure of the projective modules. For a valued arrow $[A] \overset{(a,b)}{\to} [P]$, a is the number of copies of A in a decomposition of $\mathfrak{r}P$ into a sum of indecomposable modules, and b is computed from a and $l_R(T_P)$ and $l_R(T_A)$ where we recall that $T_X = \operatorname{End}_\Lambda(X)/\mathfrak{r}_{\operatorname{End}_\Lambda(X)}$. This way we obtain all valued arrows starting at $[A]$. The corresponding irreducible morphisms from A are induced from the almost split sequences starting at the immediate predecessors. Based on this information we compute the minimal left almost split morphism $A \to E$. If A is not injective, we get the almost split sequence $0 \to A \to E \to \operatorname{Tr} DA \to 0$ and hence induced irreducible morphisms from the summands of E in a sum decomposition into indecomposable modules, and corresponding valued arrows.

Assume that we obtain a full subquiver where each vertex $[A]$ has all its immediate predecessors in the subquiver, together with the part corresponding to the minimal left almost split morphism from A and the whole almost split sequence with left hand term A if A is not injective. Then we have a whole component of the AR-quiver.

We can sometimes get started with a full subquiver having the desired properties in the following way. If S is a simple projective noninjective Λ-module, we know from V Section 3 that in the almost split sequence $0 \to S \to E \to \operatorname{Tr} DS \to 0$ the middle term E is projective. Hence we can compute this almost split sequence from knowing the structure of the projective modules, and get the corresponding valued arrows in the AR-quiver. For an indecomposable summand P of E we can compute the immediate predecessors $[X]$ of $[P]$ since they are determined by $\mathfrak{r}P$. If we can compute the almost split sequence with each possible X as left hand term, we can proceed as above with $A = P$. Note that the main problem in carrying out this procedure arises each time we reach a new projective module, since it may not be so easy to deal with its (immediate) predecessors.

Example Let k be a field and let Λ be the k-algebra given by the quiver

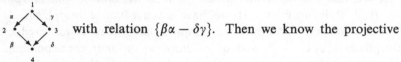 with relation $\{\beta\alpha - \delta\gamma\}$. Then we know the projective

modules and their radicals. We have $P_4 = S_4$, P_2 with $\mathfrak{r}P_2 \simeq S_4$, P_3 with $\mathfrak{r}P_3 \simeq S_4$ and P_1 with $\mathfrak{r}P_1$ given as the representation

$$\begin{array}{ccc} & 0 & \\ k & \diagup\;\diagdown & k \\ & \diagdown\;\diagup & \\ 1 & 1 & \\ & k & \end{array}$$

of the quiver. Since S_4 is simple projective we have an almost split sequence $0 \to S_4 \to P_2 \coprod P_3 \to \mathfrak{r}P_1 \to 0$. Since no indecomposable module $X \not\simeq S_4$ has an irreducible morphism to P_2 or to P_3, and neither P_2 nor P_3 is a summand of some $\mathfrak{r}Q$ for Q projective, we get almost split sequences $0 \to P_2 \to \mathfrak{r}P_1 \to \operatorname{Tr} DP_2 \to 0$ and $0 \to P_2 \to \mathfrak{r}P_1 \to \operatorname{Tr} DP_3 \to 0$. Since $\mathfrak{r}P_1$ is a summand of the radical of exactly one indecomposable projective module, namely P_1, we have an almost split sequence $0 \to \mathfrak{r}P_1 \to \operatorname{Tr} DP_2 \coprod \operatorname{Tr} DP_3 \coprod P_1 \to \operatorname{Tr} D(\mathfrak{r}P_1) \to 0$. We see that $\operatorname{Tr} DP_2 \simeq S_3$, $\operatorname{Tr} DP_3 \simeq S_2$ and $\operatorname{Tr} D(\mathfrak{r}P_1) \simeq P_1/\operatorname{soc} P_1$. We further calculate $\operatorname{Tr} DS_3 \simeq I_2$, $\operatorname{Tr} DS_2 \simeq I_3$ and $\operatorname{Tr} D(P_1/\operatorname{soc} P_1) \simeq S_4$ to get the almost split sequence $0 \to P_1/\operatorname{soc} P_1 \to I_2 \coprod I_3 \to S_1 \to 0$. There is no irreducible morphism from I_2 or I_3 to any module X not isomorphic to S_1. Hence we now have the AR-quiver

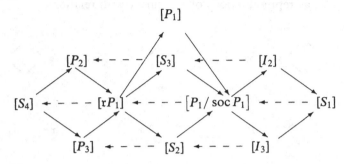

Note that we here omit the valuation since it is $(1,1)$ for each arrow.

Example For the triangular matrix algebra $\Lambda = \begin{pmatrix} \mathbb{C} & 0 \\ \mathbb{C} & \mathbb{R} \end{pmatrix}$ where $\mathbb{C}$ is the complex numbers and $\mathbb{R}$ the real numbers we have two indecomposable projective modules $P_1 = (0, \mathbb{R}, 0)$ and $P_2 = (\mathbb{C}, \mathbb{C}, 1_{\mathbb{C}})$ with $\mathbb{R}$ and $\mathbb{C}$ as endomorphism rings. Since $\mathrm{r}P_2 \simeq P_1 \coprod P_1$, we then have the valued arrow $[P_1] \overset{(2,1)}{\to} [P_2]$. P_1 is simple projective, so we have an almost split sequence $0 \to P_1 \to P_2 \to \mathrm{Tr}\, DP_1 \to 0$. From this exact sequence we see that $\mathrm{Tr}\, DP_1 \simeq (\mathbb{C}, \mathbb{R}, f)$ where $f : \mathbb{C} \to \mathbb{R}$ is nonzero. From III Section 2 we know that we have an indecomposable injective module $(\mathrm{Hom}_{\mathbb{R}}(\mathbb{C}, \mathbb{R}), \mathbb{R}, g)$ where $g : \mathbb{C} \otimes_{\mathbb{C}} \mathrm{Hom}_{\mathbb{R}}(\mathbb{C}, \mathbb{R}) \to \mathbb{R}$ is the natural morphism. Since $\mathbb{C} \otimes_{\mathbb{C}} \mathrm{Hom}_{\mathbb{R}}(\mathbb{C}, \mathbb{R}) \simeq \mathbb{C}$ and g is not zero, we have that this injective module is isomorphic to $\mathrm{Tr}\, DP_1$, using that $\mathrm{Hom}_{\mathbb{R}}(\mathbb{C}, \mathbb{R})$ is a one-dimensional $\mathbb{C}$-module. Since we have the valued arrow $[P_2] \overset{(1,2)}{\to} [\mathrm{Tr}\, DP_1]$, we have an almost split sequence $0 \to P_2 \to 2\,\mathrm{Tr}\, DP_1 \to \mathrm{Tr}\, DP_2 \to 0$. By calculating dimensions we see that $\mathrm{Tr}\, DP_2 \simeq (\mathbb{C}, 0, 0)$, which is injective. Hence we obtain the AR-quiver

$$
\begin{array}{ccc}
 & [P_2] & \overset{\text{--- --- ---}}{\longleftarrow} \quad [\mathrm{Tr}\, DP_2] \\
(2,1)\nearrow & \searrow (1,2) \quad (2,1)\nearrow & \\
[P_1] & \overset{\text{--- --- ---}}{\longleftarrow} \quad [\mathrm{Tr}\, DP_1] &
\end{array}
$$

Example Let k be a field and Λ the path algebra of the quiver

with relation $\{\beta\alpha\}$. We list the simple modules, the indecomposable projective modules and their radicals, the indecomposable injective modules and the indecomposable injective modules modulo the socle. We view all these modules as representations of the quiver with relations.

$$P_1: \quad k \underset{0 \searrow \ \swarrow 1}{\overset{1 \nwarrow \ \nearrow 1}{\diamond}} k \qquad S_2: \quad k \diamond 0 \qquad S_3: \quad 0 \diamond k$$

with top k, bottom k for P_1; top 0, bottom 0 for S_2; top 0, bottom 0 for S_3.

$$S_1: \quad 0 \diamond 0 \qquad P_2: \quad k \diamond 0 \qquad \mathrm{r}P_2 \simeq S_4$$

with top k, bottom 0 for S_1; top 0, bottom k for P_2 (edge label 1).

$$P_3: \quad 0 \diamond k \qquad \mathrm{r}P_3 \simeq S_4 \qquad P_4: \quad 0 \diamond 0$$

$$\mathrm{r}P_1 \simeq P_3 \coprod S_2 \qquad I_1 = S_1 \qquad I_3:$$

$$I_3/\operatorname{soc} I_3 \simeq S_1 \qquad I_2: \quad k \diamond 0 \qquad I_2/\operatorname{soc} I_2 \simeq S_1$$

$$I_4: \quad k \diamond k \qquad I_4/\operatorname{soc} I_4 \simeq S_2 \coprod I_3$$

Since S_4 is simple projective, we have the almost split sequence $0 \to S_4 \to P_2 \coprod P_3 \to \operatorname{Tr} DS_4 \to 0$. From this sequence we calculate that $\operatorname{Tr} DS_4$ has to be the module corresponding to the representation $k \overset{0}{\underset{1 \searrow \swarrow 1}{\diamond}} k$ over k . If $X \not\simeq S_4$ there is no irreducible morphism from X to P_2 or P_3. Since P_2 is not a summand of the radical of any projective module, we get the almost split sequence $0 \to P_2 \to \operatorname{Tr} DS_4 \to \operatorname{Tr} DP_2 \to 0$, and hence $\operatorname{Tr} DP_2 \simeq S_3$. P_3 is a summand of the radical of P_1, and $\dim_k \operatorname{Hom}_\Lambda(P_3, P_1) = 1$. Hence we get the almost split sequence $0 \to P_3 \to \mathrm{Tr}\, DS_4 \coprod P_1 \to \mathrm{Tr}\, DP_3 \to 0$. A not too hard calculation then shows that $\mathrm{Tr}\, DP_3$ is given as the representation $k \amalg k$ 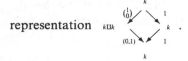.

Since $\mathfrak{r}P_1 = P_3 \coprod S_2$, and S_2 is not a composition factor of $\mathfrak{r}/\mathrm{soc}\,\mathfrak{r}$, it follows from V Theorem 3.3 that the middle term of the almost split sequence starting at S_2 is projective. Since S_2 is a summand of $\mathfrak{r}P$ with P indecomposable projective only for $P \simeq P_1$, we have the almost split sequence $0 \to S_2 \to P_1 \to \mathrm{Tr}\, DS_2 \to 0$ where $\mathrm{Tr}\, DS_2$ is given by the representation

$$0 \diagdown k \diagup 1 \diagdown k$$

Further we have an almost split sequence $0 \to P_1 \to \mathrm{Tr}\, DP_3 \coprod \mathrm{Tr}\, DS_2 \to \mathrm{Tr}\, DP_1 \to 0$, and from this we can compute $\mathrm{Tr}\, DP_1 \simeq I_4$. And we have the almost split sequence $0 \to \mathrm{Tr}\, DS_2 \to I_4 \to (\mathrm{Tr}\, D)^2 S_2 \to 0$, so that $(\mathrm{Tr}\, D)^2 S_2 \simeq S_2$. Similarly we get that $(\mathrm{Tr}\, D)^2 S_4$ corresponds to the representation

and that $\mathrm{Tr}\, DS_3 \simeq I_2$, $(\mathrm{Tr}\, D)^2 P_3 \simeq I_3$ and $(\mathrm{Tr}\, D)^3 S_4 \simeq S_1$. Collecting our information we get that the entire AR-quiver of Λ is

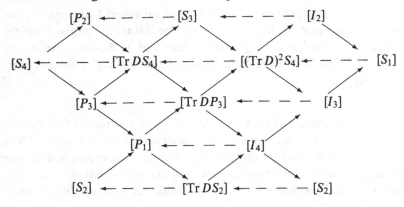

where the two copies of $[S_2]$ are to be identified.

We now give some conditions on the structure of an AR-quiver.

Proposition 2.2 *If Λ is an artin algebra of finite representation type then no arrow in the Auslander–Reiten-quiver of Λ has valuation (n, m) with $n \geq 2$ and $m \geq 2$.*

Proof Assume A and B are indecomposable and that there is an arrow in Γ_Λ from $[A]$ to $[B]$ of valuation (n, m) with $n \geq 2$ and $m \geq 2$. We want to prove that then there is no bound on the length of the indecomposable modules in mod Λ.

If $f : A \to B$ is irreducible, then the lengths of A and B cannot be the same. By duality we may assume that the length of A is bigger than the length of B. Hence B is not projective. From the almost split sequence $0 \to D\operatorname{Tr} B \to M \to B \to 0$ we get $l(D\operatorname{Tr} B) > l(A)$ by using that the multiplicity of A as a summand of M is $n \geq 2$. Hence A is not projective and we have the valued arrow $[D\operatorname{Tr} B] \overset{(m,n)}{\to} [A]$. Since $m \geq 2$ we can repeat the argument to get $l(D\operatorname{Tr} A) > l(D\operatorname{Tr} B)$. By induction we then conclude that there is no bound on the length of the indecomposable modules $(D\operatorname{Tr})^i A$, showing that Λ is not of finite representation type. $\square$

We get the following direct consequence of Proposition 2.2.

Corollary 2.3 *If Λ is an artin algebra of finite representation type over an algebraically closed field, then all arrows in the Auslander–Reiten-quiver have trivial valuation.*

Proof Since k is algebraically closed, all the division k-algebras T_A for A in ind Λ are isomorphic to k. Since the valuation is given by the dimensions of $\operatorname{Irr}(A, B)$ as T_A^{op}-vector space and as T_B-vector space, we get that they are all of the type (n, n). Hence our claim follows from Proposition 2.2. $\square$

We now give some sufficient conditions for a composition of irreducible morphisms between indecomposable modules to be nonzero. These conditions may be read off directly from the AR-quiver and will also be used to give more information on the structure of the AR-quiver.

Let I be the integers in one of the intervals $(-\infty, n]$, $[n, \infty)$, $[m, n]$ for $m < n$ or $\{1, \ldots, n\}$ modulo n. Let $\cdots \to A_i \overset{f_i}{\to} A_{i+1} \to \cdots \to A_j \overset{f_j}{\to} A_{j+1} \to \cdots$ be a sequence of irreducible morphisms between indecomposable modules with each index in I. The sequence is said to be **sectional** if $D \operatorname{Tr} A_{i+2} \not\simeq A_i$ whenever i and $i + 2$ are both in I, and the corresponding path in the AR-quiver is said to be a **sectional path**.

We have the following main result about composition of a sectional sequence of irreducible morphisms.

Theorem 2.4 *If* $A_1 \xrightarrow{f_1} A_2 \to \cdots \to A_n \xrightarrow{f_n} A_{n+1}$ *is a sectional sequence of irreducible morphisms between indecomposable modules, then the composition* $f = f_n \cdots f_1$ *is nonzero.*

This is a direct consequence of the following.

Lemma 2.5 *Let* Λ *be an artin algebra and* $A_1 \xrightarrow{f_1} A_2 \to \cdots \to A_n \xrightarrow{f_n} A_{n+1}$ *a sequence of irreducible morphisms between indecomposable modules. Suppose the composition* $f_n \cdots f_1$ *either is 0 or factors through a morphism* $g : B \to A_{n+1}$ *with* B *indecomposable, such that* $(f_n, g) : A_n \coprod B \to A_{n+1}$ *is irreducible. Then we have* $D \operatorname{Tr} A_i \simeq A_{i-2}$ *for some* i *with* $3 \le i \le n + 1$.

Proof We prove this by induction on n, and start with the case $n = 2$. Let $A_1 \xrightarrow{f_1} A_2 \xrightarrow{f_2} A_3$ be a sequence of irreducible morphisms between indecomposable modules such that $f_2 f_1 = 0$ or $f_2 f_1 = gh$ where $g : B \to A_3$ is such that $A_2 \coprod B \xrightarrow{(f_2, g)} A_3$ is irreducible. If $f_2 f_1 = 0$, then f_2 cannot be a monomorphism. Hence it is an epimorphism, and therefore A_3 is not projective. Consequently there exists an almost split sequence

$$0 \to D \operatorname{Tr} A_3 \xrightarrow{\binom{f'_2}{t'}} A_2 \coprod X \xrightarrow{(f_2, t)} A_3 \to 0.$$

Therefore there is some $h : A_1 \to D \operatorname{Tr} A_3$ with $\binom{f'_2}{t'} h = \binom{f_1}{0}$. But this implies that $f'_2 h = f_1$. Now f_1 and f'_2 being irreducible implies that h is a split monomorphism, and therefore h is an isomorphism since $D \operatorname{Tr} A_3$ is indecomposable.

In case $f_2 f_1 = gh$ we consider $A_1 \xrightarrow{\binom{f_1}{-h}} A_2 \coprod B \xrightarrow{(f_2, g)} A_3$ and get that $(f_2, g)\binom{f_1}{-h} = 0$. Hence (f_2, g) is not a monomorphism, and consequently an epimorphism. Therefore A_3 is not projective. Consider an almost split sequence

$$0 \to D \operatorname{Tr} A_3 \xrightarrow{\binom{f'_2}{g'}{t'}} A_2 \coprod B \coprod C \xrightarrow{(f_2, g, t)} A_3 \to 0.$$

We get a morphism $h': A_1 \to D \operatorname{Tr} A_3$ with $\begin{pmatrix} f_2' \\ g' \\ t' \end{pmatrix} h' = \begin{pmatrix} f_1 \\ \frac{-h}{0} \end{pmatrix}$, i.e. $f_1 = f_2' h'$. Again by the irreducibility of f_1 and f_2', we have that h' is an isomorphism.

For the inductive step assume the claim holds for $n \geq 2$. We want to prove that the claim holds for $n + 1$. Let $\tilde{f} = f_n \cdots f_1$, and assume first $f_{n+1} \tilde{f} = 0$. If $\tilde{f} = 0$ we are done by induction. But $\tilde{f} \neq 0$ implies that f_{n+1} is not a monomorphism, hence it is an epimorphism since it is irreducible, and therefore A_{n+2} is not projective. Considering an almost split sequence

$$0 \to D \operatorname{Tr} A_{n+2} \xrightarrow{\binom{g}{s}} A_{n+1} \coprod E \xrightarrow{(f_{n+1}, t)} A_{n+2} \to 0$$

and using that $f_{n+1} \tilde{f} = 0$, we see that there exists some $h: A_1 \to D \operatorname{Tr} A_{n+2}$ and an irreducible morphism $g: D \operatorname{Tr} A_{n+2} \to A_{n+1}$ with $gh = \tilde{f}$. Now if $D \operatorname{Tr} A_{n+2} \simeq A_n$ we are done. Otherwise $(f_n, g): A_n \coprod D \operatorname{Tr} A_{n+2} \to A_{n+1}$ is irreducible and we are done by induction.

If $f_{n+1} \tilde{f} = gh$ for some $h: A_1 \to B$ and $g: B \to A_{n+2}$ with $(f_{n+1}, g): A_{n+1} \coprod B \to A_{n+2}$ irreducible, then $\begin{pmatrix} \tilde{f} \\ -h \end{pmatrix}: A_1 \to A_{n+1} \coprod B$ composed with (f_{n+1}, g) is zero. Hence there exist a morphism $h': A_1 \to D \operatorname{Tr} A_{n+2}$ and an irreducible morphism $g': D \operatorname{Tr} A_{n+2} \to A_{n+1}$ with $\tilde{f} = g'h'$. If A_n is isomorphic to $D \operatorname{Tr} A_{n+2}$ we are done. If A_n is not isomorphic to $D \operatorname{Tr} A_{n+2}$, then $(f_n, g'): A_n \coprod D \operatorname{Tr} A_{n+2} \to A_{n+1}$ is irreducible and we are done by induction. □

We have the following consequence of Theorem 2.4 on the structure of the AR-quiver.

Corollary 2.6 *There is no sectional cycle in the Auslander–Reiten-quiver of an artin algebra Λ.*

Proof Assume that $A_1 \xrightarrow{f_1} A_2 \to \cdots \to A_n \xrightarrow{f_n} A_1 \xrightarrow{f_1} A_2$ is a cycle of irreducible maps between indecomposable modules. The composition $f = f_n \cdots f_1$ is clearly not an isomorphism, and is hence nilpotent in $\operatorname{End}_\Lambda(A_1)$ since $\operatorname{End}_\Lambda(A_1)$ is a local ring. Since $(f_n \cdots f_1)^t = 0$ for some t, it follows from Theorem 2.4 that the cycle $A_1 \xrightarrow{f_1} A_2 \to \cdots \to A_n \xrightarrow{f_n} A_1 \xrightarrow{f_1} A_2$ is not a sectional sequence of irreducible morphisms and hence there are no sectional cycles in the AR-quiver. □

3 Cartan matrices and subadditive functions

In the last section we gave some constraints on the structure of the AR-quiver. In this section we characterize Dynkin diagrams and Euclidean diagrams in terms of the existence of subadditive and additive functions and use this to get more information on the shape of the AR-quiver in the next section.

A handy way of representing a valued graph without loops and multiple edges is through the associated Cartan matrix, a notion we now introduce.

Let I be a set. A **Cartan matrix** C on I is a function $C: I \times I \to \mathbb{Z}$ satisfying the following properties, where we write $C(i, j) = c_{ij}$.

(a) $c_{ii} = 2$.

(b) $c_{ij} \leq 0$ for all i, j with $i \neq j$, and for each i we have that $c_{ij} < 0$ for only finitely many $j \in I$.

(c) $c_{ij} \neq 0$ if and only if $c_{ji} \neq 0$.

There is a close connection between Cartan matrices and valued graphs. Here a **valued graph** ∇ on an index set I is given by a function $d: I \times I \to \mathbb{N}$ such that (where we write $d(i, j) = d_{ij}$)

(i) $d_{ii} = 0$ for all $i \in I$,

(ii) $d_{ij} \neq 0$ if and only if $d_{ji} \neq 0$,

(iii) for each $i \in I$ there is only a finite number of j with $d_{ij} \neq 0$.

When representing a graph ∇ in the plane, the set I is identified with vertices called ∇_0 and there is an edge connecting i and j if $d_{ij} \neq 0$. We then write $i \cdot \xrightarrow{(d_{ij}, d_{ji})} \cdot j$.

If C is a Cartan matrix on a set I there is associated a graph ∇ with $\nabla_0 = I$, such that $d: I \times I \to \mathbb{N}$ is defined by $d_{ii} = 0$ and $d_{ij} = |c_{ij}|$ for $i \neq j$. Conversely, a valued graph ∇ gives rise to a Cartan matrix C on the set ∇_0 where $c_{ii} = 2$ for $i \in \nabla_0$ and $c_{ij} = -d_{ij}$ for $i \neq j$ in ∇_0.

A Cartan matrix C on a set I is said to be **indecomposable** if for all proper partitions of I as $I_1 \cup I_2$ there exist i in I_1 and j in I_2 with $c_{ij} \neq 0$. This is clearly the same as saying that the associated valued graph is connected. Observe also that the transpose C^{tr} of a Cartan matrix C on a set I given by $c_{ij}^{\text{tr}} = c_{ji}$ for all $(i, j) \in I \times I$ is also a Cartan matrix on I.

Let C be a Cartan matrix on a set I. A **subadditive function** for C is a function $d: I \to \mathbb{Z}^+$ such that $dC: I \to \mathbb{Z}$ given by $dC(j) = \sum_{i \in I} d_i c_{ij}$ has its image in $\mathbb{N}$, where we write d_i for $d(i)$, and d is said to be **additive** if $dC = 0$. Note that if the index set I is finite and we fix a total ordering of

I, we can write C as an ordinary matrix and d as a vector. The product dC can then be interpreted as a product of matrices and viewed as a vector.

We shall see that the existence of a subadditive or additive function for a Cartan matrix C imposes heavy restrictions on the valued graph associated with C. In order to describe exactly when such functions exist we need to list the diagrams known as **Dynkin** diagrams and the diagrams known as **Euclidean** diagrams. In the second case it is easy to see that there is always an additive function, and we list one for each diagram.

Dynkin diagrams

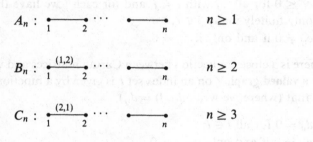

A_n : $n \geq 1$

B_n : $n \geq 2$

C_n : $n \geq 3$

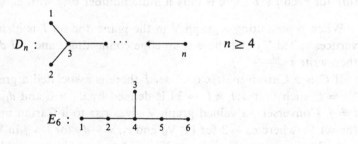

D_n : $n \geq 4$

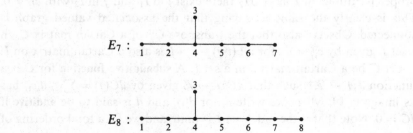

E_6 :

E_7 :

E_8 :

$F_4 :$

$G_2 :$

Euclidean diagrams   Additive functions

$\widetilde{A}_n :$

$\widetilde{B}_n :$

$\widetilde{C}_n :$

$\widetilde{D}_n :$

$\widetilde{E}_6 :$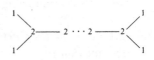

$\widetilde{E}_7 :$

$\widetilde{E}_8 :$

$\widetilde{A}_{11} :$

$\widetilde{A}_{12} :$

$\widetilde{BC}_n :$

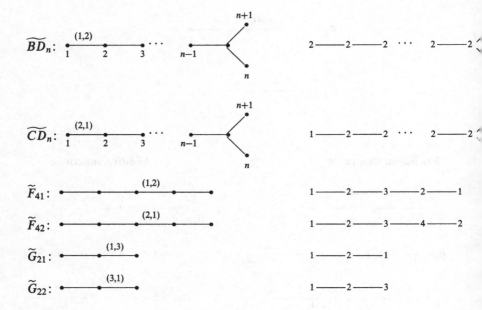

We start with the following result on subadditive functions.

Lemma 3.1 *Let C be the Cartan matrix of a Euclidean diagram. Then any subadditive function for C is additive.*

Proof We have listed an additive function for each Cartan matrix of the Euclidean diagrams in the table. Observe that if C is the Cartan matrix of a Euclidean diagram, then also C^{tr} is the Cartan matrix of a Euclidean diagram. Now let ∇ be a Euclidean diagram with Cartan matrix C and let d be a subadditive function for C. Let ε be the additive function for C^{tr} in the table and consider $\varepsilon(dC)^{\mathrm{tr}} = (\varepsilon C^{\mathrm{tr}})d^{\mathrm{tr}} = 0$. Since all values of ε are strictly positive and all values of $(dC)^{\mathrm{tr}}$ are nonnegative, we conclude that $(dC)^{\mathrm{tr}} = 0$. This shows that d is an additive function for C. $\square$

Given two indecomposable Cartan matrices C' and C on the index sets I' and I respectively we say that C' is **smaller** than C if there exists an injection $\sigma : I' \to I$ such that $|c'_{ij}| \leq |c_{\sigma(i)\sigma(j)}|$ for all $i, j \in I'$. We say that C' is strictly smaller than C if either $\sigma(I') \neq I$ or there exist i and j in I' with $|c'_{ij}| < |c_{\sigma(i)\sigma(j)}|$. We investigate the relationship between subadditive functions for two Cartan matrices when one is strictly smaller than the other.

Lemma 3.2 *Let C' and C be indecomposable Cartan matrices on I' and I respectively and let $\sigma : I' \to I$ be an injection such that C' is strictly smaller than C.*

(a) *Let $d : I \to \mathbb{Z}^+$ be a subadditive function for C. Then $d\sigma : I' \to \mathbb{Z}^+$ is a subadditive function for C' which is not additive.*

(b) *Let $d' : I' \to \mathbb{Z}^+$ be an additive function for C' and let $d : I \to \mathbb{N}$ be given by*

$$d_i = \begin{cases} 2d_j & \text{if } i = \sigma(j), \\ 1 & \text{if } i \notin \sigma(I') \text{ with } c_{ij} \neq 0 \text{ and } j \in \sigma(I'), \\ 0 & \text{otherwise.} \end{cases}$$

Then $(dC)_i \leq 0$ for all $i \in I$, with $(dC)_i < 0$ for at least one i with $d_i \neq 0$.

Proof We first prove (a). Letting $d' = d\sigma$ we have

$$
\begin{aligned}
(d'C')_i &= 2d_{\sigma(i)} + \sum_{\substack{j \neq i \\ j \in I'}} d_{\sigma(j)} c'_{ij} \\
&\geq 2d_{\sigma(i)} + \sum_{\substack{j \neq i \\ j \in I'}} d_{\sigma(j)} c_{\sigma(i)\sigma(j)} \\
&\geq 2d_{\sigma(i)} + \sum_{\substack{j \neq \sigma(i) \\ j \in I}} d_j c_{\sigma(i)j} \geq 0.
\end{aligned}
$$

Hence d' is a subadditive function for C'. Further, using that C is indecomposable we have that if $\sigma : I' \to I$ makes C' strictly smaller than C, either there exists $i \in I'$ and $j \in I$ such that $c_{j\sigma(i)} \neq 0$ and j not in the image of σ, or there exist $i, j \in I'$ with $|c'_{ji}| < |c_{\sigma(j)\sigma(i)}|$. In the first case the second inequality above is strict and in the second case the first inequality above is strict. This completes the proof of (a).

For part (b), clearly $(dC)_i \leq 0$ for i not in $\sigma(I')$ since C is indecomposable. For $i = \sigma(j)$ we have

$$
\begin{aligned}
0 = (2d'C')_j &= 4d'_j + 2 \sum_{\substack{l \neq j \\ l \in I'}} d'_l c'_{jl} \\
&= 2d_{\sigma(j)} + \sum_{\substack{l \neq j \\ l \in I'}} d_{\sigma(l)} c'_{jl} \\
&\geq 2d_{\sigma(j)} + \sum_{\substack{l \neq i \\ l \in I}} d_l c_{il} \\
&= (dC)_i .
\end{aligned}
$$

The inequality is clearly strict if $c_{\sigma(m)\sigma(j)} > c'_{mj}$ for some $(m, j) \in I' \times I'$

or if there exists an $m \in I$ with $m \notin \sigma(I')$ and $c_{mi} \neq 0$. Hence the proof of part (b) of the lemma is complete.   □

From these two lemmas we get the following conclusion.

Theorem 3.3 *Let C be an indecomposable Cartan matrix on the finite index set I with underlying valued graph ∇ and assume d is a subadditive function for C. Then we have the following.*

(a) *∇ is either a Dynkin diagram or a Euclidean diagram.*

(b) *If d is not additive, then ∇ is a Dynkin diagram.*

(c) *If d is additive, then ∇ is a Euclidean diagram.*

Proof If ∇ is a finite connected valued graph without loops which is neither a Dynkin nor a Euclidean diagram, it is not hard to see that ∇ will contain a proper valued subgraph ∇' which is a Euclidean diagram. But then the associated Cartan matrix C' of ∇' will be strictly smaller than C. Let d be a subadditive function for C and let $d' = d|_{\nabla'}$. Hence d' will be a subadditive function for C' which can not be additive according to Lemma 3.2 (a). However, Lemma 3.1 gives that all subadditive functions for the Cartan matrices with underlying valued graph a Euclidean diagram are additive. This is our desired contradiction.

Part (b) follows directly from (a) and Lemma 3.1 which says that all subadditive functions on a Euclidean diagram are additive.

To verify (c) we observe that the Cartan matrices C for the Dynkin diagrams are all nonsingular, and hence they have no additive functions.   □

We shall also investigate subadditive and additive functions for the following list of five infinite diagrams closely related to the Dynkin diagrams. We also give an additive function for each of them.

A_∞: $\bullet\!\!-\!\!-\!\!\bullet\!\!-\!\!-\!\!\bullet \cdots \bullet\!\!-\!\!-\!\!\bullet \cdots$ $1\!\!-\!\!-\!\!2\!\!-\!\!-\!\!3 \cdots n\!\!-\!\!-\!\!n+1 \cdots$

B_∞: $\overset{(1,2)}{\bullet\!\!-\!\!-\!\!\bullet}\!\!-\!\!-\!\!\bullet \cdots \bullet\!\!-\!\!-\!\!\bullet \cdots$ $1\!\!-\!\!-\!\!2\!\!-\!\!-\!\!2 \cdots 2\!\!-\!\!-\!\!2 \cdots$

C_∞: $\overset{(2,1)}{\bullet\!\!-\!\!-\!\!\bullet}\!\!-\!\!-\!\!\bullet \cdots \bullet\!\!-\!\!-\!\!\bullet \cdots$ $1\!\!-\!\!-\!\!1\!\!-\!\!-\!\!1 \cdots 1\!\!-\!\!-\!\!1 \cdots$

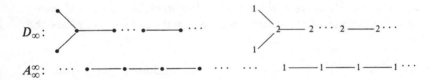

D_∞:

A_∞^∞:

The Cartan matrix of A_∞ admits subadditive functions which are not additive, but in the four other cases all subadditive functions are additive, and they are in fact bounded.

We start by considering A_∞^∞.

Lemma 3.4 *Let* $d:(A_\infty^\infty)_0 \to \mathbb{Z}^+$ *be a subadditive function. Then* d *is constant and therefore additive.*

Proof Let $a \in (A_\infty^\infty)_0$ be a vertex such that the minimum of d is obtained. Then $2d_a - d_{a+1} - d_{a-1} \geq 0$. But we have $d_{a+1} \geq d_a$ and $d_{a-1} \geq d_a$. Hence $d_a = d_{a+1} = d_{a-1}$, and therefore d has to be a constant and is hence additive. $\qquad\square$

Lemma 3.5 *If* $d:\Delta_0 \to \mathbb{Z}^+$ *is a subadditive function with* $\Delta = B_\infty$, C_∞ *or* D_∞, *then* d *is a multiple of the function listed in the table.*

Proof We consider each case separately. One checks that if

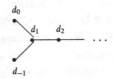

is a subadditive function for D_∞, then

$$\cdots \underset{d_2}{\bullet} \underline{\qquad} \underset{d_1}{\bullet} \underline{\qquad} \underset{d_0+d_{-1}}{\bullet} \underline{\qquad} \underset{d_1}{\bullet} \underline{\qquad} \underset{d_2}{\bullet} \cdots$$

is a subadditive function for A_∞^∞, which by Lemma 3.4 is constant.

Further

$$
\begin{aligned}
(dC)_0 &= & 2d_0 - d_1 & \geq & 0, \\
(dC)_{-1} &= & 2d_{-1} - d_1 & \geq & 0, \\
(dC)_1 &= & 2d_1 - d_0 - d_{-1} - d_2 & = & 0.
\end{aligned}
$$

But $d_1 = d_2$, so from the third equality we obtain $d_1 = d_0 - d_{-1}$.

Substituting this value for d_1 in the two inequalities gives $d_{-1} - d_0 \geq 0$ and $d_0 - d_{-1} \geq 0$. Hence $d_0 = d_{-1}$ and therefore $d_i = 2d_0$ for $i \geq 1$ showing that d is a multiple of the given additive function. This completes the case D_∞.

If $\overset{d_0}{\bullet}\overset{(1,2)}{\rule{1cm}{0.4pt}}\overset{d_1}{\bullet}\ \cdots\ \overset{d_n}{\bullet}\overset{d_{n+1}}{\rule{1cm}{0.4pt}}\bullet\ \cdots$ is a subadditive function for B_∞

then $\cdots\ \overset{d_2}{\bullet}\overset{d_1}{\rule{0.7cm}{0.4pt}}\overset{2d_0}{\bullet}\overset{d_1}{\rule{0.7cm}{0.4pt}}\overset{d_2}{\bullet}\ \cdots$ is a subadditive function for A^∞_∞ and the same analysis as above applies to conclude that d is a multiple of the additive function given in the table above.

To treat the last case C_∞ observe that if $\overset{d_0}{\bullet}\overset{d_1}{\rule{0.7cm}{0.4pt}}\overset{d_2}{\bullet}\ \cdots\ \overset{d_n}{\bullet}\overset{d_{n+1}}{\rule{0.7cm}{0.4pt}}\ \cdots$ is a subadditive function for C_∞ then $\cdots\ \overset{d_2}{\bullet}\overset{d_1}{\rule{0.7cm}{0.4pt}}\overset{d_0}{\bullet}\overset{d_1}{\rule{0.7cm}{0.4pt}}\bullet\ \cdots$ is a subadditive function for A^∞_∞ and hence we get that this is a constant by Lemma 3.4. $\qquad\square$

4 Translation quivers

In Section 1 we introduced the AR-quiver associated with an artin algebra. The AR-quiver often consists of several components, and considering each component leads us to the more general notion of a translation quiver, which we introduce in this section. Combinatorial aspects of these translation quivers will be explored and used to give results about full subquivers of the AR-quiver consisting of vertices where D Tr is everywhere defined. These combinatorial results will also be used in the next chapter dealing with hereditary artin algebras.

Let Γ be a valued quiver with vertex set Γ_0 and arrow set Γ_1 where Γ is **locally finite**, that is for each $i \in \Gamma_0$ there is only a finite number of arrows entering or leaving i. Let τ be an injective map from a subset of Γ_0 into Γ_0. For $x \in \Gamma_0$ let x^- denote the set of immediate predecessors of x. It is defined by $x^- = \{y \in \Gamma_0|$ there is an arrow $y \to x\}$. Dually let x^+ denote the set of immediate successors of x. It is defined by $x^+ = \{y \in \Gamma_0|$ there is an arrow $x \to y\}$. The pair (Γ, τ) is called a valued **translation quiver** if the following three conditions are satisfied.

(a) Γ has no loops and no multiple arrows.

(b) Whenever $x \in \Gamma_0$ is such that $\tau(x)$ is defined then $x^- = \tau(x)^+$.

(c) If α is an arrow from x to y with valuation (a, b) and $\tau(y)$ is defined then the arrow from $\tau(y)$ to x has valuation (b, a).

We say that a translation quiver (Γ, τ) is a **proper translation quiver** if in addition we have

(d) If $x \in \Gamma_0$ is such that $\tau(x)$ is defined, then x^- is nonempty.

Note that associated with a translation quiver (Γ, τ) there is always a proper translation quiver $(\widetilde{\Gamma}, \tau)$ obtained by restricting τ to the vertices x where x^- is not empty.

The partially defined map $\tau : \Gamma_0 \to \Gamma_0$ is called the **translation** of the translation quiver (Γ, τ).

Any AR-quiver of an artin algebra is a translation quiver when we let τ be the map defined by DTr.

Other finite concrete examples are

(i) 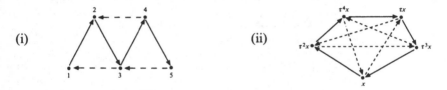 (ii)

where we have inserted a broken arrow for the translation.

Neither the example represented in figure (i) nor the example represented in figure (ii) is the AR-quiver of an artin algebra. It is easy to see this for case (ii) since there are no projective vertices.

If case (i) were an AR-quiver, vertex 1 would correspond to a simple projective module. Then vertex 2 would correspond to an indecomposable projective module, which would have length 2. But then vertex 3 would correspond to a simple module, and also to the middle term of an almost split sequence, which is impossible.

Let (Γ, τ) be a translation quiver. In analogy with the situation for the AR-quiver of an artin algebra, we call a vertex $x \in \Gamma_0$ a **projective vertex** if x is not in the domain of τ and we call a vertex $x \in \Gamma_0$ an **injective vertex** if x is not in the image of τ. As for AR-quivers, τ also induces a partially defined map on Γ_1. More precisely, if $\alpha : y \to x$ is an arrow with x not a projective vertex, there is a unique arrow $\tau(x) \to y$ in Γ_1 which we denote by $\sigma(\alpha)$. The partially defined map σ on Γ_1 obtained in this way is called the **semitranslation** of the translation quiver. It is completely determined by τ.

We now give a way of constructing families of translation quivers. Let Δ be a valued quiver without loops (possibly infinite). We now construct the translation quiver $\mathbb{Z}\Delta$. The set of vertices in $\mathbb{Z}\Delta$ is $(\mathbb{Z}\Delta)_0 = \mathbb{Z} \times \Delta_0$. The translation in $\mathbb{Z}\Delta$ is given by $\tau(n, x) = (n - 1, x)$. The arrows in $\mathbb{Z}\Delta$ are as follows. For each arrow α with valuation (a, b) from x to y in Δ form arrows α_n from (n, x) to (n, y) with valuation (a, b) and arrows $\sigma(\alpha_n)$ from $(n - 1, y)$ to (n, x) with valuation (b, a). In this way $\mathbb{Z}\Delta$ becomes a translation quiver.

If the underlying graph of Δ is a tree and the valuation is trivial then the translation quiver $\mathbb{Z}\Delta$ is independent of the orientation in Δ. Otherwise it may depend on the orientation.

Example

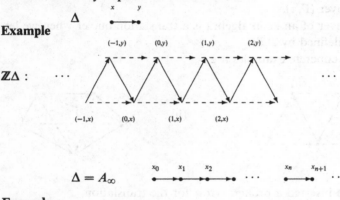

$$\Delta = A_\infty \qquad \overset{x_0}{\bullet} \longrightarrow \overset{x_1}{\bullet} \longrightarrow \overset{x_2}{\bullet} \longrightarrow \bullet \cdots \overset{x_n}{\bullet} \longrightarrow \overset{x_{n+1}}{\bullet} \cdots$$

Example

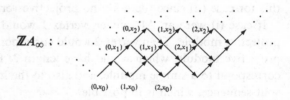

We now consider maps between translation quivers (Γ', τ') and (Γ, τ). A **translation quiver morphism** $f : (\Gamma', \tau') \to (\Gamma, \tau)$ is a pair of maps $f_0 : \Gamma_0' \to \Gamma_0$ and $f_1 : \Gamma_1' \to \Gamma_1$ such that the following two conditions are satisfied.

(a) f is a morphism of valued quivers, i.e. if α in Γ_1' is an arrow from x to y with valuation (a, b), then $f_1(\alpha)$ in Γ_1 is an arrow from $f_0(x)$ to $f_0(y)$ with valuation (a, b).

(b) $f_0(\tau'(x)) = \tau(f_0(x))$ for all nonprojective vertices x of Γ_0'.

A translation quiver morphism $f : (\Gamma', \tau') \to (\Gamma, \tau)$ is called a **covering** if the following conditions hold.

(a) If x is projective in Γ', then $f(x)$ is projective in Γ.

(b) If x is injective in Γ', then $f(x)$ is injective in Γ.

(c) For each vertex x in Γ'_0, f induces a bijection from x^- to $f_0(x)^-$ and from x^+ to $f_0(x)^+$ preserving the valuation of the arrows.

Example Let (Γ, τ) be the translation quiver

(*)

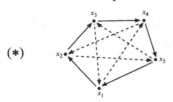

Then there is a covering $f: (\Gamma', \tau') \to (\Gamma, \tau)$, where (Γ', τ') is the following quiver.

(**)

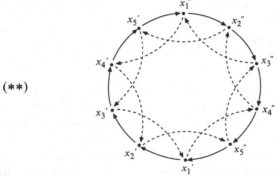

There is a theory of coverings of translation quivers, but we will not discuss this here. We will mainly use it for illustrational purposes and to deduce some more structural properties of AR-quivers.

A group G of translation quiver morphisms on a translation quiver (Γ, τ) is said to act **admissibly** if each orbit of G meets $\{x\} \cup x^-$ in at most one vertex and meets $\{x\} \cup x^+$ in at most one vertex for each x in Γ_0.

Let $\mathbb{Z}\Delta$ be as in the first example and let G be the group operating on $\mathbb{Z}\Delta$ by the automorphism ϕ where $\phi(n, x) = (n + 2, x)$. Then G is a cyclic group which acts admissibly.

Having a translation quiver Γ and a group G of translation quiver morphisms acting admissibly on Γ one can form the quotient Γ/G, which is a translation quiver. We have the natural morphism $\Gamma \to \Gamma/G$ which is a covering of translation quivers.

If Λ is an artin algebra of infinite representation type and $\mathscr{C}$ is a component of the AR-quiver containing neither projective nor injective vertices, then the translation and its inverse are everywhere defined.

Generalizing this to arbitrary translation quivers we call a translation quiver where the translation and its inverse are everywhere defined a **stable translation quiver**. Examples of stable translation quivers are the translation quivers $\mathbb{Z}\Delta$ where Δ is a quiver with no loops, and quotients of such quivers by an admissible group of automorphisms. We show that there is a strong limitation on stable subtranslation quivers of the AR-quiver of an artin algebra of the form $\mathbb{Z}\Delta/G$ where G is a group of automorphisms acting admissibly and containing some power of the translation τ. Here we use the results on additive and subadditive functions from Section 3.

Theorem 4.1 *Let Δ be a quiver without loops. If the Auslander–Reiten-quiver of an artin algebra contains a subtranslation quiver of the form $\mathbb{Z}\Delta/G$ where G is a group of automorphisms of $\mathbb{Z}\Delta$ acting admissibly and containing τ^n for some n, then either*

(i) *Δ is a Dynkin diagram and hence $\mathbb{Z}\Delta/G$ is finite or*
(ii) *Δ is A_∞.*

Proof Let Δ be a quiver without loops, n a natural number and assume $\mathbb{Z}\Delta/G$ is contained in the AR-quiver of an artin algebra Λ where G is a group of automorphisms of $\mathbb{Z}\Delta$ acting admissibly and containing τ^n. For each $a \in \Delta_0$ let $[X_0], [D \operatorname{Tr} X_0], \ldots, [D \operatorname{Tr}^{n-1} X_0]$ be the vertices, possibly with repetition, in the AR-quiver corresponding to a. Then define $d : \Delta_0 \to \mathbb{N}$ by $d(a) = \sum_{i=0}^{n-1} l(D \operatorname{Tr}^i X_0)$ where $l(Y)$ as usual denotes the length of the Λ-module Y. Using that if X is a summand of B in the exact sequence $0 \to A \to B \to C \to 0$, then $l(X) \leq l(A) + l(C)$ we obtain that d is a subadditive function for Δ. Theorem 3.3 implies that if Δ is finite then Δ is either Euclidean or Dynkin. By the same theorem, if Δ is Euclidean then d is additive. But then for each $[U]$ in $\mathbb{Z}\Delta/G$ there can be no additional neighbors in the AR-quiver than those already in $\mathbb{Z}\Delta/G$. Hence $\mathbb{Z}\Delta/G$ is a complete finite stable component of the AR-quiver of Λ. Then $\mathbb{Z}\Delta/G$ would be the whole AR-quiver by Theorem 2.1, which is impossible since there are no projective vertices in $\mathbb{Z}\Delta/G$. This finishes the proof when Δ is finite.

Assume next that Δ is infinite. Then the underlying graph $\overline{\Delta}$ cannot contain any Euclidean diagrams and hence $\overline{\Delta}$ has to be either A_∞, A_∞^∞, B_∞, C_∞ or D_∞. But A_∞^∞, B_∞, C_∞ and D_∞ only admit subadditive functions which are additive and bounded. Additivity again implies that $\mathbb{Z}\Delta/G$ is a whole component of the AR-quiver of Λ. It follows as before that $\mathbb{Z}\Delta/G$ can not be finite. But in any infinite component of the AR-quiver of an artin algebra Λ there is no bound on the length of the indecomposable modules. This excludes A_∞^∞, B_∞, C_∞ and D_∞. We are then left with the diagram A_∞. $\qquad\qquad\qquad\qquad\qquad\qquad\qquad\qquad\qquad\qquad\qquad\qquad\qquad\qquad$ □

We shall see in the next chapter that translation quivers of the form $\mathbb{Z}A_\infty/G$ actually occur as components of an AR-quiver, by showing that for the path algebra of the quiver $\cdot \rightrightarrows \cdot$ over a field k we have components of the form $\mathbb{Z}A_\infty/\langle\tau\rangle$.

If Λ is a selfinjective algebra, then we get a stable subtranslation quiver denoted by Γ_Λ^s of the AR-quiver Γ_Λ of Λ by removing the projective vertices. For applying $D\operatorname{Tr}$ or $\operatorname{Tr}D$ to an indecomposable nonprojective module we always obtain an indecomposable nonprojective module. We get the following consequence of Theorem 4.1.

Corollary 4.2 *Let Λ be an indecomposable selfinjective algebra of finite representation type. If the stable Auslander–Reiten-quiver Γ_Λ^s is of the form $\mathbb{Z}\Delta/G$ where the quiver Δ has no loops and G is a group of automorphisms of $\mathbb{Z}\Delta$ acting admissibly and containing τ^n for some n, then $\overline{\Delta}$ is a Dynkin diagram.*

Proof By Theorem 4.1 it is sufficient to show that $\overline{\Delta}$ is not A_∞. Since each $g \in G$ is an automorphism of $\mathbb{Z}\Delta$, we must have that for x and $g(x)$ in $\mathbb{Z}\Delta$ the minimal length of a path from a vertex y with only one arrow leaving y must be the same for x and $g(x)$. Hence $\overline{\Delta}$ can not be A_∞ when $\mathbb{Z}\Delta/G$ is finite. $\qquad\qquad\qquad\qquad\qquad\qquad\qquad\qquad\qquad\qquad\qquad$ □

Note that if Λ is an indecomposable selfinjective Nakayama algebra with n nonisomorphic simple modules and of Loewy length $m + 1$, then it follows from our computations of almost split sequences for Nakayama algebras that Γ_Λ^s is $\mathbb{Z}A_\infty/\langle\tau^n\rangle$. Actually it can be shown that for indecomposable selfinjective algebras of finite representation type, Γ_Λ^s must be of the form $\mathbb{Z}\Delta/G$.

Exercises

1. Let $\nabla = (\nabla_0, \nabla_1)$ be a finite graph without loops. Define $q: \mathbb{Z}^{\nabla_0} \to \mathbb{Z}$ by $q(f) = \sum_{i\in\nabla_0} f_i^2 - \sum_{l\in\nabla_1} f_{\alpha(l)}f_{\beta(l)}$ where $\alpha(l)$ and $\beta(l)$ are the end vertices of $l \in \nabla_1$. Recall that q is positive definite if $q(x) > 0$ for $x \neq 0$ and positive semidefinite if $q(x) \geq 0$ for all x.

(a) Prove that q is not positive definite if ∇ contains a cycle.
(b) Prove that q is not positive definite if ∇ contains a subgraph of the

form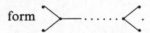

(c) Prove that if ∇ contains a subgraph of the form

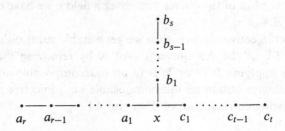

then q is not positive definite if $\frac{1}{r+1} + \frac{1}{s+1} + \frac{1}{t+1} \leq 1$.
(d) Prove that q is positive definite if and only if ∇ is one of the Dynkin
diagrams A_n, D_n, E_6, E_7 or E_8.
(e) Prove that q is positive semidefinite but not positive definite if and
only if ∇ is one of the Euclidean diagrams $\tilde{A}_n$, $\tilde{D}_n$, $\tilde{E}_6$, $\tilde{E}_7$ or $\tilde{E}_8$.

2. Let Λ and Λ' be selfinjective algebras such that $\Lambda/\operatorname{soc}\Lambda \simeq \Lambda'/\operatorname{soc}\Lambda'$.

(a) Prove that Λ is of finite representation type if and only if Λ' is of
finite representation type.
(b) Prove that $\Gamma_\Lambda \simeq \Gamma_{\Lambda'}$.
(c) Prove that the algebras given in III Exercise 8(b) are selfinjective and
that $(k\Gamma/\langle\rho\rangle)/\operatorname{soc}(k\Gamma/\langle\rho\rangle) \simeq (k\Gamma/\langle\rho'\rangle)/\operatorname{soc}(k\Gamma/\langle\rho'\rangle)$ for all fields k.
(d) Prove that $k\Gamma/\langle\rho\rangle$ from III Exercise 8(b) is of finite representation
type.
(e) Prove that $k\Gamma/\langle\rho'\rangle$ and $k\Gamma/\langle\rho\rangle$ from III Exercise 8(b) have the same
AR-quiver but that the Auslander algebra of these two examples are
not isomorphic in characteristic 2.
(f) One of the Auslander algebras in (e) is given by the mesh relations.
Determine which one.

3. Let Λ be an artin algebra and A and B indecomposable Λ-modules. Let
$f_1,\ldots,f_n : A \to B$ be irreducible morphisms. Prove that $(f_1,\ldots,f_n): nA \to B$ is irreducible if and only if $\bar{f}_1,\ldots,\bar{f}_n$ are T_A^{op}-linearly independent

elements of Irr(A, B) and that $(f_1, \ldots, f_n)^{\mathrm{tr}} \colon A \to nB$ is irreducible if and only if $\overline{f}_1, \ldots, \overline{f}_n$ are T_B-linearly independent elements of Irr(A, B).

4. Consider the translation quivers

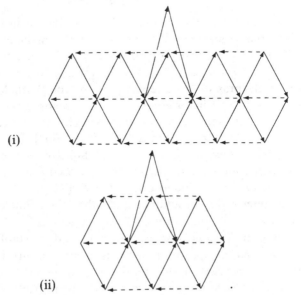

(i) .

(ii) .

Prove that the translation quiver in (i) is not the AR-quiver of an artin algebra and that the one in (ii) is the AR-quiver of an artin algebra.

5. Consider the translation quivers

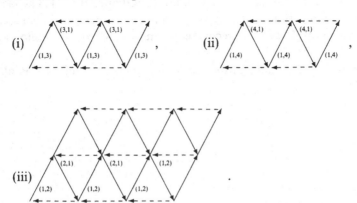

(i) , (ii) ,

(iii) .

Decide for each of the translation quivers above whether it is the AR-quiver of an artin algebra or not.

Notes

Auslander–Reiten quivers were considered by Ringel in the mid seventies. Since then they have become one of the main objects of study in the representation theory of artin algebras. Translation quivers, an abstraction of AR-quivers, played a central role in the classification of selfinjective algebras of finite representation type [Rie]. More generally, they play a basic role in covering theory as developed in [BoG].

The result that there exist no sectional cycles in an AR-quiver is due to [BauS]. The proof given here is taken from [Bo1]. The fact that there are no double arrows in the AR-quiver of finite dimensional algebras of finite representation type over algebraically closed fields is due to [Bau2]. The proof given here is due to Bongartz (see [Rie]).

The characterization of additive and subadditive functions on graphs is found in [V]. The importance of such functions on AR-quivers was first observed in [To] and subsequently exploited in [HapPR].

AR-quivers have appeared in settings other than artin algebras. For example, they are used in an essential way in the classification of two-dimensional orders of finite representation type in [ReV] and are closely related to the resolution graphs of rational double points as shown in [Au9], [AuR9]. Other examples can be found in [Yoshin].

The idea of sectional paths appeared in [Rie] and [Bau1], and several related concepts have been investigated. Information on which translation quivers occur as AR-quivers is found in [BoG] and [Bren1]. The fact that any basic k-algebra over an algebraically closed field k of characteristic different from 2 can be recovered from its AR-quiver follows from [BauGRS].

For more illustrations of the use of AR-quivers we refer to [Ben], [Er], [Rin3].

VIII
Hereditary algebras

The representation theory of hereditary artin algebras is one of the most extensively studied and best understood theories developed to date. The theory has served not only as a model for what might be expected of the representation theory of other types of artin algebras, but also as a source of conceptual insights and technical information which have often proven useful in dealing with questions about the representation theory of related but not necessarily hereditary artin algebras, for example tilted algebras.

Our purpose in this chapter is to give an introduction to this extensive theory. While some of the general theory of preprojective, preinjective and regular components of hereditary algebras is developed, we concentrate mainly on hereditary algebras of finite representation type. A bilinear form and its associated quadratic form are given on the Grothendieck group of artin algebras of finite global dimension which is used to give not only a classification of the hereditary artin algebras of finite representation over an algebraically closed field, but also a classification of their modules.

By way of an introduction to the theory of hereditary artin algebras of infinite representation type, the last part of the chapter is devoted to the study of the representation theory of the Kronecker algebra which is the simplest, at least in some respects, of the hereditary artin algebras of infinite representation type. A classification of all modules over the Kronecker algebra is given using the Auslander–Reiten-quiver of the algebra.

1 Preprojective and preinjective modules

This section is devoted to deriving basic properties of two important types of modules over hereditary algebras. These results are then used to give a description of the components of the AR-quiver of a hereditary artin algebra which contain either projective or injective modules.

We begin by pointing out some distinctive features of modules over hereditary algebras which play a crucial role in our discussion of preprojective and preinjective modules. In this connection we remind the reader that an artin algebra Λ is hereditary if and only if Λ^{op} is hereditary. We now recall and elaborate on the properties of modules over hereditary algebras given in IV Section 4.

Let Λ be a hereditary artin algebra. As usual, $\text{mod}_{\mathscr{P}}\Lambda$ denotes the subcategory of $\text{mod}\,\Lambda$ consisting of those Λ-modules with no indecomposable projective summands and $\text{mod}_{\mathscr{I}}\Lambda$ denotes the subcategory of $\text{mod}\,\Lambda$ consisting of those Λ-modules with no indecomposable injective summands. Since $\mathscr{P}(A, B) = 0$ for all A and B in $\text{mod}_{\mathscr{P}}\Lambda$ by IV Proposition 1.15, we have that $\underline{\text{mod}}_{\mathscr{P}}\Lambda = \text{mod}_{\mathscr{P}}\Lambda$. Since $\mathscr{I}(A, B) = 0$ for all A and B in $\text{mod}_{\mathscr{I}}\Lambda$, we have that $\overline{\text{mod}}_{\mathscr{I}}\Lambda = \text{mod}_{\mathscr{I}}\Lambda$. Therefore the inverse equivalences $D\,\text{Tr}:\underline{\text{mod}}_{\mathscr{P}}\Lambda \to \overline{\text{mod}}_{\mathscr{I}}\Lambda$ and $\text{Tr}\,D:\overline{\text{mod}}_{\mathscr{I}}\Lambda \to \underline{\text{mod}}_{\mathscr{P}}\Lambda$ give inverse equivalences $D\,\text{Tr}:\text{mod}_{\mathscr{P}}\Lambda \to \text{mod}_{\mathscr{I}}\Lambda$ and $\text{Tr}\,D:\text{mod}_{\mathscr{I}}\Lambda \to \text{mod}_{\mathscr{P}}\Lambda$. We now point out some other properties of these functors on which the theory of preprojective and preinjective modules is built.

For each nonnegative integer n define $\text{mod}_{\mathscr{P}}^n\Lambda$ to be the subcategory of $\text{mod}_{\mathscr{P}}\Lambda$ consisting of all Λ-modules A such that for all indecomposable summands A' of A we have that $(D\,\text{Tr})^n A'$ is not projective. Similarly, define $\text{mod}_{\mathscr{I}}^n\Lambda$ to be the subcategory of $\text{mod}_{\mathscr{I}}\Lambda$ consisting of all Λ-modules B such that for all indecomposable summands B' of B we have that $(\text{Tr}\,D)^n B'$ is not injective. It should be noted that we have $\text{mod}_{\mathscr{P}}\Lambda = \text{mod}_{\mathscr{P}}^0\Lambda \supset \text{mod}_{\mathscr{P}}^1\Lambda \supset \cdots \supset \text{mod}_{\mathscr{P}}^n\Lambda \supset \text{mod}_{\mathscr{P}}^{n+1}\Lambda \supset \cdots$ and $\text{mod}_{\mathscr{I}}\Lambda = \text{mod}_{\mathscr{I}}^0\Lambda \supset \text{mod}_{\mathscr{I}}^1\Lambda \supset \cdots \supset \text{mod}_{\mathscr{I}}^n\Lambda \supset \text{mod}_{\mathscr{I}}^{n+1}\Lambda \supset \cdots$. From the easily verified fact that $D((D\,\text{Tr})^n A) \simeq (\text{Tr}\,D)^n(DA)$ for all A in $\text{mod}\,\Lambda$, it follows that a Λ-module A is in $\text{mod}_{\mathscr{P}}^n\Lambda$ if and only if the Λ^{op}-module DA is in $\text{mod}_{\mathscr{I}}^n\Lambda$. Therefore the duality $D:\text{mod}\,\Lambda \to \text{mod}(\Lambda^{op})$ induces a duality $D:\text{mod}_{\mathscr{P}}^n\Lambda \to \text{mod}_{\mathscr{I}}^n(\Lambda^{op})$. We will usually state our results for the categories $\text{mod}_{\mathscr{P}}^n\Lambda$ and leave it to the reader to give the dual results for the categories $\text{mod}_{\mathscr{I}}^n\Lambda$.

The following result is our main reason for introducing the subcategories $\text{mod}_{\mathscr{P}}^n\Lambda$ of $\text{mod}\,\Lambda$.

Lemma 1.1 *Let Λ be a hereditary artin algebra and n a nonnegative integer. Then the functor $(D\,\mathrm{Tr})^{n+1} : \mathrm{mod}_{\mathscr{P}}^{n}\,\Lambda \to \mathrm{mod}_{\mathscr{I}}\,\Lambda$ has the following properties.*

(a) *The functor $(D\,\mathrm{Tr})^{n+1}$ is full and faithful.*

(b) *A module A in $\mathrm{mod}_{\mathscr{P}}^{n}\,\Lambda$ is indecomposable if and only if $(D\,\mathrm{Tr})^{n+1}A$ in $\mathrm{mod}_{\mathscr{I}}\,\Lambda$ is indecomposable.*

Proof (a) We proceed by induction on n. If $n = 0$, then $\mathrm{mod}_{\mathscr{P}}^{0}\,\Lambda = \mathrm{mod}_{\mathscr{P}}\,\Lambda$ and then we have the functor $D\,\mathrm{Tr} : \mathrm{mod}_{\mathscr{P}}\,\Lambda \to \mathrm{mod}_{\mathscr{I}}\,\Lambda$ which we know is full and faithful. Suppose the statement in (a) is true for $n - 1 \geq 0$. Let A and B be in $\mathrm{mod}_{\mathscr{P}}^{n}\,\Lambda$. Then A and B are in $\mathrm{mod}_{\mathscr{P}}^{n-1}\,\Lambda$ and so $(D\,\mathrm{Tr})^{n} : \mathrm{Hom}_{\Lambda}(A, B) \to \mathrm{Hom}_{\Lambda}((D\,\mathrm{Tr})^{n}A, (D\,\mathrm{Tr})^{n}B)$ is an isomorphism. But $(D\,\mathrm{Tr})^{n}(A)$ and $(D\,\mathrm{Tr})^{n}(B)$ are in $\mathrm{mod}_{\mathscr{P}}\,\Lambda$ by definition of $\mathrm{mod}_{\mathscr{P}}^{n}\,\Lambda$. Therefore $D\,\mathrm{Tr} : \mathrm{Hom}_{\Lambda}((D\,\mathrm{Tr})^{n}A, (D\,\mathrm{Tr})^{n}B)) \to \mathrm{Hom}_{\Lambda}((D\,\mathrm{Tr})^{n+1}A, (D\,\mathrm{Tr})^{n+1}B)$ is an isomorphism. This implies that the composition $(D\,\mathrm{Tr})^{n+1} : \mathrm{Hom}_{\Lambda}(A, B) \to \mathrm{Hom}_{\Lambda}((D\,\mathrm{Tr})^{n+1}A, (D\,\mathrm{Tr})^{n+1}B)$ is an isomorphism.

(b) This is a trivial consequence of (a). $\square$

An indecomposable module A over a hereditary artin algebra Λ is said to be **preprojective** if there is a nonnegative integer n such that $(D\,\mathrm{Tr})^{n}A$ is a nonzero projective module.

It is not difficult to see that for an indecomposable preprojective module A there is only one nonnegative integer n such that $(D\,\mathrm{Tr})^{n}A$ is a nonzero projective module. We denote this uniquely determined integer by $v(A)$. If A is not preprojective, we define $v(A)$ to be ∞. We also remark that if A is an indecomposable preprojective module, then $(D\,\mathrm{Tr})^{v(A)}A$ is an indecomposable projective module. We denote by $\widetilde{\mathscr{P}}(\Lambda)$ (or $\widetilde{\mathscr{P}}$ for short) the indecomposable objects of $\mathrm{ind}\,\Lambda$ which are preprojective, and sometimes we also view $\widetilde{\mathscr{P}}(\Lambda)$ as the corresponding full subcategory of $\mathrm{ind}\,\Lambda$. That a Λ-module A is an indecomposable preprojective module will often be denoted by saying that A is in $\widetilde{\mathscr{P}}$. An arbitrary Λ-module is said to be a **preprojective module** if it is isomorphic to a sum of modules in $\widetilde{\mathscr{P}}$. It is easy to see that an arbitrary module A is preprojective if and only if $(D\,\mathrm{Tr})^{n}A = 0$ for some nonnegative integer n.

An indecomposable module B over a hereditary artin algebra Λ is said to be **preinjective** if there is a nonnegative integer n such that $(\mathrm{Tr}\,D)^{n}B$ is a nonzero injective module. It is not difficult to see that for an indecomposable preinjective module B there is only one nonnegative integer n such that $(\mathrm{Tr}\,D)^{n}B$ is a nonzero injective module. We denote this uniquely determined integer by $\mu(B)$. If B is not preinjective we define $\mu(B)$ to be ∞. We also remark that if B is an indecomposable preinjective module, then $(\operatorname{Tr} D)^{\mu(B)}B$ is an indecomposable injective Λ-module. The set of objects in $\operatorname{ind}\Lambda$ which are preinjective is denoted by $\widetilde{\mathscr{I}}(\Lambda)$, or by $\widetilde{\mathscr{I}}$ when it is clear with which hereditary artin algebra Λ we are dealing. Sometimes $\widetilde{\mathscr{I}}(\Lambda)$ is viewed as the corresponding full subcategory of $\operatorname{ind}\Lambda$. That a module B is an indecomposable preinjective module will often be denoted by saying that B is in $\widetilde{\mathscr{I}}$. An arbitrary Λ-module is said to be a **preinjective** module if it is isomorphic to a sum of modules in $\widetilde{\mathscr{I}}$. It is easy to see that an arbitrary module B is preinjective if and only if $(\operatorname{Tr} D)^n B = 0$ for some nonnegative integer n.

We now show that preprojective and preinjective modules are dual notions.

Proposition 1.2 *Let Λ be a hereditary artin algebra. Then the duality $D : \operatorname{mod}\Lambda \to \operatorname{mod}(\Lambda^{\mathrm{op}})$ induces a duality between the full subcategory of $\operatorname{mod}\Lambda$ consisting of the preprojective Λ-modules and the full subcategory of $\operatorname{mod}(\Lambda^{\mathrm{op}})$ consisting of the preinjective Λ^{op}-modules.*

Proof We only have to show that if A is a Λ-module, then A is preprojective if and only if the Λ^{op}-module DA is preinjective. But we have already observed that $D(D\operatorname{Tr})^n A \simeq (\operatorname{Tr} D)^n(DA)$. Therefore $(D\operatorname{Tr})^n A = 0$ if and only if $(\operatorname{Tr} D)^n(DA) = 0$. This gives our desired result since A is preprojective if and only if $(D\operatorname{Tr})^n A = 0$ for some nonnegative integer n and a Λ^{op}-module B is preinjective if and only if $(\operatorname{Tr} D)^n B = 0$ for some nonnegative integer n. $\square$

In light of this result, we will usually concentrate on the preprojective modules and leave it to the reader to give the dual statements for preinjective modules.

We begin our study of the preprojective modules over a hereditary artin algebra Λ with the following basic facts about $\widetilde{\mathscr{P}}(\Lambda)$.

Proposition 1.3 *Let Λ be a hereditary artin algebra.*

(a) *A and B in $\widetilde{\mathscr{P}}(\Lambda)$ are isomorphic if and only if $v(A) = v(B)$ and the indecomposable projective modules $(D\operatorname{Tr})^{v(A)}A$ and $(D\operatorname{Tr})^{v(B)}B$ are isomorphic.*

(b) *Let n be a nonnegative integer. Then the subset of $\widetilde{\mathscr{P}}(\Lambda)$ consisting of those A with $v(A) \leq n$ is finite.*

(c) *The cardinality of $\widetilde{\mathscr{P}}(\Lambda)$ is at most $\aleph_0$.*

Proof (a) We only have to show that if $v(A) = v(B)$ and $(D\,\mathrm{Tr})^{v(A)}A \simeq (D\,\mathrm{Tr})^{v(B)}B$, then $A \simeq B$. But this follows from the fact that $(\mathrm{Tr}\,D)^{v(A)}((D\,\mathrm{Tr})^{v(A)}A) \simeq A$ and $(\mathrm{Tr}\,D)^{v(B)}((D\,\mathrm{Tr})^{v(B)}B) \simeq B$.

(b) Let n be a fixed nonnegative integer. Suppose P is an indecomposable projective module. Then by (a) there are at most n nonisomorphic modules A in $\widetilde{\mathscr{P}}$ such that $v(A) \leq n$ and $(D\,\mathrm{Tr})^{v(A)}A \simeq P$. Since there are only a finite number of nonisomorphic indecomposable projective modules, we have our desired result.

(c) This follows directly from (b). □

As an easy consequence of Proposition 1.3 we have the following.

Corollary 1.4 *Let A be in $\widetilde{\mathscr{P}}$.*

(a) *If B is in $\mathrm{Supp}\,\mathrm{Hom}_\Lambda(\ ,A)$, then B is in $\widetilde{\mathscr{P}}$ with $v(B) \leq v(A)$.*
(b) *$\mathrm{Supp}\,\mathrm{Hom}_\Lambda(\ ,A)$ is finite.*

Proof (a) To prove (a) it suffices to show the following. If B is in $\mathrm{Supp}\,\mathrm{Hom}_\Lambda(\ ,A)$ with $(D\,\mathrm{Tr})^iB$ not projective for $i \leq v(A) - 1$, then $(D\,\mathrm{Tr})^{v(A)}B$ is projective. Since B and A are in $\mathrm{mod}_{\mathscr{P}}^{v(A)-1}\Lambda$, we have by Lemma 1.1 that $(D\,\mathrm{Tr})^{v(A)}: \mathrm{Hom}_\Lambda(B,A) \to \mathrm{Hom}_\Lambda((D\,\mathrm{Tr})^{v(A)}B, (D\,\mathrm{Tr})^{v(A)}A)$ is an isomorphism. Therefore there is a nonzero morphism $h:$ $(D\,\mathrm{Tr})^{v(A)}B \to (D\,\mathrm{Tr})^{v(A)}A$. Since $(D\,\mathrm{Tr})^{v(A)}A$ is projective and $(D\,\mathrm{Tr})^{v(A)}B$ is indecomposable, it follows from III Lemma 1.12 that $(D\,\mathrm{Tr})^{v(A)}B$ is projective.

(b) By part (a) we have that $\mathrm{Supp}\,\mathrm{Hom}_\Lambda(\ ,A)$ is contained in the subset of $\widetilde{\mathscr{P}}$ consisting of all X with $v(X) \leq v(A)$, which we know is finite by Proposition 1.3. □

We now apply our previous observations to show that $\widetilde{\mathscr{P}}(\Lambda)$ has a natural structure as a partially ordered set. Consider the relation on $\widetilde{\mathscr{P}}(\Lambda)$ given by $B \leq A$ if there is a finite sequence $B = A_0 \xrightarrow{f_1} A_1 \xrightarrow{f_2} \cdots \xrightarrow{f_n} A_n = A$ of nonzero morphisms between modules in $\widetilde{\mathscr{P}}(\Lambda)$. Since this relation is obviously transitive and reflexive, it defines a partial order on $\widetilde{\mathscr{P}}(\Lambda)$ if it is antisymmetric. That this is indeed the case is shown in the following.

Proposition 1.5 *Let A be in $\widetilde{\mathscr{P}}(\Lambda)$ for some hereditary artin algebra Λ.*

(a) *Suppose $A = A_0 \xrightarrow{f_1} A_1 \xrightarrow{f_2} \cdots \xrightarrow{f_m} A_m = A$ is a sequence of nonzero morphisms between indecomposable modules. Then all the f_i are isomorphisms.*

(b) $\text{End}_\Lambda(A)$ *is a division algebra isomorphic to* $\text{End}_\Lambda((D\,\text{Tr})^{v(A)}A)$.

Proof (a) Since A_{m-1} is in $\text{Supp}\,\text{Hom}_\Lambda(\ ,A)$ we have by Corollary 1.4 that $v(A_{m-1}) \leq v(A)$. Therefore it follows by induction on m that $v(A) \leq v(A_1) \leq \cdots \leq v(A_i) \leq \cdots \leq v(A)$. Letting $n = v(A)$, we have that $v(A_i) = n$ for all $i = 0, 1, \ldots, m$. If $n = 0$, i.e. A is projective, then by III Lemma 1.2 we have that all the A_i are indecomposable projective modules and all the f_i are monomorphisms. But that means all the f_i are isomorphisms since $A_0 = A = A_m$.

Suppose $n > 0$. Then all the $D\,\text{Tr}^n A_i$ are indecomposable projective Λ-modules and so the A_i are in $\text{mod}_{\mathscr{P}}^{n-1} \Lambda$. Since by Lemma 1.1 the functor $D\,\text{Tr}^n : \text{mod}_{\mathscr{P}}^{n-1} \Lambda \to \text{mod}_{\mathscr{J}} \Lambda$ is full and faithful, we have that $(D\,\text{Tr})^n A = (D\,\text{Tr})^n A_0 \overset{(D\,\text{Tr})^n f_1}{\to} (D\,\text{Tr})^n A_1 \to \cdots \overset{(D\,\text{Tr})^n f_m}{\to} (D\,\text{Tr})^n A_m = (D\,\text{Tr})^n A$ is a sequence of nonzero morphisms between indecomposable projective modules. Hence all the $(D\,\text{Tr})^n f_i$ are isomorphisms as in the case $n = 0$. Therefore all the f_i are isomorphisms since $(D\,\text{Tr})^n : \text{mod}_{\mathscr{P}}^{n-1} \Lambda \to \text{mod}_{\mathscr{J}} \Lambda$ is a full and faithful functor.

(b) This is a trivial consequence of part (a). $\square$

From now on we consider $\widetilde{\mathscr{P}}(\Lambda)$ for a hereditary artin algebra Λ as a partially ordered set by means of the ordering given above, i.e. $B \leq A$ if there is a finite sequence of nonzero morphisms $B = A_0 \to \cdots \to A_n = A$ in $\widetilde{\mathscr{P}}(\Lambda)$. The following is an easily verified but nonetheless important property of the ordering on $\widetilde{\mathscr{P}}(\Lambda)$.

Proposition 1.6 *Let A be in $\widetilde{\mathscr{P}}(\Lambda)$. Then we have the following.*

(a) *If $B \leq A$, then $v(B) \leq v(A)$.*

(b) *The set of all B in $\widetilde{\mathscr{P}}(\Lambda)$ with $B \leq A$ is finite.*

Proof (a) Suppose $B \leq A$ and let $B = A_0 \overset{f_1}{\to} \cdots \overset{f_m}{\to} A_m = A$ be a sequence of nonzero morphisms in $\widetilde{\mathscr{P}}(\Lambda)$. Since $f_m : A_{m-1} \to A_m$ is not zero, we have that A_{m-1} is in $\text{Supp}\,\text{Hom}_\Lambda(\ ,A_m)$ and so $v(A_{m-1}) \leq v(A_m) = v(A)$ by Corollary 1.4. Proceeding by induction on m, we have that $v(B) \leq v(A)$.

(b) By (a) we have that the set of all B in $\widetilde{\mathscr{P}}(\Lambda)$ such that $B \leq A$ is a subset of the set of all X in $\widetilde{\mathscr{P}}(\Lambda)$ with the property $v(X) \leq v(A)$. This is a finite set by Proposition 1.3(b). $\square$

We now apply Corollary 1.4 to obtain the following criterion for when

$\operatorname{Ext}^1_\Lambda(C, A) \neq 0$ for C in $\widetilde{\mathscr{P}}(\Lambda)$. Recall that $\operatorname{Supp} \operatorname{Ext}^1_\Lambda(C, \ )$ consists of all indecomposable Λ-modules A in $\operatorname{ind} \Lambda$ such that $\operatorname{Ext}^1_\Lambda(C, A) \neq 0$.

Proposition 1.7 *Let Λ be a hereditary artin algebra and let C be in $\widetilde{\mathscr{P}}$. Then we have the following.*

(a) $\operatorname{Supp} \operatorname{Ext}^1_\Lambda(C, \ ) = \operatorname{Supp} \operatorname{Hom}_\Lambda(\ , D \operatorname{Tr} C)$.

(b) $\operatorname{Supp} \operatorname{Ext}^1_\Lambda(C, \ )$ *is a finite subset of $\widetilde{\mathscr{P}}$ with the property that if A is in $\operatorname{Supp} \operatorname{Ext}^1_\Lambda(C, \ )$, then $v(A) \leq v(C) - 1$.*

(c) $\operatorname{Ext}^1_\Lambda(C, C) = 0$.

Proof Since the proposition is clearly true for C projective, we may assume that C is not projective.

(a) Let A be an indecomposable Λ-module. Since the projective dimension of C is less than or equal to 1 we have by IV Corollary 4.7 that $l_R \operatorname{Ext}^1_\Lambda(C, A) = l_R \operatorname{Hom}_\Lambda(A, D \operatorname{Tr} C)$. Therefore we have that $\operatorname{Ext}^1_\Lambda(C, A) \neq 0$ if and only if $\operatorname{Hom}_\Lambda(A, D \operatorname{Tr} C) \neq 0$.

(b) Since C is in $\widetilde{\mathscr{P}}$ and is not projective, $D \operatorname{Tr} C$ is in $\widetilde{\mathscr{P}}$ with $v(D \operatorname{Tr} C) = v(C) - 1$. Therefore by Corollary 1.4 we have that $\operatorname{Supp} \operatorname{Hom}_\Lambda(\ , D \operatorname{Tr} C)$ is finite and if A is in $\operatorname{Supp} \operatorname{Hom}_\Lambda(\ , D \operatorname{Tr} C)$, then $v(A) \leq v(D \operatorname{Tr} C) = v(C) - 1$. Our desired result now follows from the equality $\operatorname{Supp} \operatorname{Ext}^1_\Lambda(C, \ ) = \operatorname{Supp} \operatorname{Hom}_\Lambda(\ , D \operatorname{Tr} C)$.

(c) This is a direct consequence of part (b). $\qquad\square$

We now apply our results about preprojective and preinjective modules to obtain information about the components of the AR-quiver of a hereditary artin algebra containing projective or injective vertices. Since the structure of the AR-quiver of an artin algebra Λ is determined by the irreducible morphisms in $\operatorname{mod} \Lambda$, it is natural to first investigate how irreducible morphisms and preprojective and preinjective modules are connected. The following is our first result in this direction.

Lemma 1.8 *Let Λ be a hereditary artin algebra. Suppose $B \to A$ is an irreducible morphism between indecomposable Λ-modules. Then we have the following.*

(a) *A is in $\widetilde{\mathscr{P}}$ if and only if B is in $\widetilde{\mathscr{P}}$.*

(b) *Suppose B is in $\widetilde{\mathscr{P}}$.*

    (i) *If A is projective, then B is projective.*

    (ii) *If A is not projective, then $0 \leq v(A) - 1 \leq v(B) \leq v(A)$.*

Proof Suppose A is in $\widetilde{\mathscr{P}}$. Then B is in $\operatorname{Supp} \operatorname{Hom}_\Lambda(\ ,A)$ and so B is in $\widetilde{\mathscr{P}}$ with $v(B) \leq v(A)$ by Corollary 1.4.

Suppose B is in $\widetilde{\mathscr{P}}$. If A is projective, then A is in $\widetilde{\mathscr{P}}$ and B is projective by III Lemma 1.12. Suppose A is not projective. Then there is an irreducible morphism $D \operatorname{Tr} A \to B$. So we have by our previous argument that $D \operatorname{Tr} A$ is in $\widetilde{\mathscr{P}}$ with $v(D \operatorname{Tr} A) \leq v(B)$. Therefore A is in $\widetilde{\mathscr{P}}$ and we have $0 \leq v(A) - 1 \leq v(B) \leq v(A)$ since $v(D \operatorname{Tr} A) = v(A) - 1$ when A is not projective. This finishes the proof of the lemma. □

For an indecomposable preprojective module A we call the corresponding vertex $[A]$ in the AR-quiver a preprojective vertex, and we identify the set of objects $\widetilde{\mathscr{P}}$ with the preprojective vertices in the AR-quiver. Then Lemma 1.8 shows that if one vertex of a component of the AR-quiver of a hereditary artin algebra is preprojective, then all the vertices in the component are preprojective. Such components are called the **preprojective components** of the AR-quiver. We also have that $\widetilde{\mathscr{P}}$ is the union of the vertices in the preprojective components of the AR-quiver.

Dually we identify the objects $\widetilde{\mathscr{I}}$ with the corresponding vertices in the AR-quiver which we call preinjective vertices. If one vertex of a component of the AR-quiver is preinjective, then all the vertices in the component are preinjective. Such components are called the **preinjective components** of the AR-quiver. We also have that $\widetilde{\mathscr{I}}$ is the union of the vertices in the preinjective components of the AR-quiver.

We now want to identify the preprojective components more precisely as the components containing projective modules. Dually, the preinjective components are the components containing injective modules. This follows from the following more general result.

Proposition 1.9 *Let A be an indecomposable module over a hereditary artin algebra Λ. Then the following are equivalent.*

(a) *A is preprojective.*

(b) *$\operatorname{Supp} \operatorname{Hom}_\Lambda(\ ,A)$ is finite.*

(c) *There exist a natural number m and a chain of irreducible morphisms between indecomposable modules $P = C_0 \xrightarrow{f_1} C_1 \xrightarrow{f_2} \cdots \xrightarrow{f_m} C_m = A$ with P projective and the composition $f_m \cdots f_2 f_1$ not zero.*

(d) *There exist a natural number m and a chain of irreducible morphisms between indecomposable modules $P = C_0 \xrightarrow{f_1} C_1 \xrightarrow{f_2} \cdots \xrightarrow{f_m} C_m = A$ with P projective.*

Proof (a) ⇒ (b) This is proven in Corollary 1.4(b).

(b) ⇒ (c) Since $\operatorname{Supp} \operatorname{Hom}_\Lambda(\ , A)$ is finite we know by V Lemma 7.6(b) that there is some n in $\mathbb{N}$ such that $\operatorname{rad}_\Lambda^n(\ , A) = 0$. This implies by V Proposition 7.4(b) that each nonzero morphism $f: B \to A$ with B indecomposable can be written as a sum of nonzero compositions of irreducible morphisms between indecomposable modules. Since $A \neq 0$ there is some indecomposable projective module P in $\operatorname{Supp} \operatorname{Hom}_\Lambda(\ , A)$. Therefore there is a chain of irreducible morphisms between indecomposable modules $P = C_0 \xrightarrow{f_1} C_1 \xrightarrow{f_2} \cdots \xrightarrow{f_m} C_m = A$ with nonzero composition.

(c) ⇒ (d) This is trivial.

(d) ⇒ (a) If A is projective, there is nothing to prove. Suppose A is not projective and proceed by induction on m. If $m = 1$, then there is an irreducible morphism $P \to A$. Then by Lemma 1.8 we have that A is in $\widetilde{\mathscr{P}}$.

Suppose $m > 1$. Then by the induction hypothesis we know that C_{m-1} is in $\widetilde{\mathscr{P}}$. Since there is an irreducible morphism $C_{m-1} \to A$ and A is not projective, we have by Lemma 1.8 that A is in $\widetilde{\mathscr{P}}$. This shows that (d) implies (a). □

As an immediate consequence of this proposition we have the following.

Corollary 1.10 *Let Λ be a hereditary artin algebra.*

(a) *A component of the AR-quiver of Λ is a preprojective component if and only if it contains some projective vertices.*

(b) *A component of the AR-quiver of Λ is a preinjective component if and only if it contains some injective vertices.* □

From this it follows that a hereditary algebra has only a finite number of preprojective components and a finite number of preinjective components. Naturally, one would like to know their precise number. The answer to this question is based on the description given in II Proposition 5.2 of when an arbitrary artin algebra is indecomposable as an algebra.

Suppose that Λ is a hereditary artin algebra and Q is an indecomposable projective module. Since $\mathfrak{r}Q$ is a projective module, an indecomposable projective module P is a summand of the projective cover of $\mathfrak{r}Q$ if and only if it is a summand of $\mathfrak{r}Q$, or equivalently, there is an irreducible morphism from P to Q.

Combining this remark with II Proposition 5.2 we obtain the following.

Proposition 1.11 *For a hereditary artin algebra Λ the following are equivalent.*

(a) *The algebra Λ is indecomposable as an algebra.*

(b) *All the indecomposable projective Λ-modules belong to the same component of the AR-quiver of Λ.*

(c) *The AR-quiver of Λ has only one preprojective component.* □

As an immediate consequence of this result we obtain the following answer to the question about the number of preprojective components of the AR-quiver of a hereditary artin algebra.

Corollary 1.12 *Let Λ be a hereditary artin algebra. Then we have the following.*

(a) *Two indecomposable projective Λ-modules belong to the same block of Λ if and only if they belong to the same preprojective component of the AR-quiver of Λ.*

(b) *The number of preprojective components of the AR-quiver of Λ is the same as the number of blocks of Λ.* □

We now describe in terms of preprojective and preinjective modules when a hereditary artin algebra is of finite representation type.

Proposition 1.13 *The following are equivalent for a hereditary artin algebra Λ.*

(a) *The algebra Λ is of finite representation type.*

(b) *All Λ-modules are preprojective.*

(c) *All injective Λ-modules are preprojective.*

(d) *All Λ-modules are preinjective.*

(e) *All projective Λ-modules are preinjective.*

(f) *All Λ-modules are both preinjective and preprojective.*

Proof (a) $\Leftrightarrow$ (b) We know by V Theorem 7.7 that Λ is of finite representation type if and only if $\operatorname{Supp}\operatorname{Hom}_\Lambda(\ ,A)$ is finite for all indecomposable Λ-modules A. But by Proposition 1.9 we know that an indecomposable module A is preprojective if and only if $\operatorname{Supp}\operatorname{Hom}_\Lambda(\ ,A)$ is finite. This gives the equivalence of (a) and (b).

(b) $\Leftrightarrow$ (c) That (b) implies (c) is trivial. Assume now that $D(\Lambda)$ is preprojective. By Corollary 1.4 we have that $\operatorname{Supp}\operatorname{Hom}_\Lambda(\ ,D(\Lambda))$ consists only of preprojective modules. But every indecomposable module is in $\operatorname{Supp}\operatorname{Hom}_\Lambda(\ ,D(\Lambda))$, so (c) implies (b).

(a) $\Leftrightarrow$ (d) This is the dual of (a) $\Leftrightarrow$ (b).

(d) $\Leftrightarrow$ (e) This is the dual of (b) $\Leftrightarrow$ (c).

(a) $\Leftrightarrow$ (f) This is a trivial consequence of previous implications. $\square$

We now prove a stronger version of this proposition.

Proposition 1.14 *Let Λ be a hereditary artin algebra which is indecomposable as an algebra. Then Λ is of finite representation type if and only if there is an indecomposable Λ-module A which is both preprojective and preinjective.*

Proof If Λ is of finite type then all preinjective modules are preprojective so then there exist indecomposable modules which are both preprojective and preinjective. Assume next that there exists an indecomposable preinjective module A which is preprojective. Since Λ is indecomposable as an algebra there is only one component of the AR-quiver containing preprojective modules and only one component containing preinjective modules. Since they have $[A]$ in common, they have to coincide. Therefore all injective Λ-modules are preprojective, which by Proposition 1.13 implies that Λ is of finite representation type. $\square$

Having described the preprojective components and the preinjective components of the AR-quiver of a hereditary artin algebra abstractly we now want to describe them geometrically.

Recall from Chapter III that for any artin algebra Λ we defined the valued quiver of Λ in the following way. Let $\{S_1, \ldots, S_n\}$ be a complete list of the simple Λ-modules up to isomorphism. Let $\{1, \ldots, n\}$ be the vertices of the quiver of Λ, and let there be an arrow from i to j if and only if $\mathrm{Ext}^1_\Lambda(S_i, S_j) \neq 0$ and give such an arrow valuation (a, b) where $b = \dim_{\mathrm{End}_\Lambda(S_i)^{\mathrm{op}}} \mathrm{Ext}^1_\Lambda(S_i, S_j)$ and $a = \dim_{\mathrm{End}_\Lambda(S_j)} \mathrm{Ext}^1_\Lambda(S_i, S_j)$. Let P_i be the projective cover of S_i for $i = 1, \ldots, n$. We know that if there is an arrow from i to j there is a nonzero morphism from P_j to P_i, and if Λ is hereditary there are no oriented cycles in the quiver by III Lemma 1.12. Recall also that $\mathrm{Ext}^1_\Lambda(S_i, S_j) \neq 0$ if and only if P_j is a summand of the projective cover of rP_i, which for a hereditary algebra is equivalent to the existence of an irreducible morphism from P_j to P_i. Hence if Λ is hereditary, there is an arrow from i to j in the quiver of Λ if and only if there is an irreducible morphism from P_j to P_i in the AR-quiver.

Consider the valued arrows $\underset{i}{\cdot} \overset{(a,b)}{\longrightarrow} \underset{j}{\cdot}$ in the quiver of Λ and $P_i \overset{(a',b')}{\longleftarrow} P_j$ in the AR-quiver of Λ when Λ is hereditary. By definition we have $b = $

$\dim_{\operatorname{End}_\Lambda(S_i)^{\operatorname{op}}} \operatorname{Ext}^1_\Lambda(S_i, S_j)$ and $a = \dim_{\operatorname{End}_\Lambda(S_j)} \operatorname{Ext}^1_\Lambda(S_i, S_j)$, and from Chapter VII we have $a' = \dim_{\operatorname{End}_\Lambda(S_i)^{\operatorname{op}}} \operatorname{Irr}(P_j, P_i)$ and $b' = \dim_{\operatorname{End}_\Lambda(S_j)} \operatorname{Irr}(P_j, P_i)$. Since both a and b' give the multiplicity of P_j as a summand of $\mathfrak{r}P_i$ by III Proposition 1.14 and VII Proposition 1.3, and a/b must be equal to a'/b', we have $(b, a) = (a', b')$. Hence the preprojective component of an indecomposable hereditary artin algebra Λ contains a full subquiver isomorphic to the transpose of the quiver of the opposite algebra of Λ. Since the valued quiver Δ of a hereditary artin algebra contains no oriented cycles we can form the translation quiver $\mathbb{Z}\Delta$ which also has no oriented cycles. We denote by $\mathbb{N}\Delta$ the subtranslation quiver of $\mathbb{Z}\Delta$ with vertices (n, i) with $n \geq 0$ and $i \in \Delta_0$. Identifying the set of isomorphism classes of indecomposable projective modules $\{[P_1], \ldots, [P_n]\}$ and the set of vertices in Δ we obtain the following result.

Proposition 1.15 *Let Λ be an indecomposable hereditary artin algebra and Δ the transpose of the valued quiver of $\Lambda^{\operatorname{op}}$. Then $\phi \colon \widetilde{\mathscr{P}} \to (\mathbb{N}\Delta)_0$ defined by $\phi([A]) = (v(A), [(D \operatorname{Tr})^{v(A)} A])$ gives an injective translation quiver morphism from the preprojective component of Λ to $\mathbb{N}\Delta$.*

Proof We only have to verify that ϕ behaves nicely with respect to the valued arrows in the preprojective component. So let $[A] \overset{(a,b)}{\to} [B]$ be a valued arrow in the preprojective component of Λ. If $v(A) = v(B)$ we have the valued arrow $[(D \operatorname{Tr})^{v(A)} A] \overset{(a,b)}{\to} [(D \operatorname{Tr})^{v(B)} B]$ between projective vertices in $\widetilde{\mathscr{P}}$. Hence there is an arrow from $(v(A), [(D \operatorname{Tr})^{v(A)} A])$ to $(v(B), [(D \operatorname{Tr})^{v(B)} B])$ in $\mathbb{N}\Delta$. If $v(A) = v(B) - 1$ we have an arrow from $(D \operatorname{Tr})^{v(B)} B$ to $(D \operatorname{Tr})^{v(A)} A$ with valuation (b, a). Hence there is a unique arrow from $(v(A), [(D \operatorname{Tr})^{v(A)} A])$ to $(v(B), [(D \operatorname{Tr})^{v(B)} B])$ in $\mathbb{N}\Delta$ with valuation (a, b). $\qquad\qquad\qquad\Box$

We now use the translation quiver morphism ϕ to give a different characterization of finite representation type.

Proposition 1.16 *Let Λ be an indecomposable hereditary artin algebra and let ϕ be as above. Then Λ is of infinite representation type if and only if ϕ is a bijection.*

Proof If ϕ is surjective then there are infinitely many preprojective modules and hence Λ is of infinite representation type.

Conversely if Λ is of infinite representation type, then $(\operatorname{Tr} D)^n P_i \neq 0$ for all n. Hence for each $n \in \mathbb{N}$ and P_i projective we have that

$\phi[(\operatorname{Tr} D)^n P_i] = (n, [P_i])$, which shows that ϕ is surjective. The map ϕ is always injective by Proposition 1.15 so we have that ϕ is a bijection. $\square$

The quiver Δ of the opposite algebra of an indecomposable hereditary algebra Λ is called the **type** of the preprojective component of Λ. It follows from III Proposition 1.4 that any finite nonvalued quiver without oriented cycles is realized as the type of a preprojective component of a hereditary algebra. In Section 6 we describe exactly which valued quivers are realized as the type of a preprojective component for an indecomposable hereditary artin algebra.

2 The Coxeter transformation

In this section we introduce the Coxeter transformation on the Grothendieck group $K_0(\operatorname{mod}\Lambda)$ of Λ when Λ is an artin algebra of finite global dimension. We show that this transformation is closely related to $D\operatorname{Tr}$ for hereditary algebras and use this to prove that the indecomposable preprojective and preinjective modules over a hereditary artin algebra are determined by their composition factors.

Recall from I Section 1 that the Grothendieck group $K_0(\operatorname{mod}\Lambda)$ of an artin algebra Λ is the abelian group $F(\operatorname{mod}\Lambda)/R(\operatorname{mod}\Lambda)$, where $F(\operatorname{mod}\Lambda)$ is the free abelian group on the isomorphism classes $[M]$ of finitely generated Λ-modules M and $R(\operatorname{mod}\Lambda)$ is the subgroup generated by expressions $[A] + [C] - [B]$ whenever there is an exact sequence of Λ-modules $0 \to A \to B \to C \to 0$.

Throughout this section we let $\{S_1, \ldots, S_n\}$ be a complete set of nonisomorphic simple Λ-modules. Then $[S_1], \ldots, [S_n]$ form a basis for $K_0(\operatorname{mod}\Lambda)$ by I Theorem 1.7. Let P_j be the projective cover of S_j and I_j the injective envelope of S_j for $j = 1, \ldots, n$. Recall that $I_j \simeq D\operatorname{Hom}_\Lambda(P_j, \Lambda)$ for all $j = 1, \ldots, n$ by II Proposition 4.6.

The definition of the Coxeter transformation depends on the following.

Lemma 2.1 *Let Λ be an artin algebra of finite global dimension and S_j, P_j and I_j as above. Then the sets $\{[P_j]\}_{j=1}^n$ and $\{[I_j]\}_{j=1}^n$ are both bases for $K_0(\operatorname{mod}\Lambda)$.*

Proof We prove that the set $\{[P_j]\}_{j=1}^n$ is a basis for $K_0(\operatorname{mod}\Lambda)$. The other half follows by duality. Since this set has n elements it is enough to prove that it generates $K_0(\operatorname{mod}\Lambda)$, since $K_0(\operatorname{mod}\Lambda)$ is a free abelian group of rank n. Let S be a simple Λ-module. Since Λ has finite global dimension there exists a finite projective resolution of S,

$$0 \to Q_m \to \cdots \to Q_2 \to Q_1 \to Q_0 \to S \to 0.$$

Then $[S] = \sum_{i=0}^{m} (-1)^i [Q_i]$ in $K_0(\text{mod } \Lambda)$, hence $[S]$ is in the subgroup of $K_0(\text{mod } \Lambda)$ generated by $\{ [P_j] \}_{j=1}^{n}$ and therefore $K_0(\text{mod } \Lambda)$ is generated by $\{ [P_j] \}_{j=1}^{n}$. $\square$

Let Λ be an artin algebra of finite global dimension and define $c: K_0(\text{mod } \Lambda) \to K_0(\text{mod } \Lambda)$ by $c([P_j]) = -[I_j]$, where S_j, P_j and I_j are as before. This is clearly an isomorphism since c takes a basis to a basis according to Lemma 2.1, and it is called the **Coxeter transformation**. Observe that $c[P] = -[D \operatorname{Hom}_\Lambda(P, \Lambda)]$ for any projective Λ-module P.

We now specialize to Λ being a hereditary artin algebra. We fix the basis $\mathscr{B} = \{ [S_1], \ldots, [S_n] \}$ of $K_0(\text{mod } \Lambda)$. A nonzero element $x \in K_0(\text{mod } \Lambda)$ is called **positive** if all its coordinates with respect to $\mathscr{B}$ are nonnegative, or equivalently $x = [M]$ for a module M in $\text{mod } \Lambda$. A nonzero element $x \in K_0(\text{mod } \Lambda)$ is called **negative** if all its coordinates with respect to $\mathscr{B}$ are nonpositive, or equivalently $x = -[M]$ for a module M in $\text{mod } \Lambda$. We have the following result.

Proposition 2.2 *Let Λ be a hereditary artin algebra and let c be the Coxeter transformation. We then have the following.*

(a) *If M is in $\text{mod } \Lambda$, then $c[M] = [D\operatorname{Ext}_\Lambda^1(M, \Lambda)] - [D \operatorname{Hom}_\Lambda(M, \Lambda)]$.*

(b) *If M is an indecomposable nonprojective Λ-module, then $c[M] = [D \operatorname{Tr} M]$.*

(c) *Let M be an indecomposable Λ-module. Then M is projective if and only if $c[M]$ is negative.*

(d) *If M is an indecomposable module, then $c[M]$ is either positive or negative.*

(e) *If M is in $\text{mod } \Lambda$, then $c^{-1}[M] = [\operatorname{Ext}_{\Lambda^{\text{op}}}^1(DM, \Lambda)] - [\operatorname{Hom}_{\Lambda^{\text{op}}}(DM, \Lambda)]$.*

(f) *If M is an indecomposable noninjective Λ-module then $c^{-1}[M] = [\operatorname{Tr} DM]$.*

(g) *Let M be an indecomposable Λ-module. Then M is injective if and only if $c^{-1}[M]$ is negative.*

(h) *If M is an indecomposable Λ-module, then $c^{-1}[M]$ is either positive or negative.*

Proof We only prove statements (a), (b), (c) and (d) since statements (e), (f), (g) and (h) are duals of (a), (b), (c) and (d) respectively.

In order to prove statement (a), let M be in $\text{mod } \Lambda$ and let $0 \to$

$P_1 \to P_0 \to M \to 0$ be a minimal projective resolution of M. Applying $\operatorname{Hom}_\Lambda(\ ,\Lambda)$ and writing X^* for $\operatorname{Hom}_\Lambda(X,\Lambda)$ we obtain the exact sequence

$$0 \to M^* \to P_0^* \to P_1^* \to \operatorname{Ext}_\Lambda^1(M,\Lambda) \to 0.$$

Next, applying the duality we get the exact sequence

$$0 \to D\operatorname{Ext}_\Lambda^1(M,\Lambda) \to DP_1^* \to DP_0^* \to DM^* \to 0.$$

From this we deduce that $[D\operatorname{Ext}_\Lambda^1(M,\Lambda)] - [DM^*] = [DP_1^*] - [DP_0^*] = -c[P_1] + c[P_0] = c([P_0] - [P_1]) = c[M]$.

(b) We know that if M is an indecomposable nonprojective module, then $\operatorname{Hom}_\Lambda(M,\Lambda) = 0$ since Λ is hereditary. Therefore we have $c[M] = [D\operatorname{Ext}_\Lambda^1(M,\Lambda)]$ according to (a). But for a hereditary artin algebra Λ, the functor Tr is isomorphic to $\operatorname{Ext}_\Lambda^1(\ ,\Lambda)$, and hence $c[M] = [D\operatorname{Tr} M]$.

The statements (c) and (d) are direct consequences of (a) and (b). □

As an immediate consequence of this proposition we have the following results about indecomposable modules over a hereditary artin algebra Λ.

Corollary 2.3 *Let M and N be indecomposable modules over a hereditary artin algebra Λ with $[M] = [N]$ in $K_0(\operatorname{mod}\Lambda)$ and let c be the Coxeter transformation.*

(a) *M is projective if and only if N is projective.*
(b) *If M is projective, then $M \simeq N$.*
(c) *M is preprojective if and only if there exists a natural number n with $c^n[M]$ negative.*
(d) *If M is preprojective, then $M \simeq N$.*
(e) *M is injective if and only if N is injective.*
(f) *If M is injective, then $M \simeq N$.*
(g) *M is preinjective if and only if there exists a natural number m with $c^{-m}[M]$ negative.*
(h) *If M is preinjective, then $M \simeq N$.*

Proof Again we only prove the statements (a), (b), (c) and (d). The other statements are duals of these statements.

(a) Let M and N be indecomposable modules. From Proposition 2.2 (c) we have that M is an indecomposable projective module if and only if $c[M]$ is negative. But since $c[M] = c[N]$ and N is indecomposable, we get that N is projective if and only if M is projective.

(b) Let M and N be indecomposable modules with $[M] = [N]$ and M projective. From (a) it follows that both M and N are projective. Since

M is projective and $[M] = [N]$ there exists a nonzero homomorphism $f: M \to N$. But since Λ is hereditary and N is an indecomposable projective module and M is indecomposable, f has to be a monomorphism. Finally $[M] = [N]$ implies that M and N have the same length and therefore f has to be an isomorphism.

(c) Let M be an indecomposable module. We know that M is preprojective if and only if there exists a nonnegative integer n such that $(D \operatorname{Tr})^n M$ is projective. Using Proposition 2.2(b) and (c) we have that the last statement is equivalent to the existence of a nonnegative integer m such that $c^m[M]$ is negative.

(d) Let M and N be two indecomposable modules with $[M] = [N]$ and let M be preprojective. We know from (c) that there exists a positive integer m such that $c^m[M] = c^m[N]$ is negative. Let m be the least such number. Then $[(D \operatorname{Tr})^{m-1} M] = c^{m-1}[M] = c^{m-1}[N] = [(D \operatorname{Tr})^{m-1} N]$ is positive and hence $(D \operatorname{Tr})^{m-1} M$ and $(D \operatorname{Tr})^{m-1} N$ are both indecomposable projective and therefore isomorphic according to (b). Hence M is isomorphic to N.   □

In the next chapter we will come back to the problem of when over arbitrary artin algebras indecomposable modules M are determined by $[M]$ in $K_0(\operatorname{mod} \Lambda)$, as we have just shown for preprojective and preinjective modules over a hereditary artin algebra.

Since by Proposition 1.13 we have that all modules over an hereditary artin algebra of finite representation type are preprojective, the following is an immediate consequence of the above result.

Corollary 2.4 *Suppose Λ is a hereditary artin algebra of finite representation type. Then two indecomposable Λ-modules M and N are isomorphic if and only if $[M] = [N]$ in $K_0(\operatorname{mod} \Lambda)$.*   □

3 The homological quadratic form

When a hereditary artin algebra Λ is of finite representation type, then all indecomposable Λ-modules are preprojective and preinjective. We shall see however that if Λ is of infinite representation type, there is always some indecomposable Λ-module which is neither preprojective nor preinjective, and such modules and their components of the AR-quiver will be investigated in the next section. We obtain this result as a consequence of introducing a homological quadratic form associated with any artin algebra of finite global dimension, and showing that for a hereditary algebra Λ this form is positive definite if and only if Λ is of finite representation type.

Let Λ be an artin R-algebra of finite global dimension n. For each short exact sequence $0 \to N' \to N \to N'' \to 0$ in mod Λ and each M in mod Λ we obtain the long exact sequence $0 \to \mathrm{Hom}_\Lambda(M, N') \to \mathrm{Hom}_\Lambda(M, N) \to \mathrm{Hom}_\Lambda(M, N'') \to \mathrm{Ext}^1_\Lambda(M, N') \to \cdots \to \mathrm{Ext}^n_\Lambda(M, N') \to \mathrm{Ext}^n_\Lambda(M, N) \to \mathrm{Ext}^n_\Lambda(M, N'') \to 0$. Hence by counting lengths we get

$$\sum_{i=0}^{n} (-1)^i l_R(\mathrm{Ext}^i_\Lambda(M, N)) = \sum_{i=0}^{n} (-1)^i l_R(\mathrm{Ext}^i_\Lambda(M, N' \amalg N'')).$$

Similarly, if $0 \to M' \to M \to M'' \to 0$ is an exact sequence and N is a Λ-module we obtain

$$\sum_{i=0}^{n} (-1)^i l_R(\mathrm{Ext}^i_\Lambda(M, N)) = \sum_{i=0}^{n} (-1)^i l_R(\mathrm{Ext}^i_\Lambda(M' \amalg M'', N)).$$

Therefore defining

$$B(M, N) = \sum_{i=0}^{n} (-1)^i l_R(\mathrm{Ext}^i_\Lambda(M, N))$$

for each pair of modules M and N in mod Λ we get a bilinear form from $K_0(\mathrm{mod}\,\Lambda) \times K_0(\mathrm{mod}\,\Lambda)$ to $\mathbb{Z}$, which we also denote by B. Associated with this bilinear form is the quadratic form $q: K_0(\mathrm{mod}\,\Lambda) \to \mathbb{Z}$ where $q(x) = B(x, x)$ for x in $K_0(\mathrm{mod}\,\Lambda)$, and which we call the **homological quadratic form**. Recall that q is *positive definite* if $q(x) > 0$ for all $x \neq 0$..

We now restrict our consideration to hereditary algebras and prove that q is positive definite if and only if Λ is of finite representation type.

We first prove the following.

Proposition 3.1 *Let Λ be a hereditary artin algebra and q the associated homological quadratic form. If Λ is of infinite representation type, then q is not positive definite.*

Proof Let Λ be a hereditary artin R-algebra of infinite representation type which we without loss of generality may assume is indecomposable as an algebra. Then the preprojective component of the AR-quiver Γ_Λ is infinite. Hence for each indecomposable projective Λ-module P and each natural number n we get that

$$q([(\operatorname{Tr} D)^n P]) = l_R(\operatorname{End}_\Lambda((\operatorname{Tr} D)^n P)) - l_R(\operatorname{Ext}^1_\Lambda((\operatorname{Tr} D)^n P, (\operatorname{Tr} D)^n P))$$
$$= l_R(\operatorname{End}_\Lambda((\operatorname{Tr} D)^n P))$$

since $\operatorname{Ext}^1_\Lambda((\operatorname{Tr} D)^n P, (\operatorname{Tr} D)^n P) = 0$ by Proposition 1.7. Further, for hereditary artin algebras we have that $\operatorname{End}_\Lambda((\operatorname{Tr} D)^n P)$ is a division ring isomorphic to $\operatorname{End}_\Lambda(P)$ by Proposition 1.5. Hence q assumes the value $l_R(\operatorname{End}_\Lambda(P))$ for infinitely many different elements x in $K_0(\operatorname{mod} \Lambda)$. However, if q is positive definite q extends by linearity to a positive definite form on $\mathbb{Q} \otimes_{\mathbb{Z}} K_0(\operatorname{mod} \Lambda)$. Further, q extends to a continuous quadratic form on $\mathbb{R} \otimes_{\mathbb{Z}} K_0(\operatorname{mod} \Lambda)$. However, if q is positive definite, q extends by linearity to a positive definite form on $\mathbb{Q} \otimes_{\mathbb{Z}} K_0(\operatorname{mod} \Lambda)$ and by continuity to a quadratic form on $\mathbb{R} \otimes_{\mathbb{Z}} K_0(\operatorname{mod} \Lambda)$ which we also denote by q. Considering the associated symmetric bilinear form B' on $\mathbb{Q} \otimes_{\mathbb{Z}} K_0(\operatorname{mod} \Lambda) \times \mathbb{Q} \otimes_{\mathbb{Z}} K_0(\operatorname{mod} \Lambda)$ given by $B'(x, y) = \frac{1}{2}(B(x, y) + B(x, y))$, there is by the Gram-Schmidt process an orthogonal basis $\{v_1, \ldots, v_n\}$ of $\mathbb{Q} \otimes_{\mathbb{Z}} K_0(\operatorname{mod} \Lambda)$ relative to B'. Let $v = \sum_{i=1}^n \alpha_i v_i \neq 0$ be an element of $\mathbb{R} \otimes_{\mathbb{Z}} K_0(\operatorname{mod} \Lambda)$. Then we have that $q(v) = \sum_{i=1}^n \alpha_i^2 B'(v_i, v_i) > 0$ and hence the extension of the quadratic form to $\mathbb{R} \otimes_{\mathbb{Z}} K_0(\operatorname{mod} \Lambda)$ is positive definite. Hence the extension of q to a quadratic form on $\mathbb{R} \otimes_{\mathbb{Z}} K_0(\operatorname{mod} \Lambda)$ is positive definite. But it is well known that for a positive definite quadratic form q on a finite dimensional vector space V over the real numbers, the set $\{x \in V | q(x) \leq r\}$ is a bounded set for each real number r. Hence for each real number r there is only a finite number of lattice points z in $\mathbb{R} \otimes_{\mathbb{Z}} K_0(\operatorname{mod} \Lambda)$ which satisfy $q(z) \leq v$. Especially this holds for $v = l_R(\operatorname{End}_\Lambda P)$. This finishes the proof of the proposition. $\qquad\square$

In order to prove the converse of Proposition 3.1 we will need the following lemma which also has some other applications.

Lemma 3.2 *Let Λ be an artin R-algebra and let N' and N'' be Λ-modules. If $0 \to N' \to N \to N'' \to 0$ is a nonsplit extension, then $l_R(\operatorname{End}_\Lambda(N)) \leq l_R(\operatorname{End}_\Lambda(N' \amalg N'')) - 1$.*

Proof Let Λ be an artin R-algebra, N' and N'' modules and assume $0 \to N' \to N \to N'' \to 0$ is a nonsplit extension. We then obtain the following exact commutative diagram with f nonzero.

$$
\begin{array}{ccccc}
& 0 & & 0 & & 0 \\
& \downarrow & & \downarrow & & \downarrow \\
0 \to & \mathrm{Hom}_\Lambda(N'',N') & \to & \mathrm{Hom}_\Lambda(N'',N) & \to & \mathrm{Hom}_\Lambda(N'',N'') \\
& \downarrow & & \downarrow & & \downarrow \\
0 \to & \mathrm{Hom}_\Lambda(N,N') & \to & \mathrm{Hom}_\Lambda(N,N) & \to & \mathrm{Hom}_\Lambda(N,N'') \\
& \downarrow & & \downarrow & & \downarrow \\
0 \to & \mathrm{Hom}_\Lambda(N',N') & \to & \mathrm{Hom}_\Lambda(N',N) & \to & \mathrm{Hom}_\Lambda(N',N'') \\
& \downarrow f & & & & \\
& \mathrm{Ext}^1_\Lambda(N'',N'). & & & &
\end{array}
$$

Since f is nonzero we have

$$l_R(\mathrm{Hom}_\Lambda(N,N')) \le l_R(\mathrm{Hom}_\Lambda(N'',N')) + l_R(\mathrm{Hom}_\Lambda(N',N')) - 1.$$

Hence we get

$$
\begin{aligned}
l_R(\mathrm{Hom}_\Lambda(N,N)) \;\le\;& l_R(\mathrm{Hom}_\Lambda(N,N')) + l_R(\mathrm{Hom}_\Lambda(N,N'')) \\
\le\;& l_R(\mathrm{Hom}_\Lambda(N'',N')) + l_R(\mathrm{Hom}_\Lambda(N',N')) - 1 \\
& + l_R(\mathrm{Hom}_\Lambda(N'',N'')) + l_R(\mathrm{Hom}_\Lambda(N',N'')) \\
=\;& l_R(\mathrm{End}_\Lambda(N' \textstyle\coprod N'')) - 1.
\end{aligned}
$$

$\square$

In order to establish the converse of Proposition 3.1 as well as proving that there exist modules which are neither preprojective nor preinjective for an artin algebra of infinite representation type we need the following.

Lemma 3.3 *Let Λ be a hereditary artin algebra with q the homological quadratic form. Then $q([M]) > 0$ for all nonzero M in $\mathrm{mod}\,\Lambda$ if and only if $q(x) > 0$ for all nonzero x in $K_0(\mathrm{mod}\,\Lambda)$.*

Proof We only have to prove that if $q([M]) > 0$ for all nonzero M in $\mathrm{mod}\,\Lambda$, then $q(x) > 0$ for all nonzero x in $K_0(\mathrm{mod}\,\Lambda)$. Assume therefore that $q([M]) > 0$ for all nonzero M in $\mathrm{mod}\,\Lambda$ and let $x \in K_0(\mathrm{mod}\,\Lambda)$ be nonzero. Then there exist M and N in $\mathrm{mod}\,\Lambda$ without common composition factors such that $x = [M] - [N]$. Calculating $q(x)$ we get

$$
\begin{aligned}
q(x) \;=\;& B(x,x) \\
=\;& B([M],[M]) + B([N],[N]) - B([M],[N]) - B([N],[M]) \\
=\;& q([M]) + q([N]) - l_R(\mathrm{Hom}_\Lambda(M,N)) \\
& + l_R(\mathrm{Ext}^1_\Lambda(M,N)) - l_R(\mathrm{Hom}_\Lambda(N,M)) + l_R(\mathrm{Ext}^1_\Lambda(N,M)).
\end{aligned}
$$

However, $\text{Hom}_\Lambda(M, N) = 0 = \text{Hom}_\Lambda(N, M)$ since M and N have no common composition factors, so $q(x) \geq q([M]) + q([N])$. Since at least one of the modules M and N is nonzero, we get $q(x) > 0$. $\qquad\square$

The next result shows that the homological quadratic form not being positive definite implies the existence of a certain type of indecomposable modules.

Proposition 3.4 *Let Λ be a hereditary artin algebra with homological quadratic form q. If q is not positive definite then the following hold.*

(a) *There exists an indecomposable module M with $\text{Ext}_\Lambda^1(M, M) \neq 0$.*

(b) *There exists an indecomposable module which is neither preprojective nor preinjective.*

(c) Λ *is of infinite representation type.*

Proof (b) follows from (a) since the indecomposable preprojective and preinjective modules have no selfextensions by Proposition 1.7 and its dual.

In order to establish (a), let Λ be an artin algebra where the homological quadratic form q is not positive definite. Then there exists a nonzero element $x \in K_0(\text{mod}\,\Lambda)$ with $q(x) \leq 0$. But this happens if and only if there is a nonzero module M in $\text{mod}\,\Lambda$ with $q([M]) \leq 0$ by Lemma 3.3. Choose now M nonzero such that $l_R(\text{End}_\Lambda(M))$ is minimal and such that $q[(M)] \leq 0$. We claim that there exists an indecomposable module N in $\text{mod}\,\Lambda$ with $l(N) \leq l(M)$ such that $\text{Ext}_\Lambda^1(N, N) \neq 0$.

If M contains an indecomposable summand N with $\text{Ext}_\Lambda^1(N, N) \neq 0$ we are finished. So without loss of generality we assume that all indecomposable summands N of M have the property that $\text{Ext}_\Lambda^1(N, N) = 0$. We show that this implies that M is indecomposable with $\text{Ext}_\Lambda^1(M, M) \neq 0$.

Since $q([M]) \leq 0$ it follows that $\text{Ext}_\Lambda^1(M, M) \neq 0$. Then there exists an indecomposable summand M_1 of M with $\text{Ext}_\Lambda^1(M_1, M) \neq 0$. Decompose M as $M_1 \coprod M_2$. Since by assumption $\text{Ext}_\Lambda^1(M_1, M_1) = 0$ we have $\text{Ext}_\Lambda^1(M_1, M_2) \neq 0$. So we have a nonsplit exact sequence $0 \to M_2 \to M' \to M_1 \to 0$. But then $l_R(\text{End}_\Lambda(M')) < l_R(\text{End}_\Lambda(M_1 \coprod M_2))$ by Lemma 3.2. Since $[M'] = [M]$, we get a contradiction to our choice of M. Hence M has an indecomposable summand N with $\text{Ext}_\Lambda^1(N, N) \neq 0$.

Part (c) follows from (b) since for a hereditary artin algebra of finite representation type every indecomposable module is preprojective or preinjective. $\qquad\square$

We have the following immediate consequence of Proposition 3.1 and Proposition 3.4.

Corollary 3.5 *If Λ is a hereditary artin algebra of infinite representation type, there is some indecomposable Λ-module which is neither preprojective nor preinjective.* □

We also get the following characterization of finite representation type in terms of the homological quadratic form using Proposition 3.1 and Proposition 3.4.

Theorem 3.6 *Let Λ be a hereditary artin algebra with homological quadratic form q. Then Λ is of finite representation type if and only if q is positive definite.* □

4 The regular components of the Auslander–Reiten-quiver

In the first section of this chapter we were considering the components of the AR-quiver of a hereditary artin algebra containing projective and injective vertices. This gave the description of the preprojective and preinjective modules and the preprojective and preinjective components of the AR-quiver. For hereditary artin algebras of finite representation type this gives a description of all indecomposable modules, and of the whole AR-quiver. However, if Λ is of infinite representation type, then there exist by Corollary 3.5 indecomposable Λ-modules which are neither preprojective nor preinjective. Such modules will be called **regular modules**, and the components of the AR-quiver containing regular modules will be the main object of study in this section. These components are called the **regular** components of the AR-quiver. Also a module M which is isomorphic to a sum of regular modules will be called a regular module.

In V Section 6 we associated to each indecomposable nonprojective Λ-module C the number of summands in a decomposition of B into indecomposable modules when there is a minimal right almost split morphism $f: B \to C$. This number was denoted by $\alpha(C)$ and is, as mentioned in V Section 6, in a sense a measure of the complexity of morphisms to C. It is also a measure of the complexity of Γ_Λ in a neighbourhood of the vertex $[C]$ associated with C. We will in this section prove that $\alpha(C) \leq 2$ for each C corresponding to a vertex in a regular component of the AR-quiver of a hereditary artin algebra.

Further we prove that if $f : B_1 \coprod B_2 \to C$ is a minimal right almost split morphism with B_1 and B_2 nonzero, then one of the morphisms $f|_{B_1}$ and $f|_{B_2}$ is an epimorphism and the other one is a monomorphism. This has the consequence that the valuation on the regular component of the AR-quiver of a hereditary artin algebra is always $(1,1)$ and may therefore be omitted.

Before starting the discussion of the regular modules we need some preliminary results. As usual, let $\text{mod}_{\mathscr{P}} \Lambda$ denote the full subcategory of $\text{mod} \Lambda$ consisting of all Λ-modules with no nonzero projective summands. We have the following description of irreducible morphisms in $\text{mod}_{\mathscr{P}} \Lambda$ when Λ is hereditary.

Proposition 4.1 *Let Λ be a hereditary artin algebra. A morphism $f : A \to B$ with A and B in $\text{mod}_{\mathscr{P}} \Lambda$ is irreducible if and only if in any commutative diagram*

$$
\begin{array}{ccc}
 & Y & \\
{\scriptstyle g}\nearrow & & \searrow{\scriptstyle h} \\
A & \xrightarrow{\ f\ } & B
\end{array}
$$

with Y in $\text{mod}_{\mathscr{P}} \Lambda$ either g is a split monomorphism or h is a split epimorphism.

Proof Let X be in $\text{mod} \Lambda$. Write X as $Y \coprod P$ with Y in $\text{mod}_{\mathscr{P}} \Lambda$ and P projective. Then any commutative diagram

$$
\begin{array}{ccc}
 & X & \\
{\scriptstyle g}\nearrow & & \searrow{\scriptstyle h} \\
A & \xrightarrow{\ f\ } & B
\end{array}
$$

can be written as

$$
\begin{array}{ccc}
 & Y \coprod P & \\
{\scriptstyle \binom{g_1}{g_2}}\nearrow & & \searrow{\scriptstyle (h_1,h_2)} \\
A & \xrightarrow{\ f\ } & B
\end{array}
$$

Since A is in $\text{mod}_{\mathscr{P}} \Lambda$ and P is projective, we have that $g_2 : A \to P$ is zero, so that the diagram

$$
\begin{array}{ccc}
 & Y & \\
{\scriptstyle g_1}\nearrow & & \searrow{\scriptstyle h_1} \\
A & \xrightarrow{\ f\ } & B
\end{array}
$$

commutes. Since A is in $\operatorname{mod}_{\mathscr{P}} \Lambda$ we have that g_1 is a split monomorphism if and only if $\binom{g_1}{g_2}$ is a split monomorphism and since B is in $\operatorname{mod}_{\mathscr{P}} \Lambda$ we have that h_1 is a split epimorphism if and only if (h_1, h_2) is a split epimorphism. Our desired result is a trivial consequence of this observation. □

Applying this lemma we obtain the following important connections between irreducible morphisms and the functors $D \operatorname{Tr}$ and $\operatorname{Tr} D$.

Corollary 4.2 *The following statements are equivalent for a morphism $f : A \to B$ in $\operatorname{mod}_{\mathscr{P}} \Lambda$ when Λ is a hereditary artin algebra.*

(a) *The Λ-morphism $f : A \to B$ is irreducible.*
(b) *The Λ^{op}-morphism $\operatorname{Tr} f : \operatorname{Tr} B \to \operatorname{Tr} A$ is irreducible.*
(c) *The Λ-morphism $D \operatorname{Tr} f : D \operatorname{Tr} A \to D \operatorname{Tr} B$ is irreducible.*

Proof (a) $\Leftrightarrow$ (b) This equivalence follows immediately from the fact that $\operatorname{Tr} : \operatorname{mod}_{\mathscr{P}} \Lambda \to \operatorname{mod}_{\mathscr{P}} \Lambda^{\mathrm{op}}$ is a duality and the criterion given in Proposition 4.1 for a morphism in $\operatorname{mod}_{\mathscr{P}} \Lambda$ to be irreducible.

(b) $\Leftrightarrow$ (c) This equivalence follows from the fact that a morphism f in $\operatorname{mod} \Lambda$ is irreducible if and only if $D(f)$ in $\operatorname{mod}(\Lambda^{\mathrm{op}})$ is irreducible. □

The following is the dual of Corollary 4.2.

Corollary 4.3 *The following statements are equivalent for a morphism $f : A \to B$ in $\operatorname{mod}_{\mathscr{I}} \Lambda$ where Λ is a hereditary artin algebra.*

(a) *The morphism $f : A \to B$ is irreducible.*
(b) *The morphism $\operatorname{Tr} D(f) : \operatorname{Tr} DA \to \operatorname{Tr} DB$ is irreducible.* □

In IV Corollary 1.14 we showed that for a hereditary algebra Λ, the functors $\operatorname{Tr} : \operatorname{mod} \Lambda \to \operatorname{mod}(\Lambda^{\mathrm{op}})$ and $\operatorname{Ext}_{\Lambda}^1(, \Lambda) : \operatorname{mod} \Lambda \to \operatorname{mod}(\Lambda^{\mathrm{op}})$ are isomorphic. So the functor Tr is right exact, i.e. if $0 \to A \to B \to C \to 0$ is an exact sequence, then $\operatorname{Tr} C \to \operatorname{Tr} B \to \operatorname{Tr} A \to 0$ is exact. Therefore the functor $D \operatorname{Tr} : \operatorname{mod} \Lambda \to \operatorname{mod} \Lambda$ is left exact and the functor $\operatorname{Tr} D : \operatorname{mod} \Lambda \to \operatorname{mod} \Lambda$ is right exact. We now want to refine these observations a bit.

Lemma 4.4 *Let $0 \to A \to B \to C \to 0$ be an exact sequence for a hereditary artin algebra Λ. Then we have the following.*

(a) *If A is in $\operatorname{mod}_{\mathscr{P}} \Lambda$, then $0 \to \operatorname{Tr} C \to \operatorname{Tr} B \to \operatorname{Tr} A \to 0$ and $0 \to D \operatorname{Tr} A \to D \operatorname{Tr} B \to D \operatorname{Tr} C \to 0$ are exact sequences.*

(b) *If C is in* $\mathrm{mod}_{\mathscr{I}}\Lambda$, *then* $0 \to \mathrm{Tr}\,DA \to \mathrm{Tr}\,DB \to \mathrm{Tr}\,DC \to 0$ *is exact.*

Proof (a) Applying the functor $\mathrm{Hom}_\Lambda(\ ,\Lambda)$ to the exact sequence $0 \to A \to B \to C \to 0$ we get the exact sequence $\mathrm{Hom}_\Lambda(A,\Lambda) \to \mathrm{Ext}^1_\Lambda(C,\Lambda) \to \mathrm{Ext}^1_\Lambda(B,\Lambda) \to \mathrm{Ext}^1_\Lambda(A,\Lambda) \to 0$. Since A is in $\mathrm{mod}_{\mathscr{P}}\Lambda$, it follows that $\mathrm{Hom}_\Lambda(A,\Lambda) = 0$, which gives the first part of (a). The second part of (a) follows trivially from the first part.

(b) Since C is in $\mathrm{mod}_{\mathscr{I}}\Lambda$, we have that $D(C)$ is in $\mathrm{mod}_{\mathscr{P}}(\Lambda^{\mathrm{op}})$. Then we have by (a) that $0 \to \mathrm{Tr}\,DA \to \mathrm{Tr}\,DB \to \mathrm{Tr}\,DC \to 0$ is exact.   □

We now apply these general remarks to the category $\mathscr{R}(\Lambda)$ of all **regular modules** over a hereditary artin algebra Λ of infinite representation type. It follows from Section 1 that $\mathscr{R}(\Lambda)$ has the following properties.

(a) If there is an irreducible morphism $A \to B$ between indecomposable modules, then A is in $\mathscr{R}(\Lambda)$ if and only if B is in $\mathscr{R}(\Lambda)$.
(b) If A is in $\mathscr{R}(\Lambda)$, then $D\,\mathrm{Tr}\,A$ and $\mathrm{Tr}\,DA$ are in $\mathscr{R}(\Lambda)$.
(c) $\mathscr{R}(\Lambda)$ is contained in $\mathrm{mod}_{\mathscr{P}}\Lambda \cap \mathrm{mod}_{\mathscr{I}}\Lambda$.

Combining these properties of $\mathscr{R}(\Lambda)$ with our previous results about $D\,\mathrm{Tr}$ and $\mathrm{Tr}\,D$ we have the following.

Proposition 4.5 *Let Λ be a hereditary artin algebra of infinite representation type. Then the functors $D\,\mathrm{Tr}:\mathscr{R}(\Lambda) \to \mathscr{R}(\Lambda)$ and $\mathrm{Tr}\,D:\mathscr{R}(\Lambda) \to \mathscr{R}(\Lambda)$ are inverse equivalences which preserve*

(a) *exact sequences,*
(b) *irreducible morphisms,*
(c) *almost split sequences.*   □

We are now ready to start our investigation of the structure of the regular components of a hereditary artin algebra Λ of infinite representation type. For the sake of brevity, we make the convention in this section that when we speak about regular Λ-modules we are assuming that Λ is a hereditary artin algebra of infinite representation type.

We start with the following general remark.

Proposition 4.6 *Let M and N be in $\mathscr{R}(\Lambda)$.*

(a) *If there is a proper monomorphism $f: M \to N$, then there is no proper monomorphism $g: N \to \mathrm{Tr}\,DM$.*
(b) *If there is a proper epimorphism $f: M \to N$ then there is no proper epimorphism $g: \mathrm{Tr}\,DN \to M$.*

Proof (a) Assume that there is a proper monomorphism $f: M \to N$. Since $D \operatorname{Tr}: \mathscr{R}(\Lambda) \to \mathscr{R}(\Lambda)$ is an exact equivalence of categories, $(D \operatorname{Tr})^n: \mathscr{R}(\Lambda) \to \mathscr{R}(\Lambda)$ is an exact equivalence of categories for all positive integers n. Therefore $(D \operatorname{Tr})^n f: (D \operatorname{Tr})^n M \to (D \operatorname{Tr})^n N$ is a proper monomorphism for all positive integers n. Suppose now that $g: N \to \operatorname{Tr} DM$ is a proper monomorphism. Then $(D \operatorname{Tr})^n g: (D \operatorname{Tr})^n N \to (D \operatorname{Tr})^{n-1} M$ is a proper monomorphism for all positive integers n. Then we get the infinite sequence of proper monomorphisms

$$\cdots \to (D \operatorname{Tr})^2 N \xrightarrow{(D \operatorname{Tr})^2 g} D \operatorname{Tr} M \xrightarrow{D \operatorname{Tr} f} D \operatorname{Tr} N \xrightarrow{D \operatorname{Tr} g} M \xrightarrow{f} N,$$

which contradicts the fact that N has finite length. Therefore g is not a proper monomorphism.

(b) This is dual of (a). $\square$

As an immediate consequence of this result, we have the following.

Corollary 4.7 *Let $f: M \to N$ and $g: N \to \operatorname{Tr} DM$ be irreducible morphisms with M and N in $\mathscr{R}(\Lambda)$. Then f is a monomorphism if and only if g is an epimorphism.* $\square$

We now apply this corollary to obtain the following result about irreducible morphisms in $\mathscr{R}(\Lambda)$.

Lemma 4.8 *Let Λ be a hereditary artin algebra and M an indecomposable module in $\mathscr{R}(\Lambda)$.*

(a) *Let $(f_1, f_2): X \coprod Y \to M$ be an irreducible morphism. If f_1 is an epimorphism, then f_2 is a monomorphism.*

(b) *Let $\begin{pmatrix} g_1 \\ g_2 \end{pmatrix}: M \to X \coprod Y$ be an irreducible morphism. If g_1 is a monomorphism, then g_2 is an epimorphism.*

Proof As usual we prove (a) and the proof dualizes to establish (b). Assume $(f_1, f_2): X \coprod Y \to M$ is irreducible with f_1 an epimorphism. Then there exist a module Z and a morphism $f_3: Z \to M$ such that

$$0 \to D \operatorname{Tr} M \xrightarrow{\begin{pmatrix} h_1 \\ h_2 \\ h_3 \end{pmatrix}} X \coprod Y \coprod Z \xrightarrow{(f_1, f_2, f_3)} M \to 0$$

is an almost split sequence. Since f_1 is an epimorphism, $\begin{pmatrix} h_2 \\ h_3 \end{pmatrix}: D \operatorname{Tr} M \to Y \coprod Z$ is an epimorphism by I Corollary 5.7. This implies that h_2 is an epimorphism. Therefore f_2 is a monomorphism by Corollary 4.7. $\square$

The next lemma is needed to prove that if M is an indecomposable module in $\mathcal{R}(\Lambda)$, then $\alpha(M) \leq 3$, which is a step towards proving our desired result that $\alpha(M) \leq 2$.

Lemma 4.9 *Let Λ be a hereditary artin algebra and M an indecomposable Λ-module in $\mathcal{R}(\Lambda)$. If there exists an epimorphism $f : N \to M$ which is irreducible and $(f, g) : N \coprod B \to M$ is irreducible for some $g : B \to M$, then either B is indecomposable or zero.*

Proof Let $f : N \to M$ be an irreducible epimorphism and let

$$0 \to D \operatorname{Tr} M \xrightarrow{\left(\begin{smallmatrix} h \\ h_1 \\ h_2 \end{smallmatrix}\right)} N \coprod B_1 \coprod B_2 \xrightarrow{(f, f_1, f_2)} M \longrightarrow 0$$

be an almost split sequence. Since f is an epimorphism, h is a monomorphism by Corollary 4.7. Hence (f_1, f_2) is a proper monomorphism and $\left(\begin{smallmatrix} h_1 \\ h_2 \end{smallmatrix}\right)$ is a proper epimorphism by I Corollary 5.7. Therefore $\operatorname{Tr} D \left(\begin{smallmatrix} h_1 \\ h_2 \end{smallmatrix}\right) : M \to \operatorname{Tr} D(B_1 \coprod B_2)$ is a proper epimorphism by Proposition 4.5. Then we have that $l(M) > l(\operatorname{Tr} DB_1) + l(\operatorname{Tr} DB_2)$ and $l(M) > l(B_1) + l(B_2)$. Hence we get $2l(M) > l(\operatorname{Tr} DB_1) + l(B_1) + l(\operatorname{Tr} DB_2) + l(B_2)$. However, if both B_1 and B_2 are nonzero they contain indecomposable summands B_i' for $i = 1, 2$ with almost split sequences

$$0 \to B_i' \to C_i \to \operatorname{Tr} DB_i' \to 0$$

such that M is a summand of C_i for $i = 1, 2$. Hence we get $l(M) \leq l(B_i') + l(\operatorname{Tr} DB_i') \leq l(B_i) + l(\operatorname{Tr} DB_i)$ for $i = 1, 2$. Therefore $2l(M) > 2l(M)$ if both B_1 and B_2 are nonzero, which gives a contradiction. Hence either B_1 or B_2 is zero, which completes the proof of the lemma.   □

We now prove that $\alpha(M) \leq 3$ for any indecomposable module M in $\mathcal{R}(\Lambda)$.

Proposition 4.10 *Let Λ be a hereditary artin algebra and M an indecomposable module in $\mathcal{R}(\Lambda)$. Then we have the following.*

(a) $\alpha(M) \leq 3$.

(b) *If $\alpha(M) = 3$ and*

$$0 \to D \operatorname{Tr} M \xrightarrow{\left(\begin{smallmatrix} g_1 \\ g_2 \\ g_3 \end{smallmatrix}\right)} B_1 \coprod B_2 \coprod B_3 \xrightarrow{(f_1, f_2, f_3)} M \to 0$$

is an almost split sequence, then each g_i is an epimorphism and each f_i is a monomorphism.

Proof Consider the almost split sequence

$$0 \to D \operatorname{Tr} M \xrightarrow{(g_i)^{\mathrm{tr}}} \coprod_{i=1}^{n} B_i \xrightarrow{(f_i)} M \to 0$$

where $B_1, B_2, \ldots, B_n$ are indecomposable modules. Assume $n > 2$. Then it follows from Lemma 4.9 that none of the f_i are epimorphisms. Consider the map $(f_1, f_2): B_1 \coprod B_2 \to M$. If (f_1, f_2) is an epimorphism, then $\coprod_{i=3}^{n} B_i$ is either indecomposable or zero by Lemma 4.9. Hence $n \leq 3$ in this case. If (f_1, f_2) is not an epimorphism, it is a monomorphism and hence $\binom{g_1}{g_2}$ is an epimorphism by Corollary 4.7. Therefore $(f_3, \ldots, f_n): B_3 \coprod \ldots \coprod B_n \to M$ is an epimorphism and consequently B_1 or B_2 is zero. This contradicts the indecomposability of B_1 and B_2, and completes the proof of (a) as well as (b). $\square$

We now want to exclude the possibility that $\alpha(M)$ may be 3. In order to do this we have to analyze more closely what happens in case $\alpha(M) = 3$ for an indecomposable module M in $\mathcal{R}(\Lambda)$.

Proposition 4.11 *Let Λ be a hereditary artin algebra, let M be an indecomposable Λ-module in $\mathcal{R}(\Lambda)$ with $\alpha(M) = 3$ and let $0 \to D \operatorname{Tr} M \xrightarrow{\binom{f_1}{f_2}{f_3}} B_1 \coprod B_2 \coprod B_3 \xrightarrow{(g_1, g_2, g_3)} M \to 0$ be an almost split sequence with B_i indecomposable for $i = 1, 2, 3$. Then we have the following.*

(a) *For $i = 1, 2, 3$ there are sequences of irreducible monomorphisms $B_{i,t_i} \to B_{i,t_i-1} \to \cdots \to B_{i,1} = B_i \to M$ where $\alpha(B_{i,t_i}) = 1$ and $\alpha(B_{i,j}) = 2$ for $j < t_i$.*

(b) *For $i = 1, 2, 3$, there are sequences of irreducible epimorphisms $M \to \operatorname{Tr} D B_i = \operatorname{Tr} D B_{i,1} \to (\operatorname{Tr} D)^2 B_{i,2} \to \cdots \to (\operatorname{Tr} D)^{t_i} B_{i,t_i}$ where $\alpha((\operatorname{Tr} D)^j B_{i,j}) = 2$ for $j < t_i$ and $\alpha((\operatorname{Tr} D)^{t_i} B_{i,t_i}) = 1$.*

(c) *Each indecomposable module A where the corresponding vertex $[A]$ is in the connected component of Γ_Λ determined by M is isomorphic to $(\operatorname{Tr} D)^s X$ for some $s \in \mathbb{Z}$ and $X \in \{ B_{i,j} \mid i = 1, 2, 3, \ 1 \leq j \leq t_i \} \cup \{ M \}$.*

Proof (a) Let $0 \to D \operatorname{Tr} M \xrightarrow{\binom{f_1}{f_2}{f_3}} B_1 \coprod B_2 \coprod B_3 \xrightarrow{(g_1, g_2, g_3)} M \to 0$ be an almost split sequence. By Proposition 4.10(b), each f_i is an epimorphism for $i = 1, 2, 3$. Hence Lemma 4.9 states that $\alpha(B_i) \leq 2$ for each $i = 1, 2, 3$. We now complete the proof for $i = 1$. If $\alpha(B_1) = 1$ we are done. Therefore assume $\alpha(B_1) = 2$ and let $0 \to D \operatorname{Tr} B_1 \xrightarrow{\binom{h_1}{h_{1,2}}} D \operatorname{Tr} M \coprod B_{1,2} \xrightarrow{(f_1, f_{1,2})} B_1 \to 0$ be an almost split sequence. Since f_1 is an epimorphism, $h_{1,2}$ is also an epimorphism, so $\alpha(B_{1,2}) \leq 2$. Further, again since f_1 is an epimorphism, it follows by Lemma 4.8(a) that $f_{1,2}$ is a monomorphism. Continuing this way, we get our desired chain of monomorphisms.

Statement (b) is the dual of (a) and follows by duality.

To prove (c) we deduce directly from (a) and (b) that the set of modules $\mathscr{C} = \{ (D \operatorname{Tr})^n(B_{i,j}), (D \operatorname{Tr})^n M \mid n \in \mathbb{Z}, 1 \leq i \leq 3, 1 \leq j \leq t_i \}$ has the property that if $f : X \to Y$ or $g : Y \to X$ is an irreducible morphism with X in $\mathscr{C}$, then Y is isomorphic to a module in $\mathscr{C}$. $\qquad \square$

Using this result in combination with results from Chapter V we deduce the following.

Proposition 4.12 *Let Λ be a hereditary artin algebra and C an indecomposable Λ-module in a regular component $\mathscr{C}$ of Γ_Λ such that there exist infinitely many n with $l((D \operatorname{Tr})^n C) < s$ for a fixed number s. Then we have the following.*

(a) *$\alpha(X) \leq 2$ for all X in the component $\mathscr{C}$.*

(b) *There exists some X in $\mathscr{C}$ with $\alpha(X) = 1$.*

Proof (a) If $\mathscr{C}$ contains some X with $\alpha(X) = 3$, we have by Proposition 4.11(b) and (c) that there exist an m and irreducible monomorphisms $C = C_t \xrightarrow{r_t} \cdots \to C_1 \xrightarrow{r_1} (D \operatorname{Tr})^m X$ where t is bounded by $\max\{t_i | i = 1, 2, 3\}$, and where t_i is as in Proposition 4.11(a). By V Proposition 6.6 we have that

$$l((D \operatorname{Tr})^m X) \leq l(C)(p^2 + 1)^t$$

where $p \leq l(\Lambda)$. Therefore $l((D \operatorname{Tr})^{m+n} X) \leq l((D \operatorname{Tr})^n C)(p^2 + 1)^t \leq s(p^2 + 1)^t = s'$ for infinitely many n by assumption. Hence we can organize it in such a way that there are $2^{s'}$ modules of length bounded by s' of the form $(\operatorname{Tr} D)^p Y$ with Y in the $D \operatorname{Tr}$-orbit of X and $p \geq 0$. From the almost split sequences $0 \to (\operatorname{Tr} D)^q Y \xrightarrow{(f_1, f_2, f_3)^{\text{tr}}} B_1 \coprod B_2 \coprod B_3 \xrightarrow{(g_1, g_2, g_3)} (\operatorname{Tr} D)^{q+1} Y \to 0$ we have that each g_i is a monomorphism. Therefore we get monomorphisms $(\operatorname{Tr} D)^q Y \to 3((\operatorname{Tr} D)^{q+1} Y)$ for all q, which produce monomorphisms $Y \to 3(\operatorname{Tr} DY) \to 9(\operatorname{Tr} DTY) \to \ldots$ where each morphism is in the radical of mod Λ. This gives rise to nonzero compositions of $2^{s'}$ nonisomorphisms between indecomposable modules of length bounded by s'. But we know from VI Corollary 1.3 that the number of morphisms in such a composition is bounded by $2^{s'} - 1$, giving the desired contradiction. This shows that $\alpha(X) \leq 2$ for all X in $\mathscr{C}$.

(b) Choose X in $\mathscr{C}$ such that $l(X) \leq l(Y)$ for any Y in $\mathscr{C}$. Suppose $\alpha(X) \neq 1$. Then by (a) we have $\alpha(X) = 2$. Let $0 \to D\operatorname{Tr} X \to X_1 \coprod X_2 \overset{(f_1, f_2)}{\to} X \to 0$ be an almost split sequence with the X_i indecomposable. Then by Lemma 4.8 we have that either f_1 or f_2 is a monomorphism. Suppose $f_1 : X_1 \to X$ is a monomorphism. Then $l(X_1) < l(X)$ which is a contradiction since X_1 is in $\mathscr{C}$, and hence $\alpha(X) = 1$. $\square$

Recall from V Section 7 that $\operatorname{rad}_\Lambda^\infty(, B) = \bigcap_{n \in \mathbb{N}} \operatorname{rad}_\Lambda^n(, B)$ is a subfunctor of $\operatorname{Hom}_\Lambda(, B)$, and by V Corollary 7.11 we have that $\operatorname{Supp} \operatorname{Hom}_\Lambda(, B) / \operatorname{rad}_\Lambda^\infty(, B)$ is infinite if and only if $\operatorname{Supp} \operatorname{Hom}_\Lambda(, B)$ is infinite. Further, if X is in $\operatorname{Supp} \operatorname{Hom}_\Lambda(, B) / \operatorname{rad}_\Lambda^\infty(, B)$ for an indecomposable Λ-module B, then according to VI Lemma 1.1 we have that X and B are in the same component of Γ_Λ.

We also have the following result.

Lemma 4.13 *Suppose that Λ is an arbitrary artin algebra, not necessarily hereditary. Let $f : X \to B$ be a morphism between indecomposable Λ-modules with $f \notin \operatorname{rad}_\Lambda^\infty(X, B)$. Then $\operatorname{Im} f$ contains a summand Y such that Y is in the same component of Γ_Λ as X and B.*

Proof Let $f : X \to B$ be a morphism with X and B indecomposable and $f \notin \operatorname{rad}_\Lambda^\infty(X, B)$. Then clearly $i : \operatorname{Im} f \to B$ is not in $\operatorname{rad}_\Lambda^\infty(\operatorname{Im} f, B)$ and therefore also at least one of the indecomposable summands Y of $\operatorname{Im} f$ has the property that $i|_Y : Y \to B$ is not in $\operatorname{rad}_\Lambda^\infty(Y, B)$ since $\operatorname{rad}_\Lambda^\infty$ is a relation on mod Λ. Hence, by VI Lemma 1.1, Y is in the same component of Γ_Λ as B. $\square$

With these preliminary remarks we are able to prove that for a hereditary artin algebra a regular component with an indecomposable module X with $\alpha(X) = 3$ satisfies the conditions of Proposition 4.12, which then leads to a contradiction.

Lemma 4.14 *Let Λ be a hereditary artin algebra and C an indecomposable Λ-module in a regular component $\mathscr{C}$ of Γ_Λ. If $\alpha(C) = 3$, then there exist a natural number s and infinitely many $n \in \mathbb{N}$ such that $l((\operatorname{Tr} D)^n X) < s$ for some module X in $\mathscr{C}$.*

Proof Let $\mathscr{C}$ denote the component containing C with $\alpha(C) = 3$. Then according to Proposition 4.11 there exist $B_{i,j}$ for $i = 1, 2, 3$, and $1 \le j \le t_i$ where t_i depends only on $\mathscr{C}$, such that each module in $\mathscr{C}$ is either of the form $(D \operatorname{Tr})^n (B_{i,j})$ for some n, i and j or of the form $(D \operatorname{Tr})^n C$ for some n.

Choose one of the $(D \operatorname{Tr})^n B_{i,t_i}$ of smallest length in $\mathscr{C}$ and denote this module by X. Since $\operatorname{Hom}_\Lambda(\Lambda, X) \ne 0$ and $\mathscr{C}$ contains no projective module, we have that $\operatorname{rad}_\Lambda^\infty(\Lambda, X) \ne 0$. But then $\operatorname{Supp} \operatorname{Hom}_\Lambda(\ , X)$ is infinite, and hence also $\operatorname{Supp} \operatorname{Hom}_\Lambda(\ , X)/\operatorname{rad}_\Lambda^\infty(\ , X)$ is infinite by V Corollary 7.11. Therefore we can find infinitely many nonisomorphic modules Y_i in $\mathscr{C}$ with a nonzero morphism $f_i \colon Y_i \to X$ with $f_i \notin \operatorname{rad}_\Lambda^\infty(Y_i, X)$. Hence each $\operatorname{Im} f_i$ has a summand in $\mathscr{C}$ according to Lemma 4.13. But since X was among the modules in $\mathscr{C}$ of minimal length, $\operatorname{Im} f_i = X$ for all i. Therefore each f_i is an epimorphism. Since $\mathscr{C}$ contains a module C with $\alpha(C) = 3$ we have from Proposition 4.11(c) that $\mathscr{C}$ contains only finitely many orbits under the translation $D \operatorname{Tr}$. We therefore conclude that there exists some $Z \in \{ B_{i,j} \mid i = 1, 2, 3, \ 1 \le j \le t \} \cup \{ C \}$ such that infinitely many of the Y_i are of form $(D \operatorname{Tr})^n Z$. Applying $(\operatorname{Tr} D)^n$ we get an epimorphism from Z to $(\operatorname{Tr} D)^n X$ for infinitely many n, showing that there are infinitely many n such that $l((\operatorname{Tr} D)^n X)$ is less than $l(Z)$. This completes the proof of the lemma. $\qquad\square$

We have now established the following result.

Theorem 4.15 *Let Λ be a hereditary artin algebra of infinite representation type and let $\mathscr{C}$ be a regular component of Γ_Λ. Then the following hold.*

(a) *There exists an infinite chain of irreducible monomorphisms $C_0 \xrightarrow{f_0} C_1 \xrightarrow{f_1} C_2 \xrightarrow{f_2} \cdots \xrightarrow{f_{n-1}} C_n \xrightarrow{f_n} \cdots$ in $\mathscr{C}$ such that $\alpha(C_0) = 1$ and $\alpha(C_i) = 2$ for $i \ge 1$.*

(b) *For each $n \in \mathbb{Z}$ and $i \in \mathbb{N}$, there is an almost split sequence $0 \to (D \operatorname{Tr})^{n+1} C_i \to (D \operatorname{Tr})^{n+1} C_{i+1} \amalg (D \operatorname{Tr})^n C_{i-1} \to (D \operatorname{Tr})^n C_i \to 0$, where $C_{-1} = 0$.*

(c) *The set $\{(D \operatorname{Tr})^n C_i \mid n \in \mathbb{Z}, i \in \mathbb{N}\}$ constitutes a complete set of indecomposable modules in $\mathscr{C}$ up to isomorphism.*

(d) If $h:(D\,\mathrm{Tr})^n C_{i+1} \to (D\,\mathrm{Tr})^{n-1} C_i$ is any irreducible morphism, then $\mathrm{Ker}\,h \simeq (D\,\mathrm{Tr})^n C_0$.

(e) If $(D\,\mathrm{Tr})^n C_i \simeq C_i$ for some $n \in \mathbb{Z}$ and $i \in \mathbb{N}$, then $(D\,\mathrm{Tr})^n C_j \simeq C_j$ for all $j \in \mathbb{N}$.

(f) The translation quiver $\mathscr{C}$ is isomorphic to $\mathbb{Z}A_\infty/\langle \tau^n \rangle$ where n is the smallest positive integer with $(D\,\mathrm{Tr})^n C_0 \simeq C_0$.

Proof (a) By Proposition 4.12 and Lemma 4.14 there exists some C in $\mathscr{C}$ with $\alpha(C) = 1$. Let $C = C_0$ and consider the almost split sequence $0 \to C_0 \xrightarrow{f_0} C_1 \xrightarrow{g_0} \mathrm{Tr}\,DC_0 \to 0$. The module C_1 is then indecomposable and in $\mathscr{C}$ and $\alpha(C_1) = 2$ since f_0 is not an epimorphism. Consider the almost split sequence $0 \to C_1 \xrightarrow{\binom{f_1}{g_0}} C_2 \coprod \mathrm{Tr}\,DC_0 \xrightarrow{(g_1, h_0)} \mathrm{Tr}\,DC_1 \to 0$. Since g_0 is an epimorphism, h_0 is a monomorphism by Corollary 4.7, and therefore f_1 is a monomorphism. Proceeding by induction on i we get our desired sequence of irreducible monomorphisms together with almost split sequences $0 \to C_i \xrightarrow{\binom{f_i}{g_{i-1}}} C_{i+1} \coprod \mathrm{Tr}\,DC_{i-1} \to \mathrm{Tr}\,DC_i \to 0$.

(b) Applying $(D\,\mathrm{Tr})^n$ to the sequences constructed in the proof of (a) we obtain (b).

(c) This is a direct consequence of (b) since (b) shows that the set of isomorphism classes of the modules $\{(D\,\mathrm{Tr})^n C_i | n \in \mathbb{Z}, i \in \mathbb{N}\}$ give rise to a whole component of the AR-quiver of Λ.

(d) Clearly (d) holds for $i = 0$. The rest of (d) follows by induction on i.

Parts (e) and (f) follow readily. $\qquad\square$

From Theorem 4.15 we get the following geometric picture of a regular component of the AR-quiver:

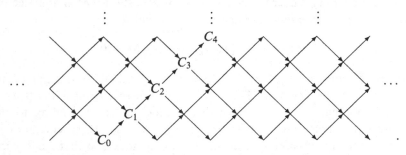

Here there is a possibility that $(D\,\mathrm{Tr})^n C_0 \simeq C_0$ for some $n \in \mathbb{N}$. Then $(D\,\mathrm{Tr})^n C_i \simeq C_i$ for all i and we obtain what is called a **stable tube**.

5 Finite representation type

In this section we consider hereditary artin algebras of finite representation type and prove that the associated AR-quivers of such algebras are all obtained from the Dynkin diagrams. We prove this by applying the results from VII Section 3 on additive functions.

Let Λ be an indecomposable hereditary artin algebra of finite representation type and let Γ_Λ be the AR-quiver of Λ. Then Γ_Λ is a finite quiver where the vertices are in one to one correspondence with the isomorphism classes of indecomposable Λ-modules. Further, since Λ is hereditary, we know from Proposition 1.15 that the preprojective component of the AR-quiver of mod Λ is a translation subquiver of $\mathbb{N}\Delta$ where Δ is the transpose of the quiver of Λ^{op}.

Now let C be the Cartan matrix of the underlying graph $\overline{\Delta}$ of the quiver Δ of Λ^{op}. We identify the vertices $\overline{\Delta}_0$ of $\overline{\Delta}$ with the D Tr-orbits in the preprojective component. Then we let $d : \overline{\Delta}_0 \to \mathbb{Z}^+$ be given by $d(x) = \sum_{M \in x} l(M)$ for each D Tr-orbit x, where $l(M)$ denotes as usual the length of the Λ-module M.

Lemma 5.1 *Let Λ be a hereditary artin algebra of finite representation type, and let $\overline{\Delta}$ be the transpose of the underlying valued graph of the valued quiver of Λ^{op}. Let further C be the Cartan matrix of $\overline{\Delta}$ and let d be the function defined above. Then each coordinate of dC is equal to 2.*

Proof Let x be a D Tr-orbit and consider the xth coordinate of dC, which is

$$2 \sum_{M \in x} l(M) - \sum_y c_{yx} \sum_{N \in y} l(N),$$

where the y ranges over the neighbors of x in $\overline{\Delta}$. But for each noninjective module M in x we have the almost split sequence

$$0 \to M \to \coprod_N c_{yx} N \to \mathrm{Tr}\, DM \to 0$$

where N runs through the modules corresponding to the immediate predecessors of $\mathrm{Tr}\, DM$ in Γ_Λ. Hence we have that $l(M) + l(\mathrm{Tr}\, DM) = \sum c_{yx} l(N)$, where N runs through the modules corresponding to the immediate predecessors of $\mathrm{Tr}\, DM$ in Γ_Λ. We use these equalities to simplify the expression for the xth coordinate above, so that what is left is the length of the projective module P_x and the injective module I_x in the D Tr-orbit x, and for the neighbors only the length of the radical of

P_x and the length of $I_x/\operatorname{soc}I_x$ are left. Hence we get

$$2\sum_{M\in x} l(M) - \sum_y c_{yx} \sum_{N\in y} l(N)$$
$$= l(P_x) - l(\mathfrak{r}P_x) + l(I_x) - l(I_x/\operatorname{soc}I_x)$$
$$= 2. \qquad\qquad \Box$$

Lemma 5.1 implies that the function d is a subadditive function on $\overline{\Delta}$ which is not additive. Hence $\overline{\Delta}$ is a Dynkin diagram by VII Theorem 3.3. As an immediate consequence we get the following.

Proposition 5.2 *If* Λ *is an indecomposable hereditary artin algebra of finite representation type, then the underlying valued graph of the quiver* Δ *of* Λ *is a Dynkin diagram.*

Proof Since the underlying valued graph of a valued quiver and its opposite quiver are the same and a valued graph is Dynkin if and only if its transpose is Dynkin, the result follows directly from the above. $\Box$

In order to prove that if the underlying valued graph of the quiver of Λ is a valued Dynkin diagram then Λ is of finite representation type, we need the following list of valued Dynkin diagrams together with their inverse Coxeter matrices and the orders of these matrices which are all finite.

Dynkin diagram:	Inverse Coxeter matrix:	Order of matrix:
A_n: •——•• $\cdots$ •——•• $\quad$ 1 $\quad$ 2 $\qquad\qquad$ n	$M_{A_n} = \begin{pmatrix} -1 & 1 & 0 & 0 & \cdots & 0 \\ -1 & 0 & 1 & 0 & \cdots & 0 \\ -1 & 0 & 0 & 1 & \cdots & 0 \\ \vdots & \vdots & \vdots & \vdots & & \vdots \\ -1 & 0 & 0 & 0 & \cdots & 1 \\ -1 & 0 & 0 & 0 & \cdots & 0 \end{pmatrix}$	$n+1$
B_n: •—$\overset{(1,2)}{}$—•• $\cdots$ •——•• $\quad$ 1 $\quad$ 2 $\qquad\qquad$ n	$M_{B_n} = \begin{pmatrix} -1 & 2 & 0 & 0 & \cdots & 0 \\ -1 & 1 & 1 & 0 & \cdots & 0 \\ -1 & 1 & 0 & 1 & \cdots & 0 \\ \vdots & \vdots & \vdots & \vdots & & \vdots \\ -1 & 1 & 0 & 0 & \cdots & 1 \\ -1 & 1 & 0 & 0 & \cdots & 0 \end{pmatrix}$	$2n$

C_n: $\overset{(2,1)}{\underset{1 \quad\quad 2}{\bullet \longrightarrow \bullet}} \cdots \underset{n}{\bullet \longrightarrow \bullet}$ $M_{C_n} = \begin{pmatrix} -1 & 1 & 0 & 0 & \cdots & 0 \\ -2 & 1 & 1 & 0 & \cdots & 0 \\ -2 & 1 & 0 & 1 & \cdots & 0 \\ \vdots & \vdots & \vdots & \vdots & & \vdots \\ -2 & 1 & 0 & 0 & \cdots & 1 \\ -2 & 1 & 0 & 0 & \cdots & 0 \end{pmatrix}$ $2n$

D_n: (diagram) $M_{D_n} = \begin{pmatrix} -1 & 0 & 1 & 0 & 0 & \cdots & 0 \\ 0 & -1 & 1 & 0 & 0 & \cdots & 0 \\ -1 & -1 & 1 & 1 & 0 & \cdots & 0 \\ -1 & -1 & 1 & 0 & 1 & \cdots & 0 \\ \vdots & \vdots & \vdots & \vdots & \vdots & & \vdots \\ -1 & -1 & 1 & 0 & 0 & \cdots & 1 \\ -1 & -1 & 1 & 0 & 0 & \cdots & 0 \end{pmatrix}$ $2(n-1)$

E_6: $\underset{1 \quad 2 \quad 4 \quad 5 \quad 6}{\bullet \to \bullet \to \overset{\overset{3}{\bullet}}{\bullet} \to \bullet \to \bullet}$ $M_{E_6} = \begin{pmatrix} -1 & 1 & 0 & 0 & 0 & 0 \\ -1 & 0 & 0 & 1 & 0 & 0 \\ 0 & 0 & -1 & 1 & 0 & 0 \\ -2 & 0 & -1 & 1 & 1 & 0 \\ -1 & 0 & -1 & 1 & 0 & 1 \\ -1 & 0 & -1 & 1 & 0 & 0 \end{pmatrix}$ 12

E_7: $\underset{1 \quad 2 \quad 4 \quad 5 \quad 6 \quad 7}{\bullet \to \bullet \to \overset{\overset{3}{\bullet}}{\bullet} \to \bullet \to \bullet \to \bullet}$ $M_{E_7} = \begin{pmatrix} -1 & 1 & 0 & 0 & 0 & 0 & 0 \\ -1 & 0 & 0 & 1 & 0 & 0 & 0 \\ 0 & 0 & -1 & 1 & 0 & 0 & 0 \\ -2 & 0 & -1 & 1 & 1 & 0 & 0 \\ -1 & 0 & -1 & 1 & 0 & 1 & 0 \\ -1 & 0 & -1 & 1 & 0 & 0 & 1 \\ -1 & 0 & -1 & 1 & 0 & 0 & 0 \end{pmatrix}$ 18

E_8: $\underset{1 \quad 2 \quad 4 \quad 5 \quad 6 \quad 7 \quad 8}{\bullet \to \bullet \to \overset{\overset{3}{\bullet}}{\bullet} \to \bullet \to \bullet \to \bullet \to \bullet}$ $M_{E_8} = \begin{pmatrix} -1 & 1 & 0 & 0 & 0 & 0 & 0 & 0 \\ -1 & 0 & 0 & 1 & 0 & 0 & 0 & 0 \\ 0 & 0 & -1 & 1 & 0 & 0 & 0 & 0 \\ -1 & 0 & -1 & 1 & 1 & 0 & 0 & 0 \\ -1 & 0 & -1 & 1 & 0 & 1 & 0 & 0 \\ -1 & 0 & -1 & 1 & 0 & 0 & 1 & 0 \\ -1 & 0 & -1 & 1 & 0 & 0 & 0 & 1 \\ -1 & 0 & -1 & 1 & 0 & 0 & 0 & 0 \end{pmatrix}$ 30

F_4: $\overset{(1,2)}{\underset{1 \quad 2 \quad 3 \quad 4}{\bullet \to \bullet \to \bullet \to \bullet}}$ $M_{F_4} = \begin{pmatrix} -1 & 1 & 0 & 0 \\ -1 & 0 & 2 & 0 \\ -1 & 0 & 1 & 1 \\ -1 & 0 & 1 & 0 \end{pmatrix}$ 12

G_2: $\overset{(1,3)}{\bullet \longrightarrow \bullet}$ $M_{G_2} = \begin{pmatrix} -1 & 3 \\ -1 & 2 \end{pmatrix}$ 6

We give a proof that the matrices M_{A_n}, M_{B_n}, M_{C_n} and M_{D_n} have the given orders but leave to the reader to verify the result for M_{E_6}, M_{E_7}, M_{E_8}, M_{F_4} and M_{G_2}, which can be done by direct calculations.

Proposition 5.3 *The matrices M_{A_n}, M_{B_n}, M_{C_n} and M_{D_n} are all diagonalizable with roots of unity as eigenvalues and have the orders $n+1$, $2n$, $2n$ and $2(n-1)$ respectively.*

Proof The characteristic polynomial of M_{A_n} is $\det(\lambda I - M_{A_n}) = \sum_{i=0}^{n} \lambda^i$. But the roots of $\sum_{i=0}^{n} \lambda^i$ are all distinct $(n+1)$-th roots of unity. Therefore M_{A_n} is diagonalizable and has order $n+1$.

Consider the matrix M_{B_n}. Conjugating with the matrix

$$G = \begin{pmatrix} 1 & 0 & 0 & \cdots & 0 \\ -1 & 1 & 0 & \cdots & 0 \\ \vdots & \vdots & \vdots & & \vdots \\ -1 & 0 & 0 & \cdots & 1 \end{pmatrix}$$

gives the matrix

$$GM_{B_n}G^{-1} = \begin{pmatrix} 1 & 2 & 0 & 0 & \cdots & 0 \\ 0 & -1 & 1 & 0 & \cdots & 0 \\ 0 & -1 & 0 & 1 & \cdots & 0 \\ \vdots & \vdots & \vdots & \vdots & & \vdots \\ 0 & -1 & 0 & 0 & \cdots & 1 \\ -1 & -1 & 0 & 0 & \cdots & 0 \end{pmatrix}.$$

Hence the characteristic polynomial of M_{B_n} is $(\lambda - 1)(\text{char.pol.}M_{A_{n-1}}) + (-1)^{n-1}2(-1)^{n-1} = \lambda^n + 1$. This polynomial has distinct roots being $2n$-th roots of unity and it has a primitive $2n$-th root of unity. Therefore M_{B_n} is diagonalizable of order $2n$.

For the matrix M_{C_n} conjugate with the same matrix G as above and obtain the matrix

$$GM_{C_n}G^{-1} = \begin{pmatrix} 0 & 1 & 0 & \cdots & 0 \\ 0 & 0 & 1 & \cdots & 0 \\ \vdots & \vdots & \vdots & & \vdots \\ 0 & 0 & 0 & \cdots & 1 \\ -1 & 0 & 0 & \cdots & 0 \end{pmatrix}.$$

Hence M_{C_n} has characteristic polynomial $\lambda^n + (-1)^{n-1}(-1)^{n-1} = \lambda^n + 1$. The same argument as for M_{B_n} applies, showing that M_{C_n} is diagonalizable of order $2n$.

For M_{D_n} conjugate with the matrix

$$G = \begin{pmatrix} 1 & 0 & 0 & \cdots & 0 \\ 1 & 1 & 0 & \cdots & 0 \\ 0 & 0 & 1 & \cdots & 0 \\ \vdots & \vdots & \vdots & & \vdots \\ 0 & 0 & 0 & \cdots & 1 \end{pmatrix}$$

to obtain

$$GM_{D_n}G^{-1} = \begin{pmatrix} -1 & 0 & 1 & 0 & \cdots & 0 \\ 0 & -1 & 2 & 0 & \cdots & 0 \\ 0 & -1 & 1 & 1 & \cdots & 0 \\ \vdots & \vdots & \vdots & \vdots & & \vdots \\ 0 & -1 & 1 & 0 & \cdots & 1 \\ 0 & -1 & 1 & 0 & \cdots & 0 \end{pmatrix}.$$

Hence the characteristic polynomial of M_{D_n} is $(\lambda + 1)(\text{char.pol.}M_{B_{n-1}}) = (\lambda + 1)(\lambda^{n-1} + 1)$. If n is odd then $n - 1$ is even and therefore -1 is not a root of $\lambda^{n-1} + 1$. Therefore M_{D_n} has distinct eigenvalues and is therefore diagonalizable with $2(n-1)$th roots of unity as eigenvalues, and it also contains a $2(n-1)$th primitive root of unity as an eigenvalue. Hence the order of M_{D_n} is $2(n-1)$ when n is odd. For n even -1 is a multiple eigenvalue. However then the vector $(1, 0, \ldots, 0)^{\text{tr}}$ and the vector $(0, 1, 0, 1, 0, \ldots, 1)^{\text{tr}}$ are eigenvectors of the matrix $GM_{D_n}G^{-1}$ corresponding to the eigenvalue -1, showing that in this case the matrix is also diagonalizable and of order $2(n-1)$. □

In order to apply the table of Dynkin quivers and Coxeter matrices we need one more notion which is also of interest in its own right. This notion is geometrically inspired as a cross section meeting each D Tr-orbit once.

Let Λ be an artin algebra with AR-quiver Γ_Λ. A **section** in Γ_Λ is a valued connected subquiver $\mathscr{S}$ of Γ_Λ such that the following two properties are satisfied.

(1) Whenever an arrow α is in $\mathscr{S}$ then $\sigma(\alpha)$ is not in $\mathscr{S}$, where σ is the semitranslation in Γ_Λ.
(2) If an arrow α in $\mathscr{S}$ has valuation (s, t) in $\mathscr{S}$ and (a, b) in Γ_Λ, then $s \leq a$ and $t \leq b$.

Example: Let k be a field and let Λ be the matrix algebra

$$\left\{ \begin{pmatrix} a\,0\,0 \\ b\,c\,0 \\ d\,e\,f \end{pmatrix} \;\middle|\; a,b,c,d,e,f \in k \right\}.$$

Then Γ_Λ is as follows.

Here $\overset{4}{\cdot} \to \overset{5}{\cdot}$ and $\overset{3}{\cdot} \to \overset{5}{\cdot} \to \overset{6}{\cdot}$ are sections, but $\overset{4}{\cdot} \to \overset{5}{\cdot} \to \overset{6}{\cdot}$ is not.

Let $\mathscr{S}$ and $\mathscr{T}$ be two sections in the AR-quiver Γ_Λ of an artin algebra Λ. We say that $\mathscr{S} \leq \mathscr{T}$ if the following two properties are satisfied.

(1) The underlying graph of $\mathscr{S}$ is a subgraph of the underlying graph of $\mathscr{T}$.
(2) If (s, s') is the valuation of an arrow α in $\mathscr{S}$ and (t, t') is the valuation of the same arrow in $\mathscr{T}$ then $s \leq t$ and $s' \leq t'$.

This relation is a partial ordering on the sections in Γ_Λ.

A section which is maximal in this ordering is called a **full section** in Γ_Λ. Two sections $\mathscr{S}$ and $\mathscr{T}$ are called **parallel** if there exists some $n \in \mathbb{Z}$ such that $\mathscr{T}$ is obtained from $\mathscr{S}$ by using the translate to the nth power, i.e. the vertices of $\mathscr{T}$ are images of the vertices in $\mathscr{S}$ by the translate in Γ_Λ to the nth power, the arrows in $\mathscr{T}$ are the images of the arrows in $\mathscr{S}$ by the semitranslate in Γ_Λ to the $2n$th power, and the valuation of each arrow in $\mathscr{T}$ is the same as the valuation of the corresponding arrow in $\mathscr{S}$.

If we now consider an indecomposable hereditary artin algebra Λ, we know from Proposition 5.2 that the preprojective component is a subtranslation quiver of $\mathbb{Z}\Delta$, where Δ is the transpose of the opposite quiver of Λ.

We are now in a situation where we can prove the converse of Proposition 5.2 and establish the following result.

Theorem 5.4 *Let Λ be an indecomposable hereditary artin algebra with associated quiver Δ. Then Λ is of finite representation type if and only if $\overline{\Delta}$ is a Dynkin diagram.*

Proof We only have to prove that if $\overline{\Delta}$ is a Dynkin diagram then Λ is of finite representation type. Let Λ be a hereditary artin algebra with quiver Δ which is Dynkin. Assume that Λ is of infinite type. Then the preprojective component $\widetilde{\mathscr{P}}$ of the AR-quiver Γ_Λ is isomorphic to $\mathbb{N}\Delta^*$ by Proposition 1.15 where Δ^* is the transpose of the opposite of the quiver Δ. Therefore $\widetilde{\mathscr{P}}$ contains infinitely many parallel neighbouring full sections of one of the valued quivers $\tilde{\Delta}$ from the list on page 289–290. So we can consider $\mathbb{N}^{\tilde{\Delta}}$ a subtranslation quiver of $\mathbb{N}^{\Delta^*}$ (up to shift). Letting $l(x)$ denote the length of the module x, we get the element $(l(x))_{x \in n \times \tilde{\Delta}}$ of $\mathbb{N}^{\tilde{\Delta}}$ corresponding to one such full section in $\widetilde{\mathscr{P}}$. Then the element $(l(x))_{x \in (n+1) \times \tilde{\Delta}}$ is given by $M_{\tilde{\Delta}}(l(x))_{x \in n \times \tilde{\Delta}}$ by just counting lengths. However, as listed on page 289–290, the matrices $M_{\tilde{\Delta}}$ all have finite order, and therefore the modules in $\widetilde{\mathscr{P}}$ are of bounded length. Since this implies that Λ is of finite type by VI Theorem 1.4 we have a contradiction. $\square$

As a direct consequence of Theorem 5.4 we have the following classification of elementary hereditary artin algebras of finite representation type. See page 65 for the definition of elementary.

Theorem 5.5 *Let Λ be an elementary hereditary artin algebra over a field k. Then the following are equivalent.*

(i) Λ *is of finite representation type.*
(ii) $\Lambda \simeq k\Delta$ *with Δ one of the Dynkin quivers A_n, D_n, E_6, E_7 or E_8.*

Proof Assume first that Λ is of finite representation type. Then the quiver Δ of Λ is one of the Dynkin quivers. Since Λ is elementary, the valuation of the quiver is symmetric. Hence the quiver Δ is in the list A_n, D_n, E_6, E_7 and E_8. From III Proposition 1.13 it follows that $\Lambda \simeq k\Delta$.

Conversely if $\Lambda \simeq k\Delta$ for Δ one of the quivers A_n, D_n, E_6, E_7 or E_8, then Theorem 5.4 implies that Λ is of finite representation type. $\square$

6 Quadratic forms and roots

In Section 2 we showed that the indecomposable preprojective and preinjective modules over a hereditary artin algebra are uniquely determined by their composition factors. In particular this is the case for all indecomposable modules over a hereditary algebra of finite representation type. We have seen that if X is indecomposable nonprojective in mod Λ we can compute $[D \operatorname{Tr} X]$ in $K_0(\operatorname{mod} \Lambda)$ as $c[X]$. In this section we give a description of which elements of the Grothendieck group $K_0(\text{mod}\,\Lambda)$ come from indecomposable modules for elementary hereditary algebras of finite representation type, in terms of the homological quadratic form. We also show how to define this quadratic form directly from the associated valued graph of the hereditary algebra, and prove that the form is positive semidefinite if and only if the valued graph is Euclidean.

Assume that Λ is a hereditary algebra of finite representation type and let q be a quadratic form from $K_0(\text{mod}\,\Lambda)$ to $\mathbb{Z}$. The elements x in $K_0(\text{mod}\,\Lambda)$ with $q(x) = 1$ are called **roots** of the quadratic form, and a root x is positive if $x = [M]$ for some M in $\text{mod}\,\Lambda$. Let now q be the homological quadratic form. If P is an indecomposable projective Λ-module, then $q([P]) = l_R(\text{End}_\Lambda(P))$ since $\text{Ext}_\Lambda^1(P,P) = 0$. Hence if there is a positive integer r such that $q([M]) = r$ for all indecomposable modules M in $\text{mod}\,\Lambda$, then $l_R(\text{End}_\Lambda(P))$ must be equal to the same number r for all indecomposable projective Λ-modules P. On the other hand, assume that $l_R(\text{End}_\Lambda(P))$ is equal to a fixed number r for each indecomposable projective Λ-module P. This is for example the case for an elementary hereditary algebra over a field k, with $r = 1$, in particular it holds when Λ is a hereditary algebra over an algebraically closed field k. We have the following preliminary results.

Lemma 6.1 *Let Λ be a hereditary algebra of finite representation type such that $l_R(\text{End}_\Lambda(P))$ is equal to a fixed integer r for each indecomposable projective Λ-module P. If q is the homological quadratic form and X is an indecomposable Λ-module, then $q([X]) = r$.*

Proof Since Λ is hereditary of finite representation type, each indecomposable Λ-module X is preprojective. Hence X is isomorphic to $(\text{Tr}\,D)^{v(X)}P$ for some indecomposable projective Λ-module P. Then we have $\text{End}_\Lambda(X) \simeq \text{End}_\Lambda(P)$ by Lemma 1.1, and therefore $q([X]) = q([P]) = r$ since $\text{Ext}_\Lambda^1(X,X) = 0$ for each indecomposable preprojective module X by Proposition 1.7. $\qquad\square$

We have the following converse of Lemma 6.1.

Lemma 6.2 *Let Λ be a hereditary artin algebra of finite representation type with homological quadratic form q and such that $l_R(\text{End}_\Lambda(P))$ is equal to a fixed integer r for all indecomposable projective Λ-modules P. Then we have the following.*

(a) *For any x in $K_0(\text{mod}\,\Lambda)$ we have that r divides $q(x)$.*

(b) *If x is a positive element of $K_0(\mathrm{mod}\,\Lambda)$ with $q(x) = r$, there exists an indecomposable Λ-module M with $[M] = x$.*

(c) *If x is an element of $K_0(\mathrm{mod}\,\Lambda)$ with $q(x) = r$, then x is either positive or negative.*

Proof Let Λ be a hereditary artin algebra of finite representation type with homological quadratic form q and such that there exists an integer r with $l_R(\mathrm{End}_\Lambda(P)) = r$ for all indecomposable projective Λ-modules P. Then we know by Lemma 6.1, since Λ is of finite type, that $l_R(\mathrm{End}_\Lambda(M)) = r$ for each indecomposable Λ-module M and therefore also r divides $l_R(X)$ for any $\mathrm{End}_\Lambda(M)$-module X, especially r divides $l_R(\mathrm{Ext}^i_\Lambda(N, M))$ for any modules M and N and all i. This gives statement (a).

In order to prove (b), let x be a positive element of $K_0(\mathrm{mod}\,\Lambda)$ with $q(x) = r$. Let M be a Λ-module such that $l_R(\mathrm{End}_\Lambda(M))$ is minimal among $l_R(\mathrm{End}_\Lambda(X))$ for all Λ-modules X with $[X] = x$. We claim that M is indecomposable.

If M is decomposable, we have that $l_R(\mathrm{End}_\Lambda(M)) \geq 2r$ since the endomorphism ring of each indecomposable module has length divisible by r. Since $q[M] = q(x) = r$ we then get that $\mathrm{Ext}^1_\Lambda(M, M) \neq 0$. But since $\mathrm{Ext}^1_\Lambda(X, X) = 0$ for each indecomposable module X, there is a decomposition of M into a sum of two modules M' and M'' such that $\mathrm{Ext}^1_\Lambda(M', M'') \neq 0$. Hence we have a nonsplit extension $0 \to M'' \to N \to M' \to 0$. But then Lemma 3.2 implies that $l_R(\mathrm{End}_\Lambda(N))$ is smaller than $l_R(\mathrm{End}_\Lambda(M))$. This gives the desired contradiction. Hence M is indecomposable, completing the proof of statement (b).

For the last statement let $x \neq 0$ be neither positive nor negative. Then there exist nonzero modules M and N without common composition factors such that $x = [M] - [N]$. Then $\mathrm{Hom}_\Lambda(M, N) = 0 = \mathrm{Hom}_\Lambda(N, M)$, and we get $q(x) = q[M] + q[N] + l_R(\mathrm{Ext}^1_\Lambda(N, M)) + l_R(\mathrm{Ext}^1_\Lambda(N, M)) \geq 2r$, which completes the proof of (c). $\square$

As a consequence of these lemmas we have the following result.

Theorem 6.3 *Let Λ be a hereditary artin algebra of finite representation type with homological quadratic form q.*

(a) *If there exists an integer r with $l_R(\mathrm{End}_\Lambda(P)) = r$ for all indecomposable projective Λ-modules P, then there is a one to one correspondence between the positive elements $x \in K_0(\mathrm{mod}\,\Lambda)$ with $q(x) = r$ and the indecomposable Λ-modules.*

(b) *If Λ is an elementary k-algebra there is a one to one correspondence between the indecomposable Λ-modules and the positive roots of q.* □

The valued quiver Δ of a hereditary artin algebra Λ contains information on the length of $\text{Ext}^1_\Lambda(S_i, S_j)$ over $\text{End}_\Lambda(S_i)$. Hence it is not so surprising that we can show that the homological quadratic form of a hereditary artin algebra Λ can be defined directly on the associated valued graph. Before doing this we discuss which valued graphs are associated with hereditary algebras.

Let C be the Cartan matrix associated with a finite valued graph ∇, which then can be viewed as an $n \times n$ matrix when n is the number of elements in ∇_0 with a fixed ordering. Then C is said to be **symmetrizable** if there exists an invertible diagonal matrix T with positive integral coefficients such that CT is symmetric. We shall show that the finite valued graphs associated with hereditary artin algebras are exactly those with symmetrizable Cartan matrices.

Proposition 6.4 *Let Λ be a hereditary artin R-algebra with valued quiver Δ. Then the associated Cartan matrix C is symmetrizable.*

Proof Let $\{S_1, \ldots, S_n\}$ be the nonisomorphic simple Λ-modules and $\{1, \ldots, n\}$ the corresponding vertices of Δ. Let $C = (c_{ij})$ be the Cartan matrix and let $t_i = l_R(\text{End}_\Lambda(S_i))$. If $\alpha: i \cdot \overset{(c_{ij}, c_{ji})}{\longrightarrow} \cdot j$ is a valued arrow, we then have

$$
\begin{aligned}
c_{ji} t_i &= l_{\text{End}_\Lambda(S_i)^{\text{op}}}(\text{Ext}^1_\Lambda(S_i, S_j)) l_R(\text{End}_\Lambda(S_i)) \\
&= l_R(\text{Ext}^1_\Lambda(S_i, S_j)) \\
&= l_{\text{End}_\Lambda(S_j)}(\text{Ext}^1_\Lambda(S_i, S_j)) l_R(\text{End}_\Lambda(S_j)) \\
&= c_{ij} t_j.
\end{aligned}
$$

This shows that if $T = (t_i)$, then CT is a symmetric matrix, and hence C is symmetrizable. □

In order to show that conversely each symmetrizable Cartan matrix is realized by a hereditary artin R-algebra we first need some preliminary observations.

Lemma 6.5 *Let ∇ be a finite connected valued graph without loops and multiple edges, and let C be the associated Cartan matrix. If T and T' are two invertible diagonal matrices with rational coefficients such that CT and CT' are symmetric, then T and T' are linearly dependent over $\mathbb{Q}$.*

Proof Let $T = (t_i)_{i \in \nabla_0}$ and $T' = (t_i')_{i \in \nabla_0}$. Since T and T' are invertible, t_i and t_i' are nonzero for all i in ∇_0. We now show that if i and j are connected by an edge in the graph ∇ then $t_i/t_i' = t_j/t_j'$, which proves the lemma. Letting $C = (c_{ij})_{i,j \in \nabla_0}$ we have the equations $c_{ij}t_j = c_{ji}t_i$ and $c_{ij}t_j' = c_{ji}t_i'$ so that $t_i/t_i' = c_{ji}t_i/c_{ji}t_i' = c_{ij}t_j/c_{ij}t_j' = t_j/t_j'$. Since ∇ is connected, this shows that T and T' are linearly dependent over $\mathbb{Q}$. $\square$

Lemma 6.6 *On any finite valued graph without loops there is an orientation such that there are no oriented cycles.*

Proof The proof goes by induction on the number of vertices. If there is one vertex, there is nothing to prove. Assume the statement holds for graphs with $n \geq 1$ vertices, and let ∇ be a graph with $n + 1$ vertices. Remove one vertex $i \in \nabla_0$ and all edges connected to i. By the induction hypothesis the remaining graph has an orientation without oriented cycles. Now add the vertex i and all edges connected to i and let i be the start of all those edges. If ∇ is given this orientation, there are no oriented cycles. $\square$

We can now give the converse of Proposition 6.4.

Proposition 6.7 *Let ∇ be a finite valued graph with the associated Cartan matrix symmetrizable. Then there exists a finite hereditary algebra Λ such that the underlying valued graph of the valued quiver of Λ is ∇.*

Proof Let ∇ be a finite valued graph with the associated Cartan matrix $C = (c_{ij})_{i,j \in \nabla_0}$ symmetrizable and let $T = (t_i)_{i \in \nabla_0}$ be a diagonal matrix with positive integral coefficients which symmetrizes C. By Lemma 6.6 choose an orientation on ∇ such that the corresponding quiver Δ has no oriented cycles. Let p be a prime number and let F be $GF(p)$, the field with p elements, and for each $i \in \Delta_0$ let $F_i = GF(p^{t_i})$, the field with p^{t_i} elements. For each arrow α in Δ let as usual $s(\alpha)$ be the start and $e(\alpha)$ the end of α. Further let $M_\alpha = GF(p^{c_{ij}t_i})$ where $i = s(\alpha)$. Since $c_{ij}t_j = c_{ji}t_i$ it follows that both F_i and F_j can be considered as subfields of M_α, and we fix such an embedding. This makes M_α a F_i-F_j-bimodule. Now let Σ be the semisimple F-algebra $\prod_{i \in \Delta_0} F_i$ with central simple idempotents e_i for $i \in \Delta_0$ and let M be the abelian group $\coprod_{\alpha \in \Delta_1} M_\alpha$. We now make M a Σ-bimodule in such a way that the tensor algebra $\Lambda = T(\Sigma, M)$ of M over Σ is finite dimensional with Δ as its valued quiver. For

$(f_i) \in \Sigma$ and $(m_\alpha) \in M$ let $(f_i)(m_\alpha) = (f_{e(\alpha)}m_\alpha)$ and $(m_\alpha)(f_i) = (m_\alpha f_{s(\alpha)})$. Now it is easy to show that this makes M a Σ-bimodule and that $0 = M^n = M \otimes_\Sigma M \otimes_\Sigma \cdots \otimes_\Sigma M$ where n is the cardinality of Δ_0. Therefore Λ is a finite dimensional algebra with $M \coprod M^2 \coprod \cdots \coprod M^{n-1}$ as the radical $\mathfrak{r}$, $\Lambda/\mathfrak{r} \simeq \Sigma$, $\mathfrak{r}/\mathfrak{r}^2 \simeq M$ as Σ-bimodules, and $\{e_i | i \in \Delta_0\}$ a complete set of primitive orthogonal idempotents of Λ. Further we have that $\mathrm{End}_\Lambda(S_i) \simeq F_i$ for the simple Λ-module $S_i = \Lambda e_i/\mathfrak{r}e_i$. Now it is easy to see that $e_j M e_i \neq 0$ if and only if there is an arrow α from i to j and that $e_j M e_i \simeq M_\alpha$. This shows that if S_i and S_j are two simple modules corresponding to the idempotents e_i and e_j then $\mathrm{Ext}^1_\Lambda(S_i, S_j)$ is isomorphic to M_α and hence has dimension c_{ij} over F_j and dimension c_{ji} over F_i. It follows that the valued quiver of Λ is Δ.

It only remains to prove that Λ is hereditary. By I Proposition 5.1 it is enough to prove that $\mathfrak{r}$ is projective. An element of $\Lambda = \Sigma \coprod M \coprod \ldots \coprod M^{n-1}$ is of the form $(m_0, m_1, \ldots, m_{n-1})$ with $m_i \in M^i$, where $M^0 = \Sigma$. Consider the map $\phi: M \otimes_\Sigma \Lambda \to M \coprod M^2 \coprod \cdots \coprod M^n$ given by $\phi(m \otimes (m_0, m_1, \ldots, m_{n-1})) = (mm_0, m \otimes m_1, \ldots, m \otimes m_{n-1})$. Then ϕ is clearly a right Λ-isomorphism. Since Σ is semisimple and M is finitely generated, M is a finitely generated projective Σ-module. Therefore $M \otimes_\Sigma \Lambda$ is a projective right Λ-module. Hence Λ is hereditary since Λ is an artin algebra. $\qquad\square$

Let now C be an $n \times n$ symmetrizable Cartan matrix on a finite valued graph ∇ with vertices $\{1, \ldots, n\}$, and let T be an invertible diagonal matrix such that CT is symmetric. Identifying the free abelian group of functions from ∇_0 to $\mathbb{Z}$ with $\mathbb{Z}^n$ we have a symmetric bilinear form $B_1 : \mathbb{Z}^n \times \mathbb{Z}^n \to \mathbb{Z}$ by defining $B_1(x, y) = xCTy^{\mathrm{tr}}$, where the elements of $\mathbb{Z}^n$ are viewed as row vectors. Let $q_1 : \mathbb{Z}^n \to \mathbb{Z}$ be the associated quadratic form given by $q_1(x) = B_1(x, x)$ which we call the **Tits form**. Recall that the quadratic form q_1 is **positive semidefinite** if $q_1(x) \geq 0$ for all x in $\mathbb{Z}^n$, and q_1 is **indefinite** if there are x and y in $\mathbb{Z}^n$ with $q_1(x) > 0$ and $q_1(y) < 0$. The question whether the quadratic form q_1 associated with CT is positive definite, positive semidefinite or indefinite is clearly independent of the choice of the diagonal matrix T with positive integer coefficients by Lemma 6.5.

Let Λ be a hereditary artin R-algebra with associated valued graph ∇. We now show that identifying $K_0(\mathrm{mod}\,\Lambda)$ with $\mathbb{Z}^n$, the homological quadratic form q coincides with the quadratic form q_1 when choosing

$T = (t_i)_{i \in \nabla_0}$ to be the diagonal matrix given by $t_i = l_R(\text{End}_\Lambda S_i)$ for S_i the simple Λ-module corresponding to the vertex i.

Proposition 6.8 *Let Λ, ∇, C and T be as above and let q be the homological quadratic form of Λ.*

(a) *The symmetric matrix (q_{ij}) of q given by $q_{ij} = \frac{1}{2}(q([S_i]+[S_j])-q([S_i]-[S_j]))$ relative to the basis $\{[S_1],\ldots,[S_n]\}$ in $K_0(\text{mod}\,\Lambda)$ is CT.*

(b) *The homological quadratic form q coincides with the Tits form q_1 given by CT.*

Proof (a) We have

$$\frac{1}{2}(q([S_i] + [S_j]) - q([S_i] - [S_j]))$$

$$= \frac{1}{2}(l_R\text{End}_\Lambda(S_i) + l_R\text{Hom}_\Lambda(S_i, S_j)$$

$$+ l_R(\text{Hom}_\Lambda(S_j, S_i)) + l_R(\text{End}_\Lambda(S_j)) - l_R(\text{Ext}^1_\Lambda(S_i, S_i)) - l_R(\text{Ext}^1_\Lambda(S_i, S_j))$$

$$- l_R(\text{Ext}^1_\Lambda(S_j, S_i)) - l_R(\text{Ext}^1_\Lambda(S_j, S_j)) - l_R(\text{End}_\Lambda(S_i)) - l_R(\text{End}_\Lambda(S_j))$$

$$+ l_R(\text{Hom}_\Lambda(S_i, S_j)) + l_R(\text{Hom}_\Lambda(S_j, S_i)) + l_R(\text{Ext}^1_\Lambda(S_i, S_i)) - l_R(\text{Ext}^1_\Lambda(S_i, S'_j))$$

$$- l_R(\text{Ext}^1_\Lambda(S_j, S_i)) + l_R(\text{Ext}^1_\Lambda(S_j, S_j))$$

$$= \begin{cases} 2l_R(\text{End}_\Lambda(S_i)) & \text{for } i = j, \\ -(l_R(\text{Ext}^1_\Lambda(S_i, S_j)) + l_R(\text{Ext}^1_\Lambda(S_j, S_i))) & \text{for } i \neq j, \end{cases} = c_{ij}l_R(\text{End}_\Lambda(S_i)).$$

(b) This is a direct consequence of (a) which shows that the quadratic forms q and q_1 have the same associated symmetric matrices.   □

If a graph ∇ is a tree, then the associated Cartan matrix is clearly symmetrizable, and this is also the case if the graph is of type $\widetilde{A}_n$. In particular the Cartan matrices associated with the Dynkin diagrams and the Euclidean diagrams are all symmetrizable. Hence they all occur as underlying valued graphs for hereditary artin R-algebras. We can now give the following characterization of the Dynkin diagrams and Euclidean diagrams in terms of the quadratic form q_1.

Theorem 6.9 *Let ∇ be a finite connected valued graph with symmetrizable Cartan matrix C, let T be an invertible diagonal integral matrix with positive coefficients such that CT is symmetric and let $q_1: \mathbb{R}^n \to \mathbb{R}$ be the quadratic form defined by $q_1(x) = xCTx^{\text{tr}}$. Then the following hold.*

(a) *q_1 is positive definite if and only if ∇ is a Dynkin diagram.*

(b) q_1 is positive semidefinite but not positive definite if and only if ∇ is a Euclidean diagram.

Proof (a) Let Λ be a hereditary artin algebra such that ∇ is the underlying valued graph of the quiver Δ of Λ. Then by Proposition 6.8 the form q_1 coincides with the homological quadratic form q. But from Theorem 3.6 we have that q is positive definite if and only if Λ is of finite representation type, which happens if and only if ∇ is a Dynkin diagram by Theorem 5.4. Hence (a) follows.

To prove (b) we start by assuming that the quadratic form is positive semidefinite but not positive definite. According to (a) ∇ is then not Dynkin. If ∇ is not Euclidean, there is a Cartan matrix C' less than C such that the valued graph of C' is Euclidean. Then according to VII Lemma 3.2(b) there is a some $d \in \mathbb{R}_+^{\nabla_0}$ where $\mathbb{R}_+^{\nabla_0}$ denotes the functions from ∇_0 to $\mathbb{R}_+$, such that $(dC)(i) \leq 0$ for all i and dC is properly negative for at least one i in the support of d. Hence we get $dCTd^{\mathrm{tr}} < 0$ showing that q is not positive semidefinite. Hence the first half of (b) is proven.

Assume now that ∇ is a Euclidean diagram. We first prove that $q_1(x) \geq 0$ for all $x \in \mathbb{R}_+^{\nabla_0}$. Let $x \in \mathbb{R}_+^{\nabla_0}$, and let d be an additive function for the Cartan matrix C. Then $q_1(x - \alpha d) = (x - \alpha d)CT(x - \alpha d)^{\mathrm{tr}} = xCT(x - \alpha d)^{\mathrm{tr}} = q_1(x) - \alpha xCTd^{\mathrm{tr}}$. However $(\alpha xCTd^{\mathrm{tr}})^{\mathrm{tr}} = \alpha d(CT)^{\mathrm{tr}}x^{\mathrm{tr}}$, but CT is symmetric, showing that $\alpha dCT = 0$. Therefore we get $q_1(x - \alpha d) = q_1(x)$. By subtracting an appropriate multiple of d we therefore get that $q_1(x) = q_1(\bar{x})$ with at least one coordinate of $\bar{x}$ being zero. However, then q_1 corresponds to the associated form of a disjoint union of Dynkin diagrams and hence $q_1(x) \geq 0$ by (a). We then get that $q_1(x) > 0$ for all $x \neq 0$ on the coordinate hyperplanes and q_1 assumes the value zero in the positive cone.

Now let $x \in \mathbb{R}^{\nabla_0}$ and assume $q_1(x) < 0$. Then $dCTx^{\mathrm{tr}} = 0$ where d is the additive function for C and hence for t and s in $\mathbb{R}$ we have $q_1(tx + sd) = (tx + sd)CT(tx + sd)^{\mathrm{tr}} = t^2 q_1(x) < 0$ for $t \neq 0$. However, since $q_1(x) < 0$, x and d are linearly independent. Therefore the subspace generated by x and d intersects any of the coordinate hyperplanes non-trivially, contradicting that $q_1(y) \leq 0$ for all y in the subspace spanned by x and d. Therefore $q_1(x) \geq 0$ for all $x \in \mathbb{R}^{\nabla_0}$. $\square$

The hereditary algebras with a positive semidefinite quadratic form are the **tame** hereditary algebras. For these algebras it is possible to classify the indecomposable modules, and we illustrate this through an important special case in the next section.

7 Kronecker algebras

In addition to algebras of finite representation type, the tame hereditary artin algebras have been studied extensively. One of the reasons for the interest in these algebras is that their indecomposable modules can be described even though there is an infinite number of isomorphism classes of such modules. This section is devoted to classifying the indecomposable modules, up to isomorphism, over the Kronecker algebras, a particular class of tame hereditary algebras. This not only gives the flavor of the general theory of tame hereditary artin algebras, but also provides results useful in the development of the general theory.

Let k be an algebraically closed field. Then the Kronecker algebra over k is the finite dimensional k-algebra Λ which is the subalgebra of $M_3(k)$ consisting of all matrices of the form

$$\begin{pmatrix} a & 0 & 0 \\ c & b & 0 \\ d & 0 & b \end{pmatrix}$$

with a, b, c and d arbitrary elements of k. It is not difficult to see that this algebra Λ is isomorphic to the path algebra of the quiver

$$\Delta : 1 \cdot \underset{\beta}{\overset{\alpha}{\rightrightarrows}} \cdot 2$$

over the field k. Hence $\operatorname{mod}\Lambda$ is equivalent to the category of finite dimensional representations of Δ over k. The associated Cartan matrix is $\begin{pmatrix} 2 & -2 \\ -2 & 2 \end{pmatrix}$, and the quadratic form is positive semidefinite.

The algebra Λ has two nonisomorphic simple modules S_1 and S_2 corresponding to the vertices 1 and 2 in the quiver Δ. Fixing the basis $\{ [S_1], [S_2] \}$ of $K_0(\operatorname{mod}\Lambda)$ we identify $K_0(\operatorname{mod}\Lambda)$ with $\mathbb{Z} \times \mathbb{Z}$ by this choice of basis. The elements in $K_0(\operatorname{mod}\Lambda)$ corresponding to the projective cover P_1 of S_1 and P_2 of S_2 and the injective envelopes I_1 of S_1 and I_2 of S_2 are the elements $(1,2), (0,1), (1,0)$ and $(2,1)$ in $\mathbb{Z} \times \mathbb{Z}$ respectively.

Direct calculations show that the Coxeter transformation c is given by the matrix $\begin{pmatrix} 3 & -2 \\ 2 & -1 \end{pmatrix} = I + \begin{pmatrix} 2 & -2 \\ 2 & -2 \end{pmatrix}$ with inverse $I - \begin{pmatrix} 2 & -2 \\ 2 & -2 \end{pmatrix}$ where I is the 2×2 identity matrix. From this we get that $c^n = I + n\begin{pmatrix} 2 & -2 \\ 2 & -2 \end{pmatrix}$ for all n

in $\mathbb{Z}$. Therefore, by this identification of $K_0(\text{mod }\Lambda)$ with $\mathbb{Z} \times \mathbb{Z}$, we get

$$[(\text{Tr }D)^n P_1] = c^{-n}[P_1] = \left(I - n \begin{pmatrix} 2 & -2 \\ 2 & -2 \end{pmatrix} \right) \begin{pmatrix} 1 \\ 2 \end{pmatrix} = \begin{pmatrix} 1 \\ 2 \end{pmatrix} + \begin{pmatrix} 2n \\ 2n \end{pmatrix},$$

$$[(\text{Tr }D)^n P_2] = \begin{pmatrix} 1 \\ 0 \end{pmatrix} + \begin{pmatrix} 2n \\ 2n \end{pmatrix},$$

$$[(D \text{ Tr})^m I_1] = \begin{pmatrix} 0 \\ 1 \end{pmatrix} + \begin{pmatrix} 2m \\ 2m \end{pmatrix}$$

and

$$[(D \text{ Tr})^m I_2] = \begin{pmatrix} 2 \\ 1 \end{pmatrix} + \begin{pmatrix} 2m \\ 2m \end{pmatrix}.$$

For each natural number n consider the following representations of the quiver Δ over the field k. Let Q_n be the representation $k^n \overset{f_\alpha}{\underset{f_\beta}{\rightrightarrows}} k^{n+1}$ where $f_\alpha = \begin{pmatrix} I \\ 0 \end{pmatrix}$ and $f_\beta = \begin{pmatrix} 0 \\ I \end{pmatrix}$ with I the $n \times n$ identity matrix. We leave as an exercise for the reader to verify that each Q_n is indecomposable. Further, for $n = 2t$, we have $[Q_n] = [(\text{Tr }D)^t P_2]$ and for $n = 2t + 1$, we have $[Q_n] = [(\text{Tr }D)^t P_1]$. Hence from Corollary 2.3(d) we get that $Q_{2t} \simeq (\text{Tr }D)^t P_2$ and $Q_{2t+1} \simeq (\text{Tr }D)^t P_1$. Similarly, for each natural number n, let J_n be the representation of the quiver Δ given by $k^{n+1} \overset{f_\gamma}{\underset{f_\delta}{\rightrightarrows}} k^n$ where $f_\gamma = (I, 0)$ and $f_\delta = (0, I)$ with I the $n \times n$ identity matrix. Then by duality we obtain the isomorphisms $J_{2t} \simeq (D \text{ Tr})^t I_1$ and $J_{2t+1} \simeq (D \text{ Tr})^t I_2$. This gives a complete list of the preprojective and preinjective Λ-modules.

Next consider the representations of the quiver Δ corresponding to the element $(1, 1)$ in $K_0(\text{mod }\Lambda)$. These correspond to the representations $R_{a,b}: k \overset{a}{\underset{b}{\rightrightarrows}} k$ with $(a, b) \in k \times k$. Then it is easy to see that $R_{a,b}$ is indecomposable if and only if $(a, b) \neq (0, 0)$. Direct calculations show that $R_{a,b}$ is isomorphic to $R_{a',b'}$ if and only there exists some $t \in k - \{0\}$ with $(ta, tb) = (a', b')$. Hence the nonisomorphic indecomposable representations of Δ over k corresponding to the element $(1, 1)$ in $K_0(\text{mod }\Lambda)$ are indexed by the projective line $\mathbb{P}^1(k)$ over k. For each $p \in \mathbb{P}^1(k)$ let R_p be the corresponding module.

We have the following result about the family $\{ R_p \mid p \in \mathbb{P}^1(k) \}$.

Proposition 7.1 *Let R_p be as above, and let p and q be in $\mathbb{P}^1(k)$. Then the following hold.*

(a) $\text{End}_\Lambda(R_p) \simeq k$ *for all p.*

(b) $\mathrm{Hom}_\Lambda(R_p, R_q) = 0$ *for* $p \neq q$.

(c) $\dim_k \mathrm{Ext}^1_\Lambda(R_p, R_p) = 1$ *for all* p.

(d) $\mathrm{Ext}^1_\Lambda(R_p, R_q) = 0$ *for* $p \neq q$.

(e) $D \mathrm{Tr}\, R_p \simeq R_p$ *for all* p.

Proof Statements (a) and (b) follow easily by direct calculations.

To prove (c) and (d), consider the minimal projective resolution of R_p which is of the form $0 \to P_2 \to P_1 \to R_p \to 0$. Applying $\mathrm{Hom}_\Lambda(\ , R_q)$ we get the exact sequence $0 \to \mathrm{Hom}_\Lambda(R_p, R_q) \to \mathrm{Hom}_\Lambda(P_1, R_q) \to \mathrm{Hom}_\Lambda(P_2, R_q) \to \mathrm{Ext}^1_\Lambda(R_p, R_q) \to 0$. From (a) and (b) we have $\dim_k \mathrm{Hom}_\Lambda(R_p, R_q) = \delta_{p,q}$, where $\delta_{p,q}$ is the Kronecker delta. In general we have $\dim_k \mathrm{Hom}_\Lambda(P_1, R_q) = 1$ and $\dim_k \mathrm{Hom}_\Lambda(P_2, R_q) = 1$, and hence $\dim_k \mathrm{Ext}^1_\Lambda(R_p, R_q) = \delta_{p,q}$.

(e) We know from Proposition 2.2(b) that $[D \mathrm{Tr}\, R_p] = c[R_p] = c \begin{pmatrix} 1 \\ 1 \end{pmatrix} = \begin{pmatrix} 1 \\ 1 \end{pmatrix}$. Hence $D \mathrm{Tr}\, R_p$ is isomorphic to one of the representations R_q. However $\mathrm{Ext}^1_\Lambda(R_p, D \mathrm{Tr}\, R_p) \neq 0$, so $D \mathrm{Tr}\, R_p \simeq R_p$ according to (d). □

Since $\mathrm{Ext}^1_\Lambda(R_p, R_p) \neq 0$, we know by Proposition 1.7 that R_p is a regular module. Therefore we can use the structure theorem for the regular components of the AR-quiver to analyze the component containing R_p. By Theorem 4.15 we have that for each $p \in \mathbb{P}^1(k)$, the component of the AR-quiver containing R_p has the form

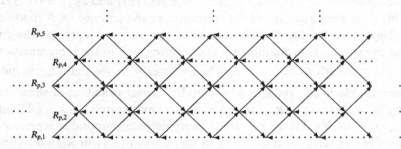

with possibly some identifications. We know that $D \mathrm{Tr}\, R_p \simeq R_p$ by Proposition 7.1(c). From this it follows that all vertices on any given dotted horizontal line in the above picture are to be identified. Such a component is called a **tube of rank** 1. In this way we obtain a list of nonisomorphic indecomposable Λ-modules which we denote by $R_{p,1}, R_{p,2}, \ldots, R_{p,j}, \ldots$. We now show that $R_p = R_{p,1}$. Since the simple module S_1 is injective and the other simple module S_2 is projective, they are not regular modules. Therefore all the modules in the components containing R_p have length at least 2. Since $l(R_p) = 2$ and by Theorem 4.15 all the morphisms going up are irreducible monomorphisms, we have that $R_p = R_{p,1}$.

Letting $R_{p,0} = 0$ we have that there is an almost split sequence $0 \to R_{p,j} \to R_{p,j+1} \coprod R_{p,j-1} \to R_{p,j} \to 0$ for $j \geq 1$. This determines the modules $R_{p,j}$ uniquely.

We have the following analogue of Proposition 7.1 for the modules $R_{p,j}$.

Proposition 7.2 *Let $R_{p,j}$ and $R_{q,i}$ for $p, q \in \mathbb{P}^1(k)$ and $i, j \in \mathbb{N}$ be as described above. Then the following hold.*

(a) $\operatorname{Hom}_\Lambda(R_{p,j}, R_{q,i}) = 0 = \operatorname{Ext}_\Lambda^1(R_{p,j}, R_{q,i})$ *for $p \neq q$.*

(b) $\dim_k \operatorname{Hom}_\Lambda(R_{p,j}, R_{p,i}) = \dim_k \operatorname{Ext}_\Lambda^1(R_{p,j}, R_{p,i}) = \min\{i, j\}$.

Proof The proof of this goes by induction on i and j separately. We start by proving the statements for $j = 1$ and i arbitrary. Proposition 7.1 states that the claims in (a) and (b) hold for $i = 1$ which is the start of the induction. Assume now the statements hold for $i \leq n$. We want to prove that they hold for $i = n + 1$.

Consider the almost split sequence $0 \to R_{q,n} \to R_{q,n+1} \coprod R_{q,n-1} \to R_{q,n} \to 0$. Applying $\operatorname{Hom}_\Lambda(R_{p,1}, \ )$ gives the long exact sequence $0 \to \operatorname{Hom}_\Lambda(R_{p,1}, R_{q,n}) \to \operatorname{Hom}_\Lambda(R_{p,1}, R_{q,n+1} \coprod R_{q,n-1}) \to \operatorname{Hom}_\Lambda(R_{p,1}, R_{q,n}) \to \operatorname{Ext}_\Lambda^1(R_{p,1}, R_{q,n}) \to \operatorname{Ext}_\Lambda^1(R_{p,1}, R_{q,n+1} \coprod R_{q,n-1}) \to \operatorname{Ext}_\Lambda^1(R_{p,1}, R_{q,n}) \to 0$. If $p \neq q$ we have by induction that the first, third, fourth and sixth groups are zero. Hence the two remaining groups in the exact sequence are also zero, showing that the claim in (a) is also valid for $j = 1$ and $i = n + 1$. Hence by induction we have that (a) holds for $j = 1$ and i arbitrary.

We now give the induction step of (b) for $j = 1$. So let $p = q$ and assume that (b) holds for $1 \leq i \leq n$. We want to show that it holds for $i = n + 1$. Since the sequence $0 \to R_{p,n} \to R_{p,n+1} \coprod R_{p,n-1} \to R_{p,n} \to 0$ is almost split, the image of the connecting homomorphism $\operatorname{Hom}_\Lambda(R_{p,1}, R_{p,n}) \to \operatorname{Ext}_\Lambda^1(R_{p,1}, R_{p,n})$ has dimension 1 if $n = 1$ and dimension zero otherwise. We get that $\dim_k \operatorname{Hom}_\Lambda(R_{p,1}, R_{p,n+1}) = 2 \dim_k \operatorname{Hom}_\Lambda(R_{p,1}, R_{p,n}) - \dim_k \operatorname{Hom}_\Lambda(R_{p,1}, R_{p,n-1}) - \delta_{1,n} = 2 - 1 = 1$ and $\dim_k \operatorname{Ext}_\Lambda^1(R_{p,1}, R_{p,n+1}) = 2 \dim_k \operatorname{Ext}_\Lambda^1(R_{p,1}, R_{p,n}) - \dim_k \operatorname{Ext}_\Lambda^1(R_{p,1}, R_{p,n-1}) - \delta_{1,n} = 2 - 1 = 1$. Hence we have established (a) and (b) for $j = 1$ and i arbitrary.

In order to complete the proof one proceeds by induction on j, fixing the index i. The inductive step here is obtained by using the functor $\operatorname{Hom}_\Lambda(\ , R_{p,i})$ on the almost split sequence $0 \to R_{p,j} \to R_{p,j+1} \coprod R_{p,j-1} \to$

$R_{p,j} \to 0$. We leave to the reader to work out the details of this induction.
$\qquad\qquad\qquad\qquad\qquad\qquad\qquad\qquad\qquad\qquad\qquad\qquad\qquad\qquad$ □

From our discussion so far we have obtained the following set of nonisomorphic indecomposable Λ-modules: the preprojective modules $\{Q_n \mid n \in \mathbb{N}\}$, the preinjective modules $\{J_n \mid n \in \mathbb{N}\}$ and the modules $\{R_{p,j} \mid p \in \mathbb{P}^1(k), j \in \mathbb{N}\}$. We now want to show that there are no other finitely generated indecomposable modules in the case when k is algebraically closed. In the proof of this we need the following result.

Lemma 7.3 *Let Λ and the notation be as before. Let X be indecomposable in* mod Λ *and let* $[X] = s[S_1] + t[S_2]$ *in* $K_0(\mathrm{mod}\,\Lambda)$.

(a) *If $s < t$ then $s = t - 1$ and X is preprojective.*

(b) *If $s > t$ then $s = t + 1$ and X is preinjective.*

(c) *If X is regular, then $s = t$ and there exists some $p \in \mathbb{P}^1(k)$ with R_p a submodule of X.*

Proof (a) If $s < t$ then applying the Coxeter transformation c to the power s gives $c^s \binom{s}{t} = \binom{s}{t} + s\begin{pmatrix}2 & -2 \\ 2 & -2\end{pmatrix}\binom{s}{t} = \binom{s}{t} + 2s\binom{s-t}{s-t}$ which is negative. Hence X is preprojective by Corollary 2.3(c). Therefore X is one of the preprojective modules Q_n and hence $s = t - 1$.

(b) If $s > t$ then applying the Coxeter transformation c to the power $-s$ gives that X is preinjective. Therefore X is isomorphic to one of the preinjective modules J_n and hence $s = t + 1$.

(c) From (a) and (b) it follows that if X is regular, then $s = t$. Considering the corresponding representations $V_1 \underset{f_\beta}{\overset{f_\alpha}{\rightrightarrows}} V_2$ in Rep($k\Delta$) we obtain that one of the following two cases has to occur: either f_α is an isomorphism or Ker $f_\alpha \neq 0$. In the first situation $f_\alpha^{-1} f_\beta : V_1 \to V_1$ has an eigenvalue λ in k since k is assumed to be algebraically closed. Let v be an associated eigenvector. Define (g, h) from $k \underset{\lambda}{\overset{1}{\rightrightarrows}} k$ to $V_1 \underset{f_\beta}{\overset{f_\alpha}{\rightrightarrows}} V_2$ by $g(x) = xv$ and $h(x) = xf_\alpha(v)$ for $x \in k$. Then $(f_\alpha g)(x) = f_\alpha(xv) = xf_\alpha(v) = h(x)$ and $(f_\beta g)(x) = f_\alpha f_\alpha^{-1} f_\beta g(x) = xf_\alpha(f_\alpha^{-1} f_\beta(v)) = xf_\alpha(\lambda v) = \lambda x f_\alpha(v) = h(\lambda x)$. Hence (g, h) is a monomorphism from $R_{\overline{(1,\lambda)}}$ to X, where $\overline{(1,\lambda)}$ is the element of $\mathbb{P}^1(k)$ corresponding to $(1, \lambda)$. This takes care of the case when f_α is an isomorphism.

If Ker $f_\alpha \neq 0$, consider the representation $R_{\overline{(0,1)}}$. Let $0 \neq v \in \mathrm{Ker}\, f_\alpha$ and define (g, h) from $k \underset{1}{\overset{0}{\rightrightarrows}} k$ to $V_1 \underset{f_\beta}{\overset{f_\alpha}{\rightrightarrows}} V_2$ by $g(x) = xv$ and $h(x) = xf_\beta(v)$

for all x in k. Then (g, h) is a nonzero morphism. Further, if (g, h) is not a monomorphism, then $\text{Im}(g, h) \simeq S_1$, which is injective. Hence S_1 is a summand of X, but then X is isomorphic to S_1 since X is indecomposable. However X is not preinjective since it is regular. From this we conclude that (g, h) is a monomorphism. This finishes the proof of the lemma. □

We now complete the classification by proving the following.

Proposition 7.4 *Let the notation be as before. Then every finitely generated indecomposable regular Λ-module is isomorphic to $R_{p,j}$ for some $p \in \mathbb{P}^1(k)$ and $j \in \mathbb{N}$.*

Proof The proof of this goes by induction on the length of the module X. We have already noticed that each indecomposable Λ-module of length 2 is in the set $\{ R_{p,1} \mid p \in \mathbb{P}^1(k) \}$. Let X be an indecomposable regular Λ-module of length $n > 2$. Assume by induction that each indecomposable regular module of length less than n is of the form $R_{p,j}$ for some $p \in \mathbb{P}^1(k)$, and $j \in \mathbb{N}$. By Lemma 7.3 there exist some $p \in \mathbb{P}^1(k)$ and an inclusion $g: R_{p,1} \to X$. Fix this p and g and let $X' = \text{Coker}\, g$. If $X' = 0$ then $X \simeq R_{p,1}$ and there is nothing to prove. We may therefore assume that $X' \neq 0$. Decompose X' into a sum of indecomposable summands X'_i and consider the short exact sequence $0 \to R_{p,1} \to X \to \coprod X'_i \to 0$. Since X is not preprojective, none of the summands X'_i of X' is preprojective. Therefore $[X_i] = s_i[S_1] + t_i[S_2]$ has by Lemma 7.3 the property that $s_i \geq t_i$ for all i. However $\sum s_i = \sum t_i$, so we have equality for all i. Therefore none of the X'_i are preinjective. Hence by the induction assumption each X'_i is isomorphic to $R_{q,j}$ for some $q \in \mathbb{P}^1(k)$ and $j \in \mathbb{N}$.

We next show that each q is equal to p. By Proposition 7.2 we have $\text{Ext}^1_\Lambda(R_{q,j}, R_{p,1}) = 0$ for $q \neq p$, so the sequence $0 \to \text{Hom}_\Lambda(R_{q,j}, R_{p,1}) \xrightarrow{(R_{q,j}, g)} \text{Hom}_\Lambda(R_{q,j}, X) \to \text{Hom}_\Lambda(R_{q,j}, X') \to 0$ is exact for $q \neq p$. Hence if $R_{q,j}$ is a summand of X' for $q \neq p$, then $R_{q,j}$ would also be a summand of X, which contradicts the indecomposability of X. Therefore each summand of X' is of the form $R_{p,j}$ for some $j \in \mathbb{N}$.

In the rest of the proof we choose j in $\mathbb{N}$ such that $R_{p,j}$ is a summand of X'. For convenience we suppress the first index p in $R_{p,i}$ for all i. From Theorem 4.15, the structure theorem for regular components for a hereditary artin algebra Λ, we have the exact sequence $\epsilon: 0 \to R_1 \to R_{j+1} \xrightarrow{\gamma} R_j \to 0$ with γ irreducible. Applying $\text{Hom}_\Lambda(\ , R_1)$ we obtain the exact sequence

$$\text{Ext}_\Lambda^1(R_j, R_1) \xrightarrow{\text{Ext}_\Lambda^1(\gamma, R_1)} \text{Ext}_\Lambda^1(R_{j+1}, R_1) \to \text{Ext}_\Lambda^1(R_1, R_1) \to 0.$$

It follows from Proposition 7.2(b) that each of these vector spaces has dimension 1, and hence $\text{Ext}_\Lambda^1(\gamma, R_1)$ is zero. Using this we want to prove that X is isomorphic to R_{j+1}. Considering the exact sequence $0 \to R_1 \xrightarrow{g} X \xrightarrow{\mu} X' \to 0$ and combining this with the sequence ϵ above, we get the following exact commutative diagram

$$
\begin{array}{ccccc}
\text{Hom}_\Lambda(R_j, X) & \xrightarrow{(R_j, \mu)} & \text{Hom}_\Lambda(R_j, X') & \xrightarrow{\delta_j} & \text{Ext}_\Lambda^1(R_j, R_1) \\
\downarrow{(\gamma, X)} & & \downarrow{(\gamma, X')} & & \downarrow{\text{Ext}_\Lambda^1(\gamma, R_1)} \\
\text{Hom}_\Lambda(R_{j+1}, X) & \xrightarrow{(R_{j+1}, \mu)} & \text{Hom}_\Lambda(R_{j+1}, X') & \xrightarrow{\delta_{j+1}} & \text{Ext}_\Lambda^1(R_{j+1}, R_1).
\end{array}
$$

Let $\sigma : R_j \to X'$ be a split monomorphism and consider $\sigma\gamma$ in $\text{Hom}_\Lambda(R_{j+1}, X')$. Since $\text{Ext}_\Lambda^1(\gamma, R_1) = 0$, we have that the connecting morphism δ_{j+1} in the bottom sequence applied to $\sigma\gamma$ is zero. Hence there exists some $v \in \text{Hom}_\Lambda(R_{j+1}, X)$ such that $\mu v = \sigma\gamma$. Now letting ρ be a left inverse of σ we get that $\rho\mu v = \gamma$. However, γ is an irreducible morphism and $\rho\mu$ is not a split epimorphism. Therefore v is a split monomorphism. Now X was assumed to be indecomposable and is therefore isomorphic to R_{j+1}.

This finishes the proof of the lemma as well as completing the classification of all finitely generated indecomposable Λ-modules. $\square$

We end this section by summarizing our findings about the finitely generated indecomposable Λ-modules and their morphisms and extensions.

Theorem 7.5 *Let k, Λ, and the notation be as before in this section. Then we have the following.*

(a) *The sets $\{ Q_n \mid n \in \mathbb{N} \}$, $\{ J_n \mid n \in \mathbb{N} \}$ and $\{ R_{p,j} \mid p \in \mathbb{P}^1(k), j \in \mathbb{N} \}$ constitute a complete set of nonisomorphic indecomposable Λ-modules.*

(b)

$$
\begin{aligned}
\max\{0, n - m + 1\} &= \dim_k \text{Hom}_\Lambda(Q_m, Q_n), \\
\max\{0, m - 1 - n\} &= \dim_k \text{Ext}_\Lambda^1(Q_m, Q_n), \\
\max\{0, m - n + 1\} &= \dim_k \text{Hom}_\Lambda(J_m, J_n), \\
\max\{0, n - 1 - m\} &= \dim_k \text{Ext}_\Lambda^1(J_m, J_n), \\
j &= \dim_k \text{Hom}_\Lambda(Q_m, R_{p,j}) = \dim_k \text{Hom}_\Lambda(R_{p,j}, J_m)
\end{aligned}
$$

$$= \dim_k \operatorname{Ext}^1_\Lambda(R_{p,j}, Q_m) = \dim_k \operatorname{Ext}^1_\Lambda(J_m, R_{p,j}),$$
$$0 = \operatorname{Hom}_\Lambda(J_m, Q_n) = \operatorname{Hom}_\Lambda(J_m, R_{p,j}) = \operatorname{Hom}_\Lambda(R_{p,j}, Q_n)$$
$$= \operatorname{Ext}^1_\Lambda(Q_n, J_m) = \operatorname{Ext}^1_\Lambda(Q_n, R_{p,j}) = \operatorname{Ext}^1_\Lambda(R_{p,j}, J_m),$$
$$\delta_{p,q} \min\{i, j\} = \dim_k \operatorname{Hom}_\Lambda(R_{p,j}, R_{q,i})$$
$$= \dim_k \operatorname{Ext}^1_\Lambda(R_{p,j}, R_{q,i}),$$

where $\delta_{p,q}$ is the Kronecker delta.

Exercises

1. Prove that the center of an indecomposable hereditary artin algebra is a field.

2. Let $F \subset K$ be a finite field extension.

(a) Show that the subring $\Lambda = \left\{ \left(\begin{smallmatrix} a & 0 \\ b & c \end{smallmatrix} \right) | a \in F, b, c \in K \right\}$ of the 2×2 matrix ring over K is hereditary.

(b) Find the preprojective component of Γ_Λ and prove that Λ is of finite type if and only if $[K : F] \leq 3$.

(c) Show that the subring of the $n \times n$ matrices over K with $n \geq 3$ given by the matrices (a_{ij}) with $a_{ij} = 0$ for $j > i$, $a_{nj} \in K$, $a_{ij} \in F$ for $i < n$ is a hereditary artin algebra.

(d) Find the preprojective component of the algebra in (c) and determine when the algebra is of finite representation type.

(e) Give examples of hereditary artin algebras of finite representation type where the underlying graphs of the valued quivers are the Dynkin diagrams different from the ones given in (a) and (c).

3. Let k be a field and Δ be the quiver $1\cdot \to \overset{\overset{\textstyle 2}{\cdot}\;\downarrow 5}{\underset{\overset{\uparrow}{\underset{\textstyle 4}{\cdot}}}{\cdot}} \leftarrow \cdot 3$ and consider the

representation M given by

$$
\begin{array}{ccc}
 & k & \\
 & \downarrow \left(\begin{smallmatrix} a_2 \\ b_2 \end{smallmatrix} \right) & \\
k \overset{\left(\begin{smallmatrix} a_1 \\ b_1 \end{smallmatrix} \right)}{\to} & k \amalg k & \overset{\left(\begin{smallmatrix} a_3 \\ b_3 \end{smallmatrix} \right)}{\leftarrow} k \\
 & \uparrow \left(\begin{smallmatrix} a_4 \\ b_4 \end{smallmatrix} \right) & \\
 & k &
\end{array}
$$

with $\binom{a_i}{b_i} \neq 0$ for $i = 1, 2, 3, 4$. Let α_{ij} be the determinant of the matrix $\begin{pmatrix} a_i & a_j \\ b_i & b_j \end{pmatrix}$ for $1 \leq i < j \leq 4$.

(a) Prove that if $\alpha_{ij} = 0$ for four sets of indices i, j with $1 \leq i < j \leq 4$ then $\alpha_{ij} = 0$ for all i, j where $1 \leq i < j \leq 4$, and the representation M decomposes into a sum of a simple projective and an injective representation.

(b) Prove that if $\alpha_{ij} = 0$ for exactly three sets of indices i, j with $1 \leq i < j \leq 4$, then there is one q with $1 \leq q \leq 4$ such that the representation M decomposes into a sum of the projective representation associated with q, and $D \operatorname{Tr} S_q$ where S_q is the simple representation corresponding to the vertex q.

(c) Prove that if $\alpha_{ij} = 0$ for exactly two sets of indices (i, j) and (p, q) then $\{i, j\} \cap \{p, q\} = \emptyset$ and the representation decomposes into a sum of two indecomposable representations N_{ij} and N_{pq} where $N_{ij}(s) = k$ for $s \in \{i, j, 5\}$ and 0 otherwise and all maps are the identity. This gives a total of six indecomposable modules which come in three pairs.

(d) Use the Coexter transformation to prove that $D \operatorname{Tr} N_{ij} \simeq N_{pq}$ when (i, j) and (p, q) are connected as in (c). Further prove that $\operatorname{End}(N_{i,j}) \simeq k$.

(e) Prove that if $\alpha_{ij} = 0$ for exactly one pair of indices i, j, then the representation M given is indecomposable and determined up to isomorphism by (i, j). Further M which we then denote by M_{ij} contains the representation N_{ij} from (c) and there is an exact sequence $0 \to N_{ij} \to M_{ij} \to N_{pq} \to 0$ where $\{i, j\} \cap \{p, q\} = \emptyset$. Prove that this is an almost split sequence (Hint: $\operatorname{End}(N_{p,q}) \simeq k$ and $N_{ij} \simeq D \operatorname{Tr}(N_{p,q})$.) Also prove that $D \operatorname{Tr} M_{ij} \simeq M_{pq}$ where $\{p, q\} \cap \{i, j\} = \emptyset$.

(f) Prove that if k has only two elements the representations given up to now give all representations of the quiver Δ of the given dimension.

(g) Prove that if $\alpha_{ij} \neq 0$ for all (i, j) then the representation M is indecomposable with $D \operatorname{Tr} M \simeq M$, $\operatorname{End}(M) \simeq k$ and is isomorphic to a representation of the form

$$
\begin{array}{ccccc}
 & & k & & \\
 & & \downarrow \binom{0}{1} & & \\
k & \overset{\binom{1}{0}}{\to} & k \amalg k & \overset{\binom{1}{1}}{\leftarrow} & k \\
 & & \uparrow \binom{1}{b_4} & & \\
 & & k & &
\end{array}
$$

where $b_4 \notin \{0, 1\}$.

(h) Conclude that the components of the AR-quiver containing the modules from (c) and (e) are of form $\mathbb{Z}A_\infty/(\tau^2)$ and that the components containing the modules from case (g) are of the form $\mathbb{Z}A_\infty/(\tau)$.

(i) Prove that if k is an algebraically closed field, then any indecomposable representation of Δ which is neither preprojective nor preinjective contains a submodule from the lists in (c) and (g) and finally show that any indecomposable module belongs to one of these components.

4. This exercise gives an alternative way of proving Theorem 6.9(a) by using Jacobi's criterion.

(a) (Jacobi's criterion) Let M be a real symmetric $n \times n$ matrix and $q : \mathbb{R}^n \to \mathbb{R}$ the quadratic form defined by $q(x) = xMx^{\text{tr}}$. Prove that q is positive definite if and only if all principal minors of M are positive, i.e. the subdeterminants formed by the i first columns and i first rows are positive for $i = 1, 2, \ldots, n$.

(b) Prove that the ith principal minor of the Cartan matrix C_{A_n} of the Dynkin diagram A_n is $i + 1$.

(c) Prove that the ith principal minor of the symmetrized Cartan matrix CT_{B_n} of the Dynkin diagram B_n is 4 (if the diagonal matrix T is chosen minimal).

(d) Show that the ith principal minor of the symmetrized Cartan matrix CT_{C_n} of the Dynkin diagram C_n is 2^i (when T is chosen minimal).

(e) Show that the ith principal minor of the Cartan matrix C_{D_n} of the Dynkin diagram D_n is 4.

(f) Show that the ith principal minors of the symmetrized Cartan matrix CT_Δ are all positive for $\Delta \in \{E_6, E_7, E_8, F_4, G_2\}$.

Notes

The starting point for the study of the representation theory of hereditary artin algebras was [Ga1] in which the hereditary algebras over algebraically closed fields of finite representation type and their modules are classified. This was done by showing that over an algebraically closed field the indecomposable representations of a connected finite quiver are of finite type if and only if the underlying graph is a Dynkin diagram of type A_n, D_n, E_6, E_7 or E_8, in which case there is a natural bijection between the positive roots of the Tits quadratic form associated with the Dynkin diagram and the isomorphism classes of the indecomposable representations of the quiver. This result was applied in [Ga1] to classifying radical square zero algebras over algebraically closed fields of finite representation type, correcting earlier work in [Yoshii]. The generalization to arbitrary hereditary and radical square zero artin algebras was given in [DlR1], [DlR2]. Other references along these lines are [BerGP], [Kru] and [AuP].

The approach to proving Gabriel's theorem in [BerGP] uses the Coxeter functor C^+, which is related to the Coxeter transformation c, and also closely related to the functor $D\,\mathrm{Tr}$ (see [BrenB1], [Ga2]). C^+ is defined as a composition of so-called partial Coxeter functors (see [BerGP]), which have a module theoretic interpretation [AuPR]. Here are the origins of tilting theory, developed in [BrenB2], [HapR].

The distinction between preprojective, preinjective and regular modules has its origin in [BerGP]. Our development of the preprojective and preinjective modules and the Coxeter transformation is based on [AuP]. In [AuP] the homological quadratic form was used, and the equivalence of the homological and Tits quadratic form was given in [Rin1]. There is a more general theory of preprojective modules over an arbitrary artin algebra developed in [AuS1], and a more general notion of preprojective components introduced in [HapR].

The structure of the regular components of a hereditary algebra was conjectured by Ringel and proved in [Rin2]. Our approach follows the independent proof in [AuBPRS].

The classification of the indecomposable modules over the Kronecker algebra was essentially done in [Kro]. The Kronecker algebra is a special case of tame hereditary algebras, and for a classification of these algebras and their indecomposable modules we refer to [DoF], [Naz], [DlR2]. A short recent proof using more machinery is given in [Rin4].

As examples of recent work on wild hereditary algebras we refer to [Ke] and [PeT].

IX

Short chains and cycles

One of the main problems in the representation theory of an artin algebra Λ is determining when two modules M and N in mod Λ are isomorphic. A completely general answer to this problem was given in Chapter VI in terms of R-lengths of the R-module of morphisms from each indecomposable module to the given module M. At first sight it seems hopeless to ever use this criterion for proving that two modules are isomorphic. Our use of it in this chapter shows that this is not entirely true. Nonetheless, it is obvious that it is desirable to have other criteria which are more manageable, even if they are not as general.

In this chapter we concentrate on giving conditions on a pair of indecomposable Λ-modules which guarantee that they are isomorphic in terms of such familiar invariants as their composition factors, their projective presentations or their tops and socles. The basic assumption is that one or both of the modules do not lie on certain types of cycles of morphisms in mod Λ, namely short cycles. We start the chapter by discussing these types of modules.

1 Short cycles

In this section we introduce and study the notions of short cycles and short chains.

A **path** from an indecomposable module M to an indecomposable module N in mod Λ is a sequence of morphisms $M \xrightarrow{f_1} M_1 \xrightarrow{f_2} M_2 \xrightarrow{f_3} \cdots \xrightarrow{f_{t-1}} M_{t-1} \xrightarrow{f_t} N$ between indecomposable modules, where $t \geq 1$ and each f_i is not zero and not an isomorphism. Note that a morphism $f: X \to Y$ between indecomposable modules is not an isomorphism if and only if $f \in \mathrm{rad}_\Lambda(X, Y)$. A path from M to M is called a **cycle** in

313

mod Λ, and the number of morphisms in the path is called the length of the cycle. M is said to **lie on a cycle** if there is a cycle from M to M.

Using results from Chapters VII and VIII we get the following relationship between cycles in mod Λ and cycles in the AR-quiver Γ_Λ of Λ. For M in ind Λ we here let M denote also the corresponding vertex in the AR-quiver.

Proposition 1.1 *Let Λ be an artin algebra and M an indecomposable Λ-module.*

(a) *If M lies on a cycle in the Auslander–Reiten-quiver Γ_Λ, then M lies on a cycle in mod Λ.*

(b) *If Λ is of finite representation type, then M lies on a cycle in mod Λ if and only if M lies on a cycle in Γ_Λ.*

Proof (a) Assume that M lies on a cycle in the AR-quiver. This means that there is a sequence of irreducible morphisms $M \xrightarrow{f_1} M_1 \xrightarrow{f_2} M_2 \xrightarrow{f_3} \cdots \xrightarrow{f_{t-1}} M_{t-1} \xrightarrow{f_t} M$ between indecomposable modules. Hence M lies on a cycle in mod Λ.

(b) If Λ is of finite representation type, it follows from V Theorem 7.8 that if $g: X \to Y$ is a nonzero morphism between indecomposable modules which is not an isomorphism, then there is a sequence of irreducible morphisms $X \xrightarrow{g_1} X_1 \xrightarrow{g_2} X_2 \xrightarrow{g_3} \cdots \xrightarrow{g_{s-1}} X_{s-1} \xrightarrow{g_s} Y$ between indecomposable modules. Hence a cycle in mod Λ gives rise to a cycle in Γ_Λ in this case. $\square$

It has been known for some time that the nature of the cycles an indecomposable module lies on has considerable impact on the properties of the module. For instance the **directing modules**, those indecomposable modules not lying on any cycle, are determined up to isomorphism by their composition factors. Further evidence along these lines is given by the fact that over the Kronecker algebra an indecomposable module is directing if and only if it is either preprojective or preinjective. In fact, this is true for any hereditary k-algebra over an algebraically closed field k. It turns out that indecomposable modules not lying on very short cycles have some properties reminiscent of directing modules. This leads to the concept of modules not lying on short cycles.

We say that a cycle of length at most 2 is a **short cycle**. Note that an indecomposable module M does not lie on a cycle of length 1 if and only if every nonzero morphism $f: M \to M$ is an isomorphism, or equivalently $\operatorname{End}_\Lambda(M)$ is a division ring.

The following example shows that not lying on a short cycle is a proper generalization of directing, and not lying on a cycle of length 1 is a proper generalization of not lying on a short cycle.

Example Let Λ be a Nakayama algebra over a field k with admissible sequence $(2,2)$. Denote by S_1 and S_2 the simple Λ-modules, and by P_1 and P_2 their projective covers. Then S_1, S_2, P_1 and P_2 are the only indecomposable Λ-modules. The module S_1 does not lie on a short cycle since P_1 is the only indecomposable module having a morphism to S_1 which is not zero and not an isomorphism, and $\mathrm{Hom}_\Lambda(S_1, P_1) = 0$. Since we have a cycle $S_1 \to P_2 \to P_1 \to S_1$, the module S_1 is not directing.

We have a short cycle $P_1 \to P_2 \to P_1$, and $\mathrm{End}_\Lambda(P_1)$ is a division ring. Hence P_1 lies on a short cycle but not on a cycle of length 1.

Closely related to the concept of short cycles is the concept of short chains for an artin R-algebra Λ. This notion is motivated by the formula $\langle X, Y \rangle - \langle Y, D\,\mathrm{Tr}\,X \rangle = \langle P_0(X), Y \rangle - \langle P_1(X), Y \rangle$ from IV Corollary 4.3, where X and Y are in $\mathrm{mod}\,\Lambda$ and $P_1(X) \to P_0(X) \to X \to 0$ is a minimal projective presentation. It is obvious that this formula is easier to deal with when one of the terms on the left hand side is zero. This observation leads to the following definition.

A sequence of morphisms $X \xrightarrow{f} Y \xrightarrow{g} D\,\mathrm{Tr}\,X$ where X is indecomposable and f and g are nonzero is said to be a **short chain**. If we have such a short chain, we say that Y is the middle of the short chain. If in addition Y is indecomposable we say that X is the start and $D\,\mathrm{Tr}\,X$ the end of the short chain. If Y is not the middle of a short chain, we have by definition $\langle X, Y \rangle = 0$ or $\langle Y, D\,\mathrm{Tr}\,X \rangle = 0$ for all indecomposable modules X in $\mathrm{mod}\,\Lambda$.

There is the following relationship between short cycles and short chains.

Theorem 1.2 *Let M be an indecomposable module over an artin algebra Λ. Then M does not lie on a short cycle if and only if M is not the middle of a short chain.*

Proof: Let M be an indecomposable Λ-module and $X \xrightarrow{f} M \xrightarrow{g} D\,\mathrm{Tr}\,X$ a short chain. We want to show that M lies on a short cycle. Consider the almost split sequence $0 \to D\,\mathrm{Tr}\,X \xrightarrow{(\alpha_i)^{\mathrm{tr}}} E_1 \amalg \cdots \amalg E_t \xrightarrow{(\beta_i)} X \to 0$, where the E_i are indecomposable. Since $(\alpha_i)^{\mathrm{tr}}$ is a monomorphism, there is some i such that the composition $M \xrightarrow{g} D\,\mathrm{Tr}\,X \xrightarrow{\alpha_i} E_i$ is not zero.

If $\beta_i: E_i \to X$ is a monomorphism, then also the composition $M \xrightarrow{g}$ $D\operatorname{Tr} X \xrightarrow{\alpha_i} E_i \xrightarrow{\beta_i} X$ is nonzero. Since the composition $h: M \to X$ is not an isomorphism we get a short cycle $M \xrightarrow{h} X \xrightarrow{t} M$, where $t = f$ if f is not an isomorphism and $t = fhf$ otherwise.

If $\beta_i : E_i \to X$ is not a monomorphism, it is an epimorphism since it is irreducible. Hence the composition $E_i \xrightarrow{\beta_i} X \xrightarrow{f} M$ is not zero, so that we get a short cycle $M \xrightarrow{\alpha_i g} E_i \xrightarrow{f\beta_i} M$. This shows that if M does not lie on a short cycle, then M is not the middle of a short chain.

In order to prove the converse we need the following.

Lemma 1.3 *Let* $g: B \to C$ *be a right minimal morphism. Then for each indecomposable summand* X *of* $\operatorname{Ker} g$ *we have* $\operatorname{Hom}_\Lambda(\operatorname{Tr} DX, C) \neq 0$.

Proof Consider the induced exact sequence $\delta: 0 \to \operatorname{Ker} g \xrightarrow{i} B \xrightarrow{g} \operatorname{Im} g \to 0$. Let X be an indecomposable summand of $\operatorname{Ker} g$ and $p: \operatorname{Ker} g \to X$ a split epimorphism. Since $g: B \to C$ is right minimal, and consequently also $g: B \to \operatorname{Im} g$ is right minimal, it is easy to see that p does not factor through $i: \operatorname{Ker} g \to B$. This shows that $\delta_*(X) \neq 0$, where δ_* is the covariant defect of δ, and hence $\delta^*(\operatorname{Tr} DX)$ is not zero by IV Theorem 4.1. It then follows that $\operatorname{Hom}_\Lambda(\operatorname{Tr} DX, \operatorname{Im} g)$ is not zero, and hence $\operatorname{Hom}_\Lambda(\operatorname{Tr} DX, C) \neq 0$.   □

We now return to the proof of Theorem 1.2. Assume that M lies on a short cycle $M \xrightarrow{f} N \xrightarrow{g} M$. We want to show that M is the middle of a short chain. We first observe that we can assume $gf = 0$. For since f and g are not isomorphisms, $gf: M \to M$ is not an isomorphism, and is hence nilpotent in $\operatorname{End}_\Lambda(M)$. If $gf \neq 0$, we can then choose i with $(gf)^i \neq 0$ and $(gf)^{i+1} = 0$, to get a short cycle $M \xrightarrow{(gf)^i} M \xrightarrow{gf} M$ where the composition of the morphisms is zero.

Assume now that we have a short cycle $M \xrightarrow{f} N \xrightarrow{g} M$ with $gf = 0$. Then we have $\operatorname{Im} f \subset \operatorname{Ker} g$ and hence $\operatorname{Hom}_\Lambda(M, \operatorname{Ker} g) \neq 0$. There is then an indecomposable summand X of $\operatorname{Ker} g$ such that $\operatorname{Hom}_\Lambda(M, X) \neq 0$. Since N is indecomposable and g is nonzero, $g: N \to M$ is right minimal. Hence we get $\operatorname{Hom}_\Lambda(\operatorname{Tr} DX, M) \neq 0$ by Lemma 1.3, so that M is the middle of a short chain.   □

Even though the concepts of lying on a short cycle and being the middle of a short chain are equivalent for an indecomposable module, it is still useful to have both concepts available. For example if $\text{End}_\Lambda(M)$ is not a division ring, it is immediate from the definition that M lies on a short cycle, but it is not obvious from the definition that M is the middle of a short chain.

Since the notions of modules being faithful or sincere play an important role in studying modules not lying on short cycles, we introduce these concepts before stating our next result. A Λ-module M is **faithful** if the annihilator $\text{ann}_\Lambda(M)$ is 0, and M is **sincere** if each simple Λ-module occurs as a composition factor of M. Choose $m_1, \ldots, m_n$ in M such that $\bigcap_{i=1}^{n} \text{ann}_\Lambda(m_i) = \text{ann}_\Lambda(M)$, and define the Λ-morphism $f: \Lambda \to nM$ by $f(1) = (m_1, \ldots, m_n)$ in nM. Then $\text{Ker} f = \text{ann}_\Lambda(M)$, so f is a monomorphism if and only if M is a faithful Λ-module. From this it follows that if M is faithful then it is sincere. For $\Lambda = k[x]/(x^2)$ the simple Λ-module is an example of an indecomposable module which is sincere and not faithful.

Our aim now is to show that even though in general an indecomposable sincere module M need not be faithful, it is faithful if it does not lie on a short cycle. This will require several steps, and we start with the following.

Lemma 1.4 *For an artin algebra Λ we have the following.*

(a) *For each right Λ-module A and left Λ-module M we have the adjointness isomorphisms $\phi_r: D(A \otimes_\Lambda M) \to \text{Hom}_{\Lambda^{op}}(A, D(M))$ and $\phi_l: D(A \otimes_\Lambda M) \to \text{Hom}_\Lambda(M, D(A))$ given by $\phi_r(f)(a)(m) = f(a \otimes m)$ and $\phi_l(f)(m)(a) = f(a \otimes m)$ for $f \in D(A \otimes_\Lambda M)$, $a \in A$ and $m \in M$.*

(b) *For each projective Λ-module P and Λ-module M we have $D(\text{Hom}_\Lambda(P, M)) \simeq \text{Hom}_\Lambda(M, D(P^*))$.*

(c) *For each projective Λ-module P and Λ-module M we have $\langle P, M \rangle = \langle M, D(P^*) \rangle$.*

Proof (a) is a standard homological fact and is therefore left to the reader.

(b) From II Proposition 4.4 we have the isomorphism $P^* \otimes_\Lambda M \simeq \text{Hom}_\Lambda(P, M)$. Using the duality D we get that $D(\text{Hom}_\Lambda(P, M)) \simeq D(P^* \otimes_\Lambda M)$ which is isomorphic to $\text{Hom}_\Lambda(M, D(P^*))$ by (a).

(c) is a direct consequence of (b). □

Before stating our next preliminary result about indecomposable modules which are not on short cycles, we state a property which the corresponding representations (V, f) have when Λ is a finite dimensional algebra over an algebraically closed field k with associated quiver Γ. Namely, for $f = (f_\alpha)_{\alpha \in \Gamma_1}$ we have that each f_α is either an epimorphism or a monomorphism. That Proposition 1.5 really implies this is left to the reader to verify (see exercise 7). Hopefully this remark will help illustrate the significance of the following.

Proposition 1.5 *Let Λ be an artin algebra and M a Λ-module which is not the middle of a short chain. If $f : P \to Q$ is a nonzero morphism between indecomposable projective Λ-modules, then $\operatorname{Hom}_\Lambda(f, M) : \operatorname{Hom}_\Lambda(Q, M) \to \operatorname{Hom}_\Lambda(P, M)$ is either a monomorphism or an epimorphism.*

Proof Assume that there is some nonzero morphism $f : P \to Q$ between indecomposable projective Λ-modules such that $\operatorname{Hom}_\Lambda(f, M)$ is neither a monomorphism nor an epimorphism. Let $X = \operatorname{Coker} f$. Since $\operatorname{Hom}_\Lambda(f, M)$ is not a monomorphism, we have $\operatorname{Hom}_\Lambda(X, M) \neq 0$. By Lemma 1.4 combined with the isomorphism $P^* \otimes_\Lambda M \simeq \operatorname{Hom}_\Lambda(P, M)$ from II Proposition 4.4 we have a commutative diagram

$$\begin{array}{ccc}
D\operatorname{Hom}_\Lambda(P, M) & \xrightarrow{D\operatorname{Hom}_\Lambda(f,M)} & D\operatorname{Hom}_\Lambda(Q, M) \\
\wr\wr & & \wr\wr \\
\operatorname{Hom}_\Lambda(P^*, DM) & \xrightarrow{\operatorname{Hom}_\Lambda(f^*,DM)} & \operatorname{Hom}_\Lambda(Q^*, DM),
\end{array}$$

where $f^* = \operatorname{Hom}_\Lambda(f, \Lambda)$. Since $\operatorname{Hom}_\Lambda(f, M)$ is not an epimorphism, $D\operatorname{Hom}_\Lambda(f, M)$ is not a monomorphism and therefore $\operatorname{Hom}_\Lambda(f^*, DM)$ is not a monomorphism. This implies that $\operatorname{Hom}_{\Lambda^{op}}(\operatorname{Coker} f^*, DM) \neq 0$. Since P and Q are indecomposable, $P \to Q \to X \to 0$ is a minimal projective presentation of X. Hence we have $\operatorname{Coker} f^* \simeq \operatorname{Tr} X$, so that $\operatorname{Hom}_{\Lambda^{op}}(\operatorname{Tr} X, DM) \neq 0$, and consequently $\operatorname{Hom}_\Lambda(M, D\operatorname{Tr} X) \neq 0$. We have already seen that $\operatorname{Hom}_\Lambda(X, M) \neq 0$. Since X is indecomposable because Q is, we get that M is the middle of a short chain. Hence if M is not the middle of a short chain, $\operatorname{Hom}(f, M)$ is either a monomorphism or an epimorphism for all nonzero $f : P \to Q$ when P and Q are indecomposable projective Λ-modules. $\square$

We are now ready to prove our promised result.

Theorem 1.6 *Let Λ be an artin algebra and M a Λ-module which is not the middle of a short chain. Then M is sincere if and only if it is faithful.*

Proof Let M be a sincere Λ-module which is not the middle of a short chain. Write $1 = e_1 + \cdots + e_n$, where the e_i are primitive orthogonal idempotents. If $\operatorname{ann} M \neq 0$, there are some i, j with $e_i(\operatorname{ann} M)e_j \neq 0$. Choose $\lambda \in \operatorname{ann} M$ such that $e_i \lambda e_j \neq 0$. Then λ induces by right multiplication a nonzero morphism $\mu \colon \Lambda e_i \to \Lambda e_j$. Since M is sincere, $\operatorname{Hom}_\Lambda(\Lambda e_i, M) \neq 0$ and $\operatorname{Hom}_\Lambda(\Lambda e_j, M) \neq 0$. By the choice of μ it is easy to see that $\operatorname{Hom}_\Lambda(\mu, M) = 0$, and $\operatorname{Hom}_\Lambda(\mu, M)$ is hence neither a monomorphism nor an epimorphism. This gives a contradiction to Proposition 1.5, so that M is faithful. This finishes the proof of the theorem since we know that M being faithful implies that M is sincere. $\square$

When an artin algebra Λ has a sincere indecomposable directing module, we shall see that there are several Λ-modules of projective dimension at most 1 and of injective dimension at most 1. We say that an indecomposable module M is **before** the indecomposable module X in $\operatorname{mod}\Lambda$ if $M = X$ or there is a path from M to X in $\operatorname{mod}\Lambda$ and M is **after** X in $\operatorname{mod}\Lambda$ if $M = X$ or there is a path from X to M in $\operatorname{mod}\Lambda$. Note that this does not in general define a partial order in $\operatorname{ind}\Lambda$.

Proposition 1.7 *Let Λ be an artin algebra and M an indecomposable Λ-module. Then we have the following.*

(a) $\operatorname{pd}_\Lambda M \leq 1$ *if and only if* $\operatorname{Hom}_\Lambda(D(\Lambda), D\operatorname{Tr} M) = 0$.

(b) $\operatorname{id}_\Lambda M \leq 1$ *if and only if* $(\operatorname{Tr} DM)^* = 0$.

(c) *Let X be a sincere indecomposable directing module. If M is before X, then $\operatorname{pd}_\Lambda M \leq 1$ and if M is after X, then $\operatorname{id}_\Lambda M \leq 1$.*

Proof (a) This is just a restatement of IV Proposition 1.16.

(b) This follows from (a) by duality.

(c) Assume that M is before the sincere indecomposable directing Λ-module X. If $\operatorname{pd}_\Lambda M > 1$ there is by (a) an indecomposable injective Λ-module I such that $\operatorname{Hom}_\Lambda(I, D\operatorname{Tr} M) \neq 0$. Since X is sincere, it is easy to see that $\operatorname{Hom}_\Lambda(X, I) \neq 0$. Since there is a path from $D\operatorname{Tr} M$ to M, it follows that X lies on a cycle. This is a contradiction, and hence we can conclude that $\operatorname{pd}_\Lambda M \leq 1$.

The second part follows by duality.   □

As an application of Theorem 1.6 we prove a similar result for sincere modules which do not lie on short cycles, which we shall need in Section 3.

Proposition 1.8 *Let* Λ *be an artin algebra and* X *an indecomposable sincere* Λ-*module which does not lie on a short cycle. Let* M *be indecomposable in* mod Λ.

(a) *If* $\mathrm{Hom}_\Lambda(M, X) \neq 0$, *then* $\mathrm{pd}_\Lambda M \leq 1$.
(b) *If* $\mathrm{Hom}_\Lambda(X, M) \neq 0$, *then* $\mathrm{id}_\Lambda M \leq 1$.

Proof (a) Assume $\mathrm{Hom}_\Lambda(M, X) \neq 0$, and let $P_1 \overset{h}{\to} P_0 \to M \to 0$ be a minimal projective presentation of M. We have the formula $\langle M, X \rangle - \langle X, D\,\mathrm{Tr}\,M \rangle = \langle P_0, X \rangle - \langle P_1, X \rangle$ by IV Corollary 4.3. Since X is not the middle of a short chain and $\mathrm{Hom}_\Lambda(M, X) \neq 0$, we have $\langle X, D\,\mathrm{Tr}\,M \rangle = 0$. Consider the exact sequence $0 \to \mathrm{Hom}_\Lambda(M, X) \to \mathrm{Hom}_\Lambda(P_0, X) \overset{\mathrm{Hom}_\Lambda(h,X)}{\to} \mathrm{Hom}_\Lambda(P_1, X)$. Then $l_R(\mathrm{Im}\,\mathrm{Hom}_\Lambda(h, X)) = \langle P_0, X \rangle - \langle M, X \rangle = \langle P_1, X \rangle$. Therefore $\mathrm{Hom}_\Lambda(h, X)$ is surjective. Since X is faithful by Theorem 1.6, we have a monomorphism $f: \Lambda \to tX$ for some $t > 0$. Hence we have a monomorphism $g: P_1 \to sX$ for some $s > 0$. Because $\mathrm{Hom}_\Lambda(h, X)$ is surjective we have a commutative diagram

$$
\begin{array}{ccc}
P_1 & \overset{h}{\to} & P_0 \\
{\scriptstyle g}\downarrow & \swarrow & \\
sX & &
\end{array}
$$

Since g is a monomorphism, h is also a monomorphism, and hence $\mathrm{pd}_\Lambda M \leq 1$.

(b) This follows from (a) by duality.   □

2 Modules determined by composition factors

We give some general sufficient conditions for indecomposable modules to be determined by their composition factors, in terms of the notions introduced in Section 1.

Theorem 2.1 *Let* M *and* N *be indecomposable modules having the same composition factors. If* M *does not lie on a short cycle, then* $M \simeq N$.

Proof Let $P_1 \to P_0 \to X \to 0$ be a minimal projective presentation for a Λ-module X, and let M and N be indecomposable Λ-modules. We have by IV Corollary 4.3 the formulas $\langle X, M \rangle - \langle M, D \operatorname{Tr} X \rangle = \langle P_0, M \rangle - \langle P_1, M \rangle$ and $\langle X, N \rangle - \langle N, D \operatorname{Tr} X \rangle = \langle P_0, N \rangle - \langle P_1, N \rangle$. Since M and N have the same composition factors, we have $\langle P_0, N \rangle = \langle P_0, M \rangle$ and $\langle P_1, N \rangle = \langle P_1, M \rangle$. Hence we get $\langle X, M \rangle - \langle M, D \operatorname{Tr} X \rangle = \langle X, N \rangle - \langle N, D \operatorname{Tr} X \rangle$. Letting $X = M$, we have $\langle M, M \rangle - \langle M, D \operatorname{Tr} M \rangle = \langle M, N \rangle - \langle N, D \operatorname{Tr} M \rangle$. Since M does not lie on a short cycle, it is not the middle of a short chain by Theorem 1.2. Therefore we get $\langle M, D \operatorname{Tr} M \rangle = 0$, so that $\langle M, N \rangle \neq 0$.

Since DM also does not lie on a short cycle in $\operatorname{mod}(\Lambda^{\operatorname{op}})$, and DM and DN have the same composition factors, we get similarly that $\langle DM, DN \rangle \neq 0$ and hence $\langle N, M \rangle \neq 0$. Since $\operatorname{Hom}_\Lambda(M, N) \neq 0$ and $\operatorname{Hom}_\Lambda(N, M) \neq 0$ and M does not lie on a short cycle, we conclude that $M \simeq N$. $\square$

We have the following immediate consequences.

Corollary 2.2 *Let M and N be indecomposable Λ-modules having the same composition factors. If M is directing, then $M \simeq N$.* $\square$

Corollary 2.3 *If there are no short cycles in $\operatorname{mod}\Lambda$, then the indecomposable Λ-modules are determined by their composition factors.* $\square$

It follows from VIII Proposition 1.5 and its dual that the indecomposable preprojective and preinjective modules over a hereditary artin algebra are directing. Hence Corollary 2.2 is a generalization of the result proven in VIII Corollary 2.3 that these modules are determined by their composition factors.

We show that $\operatorname{End}_\Lambda(M)$ being a division ring is not a sufficient condition for M to be determined by its composition factors and that M not lying on a short cycle is not a necessary condition.

Example Let Λ be a Nakayama algebra with admissible sequence $(3, 3)$ over an algebraically closed field k. Let S_1 and S_2 denote the simple Λ-modules and P_1 and P_2 their projective covers. Then $P_1/\operatorname{soc} P_1$ and $P_2/\operatorname{soc} P_2$ have the same composition factors and their endomorphism rings are division rings, but they are not isomorphic.

Further we have a short cycle $P_1 \to P_2 \to P_1$, and P_1 is determined by its composition factors.

We know that an indecomposable artin algebra is of finite representation type if we have a finite component of the AR-quiver. We have illustrated on examples in Chapter VII that we can sometimes compute the AR-quiver by starting with a simple projective module. The method is easier to carry out in practice if we can replace the indecomposable module corresponding to a vertex by its composition factors. In order to do this we need to recognize from the composition factors when we have reached an indecomposable summand X of $\mathfrak{r}P$ for some indecomposable projective module P. In particular, we need to know that such X are determined by their composition factors. We point out that we can recognize from the composition factors if we have an injective module, since for a minimal left almost split morphism $g : A \to B$ we have $l(B) < l(A)$ if and only if A is injective.

We illustrate this principle on an example already treated in Chapter VII.

Example Let k be a field and Γ the quiver $\begin{smallmatrix} & & 1 & & \\ & \alpha \nearrow & & \searrow \gamma & \\ 2 \bullet & & & & \bullet 3 \\ & \beta \searrow & & \swarrow \delta & \\ & & \bullet & & \\ & & 4 & & \end{smallmatrix}$ with relation

$\rho = \{\beta\alpha - \delta\gamma\}$. Denote by S_i and P_i the simple and projective module corresponding to the vertex i for $i = 1, 2, 3, 4$. Then $S_4 = P_4$ and $\mathfrak{r}P_2 \simeq S_4 \simeq \mathfrak{r}P_3$ and $\mathfrak{r}P_1$ is indecomposable.

It is not hard to see that $\mathfrak{r}P_1$ does not lie on a short cycle (see Exercise 5), so that $\mathfrak{r}P_1$ is determined by its composition factors by Theorem 2.1.

If $[M] = a_1[S_1] + a_2[S_2] + a_3[S_3] + a_4[S_4]$ in $K_0(\mathrm{mod}\,\Lambda)$, we denote $[M]$ by (a_1, a_2, a_3, a_4). We then have that the AR-quiver of this algebra looks like the following, using additivity.

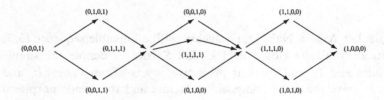

3 Sincere modules and short cycles

We have seen that for indecomposable modules, not lying on a short cycle is a proper generalization of being directing. It is sometimes possible to reduce questions about modules not lying on short cycles to questions about directing modules, as we shall illustrate in the next section. This is based on the fact that the two concepts coincide for sincere modules, which is the main result of this section.

We start with the following preliminary results.

Lemma 3.1 *Let* $\delta: 0 \to X \xrightarrow{f} Y \xrightarrow{g} Z \to 0$ *be an exact sequence in* $\operatorname{mod} \Lambda$. *If* $f \in \operatorname{rad}_\Lambda(X, Y)$ *and* $p: X \to M$ *is a split epimorphism and*

$$
\begin{array}{ccccccccc}
0 & \to & X & \xrightarrow{f} & Y & \xrightarrow{g} & Z & \to & 0 \\
 & & \downarrow p & & \downarrow q & & \| & & \\
0 & \to & M & \xrightarrow{f'} & C & \to & Z & \to & 0
\end{array}
$$

is a pushout diagram, then $f' \in \operatorname{rad}_\Lambda(M, C)$.

Proof Choose $j: M \to X$ such that $pj = 1_M$. Then we have $qfj = f'$. Since $f \in \operatorname{rad}_\Lambda(X, Y)$ we have that $f' \in \operatorname{rad}_\Lambda(M, C)$. $\square$

Lemma 3.2 *Let* M *and* N *be indecomposable* Λ-*modules with* $\operatorname{rad}^2(M, N) = 0$. *Let* $M \xrightarrow{f} X \xrightarrow{g} N$ *be a path in* $\operatorname{mod} \Lambda$ *with* $l(X)$ *smallest possible. Then* $\operatorname{Ker} g$ *is indecomposable.*

Proof Let $K = \operatorname{Ker} g$ and assume that $K = K_1 \coprod K_2$ where K_1 is indecomposable. We denote by f' the morphism $M \to K$ induced by $f: M \to X$. Since $f: M \to X$ is not zero, we can assume that $p_1 f' \neq 0$ where $p_1: K \to K_1$ is the projection map according to the above decomposition. We have the exact sequence $0 \to K \xrightarrow{s} X \xrightarrow{t} Z \to 0$ where $Z = \operatorname{Im} g$. Consider the pushout diagram

$$
\begin{array}{ccccccccc}
\delta: & 0 & \to & K & \xrightarrow{s} & X & \xrightarrow{t} & Z & \to & 0 \\
 & & & \downarrow p_1 & & \downarrow & & \| & & \\
\epsilon: & 0 & \to & K_1 & \xrightarrow{\alpha} & Y & \xrightarrow{\beta} & Z & \to & 0.
\end{array}
$$

Since $s: K \to X$ is a nonzero monomorphism which is not an epimorphism and X is indecomposable, we have that $s \in \operatorname{rad}_\Lambda(K, X)$. Then $\alpha \in \operatorname{rad}_\Lambda(K_1, Y)$ by Lemma 3.1, and hence the sequence ϵ does not split. Since $\alpha: K_1 \to Y$ is a monomorphism, there is a decomposition

$Y = Y_1 \coprod Y_2$ with Y_1 indecomposable and $q_1 \alpha p_1 f' : M \to Y_1$ nonzero, where $q_1 : Y \to Y_1$ is the projection map according to the above decomposition. If $\beta(Y_1) = 0$, we have $Y_1 \subset \alpha(K_1)$. Hence Y_1 would be a summand of $\alpha(K_1)$, and consequently $Y_1 = \alpha(K_1)$ since Y_1 is not zero and $\alpha(K_1)$ is indecomposable. But then the sequence $0 \to K_1 \overset{\alpha}{\to} Y \overset{\beta}{\to} Z \to 0$ would split. Consequently we have $\beta(Y_1) \neq 0$. Now we have a path $M \to Y_1 \to N$ with $l(Y_1) \leq l(X) - l(K_2)$. By the assumption on X we have that $K_2 = 0$. Therefore we have that $K = \operatorname{Ker} g$ is indecomposable.

$\square$

The crucial part of the main theorem is the following.

Proposition 3.3 *Let M be a sincere indecomposable module over an artin algebra Λ and assume that M does not lie on a short cycle. If N is indecomposable and there is a path $M \overset{f}{\to} X \overset{g}{\to} Y \overset{h}{\to} N$ in $\operatorname{mod}\Lambda$, then there is some path $M \overset{s}{\to} A \overset{t}{\to} N$ in $\operatorname{mod}\Lambda$.*

Proof Let M be a sincere indecomposable module over an artin algebra Λ and assume that M does not lie on a short cycle. Let $M \overset{f}{\to} X \overset{g}{\to} Y \overset{h}{\to} N$ be a path from M to N in $\operatorname{mod}\Lambda$, and assume that $l(X) + l(Y)$ is smallest possible for a path from M to N with three morphisms. We can clearly assume that $\operatorname{Hom}_\Lambda(M, Y) = 0 = \operatorname{Hom}_\Lambda(X, N)$, since otherwise we would easily get a shorter path from M to N. Since $\operatorname{Hom}_\Lambda(M, Y) = 0$ and the path $M \overset{f}{\to} X \overset{g}{\to} Y$ has $l(X)$ smallest possible, it follows from Lemma 3.2 that $\operatorname{Ker} g$ is indecomposable. Since in addition $\operatorname{Hom}_\Lambda(M, K) \neq 0$, where we write $K = \operatorname{Ker} g$, and M is a sincere indecomposable module which is not the middle of a short chain, it follows that $\operatorname{id}_\Lambda K \leq 1$ by Proposition 1.8. Writing $Z = \operatorname{Im} g$ and $C = \operatorname{Coker} g$ we have an exact sequence $\delta : 0 \to Z \overset{u}{\to} Y \overset{t}{\to} C \to 0$. The exact sequence $0 \to K \to X \overset{g'}{\to} Z \to 0$, where $g' : X \to Z$ is the morphism induced by $g : X \to Y$, gives rise to an exact sequence $\operatorname{Ext}^1_\Lambda(C, K) \to \operatorname{Ext}^1_\Lambda(C, X) \to \operatorname{Ext}^1_\Lambda(C, Z) \to 0$. Considering the element δ in $\operatorname{Ext}^1_\Lambda(C, Z)$, it follows from the surjection $\operatorname{Ext}^1_\Lambda(C, X) \to \operatorname{Ext}^1_\Lambda(C, Z)$ that we have an exact commutative diagram

$$
\begin{array}{ccccccccc}
0 & \to & X & \overset{\alpha}{\to} & B & \overset{\beta}{\to} & C & \to & 0 \\
  &   & \downarrow g' & & \downarrow v & & \| & & \\
0 & \to & Z & \overset{u}{\to} & Y & \overset{t}{\to} & C & \to & 0 \\
  &   & \downarrow & & \downarrow & & & & \\
  &   & 0 & & 0 & & & & .
\end{array}
$$

Let $B = \coprod_{i=1}^{s} B_i$ be a decomposition of B into a sum of indecomposable modules, and denote by $q_i: B_i \to B$ and $p_i: B \to B_i$ the corresponding monomorphisms and epimorphisms for $i = 1, \ldots, s$. Since $hg = 0$, g is not an epimorphism and hence $C = \operatorname{Coker} g$ is not zero. Since $g \neq 0$, we have that $Z = \operatorname{Im} g$ is not zero. It then follows that $0 \to Z \to Y \to C \to 0$ does not split since Y is indecomposable. Hence the sequence $0 \to X \xrightarrow{\alpha} B \xrightarrow{\beta} C \to 0$ does not split. If follows that $p_i\alpha: X \to B_i$ is not an isomorphism for any i.

Consider the sequence $\xi: 0 \to X \xrightarrow{\binom{\alpha}{-g'}} B \coprod Z \to Y \to 0$ induced by the above diagram which is exact by I Proposition 5.6. Since $gf = 0$ we have that $g: X \to Y$ is not a monomorphism, and hence $g': X \to Z$ is a proper epimorphism. Then $g': X \to Z$ is not a split monomorphism, so that $g' \in \operatorname{rad}_\Lambda(X, Z)$ since X is indecomposable. Since $\alpha: X \to B$ is also not a split monomorphism, we have $\alpha \in \operatorname{rad}_\Lambda(X, B)$, and consequently $\binom{\alpha}{-g'} \in \operatorname{rad}_\Lambda(X, B \coprod Z)$. Hence ξ does not split. Since both X and Y are indecomposable, $p_i\alpha: X \to B_i$ and $vq_i: B_i \to Y$ must then be nonzero nonisomorphisms for all i. Since $\alpha: X \to B$ is a monomorphism and $f: M \to X$ is not zero, there is some i with $p_i\alpha f: M \to B_i$ not zero. Since $v: B \to Y$ is an epimorphism and $h: Y \to N$ is not zero, there is some j with $hvq_j: B_j \to N$ not zero.

If $i = j$, we then have a path $M \to B_i \to N$. If $i \neq j$ consider the paths $M \to X \to B_j \to N$ and $M \to B_i \to Y \to Z$. From the exact sequence $0 \to X \to B \coprod Z \to Y \to 0$ it follows that $l(X) + l(B_j) + l(B_i) + l(Y) < 2(l(X) + l(Y))$. Hence we have $l(X) + l(B_j) < l(X) + l(Y)$ or $l(X) + l(B_i) < l(X) + l(Y)$. This contradicts our choice of path $M \to X \to Y \to N$. Hence there is some path $M \to A \to N$ in $\operatorname{mod} \Lambda$. $\qquad \square$

As a consequence we obtain the following.

Theorem 3.4 *Let M be a sincere indecomposable module over an artin algebra Λ. Then M is not directing if and only if M lies on a short cycle.*

Proof If M lies on a short cycle, then obviously M lies on a cycle and is hence not directing.

Assume conversely that M is not directing and choose a cycle $M \xrightarrow{f_0} X_1 \xrightarrow{f_1} X_2 \xrightarrow{f_2} \cdots \xrightarrow{f_{n-2}} X_{n-1} \xrightarrow{f_{n-1}} X_n = M$, where n is smallest possible. If $n \leq 2$ we have a short cycle, and we are done. If $n \geq 3$ we apply Proposition 3.3 to the path $M \xrightarrow{f_0} X_1 \xrightarrow{f_1} X_2 \xrightarrow{f_2} X_3$ to get a path $M \xrightarrow{g} Y \xrightarrow{h} X_3$.

Then we get a shorter path from M to M, which is a contradiction. Hence M lies on a short cycle. □

4 Modules determined by their top and socle

Closely related to the question of whether an indecomposable module M is determined by its composition factors is whether it is determined by the pair (top M, soc M) where top $M = M/\mathfrak{r}M$, or by the pair $(P_0(M), P_1(M))$ when $P_1(M) \to P_0(M) \to M \to 0$ is a minimal projective presentation of M. Again not lying on a short cycle plays a crucial role.

We first investigate when M is determined by its top and socle, that is by the pair of semisimple modules $(M/\mathfrak{r}M, \text{soc } M)$. We start out with M being directing, and use Theorem 3.4 to generalize to the case when M does not lie on a short cycle.

Proposition 4.1 *Let M and N be indecomposable modules over an artin algebra Λ, where M is directing. Assume that each simple Λ-module is a summand of $(M/\mathfrak{r}M) \coprod \text{soc } M$ and that $M/\mathfrak{r}M \simeq N/\mathfrak{r}N$ and $\text{soc } M \simeq \text{soc } N$.*
Then we have that $M \simeq N$.

We need the following lemmas, where M and N satisfy the assumptions of the proposition.

Lemma 4.2
(a) $\langle P(S), M \rangle \geq \langle P(S), N \rangle$ *for all simple modules S which are summands of soc M, where $P(S)$ denotes the projective cover of S. We have equality if $\text{Ext}^1_\Lambda(\text{soc } M, N) = 0$.*
(b) $\langle M, I(S) \rangle \geq \langle N, I(S) \rangle$ *for all simple modules S which are summands of soc M where $I(S)$ denotes the injective envelope of S. We have equality if $\text{Ext}^1_\Lambda(\text{soc } M, N) = 0$.*
(c) $\langle M, I(T) \rangle \geq \langle N, I(T) \rangle$ *for all simple modules T which are summands of $M/\mathfrak{r}M$, where $I(T)$ is an injective envelope of T. We have equality if $\text{Ext}^1_\Lambda(N, M/\mathfrak{r}M) = 0$.*

Proof (a) We prove this by induction on $l(P(S))$, where S is in soc M. If $l(P(S)) = 1$, we have $P(S) \simeq S$, and $\langle P(S), M \rangle = \langle S, \text{soc } M \rangle = \langle S, \text{soc } N \rangle = \langle P(S), N \rangle$.
Assume now that $l(P(S)) > 1$. Since M is sincere directing and

$\mathrm{Hom}_\Lambda(S, M) \neq 0$, we have $\mathrm{pd}_\Lambda S \leq 1$ by Proposition 1.7. Hence $\mathfrak{r}P(S)$ is the sum of indecomposable projective modules $P(S')$, where S' is simple. Since we have a chain of nonzero morphisms $S' \to \mathfrak{r}P(S)/\mathfrak{r}^2 P(S) \to P(S)/\mathfrak{r}^2 P(S) \to S \to M$ between indecomposable modules and M is directing, it follows that S' is not a summand of $M/\mathfrak{r}M$. Hence S' is a summand of $\mathrm{soc}\, M$. The exact sequence $0 \to \mathfrak{r}P(S) \to P(S) \to S \to 0$ gives rise to the exact sequences $0 \to \mathrm{Hom}_\Lambda(S, M) \to \mathrm{Hom}_\Lambda(P(S), M) \to \mathrm{Hom}_\Lambda(\mathfrak{r}P(S), M) \to \mathrm{Ext}^1_\Lambda(S, M) \to 0$ and $0 \to \mathrm{Hom}_\Lambda(S, N) \to \mathrm{Hom}_\Lambda(P(S), N) \to \mathrm{Hom}_\Lambda(\mathfrak{r}P(S), N) \to \mathrm{Ext}^1_\Lambda(S, N) \to 0$. If $\mathrm{Ext}^1_\Lambda(S, M) \neq 0$, we would have a nonsplit exact sequence $0 \to M \to E \to S \to 0$, and hence a cycle $S \to M \to E' \to S$ where E' is an indecomposable summand of E. Since M is directing, it follows that $\mathrm{Ext}^1_\Lambda(S, M) = 0$. Since $l(P(S')) < l(P(S))$, we have by the induction assumption that $\langle \mathfrak{r}P(S), M \rangle \geq \langle \mathfrak{r}P(S), N \rangle$. Hence we get $\langle P(S), M \rangle = \langle S, M \rangle + \langle \mathfrak{r}P(S), M \rangle = \langle S, N \rangle + \langle \mathfrak{r}P(S), M \rangle \geq \langle S, N \rangle + \langle \mathfrak{r}P(S), N \rangle \geq \langle P(S), N \rangle$.

If $\mathrm{Ext}^1_\Lambda(\mathrm{soc}\, M, N) = 0$, the induction assumption gives $\langle \mathfrak{r}P(S), M \rangle = \langle \mathfrak{r}P(S), N \rangle$, so that we get $\langle P(S), M \rangle = \langle S, M \rangle + \langle \mathfrak{r}P(S), M \rangle = \langle S, N \rangle + \langle \mathfrak{r}P(S), N \rangle = \langle P(S), N \rangle$.

(b) This follows directly from (a) by using Lemma 1.4(c) which implies that $\langle P(S), X \rangle = \langle X, I(S) \rangle$ for each X in $\mathrm{mod}\,\Lambda$ and each simple Λ-module S.

(c) This follows from (a) by duality. $\qquad\square$

Lemma 4.3 *We have* $\mathrm{Ext}^1_\Lambda(\mathrm{soc}\, M, N) = 0$ *or* $\mathrm{Ext}^1_\Lambda(N, M/\mathfrak{r}M) = 0$.

Proof Assume that $\mathrm{Ext}^1_\Lambda(S, N) \neq 0$ and $\mathrm{Ext}^1_\Lambda(N, T) \neq 0$ where S is a summand of $\mathrm{soc}\, M$ and T is a summand of $M/\mathfrak{r}M$. Then we would have a chain of nonzero nonisomorphisms $M \to T \to E \to N \to F \to S \to M$ between indecomposable modules, contradicting the fact that M is directing. $\qquad\square$

We are now ready to prove Proposition 4.1. By Lemma 4.3 we have that either $\mathrm{Ext}^1_\Lambda(\mathrm{soc}\, M, N) = 0$ or $\mathrm{Ext}^1_\Lambda(N, M/\mathfrak{r}M) = 0$. If $\mathrm{Ext}^1_\Lambda(N, M/\mathfrak{r}M) = 0$, then $\mathrm{Ext}^1_{\Lambda^\mathrm{op}}(D(M/\mathfrak{r}M), DN) = \mathrm{Ext}^1_{\Lambda^\mathrm{op}}(\mathrm{soc}\, DM, DN) = 0$. Further DM and DN satisfy the hypothesis of Proposition 4.1 for the artin algebra Λ^op. Hence it is enough to consider the case $\mathrm{Ext}^1_\Lambda(\mathrm{soc}\, M, N) = 0$. We will then prove that $\mathrm{Hom}_\Lambda(N, M) \neq 0$ and $\mathrm{Hom}_\Lambda(M, N) \neq 0$, giving our desired result. Since M is directing and sincere we have by Proposition 1.7

that $\mathrm{id}_\Lambda M \le 1$. Let

(*) $0 \to M \to I_0 \to I_1 \to 0$

be a minimal injective resolution. Since M is directing $\mathrm{Ext}^1_\Lambda(X, M) = 0$ for all submodules X of M, especially for $X = \mathrm{soc}\, M$. Consequently I_1 is in $\mathrm{add}\, I(M/\mathfrak{r}M)$ since all semisimple Λ-modules are in $\mathrm{add}(\mathrm{soc}\, M \coprod(M/\mathfrak{r}M))$ where $I(M/\mathfrak{r}M)$ denotes the injective envelope of $M/\mathfrak{r}M$.

Applying $\mathrm{Hom}_\Lambda(M, \ )$ and $\mathrm{Hom}_\Lambda(N, \ )$ to the exact sequence (*) gives the exact sequence $0 \to \mathrm{Hom}_\Lambda(M, M) \to \mathrm{Hom}_\Lambda(M, I_0) \to \mathrm{Hom}_\Lambda(M, I_1) \to 0$ since $\mathrm{Ext}^1_\Lambda(M, M) = 0$ and the exact sequence $0 \to \mathrm{Hom}_\Lambda(N, M) \to \mathrm{Hom}_\Lambda(N, I_0) \to \mathrm{Hom}_\Lambda(N, I_1)$. This gives that $\langle N, M \rangle \ge \langle N, I_0 \rangle - \langle N, I_1 \rangle \ge \langle N, I_0 \rangle - \langle M, I_1 \rangle = \langle M, I_0 \rangle - \langle M, I_1 \rangle = \langle M, M \rangle > 0$ where the second inequality follows by Lemma 4.2 (c) and the first equality follows since $\langle N, I_0 \rangle = \langle M, I_0 \rangle$ by Lemma 4.2 (b) using that we have assumed $\mathrm{Ext}^1_\Lambda(\mathrm{soc}\, M, N) = 0$. Hence $\mathrm{Hom}_\Lambda(N, M) \ne 0$. But then $\mathrm{Ext}^1_\Lambda(N, Y) = 0$ for all quotients Y of M since otherwise we get a path $M \to Y \to E \to N \to M$ in $\mathrm{ind}\,\Lambda$. In particular $\mathrm{Ext}^1_\Lambda(N, M/\mathfrak{r}M) = 0$. Using the duality we then get that $\mathrm{Ext}^1_\Lambda(\mathrm{soc}\, DM, DN) = 0$. Hence by our previous argument we get that $\mathrm{Hom}_\Lambda(DN, DM) \ne 0$ showing that $\mathrm{Hom}_\Lambda(M, N) \ne 0$. This then gives $M \simeq N$ completing the proof of the proposition. $\square$

Let P be a projective module over an artin algebra Λ, and let $\Gamma = \mathrm{End}_\Lambda(P)^{\mathrm{op}}$. In order to investigate $\mathrm{mod}\,\Gamma$ it will be useful to view $\mathrm{mod}\,\Gamma$ as an appropriate subcategory of $\mathrm{mod}\,\Lambda$. We have already shown in Chapter II that if $\mathrm{mod}\,P$ denotes the full subcategory of $\mathrm{mod}\,\Lambda$ whose objects are the X in $\mathrm{mod}\,\Lambda$ such that the first two terms $P_0(X)$ and $P_1(X)$ in a minimal projective presentation of X are in $\mathrm{add}\,P$, then $e_P|_{\mathrm{mod}\,P}: \mathrm{mod}\,P \to \mathrm{mod}\,\Gamma$ is an equivalence of categories. Here it is useful to deal with another subcategory of $\mathrm{mod}\,\Lambda$. Denote by $ts(P)$ the full subcategory of $\mathrm{mod}\,\Lambda$ whose objects are the M in $\mathrm{mod}\,\Lambda$ with $M/\mathfrak{r}M$ and $\mathrm{soc}\, M$ in $\mathrm{add}\, P/\mathfrak{r}P$. The following result will allow us to investigate questions about $\mathrm{mod}\,\Gamma$ by studying $ts(P)$.

Proposition 4.4 *Let P be a projective module over an artin algebra Λ, let $\Gamma = \mathrm{End}_\Lambda(P)^{\mathrm{op}}$ and $ts(P)$ be as defined above. Then the evaluation functor $e_P = \mathrm{Hom}_\Lambda(P, \ ): \mathrm{mod}\,\Lambda \to \mathrm{mod}\,\Gamma$ induces an equivalence of categories from $ts(P)$ to $\mathrm{mod}\,\Gamma$.*

Proof We know that $e_P|_{\mathrm{mod}\,P}: \mathrm{mod}\,P \to \mathrm{mod}\,\Gamma$ is an equivalence of categories by II Proposition 2.5. Let $f: X \to Y$ be a morphism in $ts(P)$. Since by assumption there is an epimorphism $h: Q \to X$ with Q in add P, it follows that if $f \neq 0$, then $e_P(f) \neq 0$. Hence $e_P|_{ts(P)}: ts(P) \to \text{mod}\,\Gamma$ is faithful.

For X in $ts(P)$ consider the exact sequence $0 \to \Omega X \to P(X) \overset{u}{\to} X \to 0$ where $u: P(X) \to X$ is a projective cover, and define $\tilde{X} = P(X)/\tau_P(\Omega X)$, where $\tau_P(\Omega X)$ denotes the submodule of ΩX generated by all images of morphisms from P to ΩX. Then $\tilde{X}$ is in mod P and we have an exact sequence $0 \to \Omega X/\tau_P(\Omega X) \to \tilde{X} \to X \to 0$. Since soc X is in add$(P/\mathfrak{r}P)$, we have that $\Omega X/\tau_P(\Omega X)$ is the largest submodule of $\tilde{X}$ where no simple composition factor is a summand of $P/\mathfrak{r}P$. We have an induced isomorphism $e_P(\tilde{X}) \to e_P(X)$ since $\text{Hom}_\Lambda(P, \Omega X/\tau_P(\Omega X)) = 0$.

Let now X and Y be in $ts(P)$ and let $\sigma: e_P(X) \to e_P(Y)$ be a morphism in mod Γ. Since $e_P|_{\text{mod}\,P}: \text{mod}\,P \to \text{mod}\,\Gamma$ is an equivalence, there is a morphism $\omega: \tilde{X} \to \tilde{Y}$ such that the diagram

$$
\begin{array}{ccc}
e_P(\tilde{X}) & \overset{\sim}{\to} & e_P(X) \\
\downarrow {\scriptstyle e_P(\omega)} & & \downarrow {\scriptstyle \sigma} \\
e_P(\tilde{Y}) & \overset{\sim}{\to} & e_P(Y)
\end{array}
$$

commutes. Using that $\Omega Y/\tau_P(\Omega Y)$ is the largest submodule of $\tilde{Y}$ where no simple composition factor is a summand of $P/\mathfrak{r}P$ and consequently $\text{Hom}_\Lambda(P, \Omega Y/\tau_P(\Omega Y)) = 0$, we get a commutative diagram

$$
\begin{array}{ccccccccc}
0 & \to & \Omega X/\tau_P(\Omega X) & \to & \tilde{X} & \to & X & \to & 0 \\
 & & \downarrow & & \downarrow {\scriptstyle \omega} & & \downarrow {\scriptstyle v} & & \\
0 & \to & \Omega Y/\tau_P(\Omega Y) & \to & \tilde{Y} & \to & Y & \to & 0.
\end{array}
$$

It then follows that $\sigma = e_P(v)$, so that $e_P|_{ts(P)}: ts(P) \to \text{mod}\,\Gamma$ is full.

Let now C be in mod Γ and let Z be in mod P with $e_P(Z) \simeq C$. Let K be the largest submodule of Z with $\text{Hom}_\Lambda(P, K) = 0$. Then soc$(Z/K)$ is in add$(P/\mathfrak{r}P)$, so that Z/K is in $ts(P)$ and $e_P(Z) \simeq e_P(Z/K)$. This shows that $e_P|_{ts(P)}: ts(P) \to \text{mod}\,\Gamma$ is dense, and finishes the proof that we have an equivalence. □

We shall need the following additional information about this equivalence.

Proposition 4.5 *Let P be a projective module over an artin algebra Λ, and let $\Gamma = \text{End}_\Lambda(P)^{\text{op}}$. Let M be in $ts(P)$.*

(a) *M is a simple Λ-module if and only if $e_P(M)$ is a simple Γ-module.*

(b) $e_P(\operatorname{soc} M) \simeq \operatorname{soc} e_P(M)$.

(c) $e_P(M/\mathfrak{r}M) \simeq e_P(M)/\mathfrak{r}_\Gamma e_P(M)$.

Proof (a) Let $f: X \to Y$ be a morphism in $ts(P)$. Since $\operatorname{soc} X$ is in $\operatorname{add}(P/\mathfrak{r}P)$, we have that $\operatorname{soc}(\operatorname{Ker} f)$ is in $\operatorname{add}(P/\mathfrak{r}P)$ and therefore $\operatorname{Hom}_\Lambda(P, \operatorname{Ker} f) \neq 0$ if $\operatorname{Ker} f \neq 0$. Further, $\operatorname{Hom}_\Lambda(P, \operatorname{Coker} f) \neq 0$ if $\operatorname{Coker} f \neq 0$ since $(\operatorname{Coker} f)/(\mathfrak{r} \operatorname{Coker} f)$ is in $\operatorname{add}(P/\mathfrak{r}P)$. Hence f is a monomorphism in $\operatorname{mod} \Lambda$ if and only if $e_P(f)$ is a monomorphism in $\operatorname{mod} \Gamma$, and f is an epimorphism in $\operatorname{mod} \Lambda$ if and only if $e_P(f)$ is an epimorphism in $\operatorname{mod} \Gamma$. It follows that a module M in $ts(P)$ is simple in $\operatorname{mod} \Lambda$ if and only if $e_P(M)$ is simple in $\operatorname{mod} \Gamma$.

(b) If we have a monomorphism $X \to e_P(M)$, then $X \simeq e_P(N)$ for some N in $ts(P)$ and we have a monomorphism $N \to M$ inducing the monomorphism $e_P(N) \to e_P(M)$. Since by (a) $e_P(N)$ is a semisimple Γ-module if and only if N is a semisimple Λ-module and $\operatorname{soc} M$ is in $ts(P)$ we get $e_P(\operatorname{soc} M) \simeq \operatorname{soc} e_P(M)$.

(c) If we have an epimorphism $e_P(M) \to Y$, then $Y \simeq e_P(N)$ for some N in $ts(P)$ and we have an epimorphism $M \to N$ inducing the epimorphism $e_P(M) \to e_P(N)$. It follows from (a) that $e_P(N)$ is a semisimple Γ-module if and only if N is a semisimple Λ-module. Since $M/\mathfrak{r}M$ is in $ts(P)$, we get $e_P(M/\mathfrak{r}_\Lambda M) \simeq e_P(M)/\mathfrak{r}_\Gamma e_P(M)$ by using that $M/\mathfrak{r}_\Lambda M$ is the largest semisimple factor module of M and $e_P(M)/\mathfrak{r}_\Gamma e_P(M)$ is the largest semisimple factor module of $e_P(M)$. $\square$

We now prove the main result of this section.

Theorem 4.6 *Let M and N be indecomposable modules over an artin algebra Λ with $M/\mathfrak{r}M \simeq N/\mathfrak{r}N$ and $\operatorname{soc} M \simeq \operatorname{soc} N$. If M does not lie on a short cycle, then $M \simeq N$.*

Proof Let P be the projective cover of $(M/\mathfrak{r}M) \coprod \operatorname{soc} M$, and consider the functor $e_P: \operatorname{mod} \Lambda \to \operatorname{mod} \Gamma$ where as usual $\Gamma = \operatorname{End}_\Lambda(P)^{\mathrm{op}}$. Then M and N are in the subcategory $ts(P)$. It follows from the equivalence between $ts(P)$ and $\operatorname{mod} \Gamma$ that since M does not lie on a short cycle in $\operatorname{mod} \Lambda$, then $e_P(M)$ does not lie on a short cycle in $\operatorname{mod} \Gamma$.

The indecomposable projective Γ-modules are those of the form $e_P(Q)$ for an indecomposable summand Q of P. Since $\operatorname{soc} e_P(M) \simeq e_P(\operatorname{soc} M)$ and $e_P(M/\mathfrak{r}M) \simeq e_P(M)/\mathfrak{r}_\Gamma(e_P(M))$, it follows from the choice of P that $\operatorname{Hom}_\Gamma(e_P(Q), \operatorname{soc}(e_P(M)) \coprod e_P(M)/\mathfrak{r}_\Gamma(e_P(M))) \neq 0$ for each indecomposable projective Γ-module $e_P(Q)$. In particular $e_P(M)$ is a sincere indecomposable Γ-module which does not lie on a short cycle, and is hence directing by Theorem 3.4. Since $\operatorname{soc} e_P(N) = e_P(\operatorname{soc} N) \simeq e_P(\operatorname{soc} M) = \operatorname{soc} e_P(M)$ and $e_P(M)/\mathfrak{r}_\Gamma e_P(N) \simeq e_P(N/\mathfrak{r}N) \simeq e_P(M/\mathfrak{r}M) \simeq e_P(M)/\mathfrak{r}_\Gamma e_P(M)$, it follows from Proposition 4.1 that $e_P(M) \simeq e_P(N)$. Hence we get $M \simeq N$ since $e_P|_{ts(P)}: ts(P) \to \operatorname{mod}\Gamma$ is an equivalence and both M and N are in $ts(P)$. $\qquad\square$

We now show that short chains also come up in investigating when an indecomposable module is determined by the first two terms in a minimal projective presentation.

Theorem 4.7 *Let M and N be indecomposable Λ-modules and $P_1(M) \to P_0(M) \to M \to 0$ and $P_1(N) \to P_0(N) \to N \to 0$ their minimal projective presentations. If $P_0(M) \simeq P_0(N)$ and $P_1(N) \simeq P_1(M)$, and M and N are not the start of any short chain, then $M \simeq N$.*

Proof Let M and N satisfy the hypothesis of the theorem. We want to prove that $\langle M, X \rangle = \langle N, X \rangle$ for all indecomposable Λ-modules X. Then it follows that $M \simeq N$ by VI Theorem 4.2.

For an indecomposable module X we have $\langle M, X \rangle - \langle X, D\operatorname{Tr} M \rangle = \langle P_0(M), X \rangle - \langle P_1(M), X \rangle = \langle N, X \rangle - \langle X, D\operatorname{Tr} N \rangle$, by IV Corollary 4.3. If $\langle M, X \rangle \neq 0$, then $\langle X, D\operatorname{Tr} M \rangle = 0$ since M is not the start of any short chain. Then we must have $\langle N, X \rangle \neq 0$, and hence $\langle X, D\operatorname{Tr} N \rangle = 0$ since N is not the start of any short chain. Hence we get $\langle M, X \rangle = \langle N, X \rangle$ when $\langle M, X \rangle \neq 0$. In the same way we get that if $\langle N, X \rangle \neq 0$ then $\langle M, X \rangle = \langle N, X \rangle$. Hence we have $\langle M, X \rangle = \langle N, X \rangle$ for all X, and this finishes the proof. $\qquad\square$

We note that for two indecomposable modules M and N, where M is directing, it may happen that $P_0(M) \simeq P_0(N)$ and $P_1(M) \simeq P_1(N)$, but M and N are not isomorphic.

Example Let $\Lambda = k(\Gamma, \rho)$ where Γ is the quiver $\underset{1}{\cdot} \longrightarrow \underset{2}{\cdot} \circlearrowleft \alpha$ and ρ the relation $\{\alpha^2\}$. The simple module S_1 is injective and is clearly directing. We let $M = S_1$ and let N be the indecomposable representation $k \xrightarrow{1} k \circlearrowleft 0$. Then $P_0(M) \simeq P_1 \simeq P_0(N)$ and $P_1(M) \simeq P_2 \simeq P_1(N)$.

We end this section by pointing out that the conditions for an indecomposable Λ-module M to be determined by its composition factors, by its top and socle or by the first two terms in a minimal projective resolution are all independent.

Let $\Lambda = k[X]/(X^2)$ where k is a field, and let S be the simple Λ-module. Then S and Λ are the only indecomposable Λ-modules. Λ is determined by its composition factors and by the first two terms in a minimal projective resolution, but not by its top and socle.

If Λ is Nakayama with admissible sequence $(2,2)$, and M is indecomposable of length 2, then M is not determined by its composition factors. But M is determined by the first two terms in a minimal projective resolution and by its top and socle.

Let Λ be Nakayama with admissible sequence $(4,4)$ and let S_1 and S_2 be the simple Λ-modules. Let M be the indecomposable Λ-module of length 3 with $M/\mathfrak{r}M \simeq S_1$. Then M has the same top and socle as S_1, and the same first two terms in a minimal projective resolution and is hence not determined by these invariants. But M is determined by its composition factors.

Exercises

1. Let Λ be an artin algebra and P a projective Λ-module. Denote by comod P the full subcategory of mod Λ whose objects are the C in mod Λ such that if $0 \to C \to I_0 \to I_1$ is a minimal injective resolution we have that soc I_0 and soc I_1 are in add$(P/\mathfrak{r}P)$.

Show that $e_P|_{\text{comod}\,P} : \text{comod}\,P \to \text{mod}\,\text{End}_\Lambda(P)^{\text{op}}$ is an equivalence of categories.

2. Show that there is a functor $F : \text{mod}\,P \to ts(P)$ sending X to $\widetilde{X}$, as defined in Section 4, which is an equivalence of categories.

3. Let Λ be an artin algebra having a sincere module which is not the middle of a short chain.

(a)   Prove that there are no oriented cycles in the quiver of Λ.

(b)   Prove that gl.dim $\Lambda \leq 2$.

(c)   Prove that if X is an indecomposable Λ-module with id$_\Lambda X = 2$, then pd$_\Lambda X \leq 1$.

4. Let Λ be a basic artin algebra and M a Λ-module which is not the middle of a short chain. Write $\Lambda = P \coprod Q$, where the simple summands of $P/\mathfrak{r}P$ are exactly the composition factors of M. Denote by $\tau_Q(\Lambda)$ the ideal of Λ generated by all images of all morphisms from Q to Λ. Prove that $\operatorname{End}_\Lambda(P)^{\mathrm{op}} \simeq \Lambda/\tau_Q(\Lambda)$ and $\tau_Q(\Lambda) = \operatorname{ann} M$.

5. Let k be a field and Γ the quiver

with relation $\rho = \{\gamma\alpha - \delta\beta\}$.

Let P_1 be the indecomposable projective injective module over the algebra $\Lambda = k(\Gamma, \rho)$.
Prove that $\mathfrak{r}P_1$ does not lie on a short cycle.

6. Let Λ be any artin algebra, A and C indecomposable Λ-modules and $A \xrightarrow{f} B \xrightarrow{g} C$ a sequence of irreducible morphisms such that $gf = 0$. Prove that $0 \to A \xrightarrow{f} B \xrightarrow{g} C \to 0$ is an almost split sequence.

7. Let (Γ, ρ) be a quiver with relations, k an algebraically closed field and $\Lambda = k(\Gamma, \rho)$. Let (V, f) be an indecomposable representation of (Γ, ρ) which is not the middle of a short chain. Prove that for $f = (f_\alpha)_{\alpha \in \Gamma_1}$ we have that each f_α is either a monomorphism or an epimorphism.

Notes

That indecomposable modules lying on cycles, but only on suitably restricted types of cycles, could have properties similar to those for directing modules was demonstrated in [AuR8] with the introduction of short chains. It was shown that indecomposable modules not in the middle of a short chain are determined up to isomorphism by their composition factors, generalizing earlier results about directing modules given in [HapR], [Hap1]. It was shown that indecomposable modules which are not the start of a short chain are determined up to isomorphism by the two projective modules occurring in the minimal projective presentation of the module. That it is not sufficient that one of the modules is directing was observed in [Bak].

The theory of short cycles initiated in [ReSS1] gives another way of viewing indecomposable modules which are not the middle of short chains. It was shown in [ReSS1] that if an indecomposable module is not the middle of a short chain, it is not on a short cycle, and that if two indecomposable modules have the same composition factors and one is not on a short cycle, then they are isomorphic. This generalized the result mentioned from [AuR8], and a similar result on directing modules in [Rin3]. That an indecomposable module is not the middle of a short chain if and only if it is not on a short cycle is given in [HapL]. The proof given here is somewhat different from the original one.

The proof that indecomposable sincere modules not lying on short cycles are directing is taken from [HapRS]. This is applied in [HapRS] to showing that indecomposable modules not lying on a short cycle are determined by their top and socle, by reducing to the case of directing modules ([BonS] and [BakS]).

Further results along these lines can be found in [ReSS2], [HapL], [Liu2] and [Sk1].

X

Stable equivalence

The category $\underline{\mathrm{mod}}\ \Lambda$ was introduced in Chapter IV in order to be able to define the functors $\mathrm{Tr} : \underline{\mathrm{mod}}\ \Lambda \to \underline{\mathrm{mod}}\ \Lambda^{\mathrm{op}}$ and $\Omega : \underline{\mathrm{mod}}\ \Lambda \to \underline{\mathrm{mod}}\ \Lambda$. Another interesting aspect of the category $\underline{\mathrm{mod}}\ \Lambda$ is that $\underline{\mathrm{mod}}\ \Lambda$ and $\underline{\mathrm{mod}}\ \Lambda'$ can be equivalent categories for seemingly rather different artin algebras Λ and Λ'. This chapter is devoted to giving various illustrations of this phenomenon. We are particularly interested in the situations where one of the algebras is either hereditary or Nakayama.

1 Stable equivalence and almost split sequences

We say that two artin algebras Λ and Λ' are **stably equivalent** if there is an equivalence $F : \underline{\mathrm{mod}}\ \Lambda \to \underline{\mathrm{mod}}\ \Lambda'$ between the associated module categories modulo projectives. Since we know by IV Proposition 1.9 that $D\,\mathrm{Tr} : \underline{\mathrm{mod}}\ \Lambda \to \overline{\mathrm{mod}}\ \Lambda$ is an equivalence of categories, it follows that Λ and Λ' are stably equivalent if and only if the categories $\overline{\mathrm{mod}}\ \Lambda$ and $\overline{\mathrm{mod}}\ \Lambda'$ of the module categories modulo injectives are equivalent. In this section we investigate properties which stably equivalent algebras have in common, including operations with which a stable equivalence $F : \underline{\mathrm{mod}}\ \Lambda \to \underline{\mathrm{mod}}\ \Lambda'$ commutes and module theoretic properties preserved by stable equivalence. A central role is played by the behavior of almost split sequences under stable equivalence.

We start with the following easy result, showing that information on when two algebras are stably equivalent is useful for classifying algebras of finite representation type. When $F : \underline{\mathrm{mod}}\ \Lambda \to \underline{\mathrm{mod}}\ \Lambda'$ is an equivalence, we also denote by F the induced correspondence between $\mathrm{mod}_{\mathscr{P}}\ \Lambda$ and $\mathrm{mod}_{\mathscr{P}}\ \Lambda'$, where as before $\mathrm{mod}_{\mathscr{P}}\ \Lambda$ denotes the full subcategory of $\mathrm{mod}\ \Lambda$ whose objects are the modules which have no nonzero projective summands.

335

Proposition 1.1 *Let Λ and Λ' be stably equivalent artin algebras. Then Λ is of finite representation type if and only if Λ' is of finite representation type.*

Proof Let $F : \underline{\mathrm{mod}}\ \Lambda \to \underline{\mathrm{mod}}\ \Lambda'$ be an equivalence. Since for M in $\mathrm{mod}_{\mathscr{P}}\ \Lambda$ we have that $\mathscr{P}(M, M) \subset \mathrm{rad}\ \mathrm{End}_{\Lambda}(M)$, it follows that $\mathrm{End}_{\Lambda}(M)$ is local if and only if $\mathrm{End}_{\Lambda'}(F(M))$ is local. Hence F gives a one to one correspondence between the nonisomorphic indecomposable nonprojective objects in $\mathrm{mod}\ \Lambda$ and $\mathrm{mod}\ \Lambda'$. Since Λ and Λ' have only a finite number of nonisomorphic indecomposable projective modules, it follows that Λ is of finite representation type if and only if Λ' is. $\quad\square$

We now study the relationship between almost split sequences, irreducible morphisms and AR-quivers for stably equivalent algebras. $D\,\mathrm{Tr}$ operates on objects in $\mathrm{mod}\,\Lambda$, and there is an induced operation from $\mathrm{mod}_{\mathscr{P}}\ \Lambda$ to $\mathrm{mod}_{\mathscr{P}}\ \Lambda$ by taking the nonprojective part $(D\,\mathrm{Tr}\,C)_{\mathscr{P}}$ of $D\,\mathrm{Tr}\,C$ for C in $\mathrm{mod}_{\mathscr{P}}\,\Lambda$. In connection with the study of almost split sequences we investigate when for an object C in $\mathrm{mod}_{\mathscr{P}}\,\Lambda$ we have $F(D\,\mathrm{Tr}_{\Lambda}\,C) \simeq D\,\mathrm{Tr}_{\Lambda'}\,F(C)$ in $\mathrm{mod}_{\mathscr{P}}\,\Lambda'$. We refer to this as F commuting with $D\,\mathrm{Tr}$. These results will also be applied to decide when F commutes with Ω for selfinjective algebras, that is when $F(\Omega C) \simeq \Omega F(C)$ in $\mathrm{mod}_{\mathscr{P}}\,\Lambda'$ for each object C in $\mathrm{mod}_{\mathscr{P}}\,\Lambda$.

We have the following connection between stable equivalence and the radical. For each f in $\mathrm{mod}\,\Lambda$ we denote by $\underline{f}$ the image of f in $\underline{\mathrm{mod}}\ \Lambda$.

Lemma 1.2 *Let $F : \underline{\mathrm{mod}}\ \Lambda \to \underline{\mathrm{mod}}\ \Lambda'$ be a stable equivalence between artin algebras and let X and Y be in $\mathrm{mod}_{\mathscr{P}}\,\Lambda$.*

(a) *Let $f : X \to Y$ be a morphism in $\mathrm{mod}\,\Lambda$ and let $f' : F(X) \to F(Y)$ in $\mathrm{mod}_{\mathscr{P}}\,\Lambda'$ be such that $F(\underline{f}) = \underline{f}'$. Then $f \in \mathrm{rad}_{\Lambda}(X, Y)$ if and only if $f' \in \mathrm{rad}_{\Lambda'}(F(X), F(Y))$.*

(b) *$F(\mathrm{rad}_{\Lambda}^{n}(X, Y) + \mathscr{P}(X, Y)) = \mathrm{rad}_{\Lambda'}^{n}(F(X), F(Y)) + \mathscr{P}(F(X), F(Y))$ for all $n \geq 1$.*

(c) *F induces an isomorphism $\mathrm{Irr}(X, Y) \simeq \mathrm{Irr}(F(X), F(Y))$ when X and Y are indecomposable in $\mathrm{mod}_{\mathscr{P}}\,\Lambda$.*

Proof (a) Assume that $f \notin \mathrm{rad}_{\Lambda}(X, Y)$. Then there is some Z in $\mathrm{mod}\,\Lambda$, which is a summand of X and is hence in $\mathrm{mod}_{\mathscr{P}}\,\Lambda$, such that $1_Z = hfg$ for some $g : Z \to X$ and $h : Y \to Z$. Then we get $1_{F(Z)} = h'f'g' + s$ with $s \in \mathscr{P}(F(X), F(Y)) \subset \mathrm{rad}_{\Lambda'}(F(X), F(Y))$, so that $f' \notin \mathrm{rad}_{\Lambda'}(F(X), F(Y))$.

It follows the same way that if $f' \notin \text{rad}_{\Lambda'}(F(X), F(Y))$, then $f \notin \text{rad}_\Lambda(X, Y)$.

(b) This is a direct consequence of (a) and the definition of rad_Λ^n.

(c) This follows from (a) and (b) using that $\mathscr{P}(X, Y) \subset \text{rad}_\Lambda^2(X, Y)$, $\mathscr{P}(F(X), F(Y)) \subset \text{rad}_{\Lambda'}^2(F(X), F(Y))$ and that F induces an isomorphism $\text{Hom}_\Lambda(X, Y)/\mathscr{P}(X, Y) \simeq \text{Hom}_{\Lambda'}(F(X), F(Y))/\mathscr{P}(F(X), F(Y))$. □

We now apply Lemma 1.2 to get the following.

Proposition 1.3 *Let* $F: \underline{\text{mod}} \ \Lambda \to \underline{\text{mod}} \ \Lambda'$ *be a stable equivalence between artin algebras. For a morphism* $f: X \to Y$ *in* $\text{mod}_{\mathscr{P}} \Lambda$ *with* X *or* Y *indecomposable, let* $f': F(X) \to F(Y)$ *be such that* $F(\underline{f}) = \underline{f}'$. *Then we have the following.*

(a) $f: X \to Y$ *is irreducible in* $\text{mod} \ \Lambda$ *if and only if* $f': F(X) \to F(Y)$ *is irreducible in* $\text{mod} \ \Lambda'$.

(b) *If* Y *is indecomposable in* $\text{mod} \ \Lambda$, *then the following are equivalent.*

   (i) *There is a morphism* $g: P \to Y$ *with* P *projective in* $\text{mod} \ \Lambda$ *such that* $(f, g): X \coprod P \to Y$ *is minimal right almost split.*

   (ii) *There is a morphism* $h: Q \to F(Y)$ *with* Q *projective in* $\text{mod} \ \Lambda'$ *such that* $(f', h): F(X) \coprod Q \to F(Y)$ *is minimal right almost split.*

(c) *If* X *is indecomposable in* $\text{mod} \ \Lambda$, *then the following are equivalent.*

   (i) *There is a morphism* $g: X \to P$ *with* P *projective such that* $\binom{f}{g}: X \to Y \coprod P$ *is minimal left almost split.*

   (ii) *There is a morphism* $h: F(X) \to Q$ *with* Q *projective in* $\text{mod} \ \Lambda'$ *such that* $\binom{f'}{h}: F(X) \to F(Y) \coprod Q$ *is minimal left almost split.*

Proof All the claims follow directly from Lemma 1.2 and VII Proposition 1.3. □

In view of Proposition 1.3 it is of interest to know how to construct the whole middle term of an almost split sequence when we know the nonprojective part. We have the following result in this direction.

Proposition 1.4 *Let* Y *be an indecomposable nonprojective module over an artin algebra* Λ, *and let* $f: X \to Y$ *in* $\text{mod}_{\mathscr{P}} \Lambda$ *be such that there is some morphism* $g: P \to Y$ *with* P *projective such that* $(f, g): X \coprod P \to Y$ *is minimal right almost split. Then we have the following.*

(a)   The composition morphism $P \to Y \to Y/\operatorname{Im} f$ is a projective cover.

(b)   If $h:Q \to Y$ is such that the composition $Q \to Y \to Y/\operatorname{Im} f$ is a projective cover, then $(f,h):X \coprod Q \to Y$ is minimal right almost split.

Proof   (a) Since (f,g) is an epimorphism, the composition $g':P \to Y \to Y/\operatorname{Im} f$ is an epimorphism. If g' is not a projective cover, there is a decomposition $P = P_1 \coprod P_2$ such that $g'|_{P_2} = 0$ and $g'|_{P_1}:P_1 \to Y/\operatorname{Im} f$ is a projective cover. Then we have $g(P_2) \subset \operatorname{Im} f$. Therefore $g|_{P_2}:P_2 \to Y$ factors through f, giving a contradiction to the right minimality of (f,g). This shows that $P_2 = 0$, and hence $g':P \to Y/\operatorname{Im} f$ is a projective cover.

(b) Since $(f,h):X \coprod Q \to Y$ is surjective, $g:P \to Y$ factors through (f,h), and hence also (f,g) factors through (f,h). This shows that (f,h) is right almost split, and it is minimal right almost split since $l(X \coprod P) = l(X \coprod Q)$. $\square$

For an artin algebra Λ with AR-quiver Γ_Λ we denote by Γ_Λ^s the full valued subtranslation quiver obtained by removing the projective vertices from Γ_Λ. When Λ is a selfinjective algebra, then Γ_Λ^s is clearly a stable translation quiver. It follows from Proposition 1.3 that if Λ and Λ' are stably equivalent algebras, then Γ_Λ^s and $\Gamma_{\Lambda'}^s$ are isomorphic as valued quivers. To investigate when they are isomorphic as translation quivers, we study what happens to almost split sequences under stable equivalence. For this we need the following.

Lemma 1.5 Let Λ be an artin algebra and $A \xrightarrow{f} B \xrightarrow{g} C$ a sequence of irreducible morphisms in $\operatorname{mod}_{\mathscr{P}} \Lambda$ with A and C indecomposable and $gf = 0$. Then we have the following.

(a) $A \simeq D\operatorname{Tr} C$.

(b)   There is a projective module Q and irreducible morphisms $g':Q \to C$ and $f':A \to Q$ and a morphism $\delta(f):A \to B$ with $\delta(f) \in \mathscr{P}(A,B)$ such that $0 \to A \xrightarrow{\binom{f+\delta(f)}{f'}} B \coprod Q \xrightarrow{(g,g')} C \to 0$ is almost split.

Proof   Let $g':B' \to C$ be such that $0 \to \operatorname{Ker}(g,g') \xrightarrow{\binom{\beta}{\beta'}} B \coprod B' \xrightarrow{(g,g')} C \to 0$ is an almost split sequence. Since $gf = 0$ there exist a projective module P and morphisms $f'':A \to P$ and $g'':P \to C$ such that $gf = g''f''$. Since (g,g') is surjective, there exists $\binom{h}{h'}:P \to B \coprod B'$ with $(g,g')\binom{h}{h'} = g''$. Hence we get $gf = (g,g')\binom{h}{h'}f''$. Consider now the morphisms $\delta(f) = -hf''$ and $f' = -h'f''$. We then have $(g,g')\binom{f+\delta(f)}{f'} = gf - ghf'' - g'h'f'' = 0$. Hence there exists a morphism $\alpha:A \to D\operatorname{Tr} C = \operatorname{Ker}(g,g')$ such that

$\binom{\beta}{\beta'}\alpha = \binom{f+\delta(f)}{f'}$. But $f + \delta(f)$ is irreducible since f is irreducible and $\delta(f) \in \mathscr{P}(A, B) \subset \operatorname{rad}_\Lambda^2(A, B)$. Hence α is a split monomorphism and consequently an isomorphism. It follows that $0 \to A \xrightarrow{\binom{f+\delta(f)}{f'}} B \coprod B' \xrightarrow{(g,g')} C \to 0$ is an almost split sequence. Since $f' = h'f'' : A \to B'$ is irreducible and $f'' : A \to P$ cannot be a split monomorphism because A is not projective, it follows that $h' : P \to B'$ must be a split epimorphism. Hence B' is a projective module. $\qquad\square$

As an immediate consequence of this lemma we get the following.

Proposition 1.6 *Let* $F : \underline{\operatorname{mod}} \Lambda \to \underline{\operatorname{mod}} \Lambda'$ *be a stable equivalence between artin algebras* Λ *and* Λ'. *Let* $0 \to A \xrightarrow{\binom{f}{h}} B \coprod P \xrightarrow{(g,t)} C \to 0$ *be an almost split sequence in* $\operatorname{mod} \Lambda$ *where A, B and C are in* $\operatorname{mod}_{\mathscr{P}} \Lambda$, *B is not zero and P is projective.*

Then for any morphism $\underline{g'} : F(B) \to F(C)$ *with* $F(\underline{g}) = \underline{g'}$ *there is an almost split sequence* $0 \to F(A) \xrightarrow{\binom{f'}{u}} F(B) \coprod P' \xrightarrow{(g',v)} F(C) \to 0$ *in* $\operatorname{mod} \Lambda'$ *where* P' *is projective and* $F(\underline{f}) = \underline{f'}$.

Proof Choose morphisms $g' : F(B) \to F(C)$ and $f'' : F(A) \to F(B)$ such that $F(\underline{g}) = \underline{g'}$ and $F(\underline{f}) = \underline{f'}$. Then $\underline{g'}$ and $\underline{f''}$ are irreducible by Lemma 1.2, and we have $\underline{g'f''} = 0$ since $\underline{gf} = 0$. Since $F(A)$ and $F(C)$ are indecomposable and in $\operatorname{mod}_{\mathscr{P}} \Lambda'$, there is by Lemma 1.5 an almost split sequence $0 \to F(A) \xrightarrow{\binom{f'}{u}} F(B) \coprod P' \xrightarrow{(g',v)} F(C) \to 0$ where P' is projective and $\underline{f'} = \underline{f''}$, and hence $F(\underline{f}) = \underline{f'}$. $\qquad\square$

Sometimes stable equivalences are induced by exact functors $F :$ $\operatorname{mod} \Lambda \to \operatorname{mod} \Lambda'$ taking projective Λ-modules to projective Λ'-modules. This occurs for example when k is a field of characteristic p and G is a finite group of order divisible by p but not p^2 and N is the normalizer in G of a subgroup of order p. Then the restriction functor $\operatorname{mod} kG \to \operatorname{mod} kN$ induces a stable equivalence. When we have such an exact functor $F : \operatorname{mod} \Lambda \to \operatorname{mod} \Lambda'$, it follows from our results that $0 \to FA \to FB \to FC \to 0$ is almost split in $\operatorname{mod} \Lambda'$ when $0 \to A \to B \to C \to 0$ is almost split in $\operatorname{mod} \Lambda$. But even if we do not have, or do not know if we have, an exact functor inducing the stable equivalence, there may be general procedures for constructing almost split sequences in $\operatorname{\underline{mod}} \Lambda'$ from almost split sequences in $\operatorname{\underline{mod}} \Lambda$ when we have an equivalence $F: \operatorname{\underline{mod}} \Lambda \to \operatorname{\underline{mod}} \Lambda'$, as we now illustrate.

For example if Λ is selfinjective we know that $\Omega: \operatorname{\underline{mod}} \Lambda \to \operatorname{\underline{mod}} \Lambda$ is an equivalence. When $0 \to A \to B \to C \to 0$ is exact, consider the exact commutative diagram

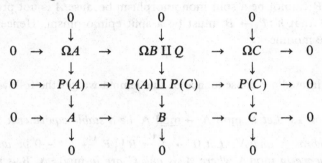

where $P(A) \to A$ and $P(C) \to C$ are projective covers. If $0 \to A \to B \to C \to 0$ is almost split it follows that $0 \to \Omega A \to \Omega B \coprod Q \to \Omega C \to 0$ is almost split, using Proposition 1.3 and the fact that ΩA and ΩC are both indecomposable.

As a consequence of our results we see that the correspondence F between the objects in $\operatorname{mod}_{\mathscr{P}} \Lambda$ and $\operatorname{mod}_{\mathscr{P}} \Lambda'$ induced by a stable equivalence $F: \operatorname{\underline{mod}} \Lambda \to \operatorname{\underline{mod}} \Lambda'$ 'usually' commutes with $D\operatorname{Tr}$.

Corollary 1.7 *Let $F: \operatorname{\underline{mod}} \Lambda \to \operatorname{\underline{mod}} \Lambda'$ be a stable equivalence between artin algebras, and let C be indecomposable in $\operatorname{mod}_{\mathscr{P}} \Lambda$.*

(a) *Assume that in the almost split sequence $0 \to A \to B \to C \to 0$ the module B is not projective. Then the following hold.*

    (i) *$D\operatorname{Tr}_\Lambda C$ is projective if and only if $D\operatorname{Tr}_{\Lambda'} F(C)$ is projective.*

    (ii) *If $D\operatorname{Tr}_\Lambda C$ is not projective then there is an isomorphism of Λ'-modules $F(D\operatorname{Tr}_\Lambda C) \simeq D\operatorname{Tr}_{\Lambda'} F(C)$.*

(b) *If $C \not\simeq \operatorname{Tr} DS$ where S is a simple module which is not a composition factor of $\mathfrak{r}I / \operatorname{soc} I$ for any injective module I, then $F(D\operatorname{Tr}_\Lambda C) \simeq D\operatorname{Tr}_{\Lambda'} F(C)$ in $\operatorname{mod}_{\mathscr{P}} \Lambda'$.*

Proof (a) Let $0 \to A \to B \to C \to 0$ be an almost split sequence where B is not projective. If A is not projective, the claim follows from Lemma 1.5. If A is projective, consider the almost split sequence $0 \to A' \to B' \to F(C) \to 0$ in $\operatorname{mod} \Lambda'$. By Proposition 1.3 B' is not projective since B is not projective, and if $A' \simeq D\operatorname{Tr}_{\Lambda'} F(C)$ was not projective, $A = D\operatorname{Tr}_\Lambda C$ would be nonprojective by Lemma 1.5. Hence A' is projective.

(b) By V Theorem 3.3(b) the middle term B of the almost split sequence $0 \to A \to B \to C \to 0$ is projective if and only if A is a simple module which is not a composition factor of $\mathfrak{r}I/\operatorname{soc}I$ for any indecomposable injective module I. Hence the claim follows from (a). $\qquad\square$

The following example shows that the correspondence F induced by a stable equivalence does not always commute with $D\operatorname{Tr}$.

Example The algebras $\Lambda = k[X]/(X^2)$ and $\Lambda' = \left(\begin{smallmatrix} k & 0 \\ k & k \end{smallmatrix}\right)$ are stably equivalent since in both cases there is only one indecomposable nonprojective module, and this module has endomorphism ring k. Then $0 \to \Lambda/\mathfrak{r} \to \Lambda \to \Lambda/\mathfrak{r} \to 0$ is the only almost split sequence in $\operatorname{mod}\Lambda$ and $0 \to \left(\begin{smallmatrix} 0 \\ k \end{smallmatrix}\right) \to \left(\begin{smallmatrix} k \\ k \end{smallmatrix}\right) \to \left(\begin{smallmatrix} k \\ k \end{smallmatrix}\right)/\left(\begin{smallmatrix} 0 \\ k \end{smallmatrix}\right) \to 0$ is the only one in $\operatorname{mod}\Lambda'$. In the first case the left hand term is not projective, whereas in the second case it is.

For selfinjective algebras we can describe when there are no almost split sequences with projective middle term.

Proposition 1.8 *Let Λ be an indecomposable nonsimple selfinjective artin algebra. Then the following are equivalent.*

(a) *There is an almost split sequence with projective middle term.*

(b) *All almost split sequences have projective middle terms.*

(c) *Λ is a Nakayama algebra of Loewy length 2.*

(d) *Λ is of Loewy length 2.*

Proof If Λ has Loewy length 2, then we know from V Section 3 that there are almost split sequences with projective middle term.

Assume that $0 \to A \to B \to C \to 0$ is almost split with B projective. Since Λ is selfinjective, B is also injective, and hence A and C must be simple by V Theorem 3.3. Since $A \simeq \Omega C$, we then get $l(B) = 2$. Since $\Omega \colon \underline{\operatorname{mod}}\,\Lambda \to \underline{\operatorname{mod}}\,\Lambda$ is an equivalence, there is an almost split sequence $0 \to K \to Q \to A \to 0$ with Q projective, and hence $l(Q) = 2$. Continuing the procedure we get an algebra summand of Λ where all indecomposable projective modules have length 2. Since Λ is indecomposable, all indecomposable projective Λ-modules have length 2. Hence Λ is a Nakayama algebra of Loewy length 2, and all almost split sequences have projective middle term. We now see that all the conditions are equivalent. □

As a consequence of Corollary 1.7 and Proposition 1.8 we see what happens to AR-quivers under stable equivalence.

Corollary 1.9 *Let $F : \underline{\mathrm{mod}}\,\Lambda \to \underline{\mathrm{mod}}\,\Lambda'$ be a stable equivalence between artin algebras.*

(a) *Let Γ_Λ be the AR-quiver of Λ and let $\mathscr{C}$ be a component of Γ^s_Λ where all simple modules are composition factors of $\tau D(\Lambda)/\operatorname{soc} D(\Lambda)$. Then F induces a translation quiver isomorphism between $\mathscr{C}$ and a component $F(\mathscr{C})$ of $\Gamma^s_{\Lambda'}$ where $\Gamma_{\Lambda'}$ is the AR-quiver of Λ'.*

(b) *If Λ and Λ' are selfinjective with no block of Loewy length 2 and Γ_Λ and $\Gamma_{\Lambda'}$ are the corresponding AR-quivers, then Γ^s_Λ and $\Gamma^s_{\Lambda'}$ are isomorphic stable translation quivers.* □

Note that $\Gamma^s_{k[X]/(X^2)}$ is the translation quiver with one vertex v and $\tau v = v$ and $\Gamma^s_{k[X]/(X^2)}$ and $\Gamma^s_{\binom{k\ 0}{k\ k}}$ are not isomorphic as translation quivers, but the associated proper translation quivers are isomorphic. Actually, we have the following general result.

Corollary 1.10 *If Λ and Λ' are stably equivalent artin algebras, then the proper translation quivers $\tilde{\Gamma}^s_\Lambda$ and $\tilde{\Gamma}^s_{\Lambda'}$ are isomorphic as translation quivers.*

Proof By definition τx is defined in $\tilde{\Gamma}^s_\Lambda$ for a vertex x if and only if x^- is not empty, that is, if and only if the middle term of the almost split sequence with the module X corresponding to x on the right is not projective. Then we get our desired result by applying Corollary 1.7. □

Note that $\left(\tilde{\Gamma}^s_\Lambda, \tau\right)$ can be defined directly in terms of the category $\underline{\mathrm{mod}}\,\Lambda$. We have seen that the quiver only depends on $\underline{\mathrm{mod}}\,\Lambda$, and the translation τ is defined by $\tau x = z$ if there is a path $\underset{z}{\cdot} \to \underset{y}{\cdot} \to \underset{x}{\cdot}$ in the quiver such that the corresponding morphisms in $\underline{\mathrm{mod}}\,\Lambda$ can be chosen to have composition zero.

The fact that the correspondence $F : \mathrm{mod}_{\mathscr{P}}\,\Lambda \to \mathrm{mod}_{\mathscr{P}}\,\Lambda'$ induced by a stable equivalence $F : \underline{\mathrm{mod}}\,\Lambda \to \underline{\mathrm{mod}}\,\Lambda'$ usually commutes with the operation $D\operatorname{Tr}$ is intimately related with the behavior of almost split sequences under stable equivalence, and such information is useful when investigating necessary conditions on algebras to be stably equivalent to a given class of algebras. We can also use this fact to show that a stable equivalence between selfinjective algebras usually commutes with Ω. This is based on the fact that the correspondence $F: \text{mod}_{\mathscr{P}} \Lambda \to \text{mod}_{\mathscr{P}} \Lambda'$ commutes with $\text{Tr} D\Omega$, when $F: \underline{\text{mod}} \Lambda \to \underline{\text{mod}} \Lambda'$ is a stable equivalence between selfinjective algebras. This is a direct consequence of the following.

Lemma 1.11 *Let Λ be a selfinjective artin algebra and C an indecomposable nonprojective Λ-module. Then the following are equivalent for an indecomposable nonprojective Λ-module X.*

(a) *X is isomorphic to $\text{Tr} D\Omega C$.*

(b) *There is a nonzero morphism $\underline{f}: X \to C$ such that if Y is indecomposable and $\underline{h}: Y \to X$ is not an isomorphism, then $\underline{f}\underline{h}$ is zero.*

Proof Let C and X be indecomposable nonprojective Λ-modules. The exact sequence $0 \to \Omega C \to P \overset{p}{\to} C \to 0$ where $p: P \to C$ is a projective cover gives rise to the long exact sequence $0 \to \text{Hom}_\Lambda(X, \Omega C) \to \text{Hom}_\Lambda(X, P) \to \text{Hom}_\Lambda(X, C) \to \text{Ext}^1_\Lambda(X, \Omega C) \to 0$. Since it is easy to see that a morphism $g: X \to C$ factors through a projective module if and only if it factors through $p: P \to C$, there is induced an isomorphism $\delta_X: \underline{\text{Hom}}_\Lambda(X, C) \to \text{Ext}^1_\Lambda(X, \Omega C)$.

Let Y be indecomposable in $\text{mod} \Lambda$ and $h: Y \to X$ a morphism such that h is not an isomorphism, or equivalently $\underline{h}$ is not an isomorphism. Then we have the commutative diagram

$$\begin{array}{ccc} \underline{\text{Hom}}_\Lambda(X, C) & \overset{\delta_X}{\to} & \text{Ext}^1_\Lambda(X, \Omega C) \\ \downarrow {\scriptstyle \underline{\text{Hom}}(h, C)} & & \downarrow {\scriptstyle \text{Ext}^1_\Lambda(h, \Omega C)} \\ \underline{\text{Hom}}_\Lambda(Y, C) & \overset{\delta_Y}{\to} & \text{Ext}^1_\Lambda(Y, \Omega C). \end{array}$$

We have that $\underline{g}$ in $\underline{\text{Hom}}_\Lambda(X, C)$ is zero if and only if $\delta_X(\underline{g})$ is zero. Hence it follows that $\underline{\text{Hom}}(h, C)(\underline{g}) = \underline{g}\underline{h}$ is zero if and only if $\text{Ext}^1_\Lambda(h, \Omega C)(\delta_X(\underline{g}))$ is zero.

Let now $X = \text{Tr} D\Omega C$ and choose $f: \text{Tr} D\Omega C \to C$ such that $\delta_X(\underline{f})$ is an almost split sequence. Then we have that $\delta_X(\underline{f})$ is not zero and $\text{Ext}^1_\Lambda(h, \Omega C)(\delta_X(\underline{f}))$ is zero for all morphisms $h: Y \to X$ where Y is indecomposable and h is not an isomorphism. It follows that $\underline{f}$ is not zero and $\underline{f}\underline{h}$ is zero for all $h: Y \to X$ where Y is indecomposable and $\underline{h}$ is not an isomorphism. This shows that (a) implies (b).

Assume now conversely that X has the property that there is some $\underline{f}: X \to C$ in $\underline{\text{mod}} \Lambda$ such that $\underline{f}$ is not zero, and $\underline{f}\underline{h}$ is zero for all $h: Y \to X$ where Y is indecomposable and $\underline{h}$ is not an isomorphism. By using the above commutative diagram again, it follows that $\delta(\underline{f}) \in \operatorname{Ext}^1_\Lambda(X, \Omega C)$ is an almost split sequence, and consequently $X \simeq \operatorname{Tr} D\Omega C$, so that (b) implies (a). $\square$

We now have the following result for selfinjective algebras.

Proposition 1.12 *Let* $F : \underline{\operatorname{mod}} \Lambda \to \underline{\operatorname{mod}} \Lambda'$ *be a stable equivalence between selfinjective artin algebras.*

(a) *If* Λ *has no algebra summand of Loewy length 2, then the correspondence* F *between objects in* $\operatorname{mod}_{\mathscr{P}} \Lambda$ *and* $\operatorname{mod}_{\mathscr{P}} \Lambda'$ *commutes with* Ω.

(b) *If* Λ *and* Λ' *are symmetric algebras, then* F *commutes with* Ω.

Proof (a) This is a direct consequence of Proposition 1.8 and Lemma 1.11.

(b) This follows from Lemma 1.11 since $\operatorname{Tr} D\Omega \simeq \Omega^{-1}$ for symmetric algebras. $\square$

2 Artin algebras with radical square zero

In this section we show that if $\mathfrak{r}^2 = 0$ for an artin R-algebra Λ, then Λ is stably equivalent to a hereditary algebra with radical square zero. The stable equivalence is induced by a functor between the module categories which takes most almost split sequences to almost split sequences. Using the classification theorem for hereditary artin algebras of finite representation type we get a similar result for algebras with radical square zero.

Let Λ be an artin algebra with $\mathfrak{r}^2 = 0$ and let Σ be the triangular matrix algebra $\begin{pmatrix} \Lambda/\mathfrak{r} & 0 \\ \mathfrak{r} & \Lambda/\mathfrak{r} \end{pmatrix}$. Σ is hereditary by III Proposition 2.7 since $\Lambda/\mathfrak{r}$ is semisimple. We define a functor $F : \operatorname{mod} \Lambda \to \operatorname{mod} \Sigma$ by $F(C) = (C/\mathfrak{r}C, \mathfrak{r}C, f)$, where $f : \mathfrak{r} \otimes_{\Lambda/\mathfrak{r}} C/\mathfrak{r}C \to \mathfrak{r}C$ is induced by the natural multiplication morphism $\mathfrak{r} \otimes_\Lambda C \to \mathfrak{r}C$, using that $\mathfrak{r}^2 = 0$. Observe that f is then an epimorphism. We then clearly have that $l(C) = l(F(C))$ for all C in $\operatorname{mod} \Lambda$. If $g : B \to C$ is a morphism in $\operatorname{mod} \Lambda$, then $g(\mathfrak{r}B) \subset \mathfrak{r}C$, so that we can define $g_2 = g|_{\mathfrak{r}B} : \mathfrak{r}B \to \mathfrak{r}C$ and there is induced a morphism $g_1 : B/\mathfrak{r}B \to C/\mathfrak{r}C$. We then define $F(g) = (g_1, g_2)$. It is easy to see that F is an R-functor. Our aim is to show that F induces an equivalence between the stable categories $\underline{\operatorname{mod}} \Lambda$ and $\underline{\operatorname{mod}}\Sigma$. For this we shall need some preliminary results on the functor F. We use the above notation in all the lemmas.

Lemma 2.1 *Let* Λ, Σ *and* F *be as above. Then we have the following.*

(a) *The functor* $F: \text{mod} \, \Lambda \to \text{mod} \, \Sigma$ *is full.*

(b) *For* E *and* E' *in* $\text{mod} \, \Lambda$, *the kernel of the epimorphism* F: $\text{Hom}_\Lambda(E, E') \to \text{Hom}_\Sigma(F(E), F(E'))$ *is* $\text{Hom}_\Lambda(E, \mathfrak{r}E')$. *Consequently the kernel of* $F: \text{End}_\Lambda(E) \to \text{End}_\Lambda(F(E))$ *is contained in* $\text{rad} \, \text{End}_\Lambda(E)$.

(c) E *and* E' *in* $\text{mod} \, \Lambda$ *are isomorphic if and only if* $F(E)$ *and* $F(E')$ *in* $\text{mod} \, \Sigma$ *are isomorphic.*

(d) E *in* $\text{mod} \, \Lambda$ *is indecomposable if and only if* $F(E)$ *in* $\text{mod} \, \Sigma$ *is indecomposable.*

(e) *Let* $X = (A, B, t)$ *be in* $\text{mod} \, \Sigma$. *Then* $X \simeq F(E)$ *for some* E *in* $\text{mod} \, \Lambda$ *if and only if* $t: \mathfrak{r} \otimes_{\Lambda/\mathfrak{r}} A \to B$ *is an epimorphism.*

Proof (a) To prove that $F: \text{mod} \, \Lambda \to \text{mod} \, \Sigma$ is full it is convenient to make some preliminary observations. Let P be a projective Λ-module. Then tensoring the exact sequence $0 \to \mathfrak{r} \to \Lambda \to \Lambda/\mathfrak{r} \to 0$ with P we obtain the exact sequence

$$0 \to \mathfrak{r} \otimes_\Lambda P \to P \to P/\mathfrak{r}P \to 0.$$

Since $\mathfrak{r}^2 = 0$, we have that the natural morphism $\mathfrak{r} \otimes_\Lambda P \to \mathfrak{r} \otimes_{\Lambda/\mathfrak{r}} (P/\mathfrak{r}P)$ is an isomorphism which we will consider as an identification. Suppose now that $f: P \to E$ is a projective cover. Then we have the exact commutative diagram

$$
\begin{array}{ccccccccc}
0 & \to & \mathfrak{r} \otimes_{\Lambda/\mathfrak{r}} (P/\mathfrak{r}P) & \to & P & \to & P/\mathfrak{r}P & \to & 0 \\
 & & \downarrow s & & \downarrow f & & \downarrow \mathcal{U}_1 & & \\
0 & \to & \mathfrak{r}E & \to & E & \to & E/\mathfrak{r}E & \to & 0 \\
 & & \downarrow & & \downarrow & & & & \\
 & & 0 & & 0 & & & &
\end{array}
$$

where the morphism s is induced by the morphism $\mathfrak{r} \otimes_\Lambda P \to E$ given by $r \otimes p \mapsto rf(p)$ for all r in $\mathfrak{r}$ and p in P. From this it follows that the Σ-modules $(P/\mathfrak{r}P, \mathfrak{r}E, s)$ and $F(E)$ are isomorphic.

Suppose that E and E' are Λ-modules and let $F(E) = (A, B, t)$ and $F(E') = (A', B', t')$. We now show that if $(u, v): (A, B, t) \to (A', B', t')$ is a morphism in $\text{mod} \, \Sigma$, then there is a morphism $w: E \to E'$ in $\text{mod} \, \Lambda$ such that $F(w) = (u, v)$.

Let $f: P \to E$ and $f': P' \to E'$ be projective covers. Then by our previous discussion we have the pushout diagrams

$$
\begin{array}{ccccccccc}
0 & \to & \mathfrak{r}\otimes_\Lambda (P/\mathfrak{r}P) & \to & P & \to & P/\mathfrak{r}P & \to & 0 \\
& & \downarrow s & & \downarrow f & & \downarrow \wr & & \\
0 & \to & B & & \to E \to & & A & \to & 0
\end{array}
$$

and

$$
\begin{array}{ccccccccc}
0 & \to & \mathfrak{r}\otimes_\Lambda (P'/\mathfrak{r}P') & \to & P' & \to & P'/\mathfrak{r}P' & \to & 0 \\
& & \downarrow s' & & \downarrow f' & & \downarrow \wr & & \\
0 & \to & B' & & \to E' \to & & A' & \to & 0
\end{array}
$$

where $(P/\mathfrak{r}P, B, s) \simeq (A, B, t)$ and $(P'/\mathfrak{r}P', B', s') \simeq (A', B', t')$. Denoting the composition $P/\mathfrak{r}P \xrightarrow{\sim} A \xrightarrow{u} A' \xrightarrow{\sim} P'/\mathfrak{r}P'$ by u' we obtain a commutative exact diagram

$$
\begin{array}{ccccccccc}
0 & \to & \mathfrak{r}\otimes_\Lambda P & \to & P & \to & P/\mathfrak{r}P & \to & 0 \\
& & \downarrow \mathfrak{r}\otimes g & & \downarrow g & & \downarrow u' & & \\
0 & \to & \mathfrak{r}\otimes_\Lambda P' & \to & P' & \to & P'/\mathfrak{r}P' & \to & 0.
\end{array}
$$

It is now easily checked that

$$
\begin{array}{ccc}
\mathfrak{r}\otimes_\Lambda P & \xrightarrow{\sim} & \mathfrak{r}\otimes_\Lambda (P/\mathfrak{r}P) \\
\downarrow \mathfrak{r}\otimes g & & \downarrow \mathfrak{r}\otimes u' \\
\mathfrak{r}\otimes_\Lambda P' & \xrightarrow{\sim} & \mathfrak{r}\otimes_\Lambda (P'/\mathfrak{r}P')
\end{array}
$$

commutes. So we obtain the commutative diagram

$$
\begin{array}{ccccccccc}
0 & \to & \mathfrak{r}\otimes_{\Lambda/\mathfrak{r}} (P/\mathfrak{r}P) & \to & P & \to & P/\mathfrak{r}P & \to & 0 \\
& & \downarrow \mathfrak{r}\otimes u' & & \downarrow g & & \downarrow u' & & \\
0 & \to & \mathfrak{r}\otimes_{\Lambda/\mathfrak{r}} (P'/\mathfrak{r}P') & \to & P' & \to & P'/\mathfrak{r}P' & \to & 0.
\end{array}
$$

Also using that

$$
\begin{array}{ccc}
\mathfrak{r}\otimes_{\Lambda/\mathfrak{r}} (P/\mathfrak{r}P) & \xrightarrow{1\otimes u'} & \mathfrak{r}\otimes_{\Lambda/\mathfrak{r}} (P'/\mathfrak{r}P') \\
\downarrow s & & \downarrow s' \\
B & \xrightarrow{v} & B'
\end{array}
$$

commutes, we obtain the commutative diagram

Using the mapping property of pushouts we obtain the commutative diagram

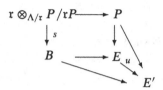

from which it is not difficult to check that $F(w) = (u', v)$. From this it follows that $F: \mathrm{Hom}_\Lambda(E, E') \to \mathrm{Hom}_\Sigma(F(E), F(E'))$ is surjective. Hence $F: \mathrm{mod}\,\Lambda \to \mathrm{mod}\,\Sigma$ is full.

(b) Let $f: E \to E'$ be a morphism in $\mathrm{mod}\,\Lambda$. Then it is easily seen that $F(f) = 0$ if and only if $f(E) \subset \mathfrak{r}E'$, that is

$$0 \to \mathrm{Hom}_\Lambda(E, \mathfrak{r}E') \to \mathrm{Hom}_\Lambda(E, E') \xrightarrow{F} \mathrm{Hom}_\Sigma(F(E), F(E')) \to 0$$

is exact. From this it follows that $\mathrm{End}_\Sigma(F(E)) \simeq \mathrm{End}_\Lambda(E)/\mathrm{Hom}_\Lambda(E, \mathfrak{r}E)$ where $\mathrm{Hom}_\Lambda(E, \mathfrak{r}E)$ is the two-sided ideal in $\mathrm{End}_\Lambda(E)$ consisting of endomorphisms $f: E \to E$ such that $f(E) \subset \mathfrak{r}E$. Since $\mathrm{Hom}_\Lambda(E, \mathfrak{r}E)^2 = 0$, it follows that $\mathrm{Hom}_\Lambda(E, \mathfrak{r}E)$ is contained in $\mathrm{rad}\,\mathrm{End}_\Lambda(E)$.

(c) Suppose E and E' are in $\mathrm{mod}\,\Lambda$ such that $F(E) \simeq F(E')$. Since F is full we know there are $f: E \to E'$ and $g: E' \to E$ such that $F(gf) = 1_{F(E)}$ and $F(fg) = 1_{F(E')}$. But then $gf: E \to E$ and $fg: E' \to E'$ are isomorphisms since the kernels of $F: \mathrm{End}_\Lambda(E) \to \mathrm{End}_\Sigma(F(E))$ and $F: \mathrm{End}_\Lambda(E') \to \mathrm{End}_\Sigma(F(E'))$ are nilpotent. Hence E and E' are isomorphic.

(d) Since the kernel of $F: \mathrm{End}_\Lambda(E) \to \mathrm{End}_\Sigma(F(E))$ is contained in $\mathrm{rad}\,\mathrm{End}_\Lambda(E)$, it follows that $\mathrm{End}_\Lambda(E)$ is a local ring if and only if $\mathrm{End}_\Sigma(F(E))$ is a local ring. Therefore E is indecomposable if and only if $F(E)$ is indecomposable.

(e) For E in $\mathrm{mod}\,\Lambda$ we have that $F(E) = (E/\mathfrak{r}E, \mathfrak{r}E, f)$ where the corresponding morphism $f: \mathfrak{r} \otimes_{\Lambda/\mathfrak{r}} (M/\mathfrak{r}M) \to \mathfrak{r}M$ induced by the canonical morphism $\mathfrak{r} \otimes M \to \mathfrak{r}M$ is obviously an epimorphism.

Let $X = (A, B, t)$ be in $\mathrm{mod}\,\Sigma$ where $t: \mathfrak{r} \otimes_{\Lambda/\mathfrak{r}} A \to B$ is an epimorphism. Let $P \to A$ be a projective cover. Identifying $P/\mathfrak{r}P$ with A we obtain as in part (a) the exact sequence $0 \to \mathfrak{r} \otimes_{\Lambda/\mathfrak{r}} A \to P \to A \to 0$. This gives the following pushout diagram.

$$
\begin{array}{ccccccccc}
 & & 0 & & 0 & & & & \\
 & & \downarrow & & \downarrow & & & & \\
 & & K & = & K & & & & \\
 & & \downarrow & & \downarrow & & & & \\
0 & \to & \mathfrak{r} \otimes_{\Lambda/\mathfrak{r}} A & \to & P & \to & A & \to & 0 \\
 & & \downarrow{\scriptstyle t} & & \downarrow & & \| & & \\
0 & \to & B & \overset{i}{\to} & E & \to & A & \to & 0 \\
 & & \downarrow & & \downarrow & & & & \\
 & & 0 & & 0 & & & &
\end{array}
$$

Since $P \to E$ is an epimorphism, we see that $i(B) = \mathfrak{r}E$ so that $F(E) \simeq (A, B, t)$.   □

Lemma 2.2 *Let Λ, Σ and F be as before. Then we have the following for C in* mod Λ.

(a)   *C is projective if and only if $F(C)$ is projective.*
(b)   *C is in* mod$_{\mathscr{P}}$ Λ *if and only if $F(C)$ is in* mod$_{\mathscr{P}}$ Σ.

Proof   (a) It is clear that $F(\Lambda) = (\Lambda/\mathfrak{r}, \mathfrak{r}, f)$, where $f: \mathfrak{r} \otimes_{\Lambda/\mathfrak{r}} (\Lambda/\mathfrak{r}) \to \mathfrak{r}$ is the natural isomorphism. Hence $F(C)$ is projective when C is projective.

Assume now that $F(C)$ is projective. To prove that C is projective we want to show that the projective cover $h: Q \to C$ in mod Λ is an isomorphism. The induced morphism $Q/\mathfrak{r}Q \to C/\mathfrak{r}C$ is an isomorphism. Writing $F(C) = (C/\mathfrak{r}C, \mathfrak{r}C, s)$ we get the commutative diagram

$$
\begin{array}{ccc}
\mathfrak{r} \otimes_{\Lambda/\mathfrak{r}} (Q/\mathfrak{r}Q) & \overset{\sim}{\to} & \mathfrak{r} \otimes_{\Lambda/\mathfrak{r}} (C/\mathfrak{r}C) \\
\downarrow{\scriptstyle \wr} & & \downarrow{\scriptstyle s} \\
\mathfrak{r}Q & \to & \mathfrak{r}C \qquad \to \quad 0.
\end{array}
$$

Then s is an isomorphism since $F(C)$ is projective, and hence the morphism $\mathfrak{r}Q \to \mathfrak{r}C$ induced by h is an isomorphism. It now follows that $h: Q \to C$ is an isomorphism, so that C is projective.

(b) Let C be indecomposable in mod Λ. Then by Lemma 2.1 we have that $F(C)$ is indecomposable, and it follows from (a) that C is not projective if and only if $F(C)$ is not projective.   □

Lemma 2.3 *Let Λ, Σ and F be as above, and let B and C be in* mod$_{\mathscr{P}}$ Λ. *Then we have $\mathscr{P}(B, C) = \mathrm{Hom}_{\Lambda}(B, \mathfrak{r}C)$.*

Proof Assume first that $f: B \to C$ in $\mathrm{mod}_{\mathscr{P}} \Lambda$ is in $\mathscr{P}(B, C)$. Then we have a commutative diagram

where Q is projective in $\mathrm{mod}\,\Lambda$. Since B is in $\mathrm{mod}_{\mathscr{P}} \Lambda$, we have $g(B) \subset \mathfrak{r}Q$. Then we have $hg(B) = f(B) \subset \mathfrak{r}C$.

Assume conversely that $f: B \to C$ is in $\mathrm{Hom}_{\Lambda}(B, \mathfrak{r}C)$. Then we have the factorization $B \to B/\mathfrak{r}B \xrightarrow{f'} \mathfrak{r}C \xrightarrow{i} C$ of $f: B \to C$. If $g: P \to C$ is a projective cover, then g induces an epimorphism $g': \mathfrak{r}P \to \mathfrak{r}C$. Using that the modules $B/\mathfrak{r}B$, $\mathfrak{r}C$ and $\mathfrak{r}P$ are semisimple, there is a morphism $h: B/\mathfrak{r}B \to \mathfrak{r}P$ such that the diagram

$$\begin{array}{ccc}
 & \mathfrak{r}P & \\
 \nearrow^{h} & & \searrow^{g'} \\
B/\mathfrak{r}B & \xrightarrow{\quad f' \quad} & \mathfrak{r}C
\end{array}$$

commutes. Hence if', and consequently f, factors through $g: P \to C$, so that $f \in \mathscr{P}(B, C)$. □

We use these lemmas to give the main result of this section.

Theorem 2.4 *Let Λ be an artin algebra with radical square zero. Let $\Sigma = \begin{pmatrix} \Lambda/\mathfrak{r} & 0 \\ \mathfrak{r} & \Lambda/\mathfrak{r} \end{pmatrix}$ and let $F: \mathrm{mod}\,\Lambda \to \mathrm{mod}\,\Sigma$ be defined as above. Then Σ is hereditary and F induces a stable equivalence $F: \underline{\mathrm{mod}}\,\Lambda \to \underline{\mathrm{mod}}\Sigma$.*

Proof Σ is hereditary since $\Lambda/\mathfrak{r}$ is semisimple. It follows from Lemma 2.2 that F induces by restriction a functor $F': \mathrm{mod}_{\mathscr{P}} \Lambda \to \mathrm{mod}_{\mathscr{P}} \Sigma$. Since $X = (A, B, t)$ is isomorphic to $F(M)$ for some M in $\mathrm{mod}\,\Lambda$ if and only if $t: \mathfrak{r} \otimes_{\Lambda/\mathfrak{r}} A \to B$ is an epimorphism by Lemma 2.1(e), the indecomposable Σ-modules not isomorphic to some $F(M)$ are the simple projective Σ-modules $(0, S, 0)$, where S is a simple projective Λ-module. Hence it follows from Lemmas 2.1 and 2.2 that F' is dense.

Since F takes projectives to projectives, F' induces a functor $F: \underline{\mathrm{mod}}\,\Lambda \to \underline{\mathrm{mod}}\Sigma$, where $\underline{\mathrm{mod}}\Sigma$ is equivalent to $\mathrm{mod}_{\mathscr{P}} \Sigma$ since Σ is hereditary. Since $\mathscr{P}(B, C) = \mathrm{Hom}_{\Lambda}(B, \mathfrak{r}C)$ for B and C in $\mathrm{mod}_{\mathscr{P}} \Lambda$ by Lemma 2.3, it follows that by Lemma 2.1(b) $F: \underline{\mathrm{mod}}\,\Lambda \to \underline{\mathrm{mod}}\Sigma$ is faithful. It is full by Lemma 2.1(a), and consequently $F: \underline{\mathrm{mod}}\,\Lambda \to \underline{\mathrm{mod}}\Sigma$ is an equivalence of categories. □

We now investigate how almost split sequences behave under the functor F. If S is a simple noninjective Λ-module, then $F(S) = (S, 0, 0)$ is an injective Σ-module, so applying F to an almost split sequence $0 \to S \to E \to \operatorname{Tr} DS \to 0$ in $\operatorname{mod} \Lambda$ cannot give an almost split sequence in $\operatorname{mod} \Sigma$.

Proposition 2.5 *Let Λ be an artin algebra with radical square zero, and let $\Sigma = \left(\begin{smallmatrix} \Lambda/\mathfrak{r} & 0 \\ \mathfrak{r} & \Lambda/\mathfrak{r} \end{smallmatrix} \right)$ and $F: \operatorname{mod} \Lambda \to \operatorname{mod} \Sigma$ be as before.*

Let $0 \to A \xrightarrow{f} B \xrightarrow{g} C \to 0$ be an exact sequence in $\operatorname{mod} \Lambda$ where A and C are indecomposable and A is not simple. Then this sequence is almost split if and only if $0 \to F(A) \xrightarrow{F(f)} F(B) \xrightarrow{F(g)} F(C) \to 0$ is almost split in $\operatorname{mod} \Sigma$.

Proof Assume first that $0 \to A \xrightarrow{f} B \xrightarrow{g} C \to 0$ is an almost split sequence. Since A is not simple, $0 \to A/\mathfrak{r}A \to B/\mathfrak{r}B \to C/\mathfrak{r}C \to 0$ is an exact sequence by V Lemma 3.2(c). Since $\mathfrak{r}A \to \mathfrak{r}B$ is a monomorphism, $\mathfrak{r}B \to \mathfrak{r}C$ is an epimorphism, $l(F(B)) = l(F(A)) + l(F(C))$ and $F(g)F(f) = 0$, it follows that $0 \to F(A) \xrightarrow{F(f)} F(B) \xrightarrow{F(g)} F(C) \to 0$ is an exact sequence. Then $F(g): F(B) \to F(C)$ is minimal right almost split by Proposition 1.3 since $g: B \to C$ is minimal right almost split. Because A is not simple, A is not a summand of $\mathfrak{r}P$ for any projective Λ-module P. Hence A, B and C are all in $\operatorname{mod}_{\mathscr{P}} \Lambda$, and consequently $F(A)$, $F(B)$ and $F(C)$ are in $\operatorname{mod}_{\mathscr{P}} \Sigma$ by Lemma 2.2. Hence $0 \to F(A) \xrightarrow{F(f)} F(B) \xrightarrow{F(g)} F(C) \to 0$ is an almost split sequence.

Assume conversely that $0 \to F(A) \to F(B) \to F(C) \to 0$ is an almost split sequence. Using the equivalence $F: \underline{\operatorname{mod}} \Lambda \to \underline{\operatorname{mod}} \Sigma$, it follows similarly that $0 \to A \to B \to C \to 0$ is almost split in $\operatorname{mod} \Lambda$. $\qquad \square$

Let Λ be an artin algebra and Γ the associated valued quiver. We associate with Γ the following quiver Γ_s called the **separated quiver** of Λ. If $\{1, \ldots, n\}$ are the vertices of Γ, then the vertices of Γ_s are $\{1, \ldots, n, 1', \ldots, n'\}$. For each valued arrow $\underset{i}{\cdot} \xrightarrow{(a,b)} \underset{j}{\cdot}$ in Γ we have by definition a valued arrow $\underset{i}{\cdot} \xrightarrow{(a,b)} \underset{j'}{\cdot}$ in Γ_s.

Using this notion and the stable equivalence between radical square zero algebras and hereditary algebras, we get the following description of the artin algebras of radical square zero of finite representation type.

Theorem 2.6 *Let* Λ *be an artin algebra with radical square zero. Then* Λ *is of finite representation type if and only if the separated quiver for* Λ *is a finite disjoint union of Dynkin quivers.*

Proof Recall that since $\mathfrak{r}^2 = 0$ the quiver of Λ is completely determined by the bimodule structure of $_{\Lambda/\mathfrak{r}}\mathfrak{r}_{\Lambda/\mathfrak{r}}$. It is then easy to see that the separated quiver of Λ coincides with the quiver of $\Sigma = \left(\begin{smallmatrix} \Lambda/\mathfrak{r} & 0 \\ \mathfrak{r} & \Lambda/\mathfrak{r} \end{smallmatrix} \right)$. Since Λ and Σ are stably equivalent by Theorem 2.4, it follows that Λ is of finite representation type if and only if Σ is of finite representation type by Proposition 1.1. Since Σ is hereditary, our claim follows from the description of hereditary algebras of finite representation type given in Chapter VIII. □

Note that for the functor $F : \operatorname{mod} \Lambda \to \operatorname{mod} \Sigma$ we have directly from the definition that $l(C) = l(F(C))$ for all C in $\operatorname{mod} \Lambda$. In particular, C is a simple Λ-module if and only if $F(C)$ is a simple Σ-module.

We have the following consequence of Theorem 2.6 about the components of the AR-quiver.

Proposition 2.7 *Let* Λ *be an algebra with radical square zero over an algebraically closed field* k. *Then each component of the Auslander–Reiten-quiver which does not contain a simple* Λ-*module is of the form* $\mathbb{Z}A_\infty$ *or* $\mathbb{Z}A_\infty / \langle \tau^n \rangle$ *for some* $n > 0$.

Proof By Theorem 2.4 there is a hereditary algebra Λ' stably equivalent to Λ. Let $\mathscr{C}$ be a component of the AR-quiver of Λ containing no simple modules. Since each indecomposable projective Λ-module has Loewy length at most 2, this means that $\mathscr{C}$ also contains no projective modules. Hence $\mathscr{C}$ is also a component of Γ^s_Λ, and consequently there is a translation quiver isomorphism between $\mathscr{C}$ and a component $F(\mathscr{C})$ of Γ^s_Λ by Corollary 1.9. Since $F(\mathscr{C})$ contains no simple Σ-module because $\mathscr{C}$ does not, $F(\mathscr{C})$ does not come from the preprojective or preinjective component, and is hence a regular component for Σ. Then we know from Chapter VIII that $F(\mathscr{C})$, and hence $\mathscr{C}$, is of the form $\mathbb{Z}A_\infty$ or $\mathbb{Z}A_\infty / \langle \tau^n \rangle$ for some $n > 0$. □

As a consequence of the functor $F : \operatorname{mod} \Lambda \to \operatorname{mod} \Sigma$ preserving length, we also have the following.

Proposition 2.8 *Let* Λ *and* Λ' *be stably equivalent artin algebras with rad-*

ical square zero. Then Λ *and* Λ' *have the same number of nonprojective simple modules.* $\square$

3 Algebras stably equivalent to symmetric Nakayama algebras

We have seen that if k is a field of characteristic p and G is a finite group whose order is divisible by p, then the group algebra kG is of finite representation type if and only if the Sylow p-subgroups of G are cyclic. If P is a cyclic p-group of order p^n, then kP is isomorphic to $k[X]/(X^{p^n})$ and is hence a Nakayama algebra. Also if $p = 3$, then the group algebra kS_3 is Nakayama where S_3 is the symmetric group on three letters. This follows by IV Theorem 2.14 once we observe that kS_3 is the skew group algebra of $k\mathbb{Z}_3$ by the action of $\mathbb{Z}_2$ obtained from the semidirect product $S_3 \simeq \mathbb{Z}_3 \rtimes \mathbb{Z}_2$. In fact it can be shown that all group algebras of finite representation type are stably equivalent to a Nakayama algebra. Motivated by this fact we investigate the algebras stably equivalent to symmetric Nakayama algebras, and show that their structure is given by Brauer trees. At the same time these results illustrate how we can get information on the projective modules over algebras stably equivalent to a given class of algebras, even though stable equivalence expresses that we have equivalence only modulo projectives.

An important role is played in this section by the algebras given by Brauer trees, so we start by introducing this class of algebras.

Let B be a finite tree, that is a finite graph with single edges and no cycles together with the following structure. To each vertex i there are associated a positive integer $m(i)$ and a circular ordering of the edges adjacent to i. This means that we have a labeling $\{\alpha_1, \ldots, \alpha_n\}$ of the edges adjacent to i such that α_{t+1} is the unique immediate successor of α_t, where the addition $t + 1$ is modulo n. We say that an artin algebra Λ is given by B if the structure of the indecomposable projective modules can be described in terms of B in the following way.

(i) The indecomposable projective Λ-modules P_α and hence the corresponding simple Λ-modules S_α, are in one to one correspondence with the edges α of B.

(ii) For an edge $\underset{i}{\bullet} \,\frac{\alpha}{\quad}\, \underset{j}{\bullet}$ in B let $(\alpha = \alpha_1, \ldots, \alpha_n)$ and $(\alpha = \beta_1, \ldots, \beta_r)$ be the circular orderings of the edges adjacent to i and j respectively. Then we have $\mathrm{r}P_\alpha = U_\alpha + V_\alpha$ where $U_\alpha \cap V_\alpha \simeq S_\alpha$, U_α is uniserial with composition factors $S_{\alpha_2}, \ldots, S_{\alpha_n}, S_{\alpha_1}$, $m(i)$ times from top to bottom, and

V_α is uniserial with composition factors $S_{\beta_2}, \ldots, S_{\beta_n}, S_{\beta_1}, m(j)$ times from top to bottom.

If at most one $m(i)$ is larger that 1, then B is a **Brauer tree** and an algebra Λ given by B is said to be given by a Brauer tree. Note that $\operatorname{soc} P_\alpha = S_\alpha$, and from this it follows that Λ is a selfinjective algebra (see IV Exercise 12). For each Brauer tree B it is easy to see that there is an algebra given by B. For let k be a field and Γ the quiver whose vertices are in one to one correspondence with the edges of B. For a vertex i of B the circular ordering $(\alpha_1, \ldots, \alpha_n)$ of the edges adjacent to i gives rise to an

oriented cycle C_i α_n •⟋•$^{\alpha_1}$ ⟍• α_2 . Denote the corresponding path with

no repeated arrows starting and ending at α_t by $p_{\alpha_t}^{(i)}$. Then the vertex α corresponding to the edge $\cdot \underset{i}{\overset{\alpha}{\longrightarrow}} \cdot$ belongs to exactly two oriented cycles C_i and C_j, and gives rise to the relations $(p_\alpha^{(i)})^{m_i} - (p_\alpha^{(j)})^{m_j}$, vu where u is the arrow in C_i with $e(u) = \alpha$ and v the arrow in C_j with $s(v) = \alpha$, or u the arrow in C_j with $e(u) = \alpha$ and v the arrow in C_i with $s(v) = \alpha$. The path algebra $k\Gamma$ modulo these relations is easily seen to be given by the original Brauer tree B.

We illustrate with the following.

Example Let B be the Brauer tree $\cdot \underset{1}{\overset{\alpha}{\longrightarrow}} \cdot \underset{2}{\overset{\beta}{\longrightarrow}} \cdot \underset{3}{\overset{\gamma}{\longrightarrow}} \cdot \underset{4}{}$ where $m(4) = 2$, and let Λ be an artin algebra given by B. Then the indecomposable projective Λ-modules P_α, P_β and P_γ have the following structure. P_α is uniserial of length 3 with composition factors S_α, S_β, S_α from top to bottom. Further we have $\mathfrak{r}P_\beta / \operatorname{soc} P_\beta \simeq S_\alpha \coprod S_\gamma$ and $\mathfrak{r}P_\gamma / \operatorname{soc} P_\gamma \simeq S_\beta \coprod S_\gamma$.

We can choose Λ to be the factor algebra of the path algebra $k\Gamma$ for Γ the quiver $^u C \cdot \underset{\alpha}{\overset{v}{\underset{w}{\leftrightarrows}}} \cdot \underset{\beta}{\overset{x}{\underset{y}{\rightleftarrows}}} \cdot \circlearrowleft^z_\gamma$ modulo the ideal $\langle u - vw, uv, wu, xw, vy, zx, yz, wv - yx, xy - z^2 \rangle$. Note that this ideal is not contained in the ideal J^2, where J is generated by the arrows. If Γ' denotes the quiver obtained by removing u, we can describe Λ in the usual way as $k\Gamma' / \langle vwv, wvw, xw, vy, zx, yz, wv - yx, xy - z^2 \rangle$.

Example Let B be the Brauer tree ⟨diagram with vertices 1, 2, 3, 4, 5, 6 and edges $\alpha, \beta, \gamma, \delta, \epsilon$⟩ with $m(i) = 1$.

for $i = 1,\ldots,6$, and where the circular ordering is given by counter-clockwise direction. Let Γ be the quiver

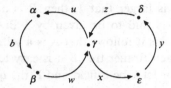

with relations $\rho = \{wvu - zyx, vuwv, wvuw, yxzy, zyxz, xw, uz\}$. If k is a field then the algebra $k(\Gamma, \rho)$ is given by B.

In order to prove that artin algebras stably equivalent to symmetric Nakayama algebras are given by Brauer trees we need a series of lemmas.

Lemma 3.1 *Let Λ be a selfinjective algebra. If $f: M \to N$ is a monomorphism or an epimorphism between indecomposable nonprojective Λ-modules, then $\underline{f}$ is nonzero in the stable category* $\underline{\mathrm{mod}}\ \Lambda$.

Proof Let Λ be selfinjective and $f: M \to N$ a monomorphism between indecomposable nonprojective Λ-modules. If $\underline{f} = 0$, there is a commutative diagram

$$
\begin{array}{ccc}
M & \xrightarrow{f} & N \\
\downarrow{g} & & \| \\
I & \xrightarrow{h} & N
\end{array}
$$

where I is injective, and it is easy to see that we can assume that $g: M \to I$ is an injective envelope. Since f is a monomorphism and g is an essential monomorphism, we conclude that $h: I \to N$ is a monomorphism. Since I is injective, h is a split monomorphism, and we get a contradiction. This shows that $\underline{f}$ is nonzero.

The proof that $\underline{f}$ is nonzero when $f: M \to N$ is an epimorphism is similar. $\square$

Assume now that Γ is an indecomposable nonsimple symmetric Nakayama algebra with nonisomorphic simple modules $T_1,\ldots,T_n$ and corresponding projective covers $Q_1,\ldots,Q_n$, where the ordering is such that $Q_{\pi(i)}$ is a projective cover of $\mathrm{r}Q_i$ when we write $\pi(i) = i + 1$ (modulo n). This means that $\{Q_1,\ldots,Q_n\}$ is a Kupisch series. Let Λ be a symmetric algebra with no semisimple block and assume Λ is stably equivalent to Γ via an equivalence $F: \underline{\mathrm{mod}}\ \Lambda \to \underline{\mathrm{mod}}\ \Gamma$, whose inverse we denote by G. Let $S_1,\ldots,S_m$ be the nonisomorphic simple Λ-modules and $P_1,\ldots,P_m$

their projective covers. We now investigate the structure of the P_i. We start out by showing that Λ and Γ have the same number of nonisomorphic simple (nonprojective) modules. But first we introduce some notation which will be useful in what follows.

Since Λ is selfinjective with no semisimple block, no simple Λ-module is projective. Denoting by F also the induced correspondence on the indecomposable nonprojective modules, we have that FS_i is an indecomposable Γ-module for all i, and is hence uniserial since Γ is Nakayama. In particular $soc(FS_i)$ and $FS_i/\mathfrak{r}(FS_i)$ are simple Γ-modules. We define the functions μ and ν from $\{1,\ldots,m\}$ to $\{1,\ldots,n\}$ by $soc(FS_i) \simeq T_{\mu(i)}$ and $FS_i/\mathfrak{r}(FS_i) \simeq T_{\nu(i)}$. We then have the following.

Lemma 3.2 *Let* $F: \underline{\mathrm{mod}}\,\Lambda \to \underline{\mathrm{mod}}\,\Gamma$ *be an equivalence where* Λ *and* Γ *are symmetric algebras, and* Γ *is an indecomposable nonsimple Nakayama algebra. Then* μ *and* ν *as defined above are bijections, so that* Λ *and* Γ *have the same number of nonisomorphic simple modules.*

Proof If $\mu(i) = \mu(j)$ for i and j in $\{1,\ldots,m\}$, then $soc\,FS_i \simeq soc\,FS_j$. Since Γ is Nakayama we can without loss of generality assume that there is a monomorphism $g: FS_i \to FS_j$ in $\mathrm{mod}\,\Gamma$. Then we have $g \neq 0$ by Lemma 3.1, so that $G(\underline{g}): S_i \to S_j$ is nonzero where $G: \underline{\mathrm{mod}}\,\Gamma \to \underline{\mathrm{mod}}\,\Lambda$ is the inverse equivalence. Since S_i and S_j are simple Λ-modules, they must then be isomorphic, and consequently $i = j$, so that μ is injective.

To show that μ is surjective, consider the simple Γ-module T_j. Let S_i be a simple Λ-module such that there is a nonzero map $g: GT_j \to S_i$. Since g is a monomorphism, we have by Lemma 3.1 that $\underline{g} \neq 0$, and hence we have a nonzero map $T_j \to FS_i$, so that $j = \mu(i)$. Hence we have shown that μ is bijective, and consequently $m = n$.

The proof that ν is bijective is dual. $\qquad\qquad\square$

It will be useful to consider permutations ρ and σ on $I = \{1,\cdots,n\}$ defined by $\rho = \mu^{-1}\nu$ and $\sigma = \nu^{-1}\pi\mu$. We define $U_i = G(T_{\nu(i)})$ and $V_i = \Omega G(T_{\mu(i)})$. Hence the U_i are the correspondents of the simple Γ-modules under the stable equivalence G and the V_i are the correspondents of the simple Γ-modules under the stable equivalence $\Omega G: \underline{\mathrm{mod}}\,\Gamma \to \underline{\mathrm{mod}}\,\Lambda$. We shall prove that these correspondents of simple modules are uniserial. The next lemma provides a step in this direction, by showing that the U_i and V_i have simple socle, and in addition gives some more specific information which will be useful in describing the structure of the P_i.

Lemma 3.3 *With the above assumptions and notation we have exact sequences*

$$0 \rightarrow U_i \rightarrow P_i \xrightarrow{\alpha_i} V_{\sigma^{-1}(i)} \rightarrow 0$$
$$\text{and} \quad 0 \rightarrow V_i \rightarrow P_i \xrightarrow{\beta_i} U_{\rho^{-1}(i)} \rightarrow 0.$$

Proof To establish the first exact sequence it is sufficient to show $\Omega(V_{\sigma^{-1}(i)}) \simeq U_i$. By definition we have $U_i = G(T_{\nu(i)})$ and $\Omega(V_{\sigma^{-1}(i)}) = \Omega^2 G(T_{\mu\sigma^{-1}(i)})$. Since the stable equivalence $G : \underline{\text{mod}}\,\Gamma \rightarrow \underline{\text{mod}}\,\Lambda$ commutes with Ω by Proposition 1.12, it is sufficient to show that $\Omega^2 T_{\mu\sigma^{-1}(i)} \simeq T_{\nu(i)}$. We know from IV Section 2 that $\Omega^2 T_{\mu\sigma^{-1}(i)} \simeq T_{\pi\mu\sigma^{-1}(i)}$, so we are done since $\pi\mu\sigma^{-1} = \pi\mu\mu^{-1}\pi^{-1}\nu = \nu$.

For the second sequence we want to show that $\Omega U_{\rho^{-1}(i)} \simeq V_i$, that is $\Omega T_{\nu\rho^{-1}(i)} \simeq \Omega T_{\mu(i)}$, which holds since $\mu = \nu\rho^{-1}$ and hence $T_{\nu\rho^{-1}(i)} \simeq T_{\mu(i)}$. □

The next result is needed for proving that U_i and V_i are uniserial, and more generally for establishing the structure of the P_i.

Lemma 3.4 *With the above notation, let U, M and N be indecomposable nonprojective Λ-modules with FU simple and also soc M and soc N simple.*

(a) *If M and N are submodules of U and $l(FM) \leq l(FN)$, then there is a monomorphism $N \rightarrow M$.*

(b) *If U has an inclusion into M and N and $l(FM) \leq l(FN)$, then there is an inclusion $M \rightarrow N$.*

Proof (a) Assume that we have inclusions $f : M \rightarrow U$ and $g : N \rightarrow U$. Then $\underline{f}$ and $\underline{g}$ are nonzero by Lemma 3.1, and hence $F(\underline{f})$ and $F(\underline{g})$ are nonzero. Therefore we have nonzero maps $f' : FM \rightarrow FU$ and $g' : FN \rightarrow FU$ with $F(\underline{f}) = \underline{f}'$ and $F(\underline{g}) = \underline{g}'$. Since FU is simple, $l(FM) \leq l(FN)$ and Γ is Nakayama, there is an epimorphism $t : FN \rightarrow FM$ such that the diagram

$$\begin{array}{ccc} FM & \xrightarrow{f'} & FU \\ \uparrow{t} & & \| \\ FN & \xrightarrow{g'} & FU \end{array}$$

commutes. Let $t' : N \rightarrow M$ be such that $G(\underline{t}) = \underline{t}'$. Then $ft' - g : N \rightarrow U$ factors through a projective module and is hence not a monomorphism by Lemma 3.1. Since soc N is simple, we have $(ft' - g)(\text{soc } N) = 0$.

Since $g(\operatorname{soc} N) \neq 0$, we must have $t'(\operatorname{soc} N) \neq 0$, so that $t' : N \to M$ is a monomorphism.

(b) Assume that we have inclusions $f : U \to M$ and $g : U \to N$. Then $\underline{f}$ and $\underline{g}$ are nonzero, and hence we get nonzero maps $f' : FU \to FM$ and $g' : FU \to FN$, which are monomorphisms since FU is simple. Since Γ is Nakayama and $l(FM) \leq l(FN)$, there is a commutative diagram

$$
\begin{array}{ccc}
FU & \overset{f'}{\to} & FM \\
\| & & \downarrow s \\
FU & \overset{g'}{\to} & FN.
\end{array}
$$

Let $s' : M \to N$ be such that $G(\underline{s}) = \underline{s}'$. Then $s'f - g : U \to N$ factors through a projective module, and is hence not a monomorphism by Lemma 3.1. Since $\operatorname{soc} U$ is simple, we must have $(s'f - g)(M) = 0$. Since $f(\operatorname{soc} U) = \operatorname{soc} M$ we would have $s'f(\operatorname{soc} U) = 0$ if s' were not a monomorphism. Then we would have $g(\operatorname{soc} U) = 0$, which contradicts the fact that g is a monomorphism. We conclude that $s' : M \to N$ is a monomorphism. $\qquad\square$

The next result is an easy consequence of the existence of a stable equivalence.

Lemma 3.5 *If X is an indecomposable Λ-module, then any simple module has multiplicity at most 1 in $\operatorname{soc} X$ and $X/\mathfrak{r}X$.*

Proof Let $f : S \to X$ and $g : S \to X$ be monomorphisms, where S is a simple Λ-module. Since Γ is Nakayama, we have a commutative diagram

$$
\begin{array}{ccc}
FS & \overset{f'}{\to} & FX \\
\downarrow t & & \| \\
FS & \overset{g'}{\to} & FX
\end{array}
$$

where $F(\underline{f}) = \underline{f}'$ and $F(\underline{g}) = \underline{g}'$. Let $t' : S \to S$ be such that $G(\underline{t}) = \underline{t}'$. Then $gt' - f : S \to X$ factors through a projective module, so that $gt' = f$ since S is simple. This shows that $\operatorname{Im} f = \operatorname{Im} g$, and hence S must occur with multiplicity 1 in $\operatorname{soc} X$.

The claim for $X/\mathfrak{r}X$ follows by duality. $\qquad\square$

As a consequence we get the following uniqueness result for submodules of Λ-modules with simple socle.

Lemma 3.6 *Let X be a Λ-module with* soc$X$ *simple. If V and W are submodules of X with $V \simeq W$, then $V = W$.*

Proof If $l(V) = 1$, then $V = W = $ soc$X$. Assume then that $l(V) > 1$, and let M be a maximal submodule of V, which by the induction assumption also is a maximal submodule of W. If $V \neq W$, we then have $(V + W)/M \simeq (V/M) \coprod (W/M)$. Since $V/M \simeq W/M$ and $V + W$ is indecomposable because it has simple socle, we get a contradiction to Lemma 3.5. Hence we conclude that $V = W$.   □

We use Lemma 3.6 to show that the correspondents of simple Γ-modules are uniserial.

Proposition 3.7 *If T is a simple Γ-module then GT is uniserial.*

Proof Let M and N be submodules of GT, and assume $l(FM) \le l(FN)$. Since $FGT \simeq T$ is simple, it follows from Lemma 3.3 that GT, and hence M and N, have simple socle. By Lemma 3.4 there is then a monomorphism $h: N \to M$. Since Im$h \simeq N$, it follows from Lemma 3.6 that Im$h = N$, and hence $N \subset M$. This shows that the submodules of GT are totally ordered by inclusion, and hence GT is uniserial.   □

We can now get explicit information on the structure of the projective Λ-modules. Note that by Lemma 3.6, U_i and V_i are uniquely determined as submodules of P_i.

Theorem 3.8 *Let P_i, U_i and V_i be as before. We then have the following.*

(a) $U_i \cap V_i = S_i$.
(b) $U_i + V_i = \mathfrak{r}P_i$.
(c) U_i/S_i *and* V_i/S_i *have no common composition factors.*

Proof (a) Since U_i is a correspondent of a simple module via the equivalence G and V_i via the equivalence $G\Omega$, it follows from Proposition 3.7 that U_i and V_i are uniserial.

Let N be a nonzero submodule of $U_i \cap V_i$, so that we have $M = $ soc$U_i \subset N \subset U_i \cap V_i \subset U_i$. If $l(FM) \le l(FN)$, we have an inclusion $N \to M$ by Lemma 3.4, and hence $M = N$. If $l(FM) \ge l(FN)$, we have $l(\Omega^{-1}F(M)) \le l(\Omega^{-1}F(N))$. Considering $M \subset N \subset U_i \cap V_i \subset V_i$, we have that V_i corresponds to a simple module via $\Omega^{-1}F$, so we apply Lemma 3.4 to get an inclusion $N \to M$. This shows that $U_i \cap V_i = S_i$.

(b) Let now M be a submodule of $\mathfrak{r}P$ with $U_i \subset U_i + V_i \subset M \subset \mathfrak{r}P_i = N$. If $l(FN) \leq l(FM)$, it follows from Lemma 3.4, using that U_i corresponds to a simple module via F, that there is an inclusion $N \to M$, which gives $M = N$. If $l(FN) \geq l(FM)$, we have $l(\Omega^{-1}FN) \leq l(\Omega^{-1}FM)$, so that by using that V_i is contained in M and N and the equivalence $\Omega^{-1}F$, again there is an inclusion $N \to M$, which gives $M = N$. This shows that $U_i + V_i = \mathfrak{r}P_i$.

(c) We have $\mathfrak{r}P_i/S_i = (U_i/S_i) \coprod (V_i/S_i)$. If U_i/S_i and V_i/S_i have a common simple composition factor T, then $T \coprod T$ would be part of the socle of a factor module of P. This contradicts Lemma 3.5. $\square$

We now get more information on the length of the U_i and V_i, and on the order in which the simple composition factors occur in a composition series.

Proposition 3.9

(a) *The simple composition factors of U_i are from top to bottom $S_{\rho(i)}, \ldots, S_{\rho^a(i)}$ and for V_i they are $S_{\sigma(i)}, \ldots, S_{\sigma^b(i)}$, where $a = l(U_i), b = l(V_i)$ and $\rho^a(i) = i = \sigma^b(i)$.*

(b) $l(U_i) = l(U_{\rho(i)})$ *and* $l(V_i) = l(V_{\sigma(i)})$.

(c) *Either $l(U_i)$ is equal to the size of the ρ-orbit containing i or $l(V_i)$ is equal to the size of the σ-orbit containing i.*

(d) *Each ρ-orbit and σ-orbit have at most one element in common.*

Proof Consider the exact sequences $0 \to U_{\sigma(i)} \to P_{\sigma(i)} \overset{\alpha_{\sigma(i)}}{\to} V_i \to 0$ and $0 \to V_{\rho(i)} \to P_{\rho(i)} \overset{\beta_{\rho(i)}}{\to} U_i \to 0$ from Lemma 3.3. Now $\text{Ker}(\alpha_{\sigma(i)}|_{V_{\sigma(i)}}) = S_{\sigma(i)}$ and $\alpha_{\sigma(i)}(V_{\sigma(i)}) = \mathfrak{r}V_i$, showing that $l(V_i) = l(V_{\sigma(i)})$ and that the order of the composition factors for V_i is as claimed. The result for U_i follows similarly. Since we know $S_i = \text{soc}\, U_i = \text{soc}\, V_i$, we have $\rho^a(i) = i = \sigma^b(i)$. This proves (a) and (b).

If $l(U_i)$ is larger than the size of the ρ-orbit, then S_i is a composition factor of U_i/S_i, and if $l(V_i)$ is larger than the size of the σ-orbit, then S_i is a composition factor of V_i/S_i. It follows from Theorem 3.8 that both cannot happen, so that (c) follows.

The elements of the ρ-orbit of i correspond by (a) exactly to the composition factors of U_i, and the elements of the σ-orbit of i correspond exactly to the composition factors of V_i. It follows from this and the fact that U_i/S_i and V_i/S_i have no common composition factors that i is the only common element of the orbits. $\square$

We use the information we have obtained to define a graph B_Λ associated with Λ which we shall show gives a Brauer tree. The set of vertices of the graph B_Λ is by definition the disjoint union of the set of all ρ-orbits and σ-orbits. We denote by $[i]_\rho$ and $[i]_\sigma$ the ρ-orbit and the σ-orbit containing i respectively. For each i there is an edge connecting $[i]_\rho$ and $[i]_\sigma$, and we write $\overset{\rho}{\cdot}\overset{i}{\rule{1.2cm}{0.4pt}}\overset{\sigma}{\cdot}$. Note that since a σ-orbit and a ρ-orbit have at most one element in common, there are no multiple edges. For each vertex representing a ρ-orbit the edges connected with this vertex are in one to one correspondence with the elements of this ρ-orbit, and correspond to the composition factors of the associated U_i. A similar comment applies to σ-orbits and V_i.

We give preliminary results leading up to proving the desired structure of the graph.

Lemma 3.10

(a) If S_j is a composition factor of U_i/S_i, there is a nonzero map $FS_i \to FS_j$ which factors through a projective module, and hence $l(\Omega FS_j) < l(FS_i)$.

(b) If S_j is a composition factor of V_i/S_i, there is a nonzero map $\Omega^{-1}FS_i \to \Omega^{-1}FS_j$ which factors through a projective module, and hence $l(FS_j) < l(\Omega^{-1}FS_i)$.

Proof (a) When S_j is a composition factor of U_i/S_i, there is a submodule W_i of U_i with $l(W_i) \geq 2$ and $W_i/\mathfrak{r}W_i \simeq S_j$. Consider the diagrams

$$
\begin{array}{ccc}
U_i & & FU_i \\
\uparrow h & \text{and} & \uparrow h' \\
S_i \xrightarrow{f} W_i \xrightarrow{g} S_j & & FS_i \xrightarrow{f'} FW_i \xrightarrow{g'} FS_j,
\end{array}
$$

where $f : S_i \to W_i$ and $h : W_i \to U_i$ are inclusion maps and $g : W_i \to S_j$ is surjective, $F(\underline{f}) = \underline{f'}$ and $F(\underline{g}) = \underline{g'}$ and $F(\underline{h}) = \underline{h'}$. Then $\underline{h'}$, $\underline{f'}$, and $\underline{g'}$ are nonzero by Lemma 3.1. Since FU_i is simple, $h' : FW_i \to FU_i$ must then be surjective. Since FW_i is uniserial, $f' : FS_i \to FW_i$ must also be surjective, and hence $g'f' : FS_i \to FS_j$ is nonzero. Further $\underline{g'f'} = 0$ because $gf = 0$. So we have a commutative diagram

$$
\begin{array}{ccc}
Q & = & Q \\
\uparrow u & & \downarrow v \\
FS_i & \xrightarrow{g'f'} & FS_j,
\end{array}
$$

where $v : Q \to FS_j$ is a projective cover. Clearly $l(FS_i) \geq l(\operatorname{Im} u)$, and by considering the exact sequence $0 \to \Omega(FS_j) \to Q \xrightarrow{v} FS_j \to 0$, we see that $\operatorname{Im} u \not\subset \Omega(FS_j)$ since $vu \neq 0$. Since Q is uniserial, we then have $\Omega(FS_j) \subset \operatorname{Im} u$, so that $l(\Omega(FS_j)) < l(\operatorname{Im} u) \leq l(FS_i)$.

(b) The proof is similar to the proof of (a), by replacing U_i by V_i and F by $\Omega^{-1}F$. □

Proposition 3.11 *Assume that in the graph B_Λ we have* $\overset{\sigma}{\cdot}\underset{i}{\_\_}\overset{\rho}{\cdot}\underset{j}{\_\_}\overset{\sigma}{\cdot}\underset{t}{\_\_}\overset{\rho}{\cdot}$ *with $i \neq j$ and $j \neq t$. Then we have $l(FS_i) > l(\Omega F S_j) > l(FS_t)$.*

Proof We have that S_j is a composition factor of U_i/S_i and S_t is a composition factor of V_j/S_j. Then we conclude by Lemma 3.10 that $l(FS_i) > l(\Omega F S_j) = l(\Omega^{-1}FS_j) > l(FS_t)$ using that the indecomposable projective Γ-modules have the same length. □

Associated with each vertex is a positive integer m in the following way. We define $m([i]_\rho)$ to be the multiplicity of S_i in the U_i and $m([i]_\sigma)$ to be the multiplicity of S_i in V_i. We shall show that for at most one vertex can the associated number m be greater than 1.

Proposition 3.12

(a) *If $m([i]_\rho) \neq 1$, then $l(FS_i) > l(\Omega F S_i)$.*
(b) *If $m([i]_\sigma) \neq 1$, then $l(\Omega F S_i) > l(FS_i)$.*

Proof (a) If $m([i]_\rho) \neq 1$, then S_i is a composition factor of U_i/S_i, so that $l(FS_i) > l(\Omega F S_i)$ by Lemma 3.10.

(b) If $m([i]_\sigma) \neq 1$, then S_i is a composition factor of V_i/S_i, so that $l(FS_i) < l(\Omega^{-1}FS_i) = l(\Omega F S_i)$, by using Lemma 3.10. □

Proposition 3.13 *The graph B_Λ associated with Λ is a tree, and we have $m(x) > 1$ for at most one vertex x.*

Proof Assume that we have a cycle

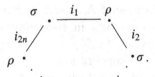

Then we have $l(FS_{i_1}) > l(FS_{i_3}) > \cdots > l(FS_{i_{2n-1}}) > l(FS_{i_1})$ by Proposition 3.11, so we have a contradiction.

Consider $\sigma \underline{\quad i_1 \quad} \rho \underline{\quad i_2 \quad} \cdots \cdot \rho \underline{\quad i_{2n} \quad} \sigma$ and assume $m([i_1]_\sigma) > 1$ and $m([i_{2n}]_\sigma) > 1$. Then we have $l(\Omega FS_{i_1}) > l(FS_{i_1}) > \cdots > l(\Omega FS_{i_{2n}}) > l(FS_{i_{2n}})$ by Propositions 3.11 and 3.12. Since for any indecomposable X and Y in $\bmod \Gamma$, we have $l(X) + l(\Omega X) = l(Y) + l(\Omega Y)$, we get a contradiction.

If we have $m([i_1]_\sigma) > 1$ and $m([i_{2n-1}]_\rho) > 1$, we get $l(\Omega FS_{i_1}) > l(FS_{i_1}) > \cdots > l(FS_{i_{2n-1}}) > l(\Omega FS_{i_{2n-1}})$, which is impossible. The rest is treated similarly. □

As a direct consequence of Theorem 3.8 and Proposition 3.13 we get the main result of this section.

Theorem 3.14 *Let* Λ *be a symmetric artin algebra stably equivalent to a symmetric Nakayama algebra. Then* Λ *is given by the Brauer tree* B_Λ. □

Exercises

1. Let Λ' be an artin algebra stably equivalent to a Nakayama algebra Λ. Prove that Λ and Λ' have the same number of nonisomorphic simple nonprojective modules.

2. Let Λ be a finite dimensional nonsemisimple algebra over an algebraically closed field k. Prove that $T_2(T_2(\Lambda))$ is of infinite representation type.

3. Let k be an algebraically closed field and $k\langle X, Y \rangle$ the free algebra over k in two noncommuting variables X and Y. For each $t \neq 0$ in k let $\Lambda_t = k\langle X, Y \rangle / I_t$ where I_t is the ideal of $k\langle X, Y \rangle$ generated by $\{X^2, Y^2, XY - tYX\}$.

(a) Prove that Λ_t is a selfinjective finite dimensional k-algebra with $\operatorname{soc} \Lambda_t = (XY) + I_t$ and that $\Lambda_t \simeq \Lambda_{t'}$ if and only if $t = t'$ or $t = \frac{1}{t'}$.

(b) Prove that $\Lambda_t / \operatorname{soc} \Lambda_t \simeq k[X, Y]/(X, Y)^2$ where $k[X, Y]$ is the polynomial ring in two commuting variables over k, and hence that $k[X, Y]/(X, Y)^2$ is stably equivalent to the Kronecker algebra.

(c) The k-algebra isomorphism $\Lambda_t / \operatorname{soc} \Lambda_t \simeq k[X, Y]/(X, Y)^2$ induces a group morphism from $\operatorname{Aut}_k(\Lambda_t)$ to $\operatorname{Aut}_k k[X, Y]/(X, Y)^2 \simeq Gl(2, k)$, where $\operatorname{Aut}_k(\Gamma)$ denotes the group of k-algebra automorphisms of a

k-algebra Γ. Prove that the image is trivial for $t \neq \pm 1$, of order 2 if $t = 1$ and char $k \neq 2$ and $GL(2, k)$ if $t = -1$.

From the stable equivalence between $k[X, Y]/(X, Y)^2$ and the Kronecker algebra we have by VIII Section 7 the following description of the AR-quiver of Λ_t. There is a $\mathbb{P}^1(k)$ family of rank 1 tubes corresponding to the regular Kronecker modules. For each $s = (1, p) \in \mathbb{P}^1(k)$ we let $M_{t,s}$ be the module $\Lambda_t/(X + pY + I_t)$ and for $s = (0, 1) \in \mathbb{P}(k)$ we let $M_{t,s}$ be the module $\Lambda_t/Y + I_t$ which constitutes the modules X with $\alpha(X) = 1$ in the tubes. For a fixed t we let M_s, $s \in \mathbb{P}_1(k)$ be the class of modules just described throughout the rest of this exercise. In addition to the tubes there is one component containing the projective Λ_t-module Λ_t. This component may be represented in the following way.

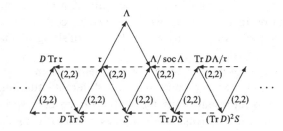

(d) Let $t \neq -1$. Prove that $\dim_k \underline{\mathrm{Hom}}(M_{(1,0)}, M_{(1,0)}) = 2$, $\dim_k \underline{\mathrm{Hom}}(M_{(0,1)}, M_{(0,1)}) = 2$ and that $\dim_k \underline{\mathrm{Hom}}(M_s, M_s) = 1$ for $(0, 1) \neq s \neq (1, 0)$.

(e) Let $t \neq -1$ and let $(1, p) \neq (1, p')$ in $\mathbb{P}^1(k)$ with $p \neq 0 \neq p'$. Prove that $\Omega M_{(1,p)} = M_{(1, -\frac{1}{t}p)}$ and that $\dim_k \underline{\mathrm{Hom}}(M_{(1,p')}, M_{(1,p)}) = \begin{cases} 0 & \text{if } p' \neq -\frac{p}{t}, \\ 1 & \text{if } p' = -\frac{p}{t}. \end{cases}$

(f) For $t = -1$ prove that $\Omega M_s \simeq M_s$ and that $\dim_k \underline{\mathrm{Hom}}(M_s, M_s) = 2$ for all $s \in \mathbb{P}^1(k)$.

(g) Prove that Λ_t and Λ_{-1} are not stably equivalent for $t \neq -1$.

(h) Now assume t is a primitive $2n$-th root of unity. Prove that there are cycles of modules $M_{s_1}, M_{s_2}, \ldots, M_{s_{2n}}$ such that $\dim_k \underline{\mathrm{Hom}}(M_{s_i}, M_{s_{i+1}}) = 1$ for $i = 1, \ldots, 2n$ when $M_{s_{2n+1}} = M_{s_1}$ and that the maximal length of such a cycle without repetition is $2n$.

(i) Deduce that if t is a primitive $2n$-th root and t' is a primitive $2m$-th root of unity with $m \neq n$, then Λ_t and $\Lambda_{t'}$ are not stably equivalent.

(j) Prove that Λ_t and $\Lambda_{t'}$ are not stably equivalent for $t \notin \{t', \frac{1}{t'}\}$. (Hint: Use the component of the AR-quiver containing the projective vertex.)

Notes

Consideration of the subgroups $\mathscr{P}(A, B)$ and factor groups $\underline{\mathrm{Hom}}(A, B)$ of $\mathrm{Hom}_\Lambda(A, B)$ predates the formal definition of the category of modules modulo projectives. The projective homotopy groups of modules [EcH], the Heller operator [He] and the notion of projective homomorphisms in group theory and orders are some instances where these notions came up, either explicitly or implicitly. But the idea of considering the category of modules modulo projectives as an object of study in its own right did not seem to gain any prominence until it arose in modular group representations in connection with the Green correspondence and the definition of the transpose as a functor establishing a duality between $\underline{\mathrm{mod}}\ \Lambda$ and $\underline{\mathrm{mod}}\ \Lambda^{\mathrm{op}}$.

The specific issue dealt with in this chapter of when two artin algebras are stably equivalent came up in two quite different situations. One was in modular group representations where the Green correspondence shows that the restriction morphism sometimes induces a stable equivalence between a group and some subgroups. It was in this connection that the problem of whether two stably equivalent artin algebras have the same number, up to isomorphism, of nonprojective simple modules first came up.

The other early case where stable equivalence came up was in [AuR1] where the artin algebras stably equivalent to hereditary algebras are described. This includes the result shown here that all radical square zero algebras are stably equivalent to hereditary algebras. The proof given here is inspired by [Au3]. There is a generalization in [M0], via the concept of a simple module being a 'node'.

The result that the artin algebras over an algebraically closed field stably equivalent to the symmetric Nakayama algebras are exactly those given by Brauer trees was proven in [GaRi]. Our approach follows [Lin].

The material in Section 1 is taken from [AuR6] [AuR7], but most proofs are different, since we do not use functor categories. For the use of functor categories in the study of stable equivalence we refer to [AuR1] and [M3].

XI

Modules determining morphisms

The basic premise of this book has been that one should study the morphisms between modules as a way of studying the modules themselves. In this enterprise, two special types of morphisms have played a particularly important role, split epimorphisms (monomorphisms) and right (left) almost split morphisms. Whether or not a morphism $f: B \to C$ is one of these types is determined by which morphisms $X \to C$ for arbitrary X can be factored through f. In fact, this situation is not as special as it seems since all right minimal morphisms $f: B \to C$ are determined by which morphisms $X \to C$ can be factored through f. A careful analysis of this observation leads to the notion of morphisms determined by modules, in terms of which a classification theorem for right minimal morphisms is given. We also get existence theorems for morphisms to an indecomposable module C which contain the existence of minimal right almost split morphisms as a special case.

In Chapter V we introduced the notion of rigid exact sequences and showed that split and almost split sequences are rigid. We use here our classification of morphisms to study further when exact sequences are rigid. In this connection, we introduce the notion of modules with waists, a notion we further exploit in studying when almost split sequences have indecomposable middle terms. This problem was also first discussed in Chapter V.

1 Morphisms determined by a module

In I Section 2 we introduced for a fixed C in $\operatorname{mod}\Lambda$ the category $\operatorname{mod}\Lambda/C$ whose objects are the morphisms $f: B \to C$ in $\operatorname{mod}\Lambda$ and whose morphisms from $f_1: B_1 \to C$ to $f_2: B_2 \to C$ are the morphisms $g: B_1 \to B_2$ such that $f_1 = f_2 g$. Our purpose there was to show that given a morphism $f: B \to C$ in $\text{mod}\,\Lambda/C$ there is a decomposition $B = B_0 \coprod B_1$ such that $f|_{B_0}: B_0 \to C$ is right minimal and $f|_{B_1}: B_1 \to C$ is the zero morphism. Our purpose now is to study the category $\text{mod}\,\Lambda/C$ by considering for each module A in $\text{mod}\,\Lambda$ the full subcategory $\text{Ker}_A(\text{mod}\,\Lambda/C)$ of $\text{mod}\,\Lambda$ whose objects are all the $f: B \to C$ with $\text{Ker}\,f$ in $\text{add}\,A$. We first consider the following problem. Given two morphisms $f_1: B_1 \to C$ and $f_2: B_2 \to C$, when is $\text{Hom}(f_1, f_2) \neq \emptyset$? Our answer leads to the notion of a morphism being determined by a module.

Suppose $f: B \to C$ is a fixed morphism. Given an arbitrary morphism $f': B' \to C$ we first show how to describe when $\text{Hom}(f', f) \neq \emptyset$ in terms of the pullback diagram

$$
\begin{array}{ccc}
B \times_C B' & \xrightarrow{\rho} & B' \\
\downarrow & & \downarrow f' \\
B & \xrightarrow{f} & C.
\end{array}
$$

We keep this notation for most of this section.

The reader should observe that a morphism $h: X \to B'$ factors through $\rho: B \times_C B' \to B'$, i.e. there is some $l: X \to B \times_C B'$ such that $\rho l = h$, if and only if the composition $f'h: X \to C$ factors through f. This is a trivial consequence of the mapping properties of pullbacks.

Lemma 1.1

(a) $\text{Im}\,f' \subset \text{Im}\,f$ if and only if $\rho: B \times_C B' \to B'$ is an epimorphism.

(b) *The morphism f' factors through f, i.e. $\text{Hom}(f', f) \neq \emptyset$, if and only if $\rho: B \times_C B' \to B'$ is a split epimorphism.*

Proof (a) Clearly ρ is an epimorphism if and only if every morphism $g: \Lambda \to B'$ factors through ρ. Now every $g: \Lambda \to B'$ factors through ρ if and only if the compositions $\Lambda \xrightarrow{g} B' \xrightarrow{f'} C$ factor through f. But this is equivalent to $\text{Im}\,f' \subset \text{Im}\,f$.

(b) The morphism ρ is a split epimorphism if and only if $1_{B'}$ factors through ρ. But $1_{B'}$ factors through ρ if and only if $f' = f'1_{B'}$ factors through f, giving our desired result. $\square$

We now give a different description of when $\rho: B \times_C B' \to B'$ is an epimorphism or a split epimorphism.

Lemma 1.2 *Let $f: B \to C$ be in $\text{mod}\,\Lambda$ and let $A = \text{Ker}\,f$.*

(a) *Suppose $\rho: B \times_C B' \to B'$ is an epimorphism. Then ρ is a split epi-morphism if and only if every morphism $\operatorname{Tr} DA \to B'$ factors through ρ.*

(b) *$\rho: B \times_C B' \to B'$ is a split epimorphism if and only if*
$$\operatorname{Hom}_\Lambda(\operatorname{Tr} DA \coprod \Lambda, B \times_C B') \xrightarrow{\operatorname{Hom}(\operatorname{Tr} DA\amalg\Lambda, \rho)} \operatorname{Hom}_\Lambda(\operatorname{Tr} DA \coprod \Lambda, B')$$
is an epimorphism.

Proof (a) By the basic properties of pullbacks we know that $A \simeq \operatorname{Ker} \rho$, so we have the exact sequence $0 \to A \xrightarrow{h} B \times_C B' \xrightarrow{\rho} B' \to 0$. Therefore we have that ρ is a split epimorphism if and only if every morphism $A \to A$ factors through h. But this is equivalent by IV Corollary 4.4 to every morphism $\operatorname{Tr} DA \to B'$ factoring through ρ.

(b) This is a trivial consequence of (a) and the fact that ρ is an epi-morphism if and only if $\operatorname{Hom}_\Lambda(\Lambda, B \times_C B') \xrightarrow{\operatorname{Hom}(\Lambda, \rho)} \operatorname{Hom}_\Lambda(\Lambda, B')$ is an epimorphism. $\square$

Combining Lemmas 1.1 and 1.2 with the mapping properties of pull-backs we have the following.

Proposition 1.3 *Let $0 \to A \to B \xrightarrow{f} C$ be an exact sequence in* $\operatorname{mod} \Lambda$.

(a) *Suppose $f': B' \to C$ is a morphism such that $\operatorname{Im} f' \subset \operatorname{Im} f$. Then f' factors through f if and only if the composition $\operatorname{Tr} DA \xrightarrow{t} B' \xrightarrow{f'} C$ factors through f for all t in $\operatorname{Hom}_\Lambda(\operatorname{Tr} DA, B')$.*

(b) *If $f': B' \to C$ is an arbitrary morphism, then f' factors through f if and only if the composition $\operatorname{Tr} DA \coprod \Lambda \xrightarrow{t} B' \xrightarrow{f'} C$ factors through f for all t in $\operatorname{Hom}_\Lambda(\operatorname{Tr} DA \coprod \Lambda, B')$.*

This basic result suggests the following definition. We say that a morphism $f: B \to C$ is **right determined by a module** X, or just deter-mined by a module X, if a morphism $f': B' \to C$ factors through f whenever $\operatorname{Im} \operatorname{Hom}_\Lambda(X, f') \subset \operatorname{Im} \operatorname{Hom}_\Lambda(X, f)$. It is easy to see that a morphism $f': B' \to C$ factors through f if and only if $\operatorname{Im} \operatorname{Hom}_\Lambda(M, f') \subset \operatorname{Im} \operatorname{Hom}_\Lambda(M, f)$ for all M in $\operatorname{mod} \Lambda$. Therefore saying that a morphism is right determined by a module is a finiteness condition which, as we now point out, all morphisms enjoy.

Corollary 1.4 *Let $0 \to A \to B \xrightarrow{f} C$ be an exact sequence. Then f is always right determined by $\operatorname{Tr} DA \coprod \Lambda$. Further, if f is an epimorphism, then it is right determined by $\operatorname{Tr} DA$.*

Proof This is an easy consequence of Proposition 1.3 and the definition of a morphism being determined by a module. □

Before moving on to develop more of the general theory of morphisms determined by modules, we discuss a few interesting but somewhat more specific questions suggested by our results so far.

Let $0 \to A \to B \xrightarrow{f} C$ be an exact sequence. Then we know that f is determined by $\text{Tr}\,DA \coprod \Lambda$. In view of this result it is natural to wonder if f is ever right determined just by Λ or just by $\text{Tr}\,DA$. We have already seen that when f is an epimorphism, then f is right determined by $\text{Tr}\,DA$. However, as we shall see, it is not necessary for f to be an epimorphism in order to be right determined by $\text{Tr}\,DA$.

We begin this discussion with the following criterion for f to be right determined by Λ.

Proposition 1.5 *Let $f : B \to C$ be a morphism in* $\text{mod}\,\Lambda$.

(a) *f is right determined by Λ if and only if the induced epimorphism $B \to \text{Im}\,f$ splits.*

(b) *If f is right minimal, then f is right determined by Λ if and only if it is a monomorphism.*

Proof (a) We have already seen that for a morphism $g : X \to C$ we have $\text{Im}\,g \subset \text{Im}\,f$ if and only if for each morphism $h : \Lambda \to X$, the composition $\Lambda \xrightarrow{h} X \xrightarrow{g} C$ factors through f. Hence if f is determined by Λ, then the inclusion $\text{Im}\,f \to C$ factors through f. But this means that the epimorphism $B \to \text{Im}\,f$ splits.

On the other hand, suppose $B \to \text{Im}\,f$ is a split epimorphism. Then there is a morphism $s : \text{Im}\,f \to B$ such that $fs(z) = z$ for all z in $\text{Im}\,f$. Let $h : X \to C$ be a morphism such that for each $t : \Lambda \to X$ the composition ht factors through f. This means that $\text{Im}\,h \subset \text{Im}\,f$. Hence we have that $h = (fs)h$, so h factors through f. This shows that f is determined by Λ.

(b) This is a trivial consequence of (a). □

We now take up the issue of when a morphism f is right determined by $\text{Tr}\,D(\text{Ker}\,f)$.

Proposition 1.6 *Let $0 \to A \to B \xrightarrow{f} C$ be an exact sequence. Then the following are equivalent.*

(a) *The morphism* $f: B \to C$ *is right determined by* $\operatorname{Tr} DA$.
(b) *A morphism* $g: \Lambda \to C$ *factors through* f *if for each morphism* $h: \operatorname{Tr} DA \to \Lambda$, *the composition* $\operatorname{Tr} DA \xrightarrow{h} \Lambda \xrightarrow{g} C$ *factors through* f.

Proof (a) $\Rightarrow$ (b) This is a trivial consequence of f being right determined by $\operatorname{Tr} DA$.

(b) $\Rightarrow$ (a) Let $s: X \to C$ be a morphism such that for each $t: \operatorname{Tr} DA \to X$ the composition st factors through f. By Proposition 1.3, in order to show that s factors through f, it suffices to show that $\operatorname{Im} s \subset \operatorname{Im} f$. Let $u: n\Lambda \to X$ be an epimorphism. Now for each $v: \operatorname{Tr} DA \to n\Lambda$ the composition $\operatorname{Tr} DA \xrightarrow{v} n\Lambda \xrightarrow{u} X \xrightarrow{s} C$ factors through f. Hence the morphism $su: n\Lambda \to C$ factors through f since for each $v: \operatorname{Tr} DA \to n\Lambda$ the composition $(su)v$ factors through f. This combined with the fact that $u: n\Lambda \to X$ is an epimorphism implies that $\operatorname{Im} s \subset \operatorname{Im} f$. $\qquad\square$

We now give an example illustrating these results.

Example Let p be a prime element in a principal ideal domain R and let $\Lambda = R/(p^n)$ for some integer $n > 0$. Then Λ is a Nakayama local ring with $A_i = R/(p^i)$, $i = 1, \ldots, n$, a complete set of nonisomorphic indecomposable Λ-modules. Let $f: A_n \to A_i$ be a morphism such that $\operatorname{Ker} f \simeq A_j$ with $1 \le j \le n$. Thus we have the exact sequence $0 \to A_j \to A_n \xrightarrow{f} A_i$. Since Λ is a local Nakayama algebra, we have that $\operatorname{Tr} DA_j \simeq A_j$. Using Proposition 1.6, it is not difficult to see that f is right determined by A_j. For it is clear that $t: A_n \to A_i$ can be factored through f if and only if $\operatorname{Ker} t \simeq A_u$ with $u \ge j$. Hence $t: A_n \to A_i$ factors through f if and only if the composition $A_j \xrightarrow{g} A_n \xrightarrow{t} A_i$ is zero for all morphisms $g: A_j \to A_n$. But zero is the only morphism $A_j \to A_i$ factoring through f. Therefore $t: A_n \to A_i$ factors through f if and only if the composition $A_j \xrightarrow{g} A_n \xrightarrow{t} A_i$ factors through f for all morphisms $g: A_j \to A_n$. This shows by Proposition 1.6 that f is right determined by $A_j \simeq \operatorname{Tr} D(\operatorname{Ker} f)$. We also have

(a) f is an epimorphism if and only if $i = n - j$ and
(b) $\operatorname{Im} \operatorname{Hom}_\Lambda(A_j, f) = 0$.

Thus for example if we choose $i = n$, then $f: A_n \to A_n$ is an example of a morphism which is right determined by $\operatorname{Tr} D(\operatorname{Ker} f)$ even though it is not an epimorphism.

Of course there is a dual theory for morphisms left determined by a module. We say a morphism $g: A \to B$ is **left determined by a module** X if a morphism $g': A' \to B$ factors through g whenever hg' factors through g for all $h: B' \to X$, i.e. there is a morphism $t: B \to B'$ such that $tg = g'$. Then, by duality, we have that a morphism $g: A \to B$ is left determined by $D\operatorname{Tr}(\operatorname{Coker} g) \coprod D(\Lambda)$ and that it is left determined by $D\operatorname{Tr}(\operatorname{Coker} g)$ if it is a monomorphism. We leave it to the reader to develop this dual theory.

Before returning to the general theory of morphisms determined by modules we introduce some notation and make some basic observations which we will use freely in the rest of the chapter. For each module X in $\operatorname{mod}\Lambda$ we denote $\operatorname{End}_\Lambda(X)^{\operatorname{op}}$ by Γ_X. Then for each Y in $\operatorname{mod}\Lambda$ we know that $\operatorname{Hom}_\Lambda(X, Y)$ is a Γ_X-module. Suppose now that $f: B \to C$ is a morphism in $\operatorname{mod}\Lambda$. We know that the morphism $\operatorname{Hom}_\Lambda(X, f): \operatorname{Hom}_\Lambda(X, B) \to \operatorname{Hom}_\Lambda(X, C)$ is functorial in X. Hence $\operatorname{Hom}_\Lambda(X, f)$ is a morphism of Γ_X-modules. Therefore $\operatorname{Im}\operatorname{Hom}_\Lambda(X, f)$ is a Γ_X-submodule of $\operatorname{Hom}_\Lambda(X, C)$ which, as we shall see, plays a critical role in the theory of morphisms determined by modules.

2 Modules determining a morphism

Let $f: B \to C$ be a morphism in $\operatorname{mod}\Lambda/C$. This section is devoted to exploring the structure of modules X which determine f. We first show that there is a unique, up to isomorphism, module $T(f)$ which determines f and such that $T(f)$ is a summand of X for all modules X determining f. We next show that there is a unique, up to isomorphism, module $U(f)$ in $\operatorname{add} T(f)$ such that $\operatorname{Hom}_\Lambda(X, U(f))$ is a Γ_X-projective cover of $\operatorname{soc}_{\Gamma_X}(\operatorname{Hom}_\Lambda(X, C)/\operatorname{Im}\operatorname{Hom}_\Lambda(X, f))$ for all modules X which determine f. Finally we show that $D\operatorname{Tr} U(f) \simeq \operatorname{Ker} f$.

We begin with the following notion which will play a crucial role in our discussion of the structure of the modules which determine a fixed morphism $f: B \to C$.

Let T be an indecomposable module. A morphism $g: T \to C$ is said to **almost factor through** f if it satisfies the following two conditions:

(a) g does not factor through f and
(b) for each module U and each morphism $h: U \to T$ in $\operatorname{rad}_\Lambda(U, T)$ the composition $U \overset{h}{\to} T \overset{g}{\to} C$ factors through f.

We say that an indecomposable module T almost factors through f if there is a morphism $T \to C$ which almost factors through f.

Even though the definition of an indecomposable module almost factoring through a morphism $f: B \to C$ has nothing to do with the modules which determine f, the concepts are nonetheless intimately related, as we now show.

Lemma 2.1 *Suppose the indecomposable module T almost factors through a morphism $f: B \to C$. Then T is a summand of X for each module X which determines f.*

Proof Since T almost factors through f, there is a morphism $g: T \to C$ which almost factors through f. Suppose T is not a summand of X where X determines f. Then each morphism $h: X \to T$ is in $\mathrm{rad}_\Lambda(X, T)$ and so the composition $X \xrightarrow{h} T \xrightarrow{g} C$ factors through $f: B \to C$ for all h in $\mathrm{Hom}_\Lambda(X, T)$. Therefore g factors through f since f is determined by X. This contradiction shows that T is a summand of X. $\square$

In view of this result, it is natural to ask which indecomposable summands of a module determining a morphism $f: B \to C$ almost factor through f. We will obtain an answer to this question as a consequence of the following result.

Proposition 2.2 *Suppose Y is an indecomposable summand of a module X which determines the morphism $f: B \to C$. Then a morphism $g: Y \to C$ almost factors through f if and only if the image of the composition of the Γ_X-morphisms $\mathrm{Hom}_\Lambda(X, Y) \xrightarrow{\mathrm{Hom}_\Lambda(X,g)} \mathrm{Hom}_\Lambda(X, C) \xrightarrow{p} \mathrm{Hom}_\Lambda(X, C)/\mathrm{Im}\,\mathrm{Hom}_\Lambda(X, f)$ is a simple Γ_X-module, where p is the canonical epimorphism.*

Proof Suppose $g: Y \to C$ almost factors through f. Since g does not factor through f which is determined by X, we have that $\mathrm{Im}(\mathrm{Hom}_\Lambda(X, Y) \xrightarrow{\mathrm{Hom}_\Lambda(X,g)} \mathrm{Hom}_\Lambda(X, C))$ is not contained in $\mathrm{Im}\,\mathrm{Hom}_\Lambda(X, f)$. Also we have that $\mathrm{Hom}_\Lambda(X, g)(\mathrm{rad}_\Lambda(X, Y)) \subset \mathrm{Im}\,\mathrm{Hom}_\Lambda(X, f)$ since for each h in $\mathrm{rad}_\Lambda(X, Y)$ the composition gh factors through f. Combining these observations with the fact that Y being an indecomposable summand of X implies that $\mathrm{Hom}_\Lambda(X, Y)/\mathrm{rad}_\Lambda(X, Y)$ is a simple Γ_X-module, we have that the image of the composition $p\mathrm{Hom}_\Lambda(X, g)$ is isomorphic to the simple Γ_X-module $\mathrm{Hom}_\Lambda(X, Y)/\mathrm{rad}_\Lambda(X, Y)$. This proves our desired result in one direction.

Suppose now that the image of the composition $p\mathrm{Hom}_\Lambda(X, g)$ is a simple Γ_X-module. Then $\mathrm{Hom}_\Lambda(X, g)(\mathrm{rad}_\Lambda(X, Y)) \subset \mathrm{Im}\,\mathrm{Hom}_\Lambda(X, f)$ and

$g: Y \to C$ does not factor through f. Let $t: A \to Y$ be in $\text{rad}_\Lambda(A, Y)$. Then for each $u: X \to A$, the composition $X \xrightarrow{u} A \xrightarrow{t} Y$ is in $\text{rad}_\Lambda(X, Y)$ which means that the composition $X \xrightarrow{u} A \xrightarrow{t} Y \xrightarrow{g} C$ factors through f. Hence for each $u: X \to A$, the composition $(gt)u$ factors through f. This implies that the composition $A \xrightarrow{t} Y \xrightarrow{g} C$ factors through f for all t in $\text{rad}_\Lambda(A, Y)$. Therefore $g: Y \to C$ almost factors through f. $\square$

As an immediate consequence of these results we have the following result whose proof we leave to the reader.

Corollary 2.3 *Let X be a Λ-module which determines a morphism $f: B \to C$ and let $\{S_1, \ldots, S_t\}$ be a complete set of nonisomorphic simple Γ_X-submodules of $\text{Hom}_\Lambda(X, C) / \text{Im} \, \text{Hom}_\Lambda(X, f)$. Then the indecomposable non-isomorphic summands $Y_1, \ldots, Y_t$ of X such that $\text{Hom}_\Lambda(X, Y_i)$ is a Γ_X-projective cover of S_i for all $i = 1, \ldots, t$ are a complete set, up to isomorphism, of nonisomorphic indecomposable modules which almost factor through f.* $\square$

Before giving our next result it is convenient to introduce the following notation. Let $\{T_1, \ldots, T_t\}$ be a complete set of nonisomorphic indecomposable modules which almost factor through a morphism $f: B \to C$. We denote by $T(f)$ any module isomorphic to $T_1 \coprod \cdots \coprod T_t$. Because of the following result, we call $T(f)$ the **minimal module determining** f.

Proposition 2.4 *Let $f: B \to C$ be a morphism and let $T(f)$ be as above. Then $T(f)$ has the following properties.*

(a) *The module $T(f)$ determines f.*

(b) *A module X determines f if and only if $T(f)$ is a summand of X.*

Proof (a) Let X be a module determining f. Then by Corollary 2.3 we know that $T(f)$ is a summand of X. Let $\{T_1, \ldots, T_t\}$ be a complete set of nonisomorphic indecomposable summands of $T(f)$. To show that $T(f)$ determines f, it suffices to show that if a morphism $g: A \to C$ does not factor through f, then there is a morphism $h: T_i \to A$ such that the composition $gh: T_i \to C$ does not factor through f.

Let $\{S_1, \ldots, S_t\}$ be a complete set of nonisomorphic simple Γ_X-submodules of $\text{Hom}_\Lambda(X, C) / \text{Im} \, \text{Hom}_\Lambda(X, f)$ labelled in such a way that $\text{Hom}_\Lambda(X, T_i)$ is a projective cover of S_i for all $i = 1, \ldots, t$. We know we can do this by Corollary 2.3. Suppose that $g: A \to C$ is a morphism which does not factor through f. Then the composition

$$\mathrm{Hom}_\Lambda(X, A) \xrightarrow{\mathrm{Hom}_\Lambda(X,g)} \mathrm{Hom}_\Lambda(X, C) \xrightarrow{p} \mathrm{Hom}_\Lambda(X, C)/\mathrm{Im}\,\mathrm{Hom}_\Lambda(X, f)$$

is not zero so there is some S_i contained in its image. Since $\mathrm{Hom}_\Lambda(X, T_i)$ is a projective cover of S_i, there is a morphism $u : \mathrm{Hom}_\Lambda(X, T_i) \to \mathrm{Hom}_\Lambda(X, A)$ such that $\mathrm{Im}(p\mathrm{Hom}_\Lambda(X, g)u) = S_i$. Because T_i is a summand of X there is a morphism $v : T_i \to A$ such that $u = \mathrm{Hom}_\Lambda(X, v)$. Then the composition $T_i \xrightarrow{v} A \xrightarrow{g} C$ does not factor through f since $\mathrm{Im}\,\mathrm{Hom}_\Lambda(X, gv)$ is not contained in $\mathrm{Im}\,\mathrm{Hom}_\Lambda(X, f)$ and f is determined by X.

(b) This is a trivial consequence of (a). $\qquad\square$

For each morphism $f : B \to C$ we have introduced the module $T(f)$ which is an invariant of f. We now introduce another module which is also an invariant of f.

Let $T(f)$ be a minimal module determining a morphism $f : B \to C$. Define $U(f)$ to be a module in $\mathrm{add}\,T(f)$ such that $\mathrm{Hom}_\Lambda(T(f), U(f))$ is a $\Gamma_{T(f)}$-projective cover of $\mathrm{soc}_{\Gamma_{T(f)}}(\mathrm{Hom}_\Lambda(T(f), C)/\mathrm{Im}\,\mathrm{Hom}_\Lambda(T(f), f))$. Since $T(f)$ is uniquely determined by f, up to isomorphism, the module $U(f)$ is also uniquely determined by f, up to isomorphism. Also we know there is a morphism $g : U(f) \to C$ such that the composition of $\Gamma_{T(f)}$-modules

$$\mathrm{Hom}_\Lambda(T(f), U(f)) \xrightarrow{\mathrm{Hom}_\Lambda(T(f),g)} \mathrm{Hom}_\Lambda(T(f), C) \to$$
$$\mathrm{Hom}_\Lambda(T(f), C)/\mathrm{Im}\,\mathrm{Hom}_\Lambda(T(f), f)$$

is a $\Gamma_{T(f)}$-projective cover of $\mathrm{soc}_{\Gamma_{T(f)}}(\mathrm{Hom}_\Lambda(T(f), C)/\mathrm{Im}\,\mathrm{Hom}_\Lambda(T(f), f))$. We call the Λ-modules $U(f)$ and the morphisms $g : U(f) \to C$ having these properties **minimal covers of the bottom** of the morphism f. It is an immediate consequence of Corollary 2.3 that $T(f)$ is a summand of $U(f)$.

It is clear that the minimal covers of the bottom of a morphism are an invariant of the morphism. We now show how this invariant of a morphism f is connected to the modules determining f.

Proposition 2.5 *Let* $g : U(f) \to C$ *be a minimal cover for the bottom of the morphism* $f : B \to C$. *If* X *is a module determining* f, *then the composition of* Γ_X-*morphisms*

$$\text{Hom}_\Lambda(X, U(f)) \xrightarrow{\text{Hom}_\Lambda(X,g)} \text{Hom}_\Lambda(X, C) \xrightarrow{p} \text{Hom}_\Lambda(X, C)/\operatorname{Im}\text{Hom}_\Lambda(X, f)$$

is a Γ_X-projective cover of $\text{soc}_{\Gamma_X}(\text{Hom}_\Lambda(X, C)/\operatorname{Im}\text{Hom}_\Lambda(X, f))$.

Proof It follows from the definition of $g: U(f) \to C$ and Proposition 2.2 that $\operatorname{Im}(p\text{Hom}_\Lambda(X, g))$ is contained in $\text{soc}_{\Gamma_X}(\text{Hom}_\Lambda(X, C)/\operatorname{Im}\text{Hom}_\Lambda(X, f))$. We now show $\operatorname{Im}(p\text{Hom}_\Lambda(X, g)) = \text{soc}_{\Gamma_X}(\text{Hom}_\Lambda(X, C)/\operatorname{Im}\text{Hom}_\Lambda(X, f))$. Suppose that S is a simple Γ_X-submodule of $\text{Hom}_\Lambda(X, C)/\operatorname{Im}\text{Hom}_\Lambda(X, f)$. Let T be an indecomposable summand of X such that $\text{Hom}_\Lambda(X, T)$ is a projective cover of S. Then there is a morphism $h: T \to C$ such that the composition $p\text{Hom}_\Lambda(X, h): \text{Hom}_\Lambda(X, T) \to \text{Hom}_\Lambda(X, C)/\operatorname{Im}\text{Hom}_\Lambda(X, f)$ has S as its image. From this it follows that $h: T \to C$ almost factors through f. Hence there is a morphism $t: T \to U(f)$ such that $gt = h$. From this it follows that S is contained in the image of the composition $\text{Hom}_\Lambda(X, U(f)) \xrightarrow{\text{Hom}_\Lambda(X,g)} \text{Hom}_\Lambda(X, C) \xrightarrow{p} \text{Hom}_\Lambda(X, C)/\operatorname{Im}\text{Hom}_\Lambda(X, f)$. Therefore the image of this composition is $\text{soc}_{\Gamma_X}(\text{Hom}_\Lambda(X, C)/\operatorname{Im}\text{Hom}_\Lambda(X, f))$. The fact that this composition is a Γ_X-projective cover of its image is an easily verified consequence of the fact that $g: U(f) \to C$ is right minimal. $\qquad\square$

Throughout the rest of this section we assume that X is a fixed Λ-module. Let $f: B \to C$ be a morphism in $\mod \Lambda$. For the sake of notational brevity we denote the Γ_X-submodule $\operatorname{Im}\text{Hom}_\Lambda(X, f)$ of $\text{Hom}_\Lambda(X, C)$ by H_f. Assume now that f is determined by X. Our aim in the rest of this section is to show that a great deal of explicit information about the exact sequence $0 \to \operatorname{Ker} f \to B \xrightarrow{f} C$ can be deduced from the associated exact sequence of Γ_X-modules $0 \to H_f \to \text{Hom}_\Lambda(X, C) \to \text{Hom}_\Lambda(X, C)/H_f \to 0$, especially when f is right minimal.

We begin our discussion with the following elementary observation.

Lemma 2.6 *Let $f_0: B_0 \to C$ be a right minimal version of a morphism $f: B \to C$ in $\mod \Lambda/C$. Then we have the following.*

(a) $H_f = H_{f_0}$.

(b) *The morphism $f: B \to C$ is right determined by the module X if and only if $f_0: B_0 \to C$ is right determined by the module X.*

Proof (a) This is an obvious consequence of the definitions involved.

(b) This follows easily from (a) and the fact that we have the commutative diagram

$$\begin{array}{ccc} B_0 & \xrightarrow{f_0} & C \\ \downarrow i & & \| \\ B & \xrightarrow{f} & C \\ \downarrow p & & \| \\ B_0 & \xrightarrow{f_0} & C \end{array}$$

where i is the inclusion and p is the projection. □

This lemma shows that there is no serious loss of generality if we restrict ourselves to considering only right minimal morphisms in mod Λ/C when studying morphisms determined by the module X. This point of view is further supported by the following.

Lemma 2.7 *Let* $f_1: B_1 \to C$ *and* $f_2: B_2 \to C$ *be two right minimal morphisms in* mod Λ/C *which are determined by the module* X. *Then* f_1 *and* f_2 *are isomorphic in* mod Λ/C *if and only if* $H_{f_1} = H_{f_2}$.

Proof Since it is clear that $H_{f_1} = H_{f_2}$ if f_1 and f_2 are isomorphic in mod Λ/C, we only show that if $H_{f_1} = H_{f_2}$, then f_1 is isomorphic to f_2 in mod Λ/C.

Suppose $H_{f_1} = H_{f_2}$. Since $H_{f_1} \subset H_{f_2}$ and f_2 is right determined by X, there is a commutative diagram

$$\begin{array}{ccc} B_1 & \xrightarrow{f_1} & C \\ \downarrow g_1 & & \| \\ B_2 & \xrightarrow{f_2} & C. \end{array}$$

Since $H_{f_1} \supset H_{f_2}$ and f_1 is right determined by X, there is a commutative diagram

$$\begin{array}{ccc} B_2 & \xrightarrow{f_2} & C \\ \downarrow g_2 & & \| \\ B_1 & \xrightarrow{f_1} & C. \end{array}$$

Since both f_1 and f_2 are right minimal, it follows that the compositions $g_2 g_1$ and $g_1 g_2$ are isomorphisms, giving our desired result. □

Let $0 \to A \xrightarrow{f} B \xrightarrow{g} C$ be an exact sequence with g a right minimal morphism determined by X. We know by Lemma 2.7 that, up to isomorphism, this sequence is uniquely determined by the Γ_X-submodule H_g

of $\text{Hom}_\Lambda(X, C)$ where $H_g = \text{Im}\,\text{Hom}_\Lambda(X, g)$. In particular, the modules A and B are determined, up to isomorphism, by the submodule H_g of $\text{Hom}_\Lambda(X, C)$. While there is no direct way known of describing B in terms of the Γ_X-submodule H_g of $\text{Hom}_\Lambda(X, C)$, there is a satisfactory answer for the module A.

Let Y be in add X such that $\text{Hom}_\Lambda(X, Y)$ is a projective cover of the Γ_X-socle of $\text{Hom}_\Lambda(X, C)/H_g$. We will prove that $A \simeq D\,\text{Tr}\,Y$. The proof goes in several steps.

For convenience of notation we make the following conventions. Let $\delta : 0 \to U \to V \to W \to 0$ be an exact sequence of Λ-modules. Then we view the induced exact sequence

$$\text{Hom}_\Lambda(X, V) \to \text{Hom}_\Lambda(X, W) \to \text{Ext}^1_\Lambda(X, U)$$

as a sequence of Γ_X-modules and consider the induced Γ_X-isomorphism $\delta^*(X) \xrightarrow{\sim} \text{Im}(\text{Hom}_\Lambda(X, W) \to \text{Ext}^1_\Lambda(X, U))$ as an identification.

Proposition 2.8 *Let $\delta : 0 \to A \to E \to \text{Tr}\,DA \to 0$ be an almost split sequence and X a module such that $\text{Tr}\,DA$ is in add X.*

(a) *Suppose $\epsilon : 0 \to A \xrightarrow{f} B \xrightarrow{g} C \to 0$ is a nonsplit exact sequence. Then we have $\epsilon^*(X) \supset \delta^*(X) \neq 0$.*

(b) *The Γ_X-module $\delta^*(X)$ is simple and is the Γ_X-socle of $\text{Ext}^1_\Lambda(X, A)$.*

(c) *$\text{Hom}_\Lambda(X, \text{Tr}\,DA)$ is a Γ_X-projective cover of the Γ_X-socle of $\epsilon^*(X)$ which is $\delta^*(X)$.*

Proof (a) Since δ is not split, $\delta^*(\text{Tr}\,DA) \neq 0$. The hypothesis that $\text{Tr}\,DA$ is in add X then implies that $\delta^*(X) \neq 0$.

Suppose now that $\epsilon : 0 \to A \to B \to C \to 0$ is a nonsplit exact sequence. Then by the basic properties of almost split sequences we have a commutative exact diagram

$$
\begin{array}{ccccccccc}
\delta : & 0 & \longrightarrow & A & \longrightarrow & E & \longrightarrow & \text{Tr}\,DA & \longrightarrow & 0 \\
& & & \| & & \downarrow & & \downarrow & & \\
\epsilon : & 0 & \longrightarrow & A & \longrightarrow & B & \longrightarrow & C & \longrightarrow & 0.
\end{array}
$$

This gives rise to the commutative diagram

$$
\begin{array}{ccc}
\text{Hom}_\Lambda(X, \text{Tr}\,DA) & \longrightarrow & \text{Ext}^1_\Lambda(X, A) \\
\downarrow & & \| \\
\text{Hom}_\Lambda(X, C) & \longrightarrow & \text{Ext}^1_\Lambda(X, A)
\end{array}
$$

which shows that $\epsilon^*(X) \supset \delta^*(X)$.

(b) Let $\epsilon: 0 \to A \to B \to X \to 0$ be a nonzero element of $\operatorname{Ext}_\Lambda^1(X, A)$. It is well known that the coboundary morphism $\operatorname{Hom}_\Lambda(X, X) \to \operatorname{Ext}_\Lambda^1(X, A)$ given by ϵ^* carries the identity to ϵ. Hence $\Gamma_X \epsilon = \epsilon^*(X) \supset \delta^*(X)$ by part (a). Therefore $\delta^*(X)$ is simple and equal to the socle of $\operatorname{Ext}_\Lambda^1(X, A)$.

(c) By definition we have an epimorphism $\operatorname{Hom}_\Lambda(X, \operatorname{Tr} DA) \to \delta^*(X)$. Since $\operatorname{Tr} DA$ is an indecomposable Λ-module in $\operatorname{add} X$, we have that $\operatorname{Hom}_\Lambda(X, \operatorname{Tr} DA)$ is an indecomposable projective Γ_X-module. Therefore it is the Γ_X-projective cover of $\delta^*(X)$, which is the socle of $\epsilon^*(X)$ by (b). $\square$

We now prove the following special case of the general theorem we are aiming for.

Proposition 2.9 *Let* $0 \to A \xrightarrow{f} B \xrightarrow{g} C \to 0$ *be an exact sequence where* g *is right minimal and determined by a module* X. *Let* Y *in* $\operatorname{add} X$ *be a minimal cover for the bottom of* g. *Then we have* $Y \simeq \operatorname{Tr} DA$.

Proof Let $A = \coprod_{i \in I} A_i$ with the A_i indecomposable modules. Since g is right minimal we have that the induced monomorphisms $A_i \to B$ are not split monomorphisms. Therefore no A_i is injective and if $0 \to A_i \to E_i \to \operatorname{Tr} DA_i \to 0$ is an almost split sequence, we have a commutative, exact diagram

$$
\begin{array}{ccccccccc}
\delta_i: & 0 & \longrightarrow & A_i & \longrightarrow & E_i & \longrightarrow & \operatorname{Tr} DA_i & \longrightarrow & 0 \\
 & & & \downarrow & & \downarrow & & \downarrow & & \\
\epsilon: & 0 & \longrightarrow & \coprod_{i \in I} A_i & \longrightarrow & B & \longrightarrow & C & \longrightarrow & 0.
\end{array}
$$

This gives rise to the exact commutative diagram

$$
\begin{array}{ccccccccc}
\coprod \delta_i: & 0 & \longrightarrow & \coprod_{i \in I} A_i & \longrightarrow & \coprod_{i \in I} E_i & \longrightarrow & \coprod_{i \in I} \operatorname{Tr} DA_i & \longrightarrow & 0 \\
 & & & \| & & \downarrow & & \downarrow & & \\
\epsilon: & 0 & \longrightarrow & \coprod_{i \in I} A_i & \longrightarrow & B & \longrightarrow & C & \longrightarrow & 0
\end{array}
$$

from which we obtain the commutative diagram

$$
\begin{array}{ccc}
\operatorname{Hom}_\Lambda(X, \coprod_{i \in I} \operatorname{Tr} DA_i) & \longrightarrow & \operatorname{Ext}_\Lambda^1(X, \coprod_{i \in I} A_i) \\
\downarrow & & \downarrow \\
\operatorname{Hom}_\Lambda(X, C) & \longrightarrow & \operatorname{Ext}_\Lambda^1(X, \coprod_{i \in I} A_i).
\end{array}
$$

Using that $\operatorname{Ext}_\Lambda^1(X, \ )$ and the socle commute with finite sums, we have by Proposition 2.8 that $(\coprod_{i \in I} \delta_i)^*(X) = \operatorname{soc} \operatorname{Ext}_\Lambda^1(X, \coprod_{i \in I} A_i)$, since $\operatorname{Tr} DA_i$ is in $\operatorname{add} X$ for all i. Our desired result now follows from the fact that $\operatorname{soc} \epsilon^*(X) = (\coprod_{i \in I} \delta_i^*)(X)$ since $\operatorname{Ext}_\Lambda^1(X, \coprod_{i \in I} A_i) \supset \epsilon^*(X) \supset \coprod_{i \in I} \delta_i^*(X)$

and that $\mathrm{Hom}_\Lambda(X, \coprod_{i \in I} \mathrm{Tr}\, DA_i)$ is a Γ_X-projective cover for $\coprod_{i \in I} \delta_i^*(X)$.

□

We are now in position to prove our announced result.

Theorem 2.10 *Let* $0 \to A \xrightarrow{f} B \xrightarrow{g} C$ *be an exact sequence with g right minimal and determined by a module* X. *Let* Y *in* $\mathrm{add}\, X$ *be a minimal cover for the bottom of g. Then* $Y \simeq \mathrm{Tr}\, DA \coprod P$ *with* P *a projective module and* $D\,\mathrm{Tr}\, Y \simeq A$.

Proof The inclusion $\mathrm{Im}\, g \to C$ induces an inclusion $\mathrm{Hom}_\Lambda(X, \mathrm{Im}\, g) \to \mathrm{Hom}_\Lambda(X, C)$. Since $H_g \subset \mathrm{Hom}_\Lambda(X, \mathrm{Im}\, g)$ we obtain the mono-morphism of Γ_X-modules $0 \to \mathrm{Hom}_\Lambda(X, \mathrm{Im}\, g)/H_g \to \mathrm{Hom}_\Lambda(X, C)/H_g$ and hence the monomorphism $0 \to \mathrm{soc}_{\Gamma_X}(\mathrm{Hom}_\Lambda(X, \mathrm{Im}\, g)/H_g) \to \mathrm{soc}_{\Gamma_X}(\mathrm{Hom}_\Lambda(X, C)/H_g)$. By Proposition 2.9 we know that $\mathrm{Hom}_\Lambda(X, \mathrm{Tr}\, DA)$ is a Γ_X-projective cover for $\mathrm{soc}_{\Gamma_X} \mathrm{Hom}_\Lambda(X, \mathrm{Im}\, g)/H_g$. Therefore to obtain our desired result it suffices to show that if a simple submodule S of $\mathrm{soc}_{\Gamma_X}(\mathrm{Hom}_\Lambda(X, C)/H_g)$ has a projective cover $\mathrm{Hom}_\Lambda(X, Z)$, with Z in $\mathrm{add}\, X$ and not projective, then $S \subset \mathrm{soc}_{\Gamma_X}(\mathrm{Hom}_\Lambda(X, \mathrm{Im}\, g)/H_g)$.

So let S be a simple submodule of $\mathrm{soc}_{\Gamma_X}(\mathrm{Hom}_\Lambda(X, C)/H_g)$. Then there is a morphism $h: Z \to C$ which almost factors through g such that S is the image of the composition $\mathrm{Hom}_\Lambda(X, Z) \xrightarrow{\mathrm{Hom}_\Lambda(X,h)} \mathrm{Hom}_\Lambda(X, C) \xrightarrow{p} \mathrm{Hom}_\Lambda(X, C)/H_g$. Now let $j: P \to Z$ be a projective cover. Since Z is not projective, we have that for each $t: X \to P$ the composition jt is in $\mathrm{rad}_\Lambda(X, Z)$. Therefore for all $t: X \to P$ the composition $X \xrightarrow{t} P \xrightarrow{j} Z \xrightarrow{h} C$ factors through g. Because f is determined by X, we have that the composition $P \to Z \to C$ factors through f. This, together with the fact that j is an epimorphism, implies that $\mathrm{Im}\, h \subset \mathrm{Im}\, g$. Therefore S is contained in $\mathrm{Hom}_\Lambda(X, \mathrm{Im}\, g)/H_g$ which shows that Z is a summand of $\mathrm{Tr}\, DA$, our desired result. □

Thus we see that we can calculate the kernel of a morphism $f: B \to C$ from the invariant of f which is the minimal cover of the bottom of f. In the next section we show how to calculate $\mathrm{Im}\, f$ knowing the Γ_X-submodule H_f of $\mathrm{Hom}_\Lambda(X, C)$ when X determines the morphism $f: B \to C$.

3 Classification of morphisms determined by a module

Let $f: B \to C$ be a morphism in $\text{mod}\,\Lambda$ with Λ an artin R-algebra and X a module determining f. In studying the structure of the module X in the previous section the Γ_X-submodule $\text{Im}\,\text{Hom}_\Lambda(X, f)$ of $\text{Hom}_\Lambda(X, C)$ played a crucial role. In view of these results it is natural to ask the following question. Let C and X be arbitrary modules. Given a Γ_X-submodule H of $\text{Hom}_\Lambda(X, C)$, is there a morphism $f: B \to C$ determined by X such that $H = \text{Im}\,\text{Hom}_\Lambda(X, f)$? The main aim of this section is to prove that this question has a positive answer.

Throughout this section we assume that X is a fixed Λ-module. Therefore there is no danger of confusion if we write Γ for Γ_X. Also, as in Section 2, given a morphism $f: B \to C$ we denote $\text{Im}\,\text{Hom}_\Lambda(X, f) \subset \text{Hom}_\Lambda(X, C)$ by H_f. It is also helpful to make the following general observation about pullbacks.

Lemma 3.1 *Let*

$$
\begin{array}{ccc}
B \times_C B' & \xrightarrow{g} & B' \\
\downarrow & & \downarrow f' \\
B & \xrightarrow{h} & C
\end{array}
$$

be a pullback diagram. Then for any module X the induced commutative diagram of Γ-modules

$$
\begin{array}{ccccccc}
\text{Hom}_\Lambda(X, B \times_C B') & \to & \text{Hom}_\Lambda(X, B') & \to & L & \to & 0 \\
\downarrow & & \downarrow & & \downarrow & & \\
\text{Hom}_\Lambda(X, B) & \to & \text{Hom}_\Lambda(X, C) & \to & U & \to & 0
\end{array}
$$

has the property that $L \to U$ is a monomorphism.

Proof Since this is an immediate consequence of the definition of a pullback diagram, we leave the proof to the reader. $\square$

As a first step in proving our announced result, we prove the following general result which is also of interest in its own right.

Suppose $\delta: 0 \to A \xrightarrow{g} B \xrightarrow{f} C \to 0$ is an arbitrary exact sequence. Then we have the exact sequence

$$
\text{Hom}_\Lambda(B, D\,\text{Tr}\,X) \to \text{Hom}_\Lambda(A, D\,\text{Tr}\,X) \to \delta_*(D\,\text{Tr}\,X) \to 0
$$

of Σ-modules, where $\Sigma = \text{End}_\Lambda(D\,\text{Tr}\,X)$. Let A_X be in $\text{add}\,D\,\text{Tr}\,X$ and $h: A \to A_X$ a morphism such that the composition

$$
\text{Hom}_\Lambda(A_X, D\,\text{Tr}\,X) \to \text{Hom}_\Lambda(A, D\,\text{Tr}\,X) \to \delta_*(D\,\text{Tr}\,X)
$$

is a Σ-projective cover of $\delta_*(D \operatorname{Tr} X)$. Using this notation we have the following.

Proposition 3.2 *Let*

$$\begin{array}{ccccccccc}
\delta: & 0 & \to & A & \xrightarrow{g} & B & \xrightarrow{f} & C & \to & 0 \\
& & & \downarrow h & & \downarrow j & & \| & & \\
\epsilon: & 0 & \to & A_X & \xrightarrow{g_X} & B_X & \xrightarrow{f_X} & C & \to & 0
\end{array}$$

be a pushout diagram with A_X and $h: A \to A_X$ as above. Then we have the following.

(a) $B_X \xrightarrow{f_X} C$ *is right determined by* X.
(b) $H_f = H_{f_X}$.
(c) f_X *is right minimal.*

Proof (a) Since $0 \to A_X \xrightarrow{g_X} B_X \xrightarrow{f_X} C \to 0$ is exact with A_X in add $D \operatorname{Tr} X$, we know by Corollary 1.4 that f_X is right determined by X.

(b) Since f factors through f_X we know that $H_f \subset H_{f_X}$. From this it follows that we have the exact commutative diagram

$$\begin{array}{ccccccc}
& & & & & 0 & \\
& & & & & \downarrow & \\
& & & & & H_{f_X}/H_f & \\
& & & & & \downarrow & \\
\operatorname{Hom}_\Lambda(X, B) & \xrightarrow{\operatorname{Hom}_\Lambda(X,f)} & \operatorname{Hom}_\Lambda(X, C) & \to & \delta^*(X) & \to & 0 \\
\downarrow & & \| & & \downarrow & & \\
\operatorname{Hom}_\Lambda(X, B_X) & \xrightarrow{\operatorname{Hom}_\Lambda(X,f_X)} & \operatorname{Hom}_\Lambda(X, C) & \to & \epsilon^*(X) & \to & 0 \\
& & & & \downarrow & & \\
& & & & 0 &
\end{array}$$

Hence $H_f = H_{f_X}$ if and only if $\delta^*(X) \to \epsilon^*(X)$ is a monomorphism.

To prove that $\delta^*(X) \to \epsilon^*(X)$ is a monomorphism it suffices to show that $l_R(\epsilon^*(X)) \geq l_R(\delta^*(X))$ since $\delta^*(X) \to \epsilon^*(X)$ is an epimorphism. Now we know by IV Theorem 4.1 that $l_R(\delta^*(X)) = l_R(\delta_*(D \operatorname{Tr} X))$ and $l_R(\epsilon^*(X)) = l_R(\epsilon_*(D \operatorname{Tr} X))$. Therefore we are done if we show that $l_R(\epsilon_*(D \operatorname{Tr} X)) \geq l_R(\delta_*(D \operatorname{Tr} X))$.

Going back to the pushout diagram, we have the following commutative diagram

$$(*) \quad \begin{array}{ccccccc}
\operatorname{Hom}_\Lambda(B_X, D \operatorname{Tr} X) & \to & \operatorname{Hom}_\Lambda(A_X, D \operatorname{Tr} X) & \to & \epsilon_*(D \operatorname{Tr} X) & \to & 0 \\
\downarrow & & \downarrow {\scriptstyle \operatorname{Hom}_\Lambda(h, D \operatorname{Tr} X)} & & \downarrow & & \\
\operatorname{Hom}_\Lambda(B, D \operatorname{Tr} X) & \to & \operatorname{Hom}_\Lambda(A, D \operatorname{Tr} X) & \to & \delta_*(D \operatorname{Tr} X) & \to & 0.
\end{array}$$

By definition the composition

$$\text{Hom}_\Lambda(A_X, D \operatorname{Tr} X) \xrightarrow{\text{Hom}_\Lambda(h, D \operatorname{Tr} X)} \text{Hom}_\Lambda(A, D \operatorname{Tr} X) \to \delta_*(D \operatorname{Tr} X)$$

is a projective cover and therefore an epimorphism. Hence $\epsilon_*(D \operatorname{Tr} X) \to \delta_*(D \operatorname{Tr} X)$ is an epimorphism and so $l_R(\epsilon_*(D \operatorname{Tr} X)) \geq l_R(\delta_*(D \operatorname{Tr} X))$, which is what we wanted to show.

(c) Suppose f_X is not right minimal. Then the exact sequence $0 \to A_X \xrightarrow{g_X} B_X \xrightarrow{f_X} C \to 0$ can be written as the sum of exact sequences

$$
\begin{array}{ccc}
A_X'' & = & A_X'' \\
\amalg & & \amalg \\
0 \quad\to\quad A_X' & \to & B_X' \quad\to\quad C \quad\to\quad 0
\end{array}
$$

with $A_X'' \neq 0$. But this implies, using the commutative exact diagram $(*)$, that there is an epimorphism of Σ-modules $\text{Hom}_\Lambda(A_X', D \operatorname{Tr} X) \to \delta_*(D \operatorname{Tr} X)$. Since $\text{Hom}_\Lambda(A_X', D \operatorname{Tr} X)$ is a projective Σ-module, this contradicts the fact that $\text{Hom}_\Lambda(A, D \operatorname{Tr} X)$ is a projective cover for $\delta_*(D \operatorname{Tr} X)$. Therefore we have shown that f_X is right minimal. $\square$

We say that any epimorphism $f_X: B_X \to C$ is a **right X-version** of the epimorphism $f: B \to C$ if

(a) f_X is right determined by X and is also right minimal and
(b) $H_{f_X} = H_f$.

Summarizing our results we have the following.

Proposition 3.3 *Let $f: B \to C$ be an epimorphism. Then we have the following.*

(a) *The epimorphism f has a right X-version.*
(b) *If $f_X: B_X \to C$ is a right X-version of f then there is a morphism $f \to f_X$ in $\operatorname{mod} \Lambda/C$, i.e. there is a commutative diagram*

$$
\begin{array}{ccc}
B & \xrightarrow{f} & C \\
\downarrow{j_X} & & \parallel \\
B_X & \xrightarrow{f_X} & C.
\end{array}
$$

(c) *If $f_X': B_X' \to C$ is another right X-version of f, then f and f' are isomorphic in $\operatorname{mod} \Lambda/C$.* $\square$

We now apply Proposition 3.3 to obtain the following.

Corollary 3.4 *Let* C *be a module and* H *a* Γ-*submodule of* $\operatorname{Hom}_\Lambda(X, C)$ *such that* $\mathscr{P}(X, C) \subset H$. *Then there is a unique, up to isomorphism in* $\operatorname{mod} \Lambda/C$, *epimorphism* $f_H : B_H \to C$ *satisfying the following.*

(a) f_H *is right minimal and right determined by* X.

(b) $H_{f_H} = H$.

Proof Since H is a finitely generated Γ-module there is some X' in $\operatorname{add} X$ and a morphism $X' \to C$ such that $\operatorname{Im}(\operatorname{Hom}_\Lambda(X, X') \to \operatorname{Hom}_\Lambda(X, C)) = H$. Let $P \to C$ be a projective cover. Then the induced morphism $B \xrightarrow{f} C$, where $B = X' \coprod P$, is an epimorphism with $\operatorname{Im}(\operatorname{Hom}_\Lambda(X, f)) = H$ because of the assumption that $\mathscr{P}(X, C)$ is contained in H. Then letting $f_H : B_H \to C$ be a right X-version of $f : B \to C$ we have our desired result by Proposition 3.3. $\qquad\qquad\square$

It is useful to have the following observations in order to show how our general result follows from the special case given in Corollary 3.4.

Lemma 3.5 *Let* $f : B \to C$ *be an arbitrary morphism. Then we have the following.*

(a) *If* f *is an epimorphism, then* $\mathscr{P}(X, C) \subset H_f$.

(b) *If* f *is determined by* X, *then* f *is an epimorphism if and only if* $\mathscr{P}(X, C) \subset H_f$.

Proof (a) This follows from the fact that if f is an epimorphism, then every morphism in $\mathscr{P}(X, C)$ factors through f.

(b) Let $g : P \to C$ be a projective cover of C. Since f is right determined by X, the hypothesis that $\mathscr{P}(X, C) \subset H_f$ means that g factors through f. Therefore f is an epimorphism. $\qquad\qquad\square$

We now generalize Corollary 3.4 to obtain our promised result.

Theorem 3.6 *Let* X *and* C *be* Λ-*modules and* H *an arbitrary* Γ-*submodule of* $\operatorname{Hom}_\Lambda(X, C)$. *Then there exists a unique, up to isomorphism in* $\operatorname{mod} \Lambda/C$, *morphism* $B_H \xrightarrow{f_H} C$ *satisfying the following conditions.*

(a) f_H *is right minimal.*

(b) f_H *is right* X-*determined.*

(c) $H_{f_H} = H$.

We show that we can reduce this general case to the case of epimorphisms studied before. This reduction is based on the following considerations.

First observe that if C' is a submodule of C, then $\operatorname{Hom}_\Lambda(X, C') \subset \operatorname{Hom}_\Lambda(X, C)$ is a Γ-submodule of $\operatorname{Hom}_\Lambda(X, C)$. Now it is not hard to see that if C' and C'' are submodules of C such that $\mathscr{P}(X, C') \subset H$ and $\mathscr{P}(X, C'') \subset H$, then $\mathscr{P}(X, C' + C'') \subset H$. Thus there is a unique submodule C' of C maximal with respect to the condition $\mathscr{P}(X, C') \subset H$. We denote this uniquely determined submodule C' of C by C_H. We now give some basic properties of C_H.

Lemma 3.7 *Let X and C be Λ-modules and suppose H is a Γ-submodule of $\operatorname{Hom}_\Lambda(X, C)$. Then the submodule C_H of C as described above has the following properties.*

(a) *C_H contains the submodule C' of C generated by $\{\operatorname{Im}(f : X \to C)\}_{f \in H}$.*
(b) *$\mathscr{P}(X, C_H) \subset H \subset \operatorname{Hom}_\Lambda(X, C_H)$.*
(c) *If $t : Z \to C$ has the property that*

$$\operatorname{Im}(\operatorname{Hom}_\Lambda(X, t) : \operatorname{Hom}_\Lambda(X, Z) \to \operatorname{Hom}_\Lambda(X, C))$$

is contained in H, then $\operatorname{Im} t \subset C_H$.

Proof (a) Since H is a finitely generated Γ-submodule of $\operatorname{Hom}_\Lambda(X, C)$, there are $h_1, \ldots, h_n$ in H such that the induced morphism $h : nX \to C$ has the property $\operatorname{Im}(\operatorname{Hom}_\Lambda(X, nX) \xrightarrow{\operatorname{Hom}_\Lambda(X,h)} \operatorname{Hom}_\Lambda(X, C)) = H$. Then we have that $\operatorname{Im} h = C'$. Since a projective cover $P \to C'$ factors through h, this shows that $\mathscr{P}(X, C') \subset H$ and hence $C' \subset C_H$.

(b) We clearly have $H \subset \operatorname{Hom}_\Lambda(X, C')$. Since $C' \subset C_H$, we have that $H \subset \operatorname{Hom}_\Lambda(X, C_H)$.

(c) Suppose $t : Z \to C$ is a morphism. Since a projective cover $P \to \operatorname{Im} t$ factors through t, we have that $\mathscr{P}(X, \operatorname{Im} t) \subset \operatorname{Im} \operatorname{Hom}_\Lambda(X, t)$. Therefore if $\operatorname{Im} \operatorname{Hom}_\Lambda(X, t) \subset H$, then $\mathscr{P}(X, \operatorname{Im} t) \subset H$ and so $\operatorname{Im} t \subset C_H$. $\square$

We now return to the proof of Theorem 3.6.

Suppose H is a Γ-submodule of $\operatorname{Hom}_\Lambda(X, C)$. Then by Lemma 3.7 the submodule C_H of C satisfies $\mathscr{P}(X, C_H) \subset H \subset \operatorname{Hom}_\Lambda(X, C_H)$. Therefore by Corollary 3.4 we have the exact sequence $0 \to A_H \to B_H \xrightarrow{f_H} C_H \to 0$ which has the properties that f_H is a right minimal morphism which is determined by X and $H = H_{f_H}$. Consider now the composition

$B_H \xrightarrow{f_H} C_H \xrightarrow{i} C$, which we also denote by f_H. Then $0 \to A_H \to B_H \xrightarrow{f_H} C$ is our desired exact sequence provided f_H is right X-determined. Suppose $t : Z \to C$ is such that $\operatorname{Im}\operatorname{Hom}_\Lambda(X,t) \subset H$. Then by Lemma 3.7(c) we have that $\operatorname{Im} t \subset C_H$. Hence t factors through f_H and we are done. $\square$

As an immediate consequence of this result we have the following.

Corollary 3.8 *Let C be a Λ-module and H any Γ-submodule of $\operatorname{Hom}_\Lambda(X, C)$. Then the submodule C_H, which is completely determined by H, is the image of any morphism $f : B \to C$ such that $H_f = H$.* $\square$

We summarize our results in the following classification theorem.

Theorem 3.9 *Let X and C be Λ-modules and let $\operatorname{Sub}\operatorname{Hom}_\Lambda(X, C)$ be the set of Γ-submodules of $\operatorname{Hom}_\Lambda(X, C)$.*

(a) *Then the map $\operatorname{mod}\Lambda/C \to \operatorname{Sub}\operatorname{Hom}_\Lambda(X, C)$ given by $f \mapsto H_f$ induces a bijection between the isomorphism classes of f which are the right minimal morphisms determined by the module X and $\operatorname{Sub}\operatorname{Hom}_\Lambda(X, C)$.*

(b) *For f in $\operatorname{mod}\Lambda/C$ we have that $\operatorname{Im} f = C_{H_f}$ and $\operatorname{Ker} f = D\operatorname{Tr} U(f)$ where $U(f)$ is the cover of the bottom of f.*

(c) *$\operatorname{Hom}_{\operatorname{mod}\Lambda/C}(f_1, f_2)$ is not empty if and only if $H_{f_1} \subset H_{f_2}$.* $\square$

As an application of this result we have the following.

Corollary 3.10 *Let A be an arbitrary Λ-module with no nonzero injective summands.*

(a) *Then the right minimal morphisms $f : B \to C$ with $\operatorname{Ker} f$ in $\operatorname{add} A$ are precisely the right minimal morphisms in $\operatorname{mod}\Lambda/C$ which are right determined by $X = \operatorname{Tr} DA \coprod \Lambda$.*

(b) *The isomorphism classes of right minimal morphisms in $\operatorname{mod}\Lambda/C$ with kernel in $\operatorname{add} A$ are classified by $\operatorname{Sub}\operatorname{Hom}_\Lambda(X, C)$, the set of Γ-submodules of $\operatorname{Hom}_\Lambda(X, C)$ by means of the bijection given by $f \mapsto H_f$.*

(c) *$\operatorname{Hom}_{\operatorname{mod}\Lambda/C}(f_1, f_2)$ is not empty if and only if $H_{f_1} \subset H_{f_2}$.* $\square$

Note that Theorem 3.6 can be viewed as a generalization of the existence theorem for minimal right almost split morphisms. For let C be an indecomposable Λ-module. Let $X = C$ and let $H = \operatorname{rad}\operatorname{End}_\Lambda(C) \subset \operatorname{Hom}_\Lambda(C, C)$. Then the morphisms $f : B \to C$ with $\operatorname{Im}\operatorname{Hom}_\Lambda(C, f) \subset H$ are those which are not split epimorphisms, and the morphism $f_H : B_H \to C$ is the minimal right almost split morphism. In the more general situation we make a different restriction on the morphisms we consider, namely for given X and $H \subset \operatorname{Hom}_\Lambda(X, C)$ we consider the morphisms $f: B \to C$ with $\operatorname{Im} \operatorname{Hom}_\Lambda(X, f) \subset H$. Then Theorem 3.6 says that we have a similar finiteness condition to what is expressed through the existence of minimal right almost split morphisms.

4 Rigid exact sequences

In this section we use the classification of morphisms to investigate when exact sequences are rigid. We have already seen in V Proposition 2.3 that almost split sequences and split sequences are rigid. It is also clear that an exact sequence $0 \to A \xrightarrow{f} B \xrightarrow{g} C \to 0$ is rigid if either $f: A \to B$ is an injective envelope or $g: B \to C$ is a projective cover. However, in general exact sequences are not rigid (see Exercise 1).

Our discussion of when exact sequences are rigid is based on the following criteria for when exact sequences are isomorphic.

Proposition 4.1 *Let* $0 \to A \xrightarrow{f} B \xrightarrow{g} C \to 0$ *be an exact sequence with* g *right minimal and let* $X = \operatorname{Tr} DA$. *An exact sequence* $0 \to A' \xrightarrow{f'} B' \xrightarrow{g'} C' \to 0$ *with* $A \simeq A'$ *and* $C \simeq C'$ *is isomorphic to* $0 \to A \xrightarrow{f} B \xrightarrow{g} C \to 0$ *if and only if there is an isomorphism* $t: C \to C'$ *such that the induced isomorphism* $\operatorname{Hom}_\Lambda(X, t): \operatorname{Hom}_\Lambda(X, C) \to \operatorname{Hom}_\Lambda(X, C')$ *has the property that* $\operatorname{Hom}_\Lambda(X, t)(H_g) = H_{g'}$, *where* $H_g = \operatorname{Im} \operatorname{Hom}_\Lambda(X, g)$ *and* $H_{g'} = \operatorname{Im} \operatorname{Hom}_\Lambda(X, g')$.

Proof Suppose $t: C \to C'$ is an isomorphism such that $\operatorname{Hom}_\Lambda(X, t)(H_g) = H_{g'}$. Since both g and g' are epimorphisms with kernels isomorphic to $D \operatorname{Tr} X$, it follows from Corollary 1.4 that they are determined by X. Since $\operatorname{Hom}_\Lambda(X, t)(H_g) = H_{g'}$ and $\operatorname{Hom}_\Lambda(X, t^{-1})(H_{g'}) = H_g$ there is a commutative exact diagram

$$
\begin{array}{ccccccccc}
0 & \longrightarrow & A & \xrightarrow{f} & B & \xrightarrow{g} & C & \longrightarrow & 0 \\
& & \downarrow{u} & & \downarrow{w} & & \downarrow{t} & & \\
0 & \longrightarrow & A' & \xrightarrow{f'} & B' & \xrightarrow{g'} & C' & \longrightarrow & 0 \\
& & \downarrow & & \downarrow{v} & & \downarrow{t^{-1}} & & \\
0 & \longrightarrow & A & \xrightarrow{f} & B & \xrightarrow{g} & C & \longrightarrow & 0.
\end{array}
$$

But then $vw: B \to B$ is an automorphism because g is right minimal. Hence w is a monomorphism and therefore u is also a monomorphism.

But then $u: A \to A'$ is an isomorphism since $A \simeq A'$. Hence $w: B \to B'$ is an isomorphism. This shows that the exact sequences $0 \to A \to B \to C \to 0$ and $0 \to A' \to B' \to C' \to 0$ are isomorphic. The proof in the other direction is trivial. $\square$

Using this result it is not difficult to prove the following.

Corollary 4.2 *Let X be a Λ-module and $g: B \to C$ an epimorphism with kernel $D \operatorname{Tr} X$. Assume that the Γ-submodule $H_g = \operatorname{Im} \operatorname{Hom}_\Lambda(X, g)$ of $\operatorname{Hom}_\Lambda(X, C)$ either is contained in or contains all Γ-submodules M of $\operatorname{Hom}_\Lambda(X, C)$ with $\mathscr{P}(X, C) \subset M$. Then the following are equivalent for an exact sequence $\epsilon: 0 \to D \operatorname{Tr} X \xrightarrow{f'} B' \xrightarrow{g'} C \to 0$.*

(i) *ϵ is isomorphic to $0 \to D \operatorname{Tr} X \to B \xrightarrow{g} C \to 0$.*

(ii) *$B \simeq B'$.*

(iii) *$l_R(\operatorname{Hom}_\Lambda(X, B)) = l_R(\operatorname{Hom}_\Lambda(X, B'))$.*

Proof (i) $\Rightarrow$ (ii) and (ii) $\Rightarrow$ (iii) are trivial. In order to prove that (iii) $\Rightarrow$ (i) observe that $l_R(\operatorname{Hom}_\Lambda(X, B)) = l_R(\operatorname{Hom}_\Lambda(X, B'))$ implies that $l_R(H_{g'}) = l_R(\operatorname{Im} \operatorname{Hom}(X, g')) = l_R(H_g)$. Since $\mathscr{P}(X, C) \subset H_{g'}$ we have $H_g = H_{g'}$, and the result follows from Proposition 4.1. $\square$

Corollary 4.3 *Let $0 \to A \xrightarrow{f} B \xrightarrow{g} C \to 0$ be an exact sequence with g right minimal and let $X = \operatorname{Tr} DA$. Suppose $H_g \subset \operatorname{Hom}_\Lambda(X, C)$ has the property that if $0 \to A \xrightarrow{f'} B' \xrightarrow{g'} C \to 0$ is exact, then $H_{g'}$ either contains or is contained in H_g. Then $0 \to A \xrightarrow{f} B \xrightarrow{g} C \to 0$ is a rigid exact sequence.* $\square$

As an example of how to apply Corollary 4.3 we give the following result.

Proposition 4.4 *Let X and C be nonprojective Λ-modules. Let $0 \to A \xrightarrow{f} B \xrightarrow{g} C \to 0$ be the uniquely determined exact sequence with g a right minimal morphism determined by X where $H_g = \mathscr{P}(X, C)$. Then $0 \to A \xrightarrow{f} B \xrightarrow{g} C \to 0$ is a rigid exact sequence.*

Proof In Lemma 3.5 we showed that if $0 \to A \xrightarrow{f'} B' \xrightarrow{g'} C \to 0$ is any exact sequence, then $H_g \supset \mathscr{P}(X, C) = H_g$ and so $0 \to A \xrightarrow{f} B \xrightarrow{g} C \to 0$ is rigid by Corollary 4.3. $\square$

The hypothesis of Corollary 4.3 suggests introducing the following notion.

Let M be a Λ-module. A submodule M' of M is said to be a **waist** of M if M' is a submodule of M with the property that a submodule M'' of M either contains M' or is contained in M'. A submodule M' of M is called a **proper waist** if $0 \neq M' \neq M$ and M' is a waist of M.

Our reason for introducing the notion of a waist of a module is apparent from the following.

Proposition 4.5 *Let* $0 \to A \xrightarrow{f} B \xrightarrow{g} C \to 0$ *be an exact sequence with* g *right minimal and let* $X = \operatorname{Tr} DA$. *Then* $0 \to A \xrightarrow{f} B \xrightarrow{g} C \to 0$ *is rigid if* H_g *is a waist of the* $\operatorname{End}_\Lambda(X)^{\mathrm{op}}$-*module* $\operatorname{Hom}_\Lambda(X, C)$.

Proof This is a trivial consequence of the definition of a waist of a module and Corollary 4.3. $\square$

Before illustrating how Proposition 4.5 can be used to give more examples of exact sequences which are rigid, it is useful to discuss some elementary properties and examples of waists of modules.

As an immediate consequence of the definition of a waist we have the following.

Proposition 4.6 *Let* M' *be a submodule of the* Λ-*module* M.

(a) M' *is a waist of* M *if and only if* $\Lambda m \supset M'$ *for all elements* m *of* M *not in* M'.

(b) *Suppose* M' *is a proper waist of* M. *Then* M *is indecomposable.*

Proof (a) It is clear that if M' is a waist of M, then Λm contains M' if m is not in M'. Suppose $\Lambda m \supset M'$ if m is not in M'. Let M'' be a submodule of M. If every m in M'' is in M', then $M'' \subset M'$. If there is an m in M'' not in M', then $M'' \supset \Lambda m \supset M'$. Hence M' is a waist of M.

(b) Suppose $M = M_1 \coprod M_2$ and that M' is a proper waist of M. Then M' is not contained in both M_1 and M_2 and does not contain both M_1 and M_2. Hence one M_i is contained in M' and the other one contains M'. This is a contradiction unless one of them is zero and therefore M is indecomposable. $\square$

We now give some examples of waists of modules.

Proposition 4.7

(a) *A nonzero Λ-module M is uniserial if and only if every submodule of M is a waist of M.*

(b) *Suppose $0 \to A \xrightarrow{f} B \xrightarrow{g} C \to 0$ is an exact sequence with $f : A \to B$ an irreducible morphism. Let X be a Λ-module. Then $\operatorname{Im}\operatorname{Hom}_\Lambda(X, g)$ is a waist of the $\operatorname{End}_\Lambda(X)^{\mathrm{op}}$-module $\operatorname{Hom}_\Lambda(X, C)$.*

Proof (a) This is left as an exercise.

(b) Let h be in $\operatorname{Hom}_\Lambda(X, C)$. Since f is irreducible we know by V Proposition 5.6 that either h factors through g, in which case h is in $\operatorname{Im}\operatorname{Hom}_\Lambda(X, g)$, or g factors through h, in which case $\operatorname{Im}\operatorname{Hom}_\Lambda(X, g) \subset \operatorname{End}_\Lambda(X)^{\mathrm{op}} h$. Hence we have that $\operatorname{Im}\operatorname{Hom}_\Lambda(X, g)$ is a waist of $\operatorname{Hom}_\Lambda(X, C)$ by Proposition 4.6. $\square$

We now apply these remarks to obtain more examples of rigid exact sequences.

Proposition 4.8 *Suppose Λ is a Nakayama algebra. If $0 \to A \xrightarrow{f} B \xrightarrow{g} C \to 0$ is an exact sequence of Λ-modules with A and C indecomposable modules, then it is a rigid exact sequence.*

Proof If A is injective, the sequence splits and is therefore rigid. Therefore we may assume that A is not injective. Since A is indecomposable, we have that $X = \operatorname{Tr} DA$ is indecomposable. Since X and C are indecomposable and Λ is a Nakayama algebra, we know by VI Corollary 2.4 that $\operatorname{Hom}_\Lambda(X, C)$ is a uniserial module over the Nakayama algebra $\operatorname{End}_\Lambda(X)^{\mathrm{op}}$. Therefore, by Proposition 4.7 we have that H_g is a waist of $\operatorname{Hom}_\Lambda(X, C)$ since it is a submodule of $\operatorname{Hom}_\Lambda(X, C)$. Hence $0 \to A \xrightarrow{f} B \xrightarrow{g} C \to 0$ is a rigid exact sequence by Corollary 4.3. $\square$

As our final example of rigid exact sequences we have the following immediate consequence of Proposition 4.7.

Corollary 4.9 *An exact sequence $0 \to A \xrightarrow{f} B \xrightarrow{g} C \to 0$ is rigid if f is irreducible.*

Proof Let $X = \operatorname{Tr} DA$. Then by Proposition 4.7, we know that $\operatorname{Im}\operatorname{Hom}_\Lambda(X, g)$ is a waist of $\operatorname{Hom}_\Lambda(X, C)$. Therefore $0 \to A \xrightarrow{f} B \xrightarrow{g} C \to 0$ is a rigid exact sequence. $\square$

5 Indecomposable middle terms

In V Section 6 a method was given for constructing almost split sequences with indecomposable middle term that was sufficiently general to show that all nonsemisimple artin algebras have such almost split sequences. However, it is easy to give examples of almost split sequences with indecomposable middle term which cannot be obtained by the method given in V Section 6. Therefore it is natural to try to find other conditions on an indecomposable module C which guarantee that $\alpha(C) = 1$, i.e. in the almost split sequence $0 \to D \operatorname{Tr} C \to E \to C \to 0$ the module E is indecomposable. This section is devoted to a discussion of this problem. Using the well known properties of pushouts and their connection with exact sequences stated in I Proposition 5.6 and Corollary 5.7 we obtain the following result about the indecomposable modules C with $\alpha(C) = 1$.

Proposition 5.1 *Let* $0 \to D \operatorname{Tr} C \to E \to C \to 0$ *be an almost split sequence with E an indecomposable Λ-module. Then the following are equivalent.*

(a) *E is not projective.*

(b) *There is an exact sequence $0 \to D \operatorname{Tr} E \xrightarrow{f} B \to C \to 0$ with f an irreducible but not left almost split morphism.*

(c) *There is an exact sequence $0 \to A \xrightarrow{f} B \to C \to 0$ with f an irreducible but not left almost split morphism, with A or B indecomposable.*

Proof (a) $\Rightarrow$ (b) Since E is indecomposable and not projective, there is an almost split sequence $0 \to D \operatorname{Tr} E \xrightarrow{\binom{f}{f'}} B \coprod D \operatorname{Tr} C \xrightarrow{(g',g)} E \to 0$ with $B \neq 0$. It then follows from I Corollary 5.7 that we have the exact commutative pushout diagram

$$
\begin{array}{ccccccccc}
0 & \to & D \operatorname{Tr} E & \xrightarrow{f} & B & \xrightarrow{h} & \operatorname{Coker} f & \to & 0 \\
 & & \downarrow{-f'} & & \downarrow{g'} & & \| h' & & \\
0 & \to & D \operatorname{Tr} C & \xrightarrow{g} & E & \to & C & \to & 0.
\end{array}
$$

The exact sequence $0 \to D \operatorname{Tr} E \xrightarrow{f} B \xrightarrow{h'h} C \to 0$ has our desired property since f is irreducible but not left almost split.

(b) $\Rightarrow$ (c) This is trivial.

(c) $\Rightarrow$ (a) Assume $0 \to A \xrightarrow{f} B \to C \to 0$ is an exact sequence with f irreducible but not left almost split and where A or B is indecomposable. Using that $0 \to D \operatorname{Tr} C \to E \to C \to 0$ is an almost split sequence we obtain the following commutative diagram.

$$
\begin{array}{ccccccccc}
0 & \to & A & \overset{f}{\to} & B & \to & C & \to & 0 \\
 & & \downarrow g & & \downarrow h & & \| & & \\
0 & \to & D\,\mathrm{Tr}\,C & \overset{f'}{\to} & E & \to & C & \to & 0
\end{array}
$$

Assume that E is projective. We then have the split exact sequence $0 \to A \overset{\binom{f}{g}}{\to} B \coprod D\,\mathrm{Tr}\,C \overset{(h,-f')}{\to} E \to 0$. Since A or B is indecomposable and f is irreducible, it follows that $f \in \mathrm{rad}_\Lambda(A, B)$. But $\binom{f}{g}$ is a split monomorphism and therefore g is a split monomorphism. Hence g is an isomorphism and therefore h is an isomorphism. But then f is a left almost split morphism which gives the desired contradiction. Hence E is not projective. □

For the sake of brevity we say that a monomorphism $f : A \to B$ is properly irreducible if A is indecomposable and f is irreducible but not left almost split. While such morphisms should really be called left properly irreducible, we use the shorter terminology properly irreducible since we do not use the dual notion of right properly irreducible morphisms in this book. In view of Proposition 5.1 it is natural to ask if for an indecomposable module C which is the cokernel of a properly irreducible morphism we have $\alpha(C) = 1$. The following example shows that this is not always the case when C is a simple module.

Example Let $k[X, Y]$ be a polynomial ring over a field k and let $\Lambda = k[X, Y]/(X^2, Y^2)$. Denoting the images of X and Y in Λ by x and y respectively, it is not difficult to see that for the ideal $\mathfrak{m} = (x, y)$ we have $\Lambda/\mathfrak{m} \simeq k$ and $\mathfrak{m}^3 = 0$. So Λ is a local k-algebra of k-dimension 4 with $\mathfrak{m} = \mathrm{rad}\,\Lambda$. It is also not difficult to see that $\mathrm{soc}\,\Lambda = (xy) = \mathfrak{m}^2$. Therefore Λ is a selfinjective algebra since it is a local commutative artin algebra with simple socle (see IV Section 3) and hence a symmetric algebra since Λ is commutative. Now by V Proposition 5.5 we know that (*) $0 \to \mathfrak{m} \to (\mathfrak{m}/\mathrm{soc}\,\Lambda) \coprod \Lambda \overset{(j,h)}{\to} \Lambda/\mathrm{soc}\,\Lambda \to 0$ is an almost split sequence where $h : \Lambda \to \Lambda/\mathrm{soc}\,\Lambda$ is the natural epimorphism and $j : \mathfrak{m}/\mathrm{soc}\,\Lambda \to \Lambda/\mathrm{soc}\,\Lambda$ is the natural inclusion. Since $\mathfrak{m}^2 = \mathrm{soc}\,\Lambda$ we have that $\mathfrak{m}/\mathrm{soc}\,\Lambda \simeq (\Lambda/\mathfrak{m})\coprod(\Lambda/\mathfrak{m})$. Since (*) is an almost split sequence with $\mathfrak{m}$ not simple we obtain the exact commutative diagram

$$
\begin{array}{ccccccccc}
& & 0 & & 0 & & 0 & & \\
& & \downarrow & & \downarrow & & \downarrow & & \\
0 & \to & \Omega^2(\Lambda/\mathfrak{m}) & \to & 2\mathfrak{m} & \to & \Lambda/\mathfrak{m} & \to & 0 \\
& & \downarrow & & \downarrow & & \downarrow & & \\
0 & \to & 2\Lambda & \to & 2\Lambda \amalg \Lambda & \to & \Lambda & \to & 0 \\
& & \downarrow f_0 & & \downarrow g_0 & & \downarrow h_0 & & \\
0 & \to & \mathfrak{m} & \to & (\mathfrak{m}/\operatorname{soc}\mathfrak{m}) \amalg \Lambda & \to & \Lambda/\operatorname{soc}\Lambda & \to & 0 \\
& & \downarrow & & \downarrow & & \downarrow & & \\
& & 0 & & 0 & & 0 & &
\end{array}
$$

where f_0, g_0 and h_0 are projective covers. Therefore the exact sequence $0 \to \Omega^2(\Lambda/m) \to 2\mathfrak{m} \to \Lambda/m \to 0$ is an almost split sequence by X Section 1. Hence we have $\alpha(\Lambda/\mathfrak{m}) = 2$. But in the exact sequence $0 \to \mathfrak{m} \xrightarrow{t} \Lambda \to \Lambda/m \to 0$, the morphism t is properly irreducible. Therefore the simple Λ-module Λ/m has $\alpha(\Lambda/\mathfrak{m}) = 2 > 1$ even though it is the cokernel of a properly irreducible morphism. Of course this also gives an example of a simple module C with $\alpha(C) > 1$.

Therefore the following is the best general theorem we can hope for concerning when indecomposable modules C which are cokernels of properly irreducible morphisms have the property $\alpha(C) = 1$.

Theorem 5.2 *Let C be a nonsimple indecomposable module which is the cokernel of a properly irreducible morphism. Then we have that $\alpha(C) = 1$.*

The proof of this theorem occupies most of the rest of this section. We start with the following.

Proposition 5.3 *Let $\eta : 0 \to A \xrightarrow{f} B \xrightarrow{g} C \to 0$ be exact with f irreducible. Then $\operatorname{End}_\Lambda(A)\eta \subset \operatorname{Ext}^1_\Lambda(C, A)$ is an $\operatorname{End}_\Lambda(A)$-waist of $\operatorname{Ext}^1_\Lambda(C, A)$.*

Proof If $\operatorname{End}_\Lambda(A)\eta = \operatorname{Ext}^1_\Lambda(C, A)$, there is nothing to prove. If $\operatorname{End}_\Lambda(A)\eta \neq \operatorname{Ext}^1_\Lambda(C, A)$, let $\delta : 0 \to A \xrightarrow{f'} B' \xrightarrow{g'} C \to 0$ be an element of $\operatorname{Ext}^1_\Lambda(C, A)$. Since f is an irreducible monomorphism, it follows from V Proposition 5.6 that there is either some $h : B' \to B$ such that $g' = gh$ or an $h' : B \to B'$ such that $g = g'h'$. If such an h exists, h induces a morphism $i : A \to A$ such that $\eta = i\delta$, and hence $\eta \in \operatorname{End}_\Lambda(A)\delta$. If such an h' exists, h' induces a morphism $j : A \to A$ such that $\delta = j\eta$, hence $\delta \in \operatorname{End}_\Lambda(A)\eta$.

Therefore either $\text{End}_\Lambda(A)\delta \subset \text{End}_\Lambda(A)\eta$ or $\text{End}_\Lambda(A)\eta \subset \text{End}_\Lambda(A)\delta$. It then follows by Proposition 4.6 that $\text{End}_\Lambda(A)\eta$ is an $\text{End}_\Lambda(A)$-waist of $\text{Ext}_\Lambda^1(C, A)$. □

From this we deduce the following result.

Corollary 5.4 *Let* A *be an indecomposable* Λ-*module and* $\eta : 0 \to A \xrightarrow{f} B \to C \to 0$ *an exact sequence with* f *irreducible. Then* $\text{End}_\Lambda(A)\eta$ *is an* $\text{End}_\Lambda(A)$-$\text{End}_\Lambda(C)^{\text{op}}$-*submodule of* $\text{Ext}_\Lambda^1(C, A)$ *with* $\mathfrak{r}_{\text{End}_\Lambda(A)}\eta$ *an* $\text{End}_\Lambda(A)$-$\text{End}_\Lambda(C)^{\text{op}}$-*submodule which is the unique maximal* $\text{End}_\Lambda(A)$-*submodule of* $\text{End}_\Lambda(A)\eta$.

Proof Generally, if M is a Γ-Γ'-bimodule and M' is a characteristic Γ-submodule of M, i.e. M' is mapped into itself by all Γ-homomorphisms from M to M, then M' is a Γ-Γ'-subbimodule of M. Clearly, any $\text{End}_\Lambda(A)$-waist X of $\text{Ext}_\Lambda^1(C, A)$ is a characteristic submodule and so is also the $\text{End}_\Lambda(A)$-radical of X. Hence $\text{End}_\Lambda(A)\eta$ and $\mathfrak{r}_{\text{End}_\Lambda(A)}\eta$ are both $\text{End}_\Lambda(A)$-$\text{End}_\Lambda(C)^{\text{op}}$-bimodules. Further, since A is indecomposable, $\text{End}_\Lambda(A)\eta$ has $\mathfrak{r}_{\text{End}_\Lambda(A)}\eta$ as a unique maximal $\text{End}_\Lambda(A)$-submodule. □

We next explain how we identify $D\text{Ext}_\Lambda^1(C, A)$ and $\overline{\text{Hom}}_\Lambda(A, D\text{Tr}\, C)$ as $\text{End}_\Lambda(A)^{\text{op}}$-$\text{End}_\Lambda(C)$-bimodules. Let C be any Λ-module and let $P_2 \xrightarrow{f_2} P_1 \xrightarrow{f_1} P_0 \to C \to 0$ be the start of a minimal projective resolution of C. We then obtain the sequence of Λ^{op}-modules $P_0^* \xrightarrow{f_1^*} P_1^* \xrightarrow{f_2^*} P_2^*$ where $\text{Coker}\, f_1^* = \text{Tr}\, C$ and the morphism f_2^* factors through $\text{Tr}\, C$. Denoting also by f_2^* the induced morphism from $\text{Tr}\, C$ to P_2^* we observe that $f_2^* : \text{Tr}\, C \to P_2^*$ has the property that for any Λ^{op}-morphism $g : \text{Tr}\, C \to Q$ with Q projective, there exists an $h : P_2^* \to Q$ with $g = hf_2^*$. Dualizing this we obtain that the Λ-morphism $Df_2^* : DP_2^* \to D\text{Tr}\, C$ has the property that for any Λ-morphism $g : I \to D\text{Tr}\, C$ with I injective, there exists an $h : I \to DP_2^*$ with $g = (Df_2^*)h$.

From this we deduce that $\text{Im}\,\text{Hom}_\Lambda(A, Df_2^*) = \mathscr{I}(A, D\text{Tr}\, C)$ for all A in $\text{mod}\,\Lambda$. Hence we have the exact sequence $\text{Hom}_\Lambda(A, DP_2^*) \to \text{Hom}_\Lambda(A, D\text{Tr}\, C) \to \overline{\text{Hom}}_\Lambda(A, D\text{Tr}\, C) \to 0$ where the last morphism is an $\text{End}_\Lambda(A)^{\text{op}}$-$\text{End}_\Lambda(D\text{Tr}\, C)$-bimodule morphism. But we also have the complex $\text{Hom}_\Lambda(A, DP_2^*) \to \text{Hom}_\Lambda(A, DP_1^*) \to \text{Hom}_\Lambda(A, DP_0^*)$. However for each projective Λ-module Q we have natural isomorphisms

$\mathrm{Hom}_\Lambda(A, DQ^\bullet) \simeq D(Q^\bullet \otimes A) \simeq D\mathrm{Hom}_\Lambda(Q, A)$ giving rise to the commuting diagram.

$$\mathrm{Hom}_\Lambda(A, DP_2^\bullet) \xrightarrow{(A, Df_2^\bullet)} \mathrm{Hom}_\Lambda(A, DP_1^\bullet) \xrightarrow{(A, Df_1^\bullet)} \mathrm{Hom}_\Lambda(A, DP_0^\bullet)$$
$$\downarrow \wr \qquad\qquad\qquad \downarrow \wr \qquad\qquad\qquad \downarrow \wr$$
$$D\mathrm{Hom}_\Lambda(P_2, A) \xrightarrow{D(f_2, A)} D\mathrm{Hom}_\Lambda(P_1, A) \xrightarrow{D(f_1, A)} D\mathrm{Hom}_\Lambda(A, P_0)$$

where all maps are $\mathrm{End}_\Lambda(A)^{\mathrm{op}}$-morphisms. By definition $\mathrm{Ker}\, D(f_1, A)/\mathrm{Im}\, D(f_2, A) \simeq D\mathrm{Ext}^1_\Lambda(C, A)$ which is then isomorphic to $\mathrm{Ker}(A, Df_1^\bullet)/\mathrm{Im}(A, Df_2^\bullet)$ as an $\mathrm{End}_\Lambda(A)^{\mathrm{op}}$-module. But we have seen above that $\mathrm{Ker}(A, Df_1^\bullet) = \mathrm{Hom}_\Lambda(A, D\,\mathrm{Tr}\, C)$ and $\mathrm{Im}(A, Df_2^\bullet) = \mathscr{I}(A, D\,\mathrm{Tr}\, C)$ so $D\mathrm{Ext}^1_\Lambda(C, A) \simeq \overline{\mathrm{Hom}}_\Lambda(A, D\,\mathrm{Tr}\, C)$ as $\mathrm{End}_\Lambda(A)^{\mathrm{op}}$-modules. Also identifying $\underline{\mathrm{End}}_\Lambda(C)$ with $\overline{\mathrm{End}}_\Lambda(D\,\mathrm{Tr}\, C)$ it is not too hard to verify that the isomorphism $D\mathrm{Ext}^1_\Lambda(C, A) \simeq \overline{\mathrm{Hom}}_\Lambda(A, D\,\mathrm{Tr}\, C)$ is an $\mathrm{End}_\Lambda(A)^{\mathrm{op}}$-$\underline{\mathrm{End}}_\Lambda(C)$-bimodule isomorphism as well.

As before let A be an indecomposable Λ-module and $\eta : 0 \to A \xrightarrow{f} B \to C \to 0$ an exact sequence with f irreducible. Consider the following exact commutative diagram of $\mathrm{End}_\Lambda(A)$-modules and $\mathrm{End}_\Lambda(A)$-morphisms with T a simple $\mathrm{End}_\Lambda(A)$-module.

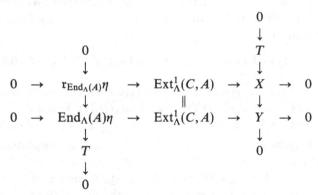

Dualizing, using that $\mathrm{End}_\Lambda(A)\eta$ is a waist of $\mathrm{Ext}^1_\Lambda(C, A)$ we get that DX and DY are the only $\mathrm{End}_\Lambda(A)^{\mathrm{op}}$-submodules of $D\mathrm{Ext}^1_\Lambda(C, A)$ of length $l_R(X)$ and $l_R(Y)$ over R respectively, with DY a maximal $\mathrm{End}_\Lambda(A)^{\mathrm{op}}$-submodule of DX. Therefore DX is an $\mathrm{End}_\Lambda(A)^{\mathrm{op}}$-waist of $D\mathrm{Ext}^1_\Lambda(C, A)$ with a unique maximal submodule. Using the $\mathrm{End}_\Lambda(A)^{\mathrm{op}}$-isomorphism $D\mathrm{Ext}^1_\Lambda(C, A) \simeq \overline{\mathrm{Hom}}_\Lambda(A, D\,\mathrm{Tr}\, C)$ we have that $\overline{\mathrm{Hom}}_\Lambda(A, D\,\mathrm{Tr}\, C)$ also has only one $\mathrm{End}_\Lambda(A)^{\mathrm{op}}$-submodule M of length $l_R(X)$ and M has a unique maximal $\mathrm{End}_\Lambda(A)^{\mathrm{op}}$-submodule M'.

We now consider the natural R-bilinear map

$$\phi : \overline{\operatorname{Hom}_\Lambda}(A, D \operatorname{Tr} C) \times \operatorname{Ext}_\Lambda^1(C, A) \to \operatorname{Ext}_\Lambda^1(C, D \operatorname{Tr} C)$$

given by pushout diagrams, i.e. if $\delta : 0 \to A \to U \to C \to 0$ is an element of $\operatorname{Ext}_\Lambda^1(C, A)$, and h is an element of $\operatorname{Hom}_\Lambda(A, D \operatorname{Tr} C)$, then $\phi(\bar{h}, \delta) = \delta'$ where δ' is obtained from the pushout diagram

$$
\begin{array}{ccccccccc}
\delta : & 0 & \to & A & \to & U & \to & C & \to & 0 \\
 & & & \downarrow h & & \downarrow & & \| & & \\
\delta' : & 0 & \to & D \operatorname{Tr} C & \to & U' & \to & C & \to & 0.
\end{array}
$$

Clearly δ' depends only on the image $\bar{h}$ of h in $\overline{\operatorname{Hom}}_\Lambda(A, D \operatorname{Tr} C)$.

Using properties of pushouts it is easy to see that ϕ satisfies the following properties: $\phi(\bar{h}, \bar{i}\delta) = \phi(\overline{hi}, \delta)$ and $\phi(\bar{h}, \delta \underline{j}) = \phi(\bar{h}, \delta)\underline{j}$ and $\phi(\bar{s}\bar{h}, \delta) = \bar{s}\phi(\bar{h}, \delta)$ for all $\bar{h} \in \overline{\operatorname{Hom}}_\Lambda(A, D \operatorname{Tr} C)$, $\delta \in \operatorname{Ext}_\Lambda^1(C, A)$, $\bar{i} \in \overline{\operatorname{End}}_\Lambda(A)$, $\underline{j} \in \underline{\operatorname{End}}_\Lambda(C)$ and $\bar{s} \in \overline{\operatorname{End}}_\Lambda(D \operatorname{Tr} C)$. Hence we obtain an $\operatorname{End}_\Lambda(D \operatorname{Tr} C)$-$\operatorname{End}_\Lambda(C)^{\operatorname{op}}$-bimodule morphism

$$\check{\phi} : \overline{\operatorname{Hom}}_\Lambda(A, D \operatorname{Tr} C) \otimes_{\overline{\operatorname{End}}_\Lambda(A)} \operatorname{Ext}_\Lambda^1(C, A) \to \operatorname{Ext}_\Lambda^1(C, D \operatorname{Tr} C).$$

Using the fact that there is an almost split sequence $\delta' : 0 \to D \operatorname{Tr} C \to E \to C \to 0$ we obtain that for each $0 \neq \delta \in \operatorname{Ext}_\Lambda^1(C, A)$ there exists some $\bar{h} \in \overline{\operatorname{Hom}}_\Lambda(A, D \operatorname{Tr} C)$ with $\phi(\bar{h}, \delta) \neq 0$. From this it follows that the adjoint morphism

$$\hat{\phi} : \operatorname{Ext}_\Lambda^1(C, A) \to \operatorname{Hom}_{\overline{\operatorname{End}}_\Lambda(D \operatorname{Tr} C)}(\overline{\operatorname{Hom}}_\Lambda(A, D \operatorname{Tr} C), \operatorname{Ext}_\Lambda^1(C, D \operatorname{Tr} C))$$

given by $\hat{\phi}(\delta)(\bar{h}) = \phi(\bar{h}, \delta)$ is an $\operatorname{End}_\Lambda(C)^{\operatorname{op}}$-monomorphism. Since $\operatorname{Ext}_\Lambda^1(C, D \operatorname{Tr} C) \simeq D \overline{\operatorname{End}}_\Lambda(D \operatorname{Tr} C)$ as $\overline{\operatorname{End}}_\Lambda(D \operatorname{Tr} C)$-modules, we have by IX Lemma 1.4 that the functor $\operatorname{Hom}_{\overline{\operatorname{End}}_\Lambda(D \operatorname{Tr} C)}(\ , \operatorname{Ext}_\Lambda^1(C, D \operatorname{Tr} C))$ preserves lengths over R and therefore $\hat{\phi}$ is an isomorphism. Hence ϕ induces a duality between the $\underline{\operatorname{End}}_\Lambda(C)^{\operatorname{op}}$-module $\operatorname{Ext}_\Lambda^1(C, A)$ and the $\underline{\operatorname{End}}_\Lambda(C)$-module $\operatorname{Hom}_\Lambda(A, D \operatorname{Tr} C)$ when $\underline{\operatorname{End}}_\Lambda(C)$ and $\overline{\operatorname{End}}_\Lambda(D \operatorname{Tr} C)$ are identified. It is easy to see that the orthogonals $(\operatorname{End}_\Lambda(A)\eta)^\perp$ and $(\operatorname{r}_{\operatorname{End}_\Lambda(A)}\eta)^\perp$ of $\operatorname{End}_\Lambda(A)\eta$ and $\operatorname{r}_{\operatorname{End}(A)}\eta$ respectively in $\overline{\operatorname{Hom}}_\Lambda(A, D \operatorname{Tr} C)$ with respect to ϕ are $\operatorname{End}_\Lambda(A)^{\operatorname{op}}$-submodules of $\overline{\operatorname{Hom}}_\Lambda(A, D \operatorname{Tr} C)$. But $(\operatorname{End}_\Lambda(A)\eta)^\perp = \bigcap\{\operatorname{Ker}(\hat{\phi}(\delta)) \mid \delta \in \operatorname{End}_\Lambda(A)\eta\}$ and $(\operatorname{r}_{\operatorname{End}_\Lambda(A)}\eta)^\perp = \bigcap\{\operatorname{Ker}(\hat{\phi}(\delta)) \mid \delta \in \operatorname{r}_{\operatorname{End}_\Lambda(A)}\eta\}$ so they are also $\overline{\operatorname{End}}_\Lambda(D \operatorname{Tr} C)$-submodules since they are intersections of kernels of $\operatorname{End}_\Lambda(D \operatorname{Tr} C)$-morphisms. Now from the duality we get that $(\operatorname{r}_{\operatorname{End}_\Lambda(A)}\eta)^\perp$ and X have the same R-length and $(\operatorname{End}_\Lambda(A)\eta)^\perp$ and Y have the same R-length. Using that $\overline{\operatorname{Hom}}_\Lambda(A, D \operatorname{Tr} C)$ has only one $\operatorname{End}_\Lambda(A)^{\operatorname{op}}$-submodule of each of these lengths, namely M and M', we get that $(\mathfrak{r}_{End_\Lambda(A)}\eta)^\perp = M$ and $(End_\Lambda(A)\eta)^\perp = M'$. Since M' is the only maximal $End_\Lambda(A)^{op}$-submodule of M we have that M is cyclic, and using that it is a waist we get that any generator α for $(\mathfrak{r}_{End_\Lambda(A)}\eta)^\perp$ has the property that for all $\bar\beta \in \overline{Hom}_\Lambda(A, D\,Tr\,C)$ with $\bar\beta \notin (End_\Lambda(A)\eta)^\perp$ there is some $\alpha' \in End_\Lambda(A)$ with $\overline{\beta\alpha'} = \bar\alpha$. Further, $\bar\alpha \notin (End_\Lambda(A)\eta)^\perp$ implies that there is some $\alpha'' \in End_\Lambda(A)$ with $\phi(\bar\alpha\alpha'', \eta) \neq 0$. But then $\overline{\alpha\alpha''} \notin (End_\Lambda(A)\eta)^\perp$. Since $\phi(\bar\alpha\alpha'', \eta) \neq 0$ there is some $\beta': D\,Tr\,C \to D\,Tr\,C$ with $\phi(\overline{\beta'\alpha\alpha''}, \eta)$ almost split.

Letting a be the element $\beta'\alpha\alpha''$ we collect our findings in the following proposition.

Proposition 5.5 *Let* $\eta: 0 \to A \xrightarrow{f} B \xrightarrow{g} C \to 0$ *be an exact sequence with* f *a properly irreducible morphism and let* $\phi: \overline{Hom}_\Lambda(A, D\,Tr\,C) \times Ext^1_\Lambda(C, A) \to Ext^1_\Lambda(C, D\,Tr\,C)$ *be the natural bilinear map. Then there exists some* $a \in Hom_\Lambda(A, D\,Tr\,C)$ *with the following properties.*

(a) $\phi(\bar a, \eta)$ *is almost split.*
(b) *For each* $\beta \in Hom_\Lambda(A, D\,Tr\,C)$ *with* $\phi(\bar\beta, \eta) \neq 0$ *there is some* $\alpha' \in End_\Lambda(A)$ *with* $\overline{\beta\alpha'} = \bar a$. $\qquad\square$

In order to prove the main theorem of this section we need the following two lemmas.

Lemma 5.6 *Let* $\eta: 0 \to A \xrightarrow{f} B \xrightarrow{g} C \to 0$ *be exact with* f *a properly irreducible morphism. Then* gh *is not irreducible for any module* B' *and morphism* $h: B' \to B$.

Proof Let η be as above and let $0 \to A \to F \xrightarrow{b} Tr\,DA \to 0$ be an almost split sequence. Then there exist morphisms a and c making the following diagram commute.

$$
\begin{array}{ccccccccc}
0 & \to & A & \to & F & \xrightarrow{b} & Tr\,DA & \to & 0 \\
  &   & \| &   & \downarrow a & & \downarrow c & & \\
0 & \to & A & \xrightarrow{f} & B & \xrightarrow{g} & C & \to & 0
\end{array}
$$

Since f is irreducible, a is a split epimorphism. Let $h: B' \to B$ be a morphism and assume $gh: B' \to C$ is irreducible and let a' be a map from B to F such that aa' is the identity on B. Then $gh = gaa'h = cba'h$. But $ba'h$ is not a split monomorphism, so c is a split epimorphism. But then a and c are isomorphisms and hence g is irreducible. Hence both f and

g are irreducible and therefore $0 \to A \xrightarrow{f} B \xrightarrow{g} C \to 0$ is an almost split sequence by V Proposition 5.9. But then f is not a properly irreducible morphism. This gives the desired contradiction.   □

Using Lemma 5.6 we obtain the following generalization of V Corollary 5.8.

Lemma 5.7 *Let* $\eta: 0 \to A \xrightarrow{f} B \xrightarrow{g} C \to 0$ *be an exact sequence with* f *a properly irreducible morphism. Let* $g': B' \to C$ *be irreducible. Then there exists* $h: B \to B'$ *with* $g = g'h$. *In particular* g' *is an epimorphism.*

Proof This follows easily from the previous lemma.   □

We are now ready to prove our promised result.

Theorem 5.8 *Let* $\eta: 0 \to A \xrightarrow{f} B \xrightarrow{g} C \to 0$ *be an exact sequence with* f *a properly irreducible morphism and let* $\delta: 0 \to D\operatorname{Tr}C \to E \to C \to 0$ *be an almost split sequence. If* E *is decomposable then* C *is simple.*

Proof Let $\eta: 0 \to A \xrightarrow{f} B \xrightarrow{g} C \to 0$ be exact with f a properly irreducible morphism and let $a \in \operatorname{Hom}_\Lambda(A, D\operatorname{Tr}C)$ be fixed such that $\bar{a} \in \overline{\operatorname{Hom}}_\Lambda(A, D\operatorname{Tr}C)$ satisfies conditions (a) and (b) in Proposition 5.5. Assume that $0 \to D\operatorname{Tr}C \to E \to C \to 0$ is an almost split sequence with E decomposable. Let $E = E_1 \coprod E_2$ with $E_i \neq 0$ for $i = 1, 2$ and consider the sequence $0 \to D\operatorname{Tr}C \xrightarrow{\binom{i}{j}} E_1 \coprod E_2 \xrightarrow{(p,q)} C \to 0$. According to Lemma 5.7, the morphisms p and q are both epimorphisms and there exist morphisms $h_1: B \to E_1$ and $h_2: B \to E_2$ with $g = ph_1$, and $g = qh_2$. Hence there exists $f_1: A \to D\operatorname{Tr}C$ with $if_1 = h_1f_1$ and $jf_1 = 0$ and there exists $f_2: A \to D\operatorname{Tr}C$ with $if_2 = 0$ and $jf_2 = h_2f$. Therefore $\phi(\bar{f_i}, \eta) \neq 0$ for $i = 1, 2$. So by Proposition 5.5(b) there exist a_1 and a_2 in $\operatorname{End}_\Lambda(A)^{\operatorname{op}}$ with $a_1\bar{f_1} = \bar{a}$ and $a_2\bar{f_2} = \bar{a}$. But then $\overline{ia} = 0$ and $\overline{ja} = 0$ and hence we get that $\binom{i}{j}a \in \mathscr{I}(A, E_1 \coprod E_2)$, the subgroup of $\operatorname{Hom}_\Lambda(A, E_1 \coprod E_2)$ consisting of morphisms factoring through injective modules. Hence there is a morphism $\tilde{\psi}: B \to E_1 \coprod E_2$ factoring through an injective module making the following diagram commute.

$$
\begin{array}{ccccccccc}
\eta: & 0 & \to & A & \xrightarrow{f} & B & \xrightarrow{g} & C & \to & 0 \\
 & & & \downarrow a & & \downarrow \tilde{\psi} & & \downarrow \psi & & \\
 & 0 & \to & D\operatorname{Tr}C & \xrightarrow{\binom{i}{j}} & E_1 \amalg E_2 & \xrightarrow{(p,q)} & C & \to & 0
\end{array}
$$

Now the pushout of η by a is an almost split sequence. Therefore ψ is an isomorphism. Suppose C is not simple and let $\gamma : \operatorname{soc} C \to C$ be the inclusion. Then there exists $\beta : \operatorname{soc} C \to B$ with $\gamma = g\beta$. Hence $\psi\gamma = (p,q)\tilde{\psi}\beta$ where $\tilde{\psi}$ factors through an injective module. This gives the desired contradiction. $\square$

The following restatement of Theorem 5.2 follows easily from Theorem 5.8.

Theorem 5.9 *Let* $0 \to A \xrightarrow{f} B \xrightarrow{g} C \to 0$ *be an exact sequence with* A *indecomposable,* f *a properly irreducible morphism and* C *not simple. Then in the almost split sequence* $0 \to D\operatorname{Tr} C \to E \to C \to 0$ *the middle term* E *is indecomposable.* $\square$

Theorem 5.9 does not answer the question of under what circumstances $\alpha(C) = 1$ when C is a simple nonprojective module. We have already given an example of a simple nonprojective module C where $\alpha(C) > 1$ even though there is an exact sequence $0 \to A \xrightarrow{f} B \to C \to 0$ with f a properly irreducible morphism. We end this discussion with an example of a nonprojective simple module C which is the cokernel of a properly irreducible monomorphism with the property $\alpha(C) = 1$. Hence more is involved in determining when a nonprojective simple module C has the property $\alpha(C) = 1$ than just whether or not C is the cokernel of a properly irreducible monomorphism. The example we now give is closely related to our previous example.

Example Let $k[X, Y]$ be the polynomial ring in two commuting variables X and Y over a field k. Define $\Lambda = k[X, Y]/(X^3, Y^3)$ and denote by $\overline{f}(X, Y)$ the image in Λ of a polynomial $f(X, Y)$ in $k[X, Y]$ and write $\overline{X} = x$ and $\overline{Y} = y$. Since (X, Y) is a maximal ideal in $k[X, Y]$ and $(X, Y)^5 \subset (X^3, Y^3)$ it follows that Λ is a finite dimensional local k-algebra with unique maximal ideal $\mathfrak{m} = (x, y)$ and that $\Lambda/\mathfrak{m} = k$. It is also not difficult to check that $\operatorname{soc} \Lambda$ is simple with generator $(xy)^2$. Therefore since Λ is commutative, it is a symmetric k-algebra by IV Section 3. Hence $\mathfrak{m}$ is an indecomposable Λ-module. Then the exact sequence $0 \to \mathfrak{m} \xrightarrow{f} \Lambda \to k \to 0$, where f is the inclusion morphism, has the property that f is an irreducible morphism without being an almost split sequence. So the simple Λ-module k is the cokernel of the properly irreducible monomorphism f. We now show that $\alpha(k) = 1$, giving our desired example.

Since Λ is symmetric we have by V Proposition 5.5 the almost split sequence $0 \to \mathfrak{m} \to \Lambda \coprod (\mathfrak{m}/\operatorname{soc}\Lambda) \overset{(f,g)}{\to} \Lambda/\operatorname{soc}\Lambda \to 0$ where $g:\Lambda \to \Lambda/\operatorname{soc}\Lambda$ is the natural epimorphism and $h:\mathfrak{m}/\operatorname{soc}\Lambda \to \Lambda/\operatorname{soc}\Lambda$ is the inclusion morphism. We have the commutative exact diagram

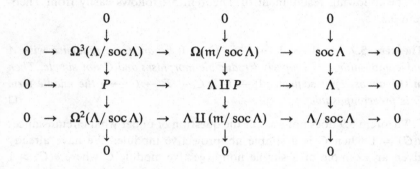

where $P \to \Omega^2(\Lambda/\operatorname{soc}\Lambda)$ is a projective cover. Then $0 \to \Omega^3(\Lambda/\operatorname{soc}\Lambda) \to \Omega(\mathfrak{m}/\operatorname{soc}\Lambda) \to \operatorname{soc}\Lambda \to 0$ is an almost split sequence by X Section 1. Since $\operatorname{soc}\Lambda$ is isomorphic to the unique simple Λ-module k, we have that $\alpha(k) = 1$ if we show that $\Omega(\mathfrak{m}/\operatorname{soc}\Lambda)$ is indecomposable. To show that $\Omega(\mathfrak{m}/\operatorname{soc}\Lambda)$ is indecomposable, it suffices to show that $\mathfrak{m}/\operatorname{soc}\Lambda$ is indecomposable since Λ is a symmetric algebra.

Since $\mathfrak{m}/\operatorname{soc}\Lambda = (X,Y)/(X^3,(XY)^2,Y^3)$ and $(X,Y)^3 \supset (X^3,(XY)^2,Y^3)$ it follows that $(X,Y)^2(X,Y)/(X^3,(XY)^2,Y^3) \simeq (X,Y)^3/(X^3,(XY)^2,Y^3)$ and so $(\mathfrak{m}/\operatorname{soc}\Lambda)/\mathfrak{m}^2(\mathfrak{m}/\operatorname{soc}\Lambda) \simeq (X,Y)/(X,Y)^3$. Therefore $\mathfrak{m}/\operatorname{soc}\Lambda$ is an indecomposable Λ-module if $(X,Y)/(X,Y)^3$ is an indecomposable Λ-module. Now $(X,Y)/(X,Y)^3$ is a Γ-module where Γ is the factor ring $k[X,Y]/(X,Y)^2$ of Λ. So it suffices to show that $(X,Y)/(X,Y)^3 = A$ is an indecomposable Γ-module. It is not difficult to see that $\operatorname{ann}_\Gamma(z) = 0$ for each z in $A - \mathfrak{m}'A$ where $\mathfrak{m}'$ is the maximal ideal of Γ. Therefore $\Gamma z \simeq \Gamma$ for all z in $A - \mathfrak{m}'A$.

Suppose now that $A = B \coprod C$ with $B \neq 0$ and $C \neq 0$. Then there are b and c in $A - \mathfrak{m}'A$ such that $\Gamma b \simeq B$ and $\Gamma c \simeq C$, so $A \simeq 2\Gamma$. But $\dim_k A = 5$ while $\dim_k(2\Gamma) = 6$. This contradiction shows that A is indecomposable, finishing the proof that the example has the desired properties.

Exercises

1. Let Δ be the quiver

,

let k be a field and consider the simple $k\Delta$-module S_4 corresponding to the vertex 4.

(a) Prove that $\operatorname{Tr} DS_4$ corresponds to the representation

and that $(\operatorname{Tr} D)^2 S_4$ corresponds to the representation

.

(b) Show that there are exact sequences $0 \to S_4 \to \operatorname{Tr} DS_4 \to (\operatorname{Tr} D)^2 S_4 \to 0$ which are not rigid.

(c) Find three nonzero rigid exact sequences in $\operatorname{Ext}^1_\Lambda(S_4, (\operatorname{Tr} D)^2 S_4)$.

2. Let k be a field and $\Lambda = k[X]/(X^3)$. Prove that there are nonrigid exact sequences in $\operatorname{Ext}^1_\Lambda((\Lambda/\mathfrak{r}) \coprod (\Lambda/\mathfrak{r}^2), (\Lambda/\mathfrak{r}) \coprod (\Lambda/\mathfrak{r}^2))$.

3. Let Λ be an artin algebra and let X be in $\operatorname{mod} \Lambda$. For each Λ-module A denote by $\tau_X(A)$ the submodule of A generated by all images of all morphisms in $\operatorname{Hom}_\Lambda(X, A)$.

(a) Prove that for each morphism g in $\operatorname{Hom}_\Lambda(A, B)$ we have $g(\tau_X(A)) \subset \tau_X(B)$ and deduce that τ_X induces a functor from $\operatorname{mod} \Lambda$ to $\operatorname{mod} \Lambda$ which is a subfunctor of the identity functor.

(b) Show that $\tau_X(\Lambda)$ is a two-sided ideal in Λ.

(c) Let $f : B \to C$ be a morphism in $\operatorname{mod} \Lambda$ such that $\operatorname{Hom}_\Lambda(X, f) = 0$. Prove that $\operatorname{Im} f \subset \{c \in C \mid \tau_X(\Lambda)c = 0\}$.

(d) Let $f : B \to C$ be a morphism in $\operatorname{mod} \Lambda$ which is right determined by X. Prove that $\operatorname{Im} f \supset \{c \in C \mid \tau_X(\Lambda)c = 0\}$.

(e) Let $f : B \to C$ be a morphism in $\operatorname{mod} \Lambda$ which is right determined by X and such that $\operatorname{Hom}(X, f) = 0$. Prove that $\operatorname{Im} f = \{c \in C \mid \tau_X(\Lambda)c = 0\}$.

4. Assume throughout this exercise that $\mathfrak{a}$ is a left ideal of the artin R-algebra Λ and that $\Gamma = \mathrm{End}_\Lambda(\Lambda/\mathfrak{a})^{\mathrm{op}}$. Recall from I Exercise 12 that the idealizer of $\mathfrak{a}$ in Λ is the subring $\Sigma = \{\lambda \in \Lambda | \mathfrak{a}\lambda \subset \mathfrak{a}\}$ of Λ and that $\phi: \Sigma \to \Gamma$, given by $\phi(\gamma)(\lambda + \mathfrak{a}) = \lambda\gamma + \mathfrak{a}$ for $\gamma \in \Sigma$ and $\lambda + \mathfrak{a} \in \Lambda/\mathfrak{a}$, is a surjective ring morphism. We consider Γ-modules as Σ-modules by means of the ring morphism ϕ and Λ-modules as Γ-modules through the inclusion $\Gamma \subset \Lambda$.

(a) Prove that Σ is an artin R-subalgebra of Λ and that ϕ is an artin R-algebra morphism.

(b) For C in $\mathrm{mod}\,\Lambda$ let $_\mathfrak{a}C = \{c \in C | \mathfrak{a}c = 0\}$. Prove that $_\mathfrak{a}C$ is a Σ-submodule of C.

(c) For C in $\mathrm{mod}\,\Lambda$ we have that $\mathrm{Hom}_\Lambda(\Lambda/\mathfrak{a}, C)$ is a Γ-module and hence a Σ-module. Prove that the map $\beta: \mathrm{Hom}_\Lambda(\Lambda/\mathfrak{a}, C) \to C$ given by $\beta(f) = f(1 + \mathfrak{a})$ for all $f \in \mathrm{Hom}_\Lambda(\Lambda/\mathfrak{a}, C)$ is a monomorphism of Γ-modules with image $_\mathfrak{a}C$.

(d) Let $f: B \to C$ be a morphism in $\mathrm{mod}\,\Lambda$. Prove that $f(_\mathfrak{a}B) \subset {}_\mathfrak{a}C$ is a Γ-submodule of C.

(e) Let $f: B \to C$ be a morphism in $\mathrm{mod}\,\Lambda$. Prove that the following are equivalent.

   (i) f is right determined by $\Lambda/\mathfrak{a}$.

   (ii) Each morphism $g: Y \to C$ with $g(_\mathfrak{a}Y) \subset f(_\mathfrak{a}B)$ factors through f.

(f) Let C be in $\mathrm{mod}\,\Lambda$ and let $\mathcal{H}$ be the set of Γ-submodules of $_\mathfrak{a}C$. Prove that the map $\psi: \mathrm{mod}\,\Lambda/C \to \mathcal{H}$ given by $\psi(f) = f(_\mathfrak{a}B)$ for each $f: B \to C$ in $\mathrm{mod}\,\Lambda/C$ induces a bijection between the isomorphism classes of right $(\Lambda/\mathfrak{a})$-determined morphisms in $\mathrm{mod}\,\Lambda/C$ and $\mathcal{H}$.

(g) Let $\mathfrak{b}$ be the right ideal $_\mathfrak{a}\Lambda$ of Λ and let $f: B \to C$ be a morphism in $\mathrm{mod}\,\Lambda$ which is right determined by $\Lambda/\mathfrak{a}$. Prove that $\mathrm{Im}\,f = \{c \in C | \mathfrak{b}c \subset f(_\mathfrak{a}B)\}$.

5. Let Λ be a local artin ring with maximal ideal $\mathfrak{m}$ and denote $\Lambda/\mathfrak{m}$ by k.

(a) Prove that the following are equivalent for an indecomposable Λ-module Y.

   (i) $\mathrm{Ext}^1_\Lambda(k, Y) \simeq k$ as a k^{op}-module.

   (ii) There exists a proper left ideal $\mathfrak{a}$ in Λ with $\mathrm{Tr}\,DY \simeq \Lambda/\mathfrak{a}$.

(b) Let $\mathfrak{a}$ be a proper left ideal of Λ and let $0 \to D\operatorname{Tr}(\Lambda/\mathfrak{a}) \xrightarrow{g} B \xrightarrow{f} k \to 0$ be a nonsplit exact sequence. Show that f is a right minimal morphism determined by $\Lambda/\mathfrak{a}$ with $\operatorname{Hom}_\Lambda(\Lambda/\mathfrak{a}, f) = 0$.

(c) Let $g: \Lambda/\mathfrak{a} \to k$ be the natural morphism. Prove that in the pullback diagram

$$
\begin{array}{ccccccccc}
0 & \to & D\operatorname{Tr}(\Lambda/a) & \to & E & \to & \Lambda/\mathfrak{a} & \to & 0 \\
 & & \| & & \downarrow h & & \downarrow g & & \\
0 & \to & \operatorname{Tr}D(\Lambda/a) & \to & B & \to & k & \to & 0
\end{array}
$$

the sequence $0 \to D\operatorname{Tr}(\Lambda/a) \to E \to \Lambda/\mathfrak{a} \to 0$ is almost split.

(d) Suppose $\mathfrak{a}$ is a two-sided ideal contained in the right socle of Λ, and let $0 \to nD\operatorname{Tr}(\Lambda/\mathfrak{a}) \to A \xrightarrow{h} C$ be an exact sequence with h a right minimal morphism determined by $\Lambda/\mathfrak{a}$ and with $\operatorname{Hom}_\Lambda(\Lambda/\mathfrak{a}, h) = 0$. Show that $\operatorname{Im} h = \operatorname{soc} C$, that $n = l(\operatorname{soc} C)$ and that the induced exact sequence $0 \to nD\operatorname{Tr}(\Lambda/\mathfrak{a}) \to A \to \operatorname{Im} h \to 0$ is isomorphic to

$0 \to nD\operatorname{Tr}(\Lambda/\mathfrak{a}) \xrightarrow{\coprod g} nB \xrightarrow{\coprod f} nk \to 0$ where $0 \to D\operatorname{Tr}(\Lambda/\mathfrak{a}) \xrightarrow{g} B \xrightarrow{f} k \to 0$ is the sequence given in (b).

6. Let R be the commutative polynomial ring $k[X_1, \ldots, X_n]$ in n variables over the field k. Let $\mathfrak{m}_0$ be the maximal ideal of R generated by $X_1, X_2, \ldots, X_n$, and let $\mathfrak{a}$ be an ideal such that there is some $q \geq 2$ such that $\mathfrak{m}_0^q \subset \mathfrak{a} \subset \mathfrak{m}_0^2$. Let $\Lambda = R/\mathfrak{a}$ and let $h: R \to \Lambda$ be the natural ring morphism such that $h(r) = r + \mathfrak{a}$.

(a) Show that Λ is a commutative local artin ring with maximal ideal $\mathfrak{m} = h(\mathfrak{m}_0)$ and with $\Lambda/\mathfrak{m} \simeq k$.

(b) Denote $h(X_i)$ by x_i for $i = 1, \ldots, n$. Show that the kernel of the morphism $f: \Lambda \to n\Lambda$ given by $f(1) = (x_1, \ldots, x_n)$ is $\operatorname{soc} \Lambda$, that $\operatorname{Coker} f \simeq \operatorname{Tr} D(k)$ and that the induced exact sequence (*) $0 \to k \to n\Lambda/(\mathfrak{m}(x_1, \ldots, x_n)) \to \operatorname{Coker} f \to 0$ is an almost split sequence.

(c) Let t be the number of summands in a decomposition of $n\Lambda/(\mathfrak{m}(x_1, \ldots, x_n))$ into a sum of indecomposable modules. Show that $t = \alpha(\Lambda/\mathfrak{m})$, where α is as defined in V Section 6.

(d) Let $\mathfrak{m} = \coprod_{i=1}^s \mathfrak{m}_i$ be a decomposition of $\mathfrak{m}$ into a sum of indecomposable Λ-modules. Show that $s \leq t$.

(e) Deduce from (d) that if $\alpha(k) = 1$ then $\mathfrak{m}$ is indecomposable and give an example when $\mathfrak{m}$ is indecomposable and $\alpha(k) \geq 2$.

(f) Show that if $\mathfrak{a} \subset \mathfrak{m}_0^3$ then $\alpha(k) = 1$.

(g) Determine precise conditions for $\alpha(k) = 1$.

7. Let Λ be an artin R-algebra. Let M be in mod Λ. An R-submodule N of M is said to be of finite definition if there exists a Λ-homomorphism $f: \Lambda \to X$ such that $N = \operatorname{Im} \operatorname{Hom}_\Lambda(f, M)$ where $\operatorname{Hom}_\Lambda(\Lambda, M)$ is identified with M. Prove that $N \subset M$ is an R-submodule of M of finite definition if and only if N is an $\operatorname{End}_\Lambda(M)$-submodule of M.

8. Let M be in mod Λ where Λ is an artin algebra. Prove that there exists some X in mod Λ and an x in X such that for each Λ-module N the following are equivalent.

(i) $\operatorname{Hom}_\Lambda(N, M) = 0$.
(ii) For each n in N there exists an f in $\operatorname{Hom}_\Lambda(X, N)$ with $f(x) = n$.

9. Let Λ be a nonsemisimple artin algebra. Prove that there exists a simple Λ-module S and either an irreducible morphism $f: S \to B$ with B indecomposable and $\alpha(\operatorname{Coker} f) = 1$ or an irreducible morphism $g: B \to S$ with B indecomposable and $\alpha(D \operatorname{Ker} g) = 1$.

The following set of exercises is devoted to investigating epimorphisms $B \xrightarrow{f} C$ which are not split epimorphisms but have the property that each proper inclusion $C' \to C$ factors through f. Such morphisms $f: B \to C$ are called subsplit epimorphisms.

10. Suppose $0 \to A \to B \xrightarrow{f} C \to 0$ is an exact sequence with f a subsplit epimorphism. Show that the following statements hold.

(a) C is indecomposable.

(b) If $\begin{array}{ccc} B & \xrightarrow{f} & C \\ g \downarrow & \nearrow_h & \\ B' & & \end{array}$ commutes and h is not a split epimorphism, then h is a subsplit epimorphism.

(c) If $f_0: B_0 \to C$ is a right minimal version of f, then f_0 is also a subsplit epimorphism.

(d) If A is indecomposable and $0 \to A \to E \to \operatorname{Tr} DA \to 0$ is an almost split sequence, then there is a commutative exact diagram

$$
\begin{array}{ccccccccc}
0 & \to & A & \to & E & \to & \operatorname{Tr} DA & \to & 0 \\
  &   & \| &   & \downarrow & & \downarrow & & \\
0 & \to & A & \to & B & \to & C & \to & 0 \\
  &   &   &   & \downarrow & & \downarrow & & \\
  &   &   &   & 0 & & 0 & & .
\end{array}
$$

In particular, there is an epimorphism $\operatorname{Tr} DA \to C$.

(e) Let

$$
\begin{array}{ccccccccc}
0 & \to & A & \to & B \times_C U & \xrightarrow{h} & U & \to & 0 \\
  &   & \| &   & \downarrow &   & \downarrow{\scriptstyle g} \\
0 & \to & A & \to & B & \xrightarrow{f} & C & \to & 0
\end{array}
$$

be a pullback diagram.

(i) Every morphism $X \xrightarrow{t} U$ such that the composition $gt: X \to C$ is not an epimorphism factors through h.

(ii) If h is not a split epimorphism and $g: U \to C$ is an essential epimorphism, then h is a subsplit epimorphism.

11. Show that the following statements are equivalent for an indecomposable, nonprojective module C over an artin algebra Λ.

(a) Every subsplit epimorphism $f: B \to C$ is right almost split.

(b) If $g: U \to C$ is an irreducible epimorphism with U an indecomposable module, then U is projective.

12. Suppose $0 \to A \to B \xrightarrow{f} C \to 0$ is a nonsplit exact sequence. Let C' be a submodule of C minimal with respect to the property that the induced epimorphism $f^{-1}(C') \to C'$ does not split. Show that the following statements hold.

(a) The epimorphism $f^{-1}(C') \to C'$ is a subsplit epimorphism.
   Suppose further that A is an indecomposable module.

(b) If C' has the property that every subsplit epimorphism $t: U \to C$ is right almost split, then the induced exact sequence $0 \to A \to f^{-1}(C') \to C' \to 0$ is almost split, so $\operatorname{Tr} DA \simeq C'$.

(c) If $\operatorname{Tr} DA$ is not isomorphic to any proper submodule of C, then there is an irreducible epimorphism $V \to C'$ with V an indecomposable nonprojective Λ-module.

13. Let $0 \to \Omega(C) \to P \xrightarrow{f} C \to 0$ be a projective cover. Show that the following statements are equivalent.

(a) C is almost projective, as defined in V Section 3.
(b) $f: P \to C$ is a subsplit epimorphism.
(c) The inclusion $\Omega(C) \to P$ is irreducible.

14. Let $0 \to A \to P \xrightarrow{f} C \to 0$ be exact with P an indecomposable projective module and $A \subset \mathfrak{r}P$. Then show that the following are equivalent.

(a) The inclusion $A \to \mathfrak{r}P$ is a split monomorphism.

(b) f is a subsplit epimorphism.

(c) C is almost projective.

(d) For each indecomposable summand A' of A, then there is an epimorphism $\operatorname{Tr} DA \to C$.

15. Suppose $0 \to \Omega(C) \to P \xrightarrow{f} C \to 0$ is exact where $f : P \to C$ is a projective cover of the almost projective module C. Show that the following hold.

(a) A submodule C' of C has the property that $0 \to C'/\mathfrak{r}C' \to C/\mathfrak{r}C$ is exact if and only if there is a decomposition $P = P' \coprod P''$ such that $f(P') = C'$ and the induced morphism $P' \to C'$ is an isomorphism.

(b) If $P = P' \coprod P''$, then the submodule $C' = f(P')$ of C has the property that the induced morphism $C'/\mathfrak{r}C' \to C/\mathfrak{r}C$ is a monomorphism. Hence the induced epimorphism $P' \to C'$ is an isomorphism. So if both P' and P'' are nonzero, then $P' \cap \Omega(C) = 0 = P'' \cap \Omega(C)$.

(c) Let C' be a submodule of C such that $0 \to C'/\mathfrak{r}C' \to C/\mathfrak{r}C$ is exact. If P' is a submodule of P such that $f|_{P'} : P' \to C$ is a monomorphism with image C', then the inclusion $i : P' \to P$ is a split monomorphism.

(d) Suppose $0 \to C' \to C \xrightarrow{f} C'' \to 0$ is an exact sequence such that $C'/\mathfrak{r}C' \to C/\mathfrak{r}C$ is a proper monomorphism. Then C'' is almost projective. (Hint: First show using (c) that f is a subsplit epimorphism.)

(e) Show that if P' is an indecomposable summand of P, then $\Omega(C)$ is isomorphic to a summand of $\mathfrak{r}P'$.

(f) Show that there is an epimorphism $\operatorname{Tr} D(A) \to C$ for each indecomposable summand A of $\Omega(C)$.

(g) If E is any almost projective module, then $l_\Lambda(E) \leq \max\{l_\Lambda(\operatorname{Tr} DA) | A$ indecomposable summand of $\mathfrak{r}\}$.

16. Show for an artin algebra that the set of modules with proper waists has bounded length.

Notes

The notion of morphisms being determined by modules arose originally in [Au5] in connection with subfunctors of a functor being determined by an object in an abelian category. In [Au5] the basic facts about morphisms being determined by modules were derived as consequences of theorems about functors. The development given here is new. While less conceptual than the functorial point of view, it has the advantage of being more explicit and therefore hopefully more constructive.

The notion of the waist of a module was introduced in [AuGR]. The original inspiration for this idea came from studying modules over the Auslander algebra where nontrivial waists frequently appear.

Theorem 5.9 was first conjectured by Brenner and proved independently by Brenner and Krause. The proof of this result given here is a modified version of [Kra1]. Exercise 6 is based on [Lu]. (See also [Bren2]).

Notation

Set notation

$\mathbb{Z}$	integers
$\mathbb{Z}^+$	$\{1, 2, 3, \ldots\}$
$\mathbb{N}$	$\{0, 1, 2, \ldots\}$
$\mathbb{Q}$	rational numbers
$\mathbb{R}$	real numbers
$\mathbb{C}$	complex numbers
$\aleph_0$	countably infinite cardinal
$\subset$	inclusion (not necessarily proper)
$X-Y$	the elements in a set X which are not in a set Y

For a given ring (artin algebra) Λ

$\text{Mod}\,\Lambda$	category of (left) Λ-modules
$\text{f.l.}(\Lambda)$	category of left Λ-modules of finite length (any ring)
$\text{mod}\,\Lambda$	category of finitely generated Λ-modules (artin algebras)
$\text{ind}\,\Lambda$	category of a chosen set of representatives of nonisomorphic indecomposable finitely generated Λ-modules
$\text{mod}_{\mathscr{P}}\,\Lambda$	subcategory of $\text{mod}\,\Lambda$ consisting of modules without projective summands
$\text{mod}_{\mathscr{I}}\,\Lambda$	subcategory of $\text{mod}\,\Lambda$ consisting of modules without injective summands
$\underline{\text{mod}}\,\Lambda$	category $\text{mod}\,\Lambda$ modulo projectives
$\overline{\text{mod}}\,\Lambda$	category $\text{mod}\,\Lambda$ modulo injectives
$K_0(\text{mod}\,\Lambda)$	Grothendieck group of $\text{mod}\,\Lambda$ (modulo exact sequences)
$K_0(f.l.\Lambda)$	Grothendieck group of $f.l.\Lambda$
$\mathscr{P}(\Lambda)$	category of finitely generated projective Λ-modules
$\mathscr{I}(\Lambda)$	category of finitely generated injective Λ-modules
$\widetilde{\mathscr{P}}(\Lambda)$	category of finitely generated preprojective Λ-modules for a hereditary artin algebra Λ

406

$\widetilde{\mathscr{I}}(\Lambda)$	the category of finitely generated preinjective Λ-modules for a hereditary artin algebra Λ
gl.dim Λ	global dimension of Λ
$T_2(\Lambda)$	ring of 2×2 lower triangular matrices over Λ
$r_\Lambda = \operatorname{rad} \Lambda$	radical of Λ (also written r)
$\operatorname{rl}(\Lambda))$	Loewy length of Λ (index of nilpotency)
Λ^{op}	opposite ring of Λ
$D : \operatorname{mod} \Lambda \to$	
$\operatorname{mod}(\Lambda^{op})$	standard duality for artin algebras
1 :	identity element in rings (also identity element in groups written multiplicatively)

For a given Λ-module A (Λ artin R-algebra)

$\operatorname{pd}_\Lambda A$	projective dimension
$\operatorname{id}_\Lambda A$	injective dimension
1_A	the identity map from A to A
$\operatorname{rad} A$	radical of A
$\operatorname{add} A$	category of summands of finite sums of copies of A
dom.dim A	dominant dimension of A
$l(A)$	length of A as a Λ-module
$l_R(A) = \langle A \rangle$	length of A as R-module
$\Omega^i(A)$	ith syzygy module of A ($\Omega A = \Omega^1 A$)
$\operatorname{Tr}(A)$	transpose of A
nA	sum of n copies of A
$\operatorname{soc} A$	socle of A
$m_S(A)$	multiplicity of a simple module S in a composition series of A
$\operatorname{rl}(A)$	radical length of A
$\operatorname{sl}(A)$	socle length of A
e_A	the functor $\operatorname{Hom}_\Lambda(A, \ )$
A^*	the Λ^{op}-module $\operatorname{Hom}_\Lambda(A, \Lambda)$
$\operatorname{ann}_\Lambda A$	annihilator of A in Λ
$\operatorname{End}_\Lambda(A)$	endomorphism ring of A
T_A	$\operatorname{End}_\Lambda(A)/ \operatorname{rad} \operatorname{End}_\Lambda(A)$ (A indecomposable)

For given Λ-modules A and B (Λ artin R-algebra)

$\operatorname{rad}_\Lambda(\ , \ )$	radical of the category $\operatorname{mod} \Lambda$
$\operatorname{Irr}(A, B)$	$\operatorname{rad}_\Lambda(A, B)/ \operatorname{rad}_\Lambda^2(A, B)$
$\operatorname{rad}_\Lambda^\infty(A, B)$	$\bigcap_{m \geq 0} \operatorname{rad}_\Lambda^m(A, B)$
$A + B$	submodule generated by A and B

$\mathscr{P}(A, B)$ the morphisms from A to B which factor through a projective module

$\mathscr{I}(A, B)$ the morphisms from A to B which factor through an injective module

$\langle A, B \rangle$ the length of $\mathrm{Hom}_\Lambda(A, B)$ as an R-module

$\tau_A(B)$ submodule of B generated by the images of Λ-morphisms from A to B

For a given map $f: A \to B$ between modules

$\mathrm{Im}\, f$ image of f

$\mathrm{Ker}\, f$ kernel of f

$f|_C$ restriction to a submodule C of A

$f^{-1}(E)$ preimage in A of a submodule E of B

$\mathrm{Coker}\, f$ cokernel of F

Others

$\mathrm{Supp}\, F$ the objects A in $\mathrm{ind}\,\Lambda$ with $F(A) \neq 0$ for a functor $F: \mathrm{mod}\,\Lambda \to \mathrm{mod}\, R$

$\coprod$ categorical sum

$f \amalg g$ map from $A \amalg B$ to $X \amalg Y$ induced by $f: A \to X$ and $g: B \to Y$ given by $(f \amalg g)(a, b) = (f(a), g(b))$

Ab category of abelian groups

$T(\Sigma, M)$ tensor ring of the Σ-bimodule M over Σ

$\Pi_n(k)$ k-algebra which is the product of n copies of a field k with k acting diagonally

$A \times^C B$ pushout of diagram $\begin{array}{ccc} C & \to & A \\ \downarrow & & \\ B & & \end{array}$

$A \times_C B$ pullback of diagram $\begin{array}{ccc} & & B \\ & & \downarrow \\ A & \to & C \end{array}$

(f_{ij}) the induced map $\coprod\limits_{j=1}^{n} A_j \to \coprod\limits_{i=1}^{t} B_i$ for Λ-morphisms $f_{ij}: A_j \to B_i$ for $1 \leq j \leq n$ and $1 \leq i \leq t$

δ^* the functor given by the exact sequence $\mathrm{Hom}_\Lambda(\ , B) \overset{\mathrm{Hom}_\Lambda(\ , g)}{\to} \mathrm{Hom}_\Lambda(\ , C) \to \delta^* \to 0$ for an exact sequence $\delta: 0 \to A \overset{f}{\to} B \overset{g}{\to} C \to 0$ in $\mathrm{mod}\,\Lambda$

δ_* the functor given by the exact sequence $\mathrm{Hom}(B, \ ) \overset{\mathrm{Hom}(f, \ )}{\to} \mathrm{Hom}(A, \ ) \to \delta_* \to 0$ for an exact sequence $\delta: 0 \to A \overset{f}{\to} B \overset{g}{\to} C \to 0$ in $\mathrm{mod}\,\Lambda$

Conjectures

We now list some well known conjectures in the representation theory of artin algebras covered in this book.

(1) If Λ is an infinite artin algebra of infinite representation type, then there are infinitely many integers n with infinitely many indecomposable modules of length n.

(It is enough to find one such n (see [Sm]). The conjecture is verified for finite dimensional algebras over an algebraically closed field (see [Bau3], [BretT] and [Fi]). It would be interesting to find a more direct proof even in this case.)

(2) If the AR-quiver of an artin algebra Λ has only one component, then Λ is of finite representation type.

(3) If Λ is an artin algebra of infinite representation type, then the AR-quiver of Λ has infinitely many components. (This conjecture holds for hereditary artin algebras.)

(4) If Λ is an artin algebra and all indecomposable Λ-modules are determined up to isomorphism by their composition factors, then Λ is of finite representation type.

(5) Let Λ and Λ' be two artin algebras which are stably equivalent. Then the numbers of isomorphism classes of simple nonprojective modules are the same for Λ and Λ'. (It is enough to consider the situation when Λ is selfinjective (see [M3]). This is proven when Λ is of finite representation type (see [M2]).)

409

(6) Let Λ be an artin algebra of finite global dimension with $S_1, \ldots, S_n$ a complete list of nonisomorphic simple Λ-modules and $P_1, \ldots, P_n$ the corresponding projective covers. Let C be the Cartan matrix of Λ, i.e. $C = (c_{ij})$ where $[P_j] = \sum_{i=1}^{n} c_{ij}[S_i]$ in the Grothendieck group of Λ. Then the determinant of C is 1. (This has been proven for graded artin algebras ([W] and [FuZ]). For other cases see [Z] and [BurF].)

(7) Let Λ be an artin algebra and S a simple Λ-module with $\text{Ext}^1_\Lambda(S, S) \neq 0$. Then $\text{pd}_\Lambda(S) = \infty$. (It is known that under the given assumption we have gl.dim $\Lambda = \infty$; see [Le] and [I].)

(8) Let Λ be an artin algebra and $0 \to A \to I_0 \to I_1 \to I_2 \to \cdots$ a minimal injective resolution of Λ as a left Λ-module. If I_j is projective for all $j \geq 0$, then Λ is selfinjective. (This is known as the Nakayama Conjecture, and has been proven for graded artin algebras ([W] and [FuZ]).)

(9) Let Λ be an artin algebra and S a simple Λ-module. Then $\text{Ext}^i_\Lambda(S, \Lambda) \neq 0$ for some i. An affirmative answer to (9) implies that the conjecture in (8) also holds (see [AuR3]).

(10) A Λ-module M is projective if $\text{Ext}^i_\Lambda(M, M \amalg \Lambda) = 0$ for all $i > 0$. (This is equivalent to (9), see [AuR1], and was conjectured by Tachikawa for selfinjective algebras in [Ta].)

(11) Let Λ be an artin algebra. Then $\sup\{\text{pd}_\Lambda X | X$ in mod Λ and $\text{pd}_\Lambda X < \infty\}$ is finite. (This is proven for some classes of algebras (see [IZ], [GrZ] and [GrKK]). An affirmative answer to (11) implies that the conjecture in (9) also holds.)

(12) A Λ-module M is zero if $\text{Ext}^i_\Lambda(M, \Lambda) = 0$ for all $i \geq 0$. (This is a consequence of (11).)

(13) If $\text{id}_\Lambda \Lambda < \infty$, then $\text{id}_{\Lambda^{op}} \Lambda < \infty$. (This is a consequence of (11).)

Open problems

We now list some open problems based on the topics covered in this book.

(1) Give a method for deciding when two uniserial modules over an artin algebra are isomorphic.

(2) Which artin algebras of infinite representation type have only a finite number of pairwise nonisomorphic uniserial modules?

(3) Let A and C be uniserial modules. Give a method for deciding if there are exact sequences $0 \to A \to B \to C \to 0$ with the property that B is also uniserial.

(4) A uniserial module is said to be a maximal uniserial module if it is neither a proper submodule, or a proper factor module of a uniserial module. Which algebras Λ have the property that all maximal uniserial submodules have length equal to the Loewy length of Λ? (This has been solved for commutative local rings containing an infinite field in [Lu].)

(5) An artin algebra Λ is called a monomial algebra if Λ is isomorphic to a path algebra of a quiver with relations over a field k and where all relations can be chosen to be paths. Give an invariant description, one that is independent of generators and relations, of when an artin algebra is a monomial algebra. (See [BurFGZ] for partial results.)

(6) Describe the artin algebras with only a finite number of nonisomorphic almost projective modules. (This question is decided in [M1] for hereditary algebras.)

411

(7) Suppose Λ and Γ are artin algebras and $G:\underline{\mathrm{mod}}\,\Lambda \to \underline{\mathrm{mod}}\,\Gamma$ is an equivalence of categories.

(a) When is there a functor $F:\mathrm{mod}\,\Lambda \to \mathrm{mod}\,\Gamma$ with the property $F(\mathscr{P}(\Lambda)) \subset \mathscr{P}(\Gamma)$ such that the induced functor $F':\underline{\mathrm{mod}}\,\Lambda \to \underline{\mathrm{mod}}\,\Gamma$ is isomorphic to G?

(b) When is there such a functor $F:\mathrm{mod}\,\Lambda \to \mathrm{mod}\,\Gamma$ which is half exact, right exact, left exact or exact?

(c) Is there some functor $F:\mathrm{mod}\,\Lambda \to \mathrm{mod}\,\Gamma$ with the property $F(\mathscr{P}(\Lambda)) \subset \mathscr{P}(\Gamma)$ such that the induced functor $F':\underline{\mathrm{mod}}\,\Lambda \to \underline{\mathrm{mod}}\,\Gamma$ is an equivalence of categories.

(8) Describe the infinite artin algebras with the property that each indecomposable module has only a finite number of indecomposable factor modules up to isomorphism.

(9) Describe the artin algebras which have only a finite number of nonisomorphic modules with proper waists.

(10) Suppose that the Loewy lengths of the endomorphism ring as a module over itself or as a module over the ground ring are bounded for all indecomposable Λ-modules. Is Λ of finite representation type? It is known that if the lengths of the endomorphism ring are bounded for all indecomposable modules, then Λ is of finite representation type [SmV]. (The case when the length is 1 was proved in [Sk2].)

(11) When is a selfinjective algebra stably equivalent to a symmetric algebra also symmetric?

Bibliography

[AnF]   F. W. Anderson and K. R. Fuller. *Rings and Categories of Modules.* Graduate Texts in Math., 13, Springer-Verlag, Berlin–Heidelberg–New York (1992).

[Au1]   M. Auslander. *Modules over unramified regular local rings.* Proc. of Int. Congress of Math. Stockholm (1962) 230–233.

[Au2]   M. Auslander. *Coherent functors.* Proceedings Conference on Categorical Algebra, La Jolla, Springer-Verlag, Berlin–Heidelberg–New York (1966) 189–231.

[Au3]   M. Auslander. *Representation Dimension of artin algebras.* Queen Mary College Mathematical Notes, London (1971).

[Au4]   M. Auslander. *Representation theory of artin algebras II.* Comm. in Algebra, 2 (1974) 269–310.

[Au5]   M. Auslander. *Functors and morphisms determined by objects.* Proc. Conf. on Representation Theory, Philadelphia, Lecture Notes in Pure and Applied Math. 37, Marcel Dekker, New York–Basel (1978) 1–244.

[Au6]   M. Auslander. *Applications of morphisms determined by modules.* Proc. Conf. on Representation Theory, Philadelphia, Lecture Notes in Pure and Applied Math. 37, Marcel Dekker, New York–Basel (1978) 245–327.

[Au7]   M. Auslander. *Representation theory of finite dimensional algebras.* Contemp. Math., 13 (1982) 27–39.

[Au8]   M. Auslander. *Relations for Grothendieck groups of artin algebras.* Proc. Amer. Math. Soc., 91 (1984) 336–340.

[Au9]   M. Auslander. *Rational singularities and almost split sequences.* Trans. Amer. Math. Soc., 293 (1986) 511–531.

[AuB]   M. Auslander and M. Bridger. *Stable Module Theory.* Memoirs Amer. Math. Soc. no. 94, Providence R. I., (1969).

[AuBPRS]   M. Auslander, R. Bautista, M. I. Platzeck, I. Reiten and S. O. Smalø. *Almost split sequences whose middle term has at most two indecomposable summands.* Canad. J. Math., 31 (1979) 942–960.

[AuC]   M. Auslander and J. F. Carlson. *Almost split sequences and group algebras.* J. Algebra, 103 (1986) 122–140.

413

[AuGR]   M. Auslander, E. Green and I. Reiten. *Modules with waists*. Ill. J. Math., 19, No. 3 (1975) 467–478.

[AuP]   M. Auslander and M. I. Platzeck. *Representation theory of hereditary artin algebras*. Lecture Notes in pure and applied mathematics, 37, Marcel Dekker, New York and Basel (1978) 389–424.

[AuPR]   M. Auslander, M. I. Platzeck, and I. Reiten. *Coxeter functors without diagrams*. Trans. Amer. Math. Soc., 250 (1979) 1–12.

[AuR1]   M. Auslander and I. Reiten. *Stable equivalence of artin algebras*. Proc. Ohio Conf. on orders, group rings and related topics, Lecture Notes in Math., 353, Springer-Verlag, Berlin–Heidelberg–New York (1973) 8–71.

[AuR2]   M. Auslander and I. Reiten. *Stable equivalence of dualizing R-varieties I*. Advances in Math., 12 (1974) 306–366.

[AuR3]   M. Auslander and I. Reiten. *On a generalized version of the Nakayama conjecture*. Proc. Amer. Math. Soc., 52 (1975) 69–74.

[AuR4]   M. Auslander and I. Reiten. *Representation theory of artin algebras III*. Comm. in Algebra, 3 (1975) 239–294.

[AuR5]   M. Auslander and I. Reiten. *Representation theory of artin algebras IV. Invariants given by almost split sequences*. Comm. in Algebra, 5 (1977) 443–518.

[AuR6]   M. Auslander and I. Reiten. *Representation theory of artin algebras V. Methods for computing almost split sequences and irreducible morphisms*. Comm. in Algebra, 5 (1977) 519–554.

[AuR7]   M. Auslander and I. Reiten. *Representation theory of artin algebras VI. A functorial approach to almost split sequences*. Comm. in Algebra, 2 (1977) 279–291.

[AuR8]   M. Auslander and I. Reiten. *Modules determined by their composition factors*. Ill. J. of Math., 29 (1985) 280–301.

[AuR9]   M. Auslander and I. Reiten. *Almost split sequences for rational double points*. Trans. Amer. Math. Soc., 293 (1987) 87–97.

[AuR10]   M. Auslander and I. Reiten. *The Cohen–Macaulay type of Cohen–Macaulay rings*. Advances in Math., 73 (1989) 1–23.

[AuR11]   M. Auslander and I. Reiten. *Almost split sequences for Cohen–Macaulay modules*. Math. Ann., 277 (1987) 345–349.

[AuRo]   M. Auslander and K. W. Roggenkamp. *A characterization of orders of finite lattice type*. Invent. Math., 17 (1972) 79–84.

[AuS1]   M. Auslander and S. O. Smalø. *Preprojective modules over artin algebras*. J. Algebra, 66 (1980) 61–122.

[AuS2]   M. Auslander and S. O. Smalø. *Almost split sequences in subcategories*. J. Algebra, 69 (1981) 426–454. Addendum. J. Algebra, 71 (1981) 592–594.

[Bak]   Ø. Bakke. Sufficient condition for modules with the same first terms in their projective resolutions to be isomorphic, *Comm. in Alg.* 16(3) 437–442 (1988).

[BakS]   Ø. Bakke and S. O. Smalø. *Modules with the same socles and tops as a directing module are isomorphic*. Comm. in Algebra, 15 (1987) 1–9.

[Bau1]   R. Bautista. *Sections in Auslander–Reiten quivers.* Proc. ICRA II Ottawa 1979, Lecture Notes in Math., 832, Springer-Verlag, Berlin–Heidelberg–New York (1980) 74–96.

[Bau2]   R. Bautista. *Irreducible morphisms and the radical of a category.* An. Inst. Math., 22, Univ. Nac. Auton., Mex. (1982) 83–135.

[Bau3]   R. Bautista. *On algebras of strongly unbounded representation type.* Comm. Math. Helv., 60 (1985) 392–399.

[BauB]   R. Bautista and S. Brenner. *Replication numbers for non-Dynkin sectional subgraphs in finite Auslander–Reiten quivers and some properties of Weyl roots,* Proc. London Math. Soc., 43 (1983) 429–462.

[BauGRS] R. Bautista, P. Gabriel, A. V. Roiter and L. Salmeron. *Representation-finite algebras and multiplicative bases.* Invent. Math., 81 (1985) 277–285.

[BauS]   R. Bautista and S. O. Smalø. *Nonexistent cycles.* Comm. in Algebra, 11 (1983) 1755–1767.

[Ben]   D. J. Benson. *Representations and Cohomology I & II.* Cambridge Univ. Press, 30, 31, Cambridge (1991).

[BerGP]   I. N. Bernstein, I. M. Gelfand, and V. A. Ponomarev. *Coxeter functors and Gabriel's theorem.* Usp. Mat. Nauk 28 (1973) 19–33, Transl. Russ. Math. Surv. 28 (1973) 17–32.

[Bo0]   K. Bongartz. *Treue einfach zusammenhängende Algebren I.* Comment. Math. Helv., 57 (1982) 228–330.

[Bo1]   K. Bongartz. *On a result of Bautista and Smalø on cycles.* Comm. in Algebra, 11 (1983) 2123–2124.

[Bo2]   K. Bongartz. *Critical simply connected algebras.* Manuscr. Math., 46 (1984) 117–136.

[Bo3]   K. Bongartz. *Indecomposables are standard.* Comm. Math. Helv., 60 (1985) 400–410.

[Bo4]   K. Bongartz. *A generalization of a theorem of M. Auslander.* Bull. London Math. Soc., 21 (1989) 255–256.

[BoG]   K. Bongartz and P. Gabriel. *Covering spaces in representation theory.* Invent. Math., 65 (1982) 331–378.

[BonS]   K. Bongartz and S. O. Smalø. *Modules determined by their tops and socles.* Proc. Amer. Math. Soc., 96 (1986) 34–38.

[Bren1]   S. Brenner. *A combinatorial characterization of finite Auslander–Reiten quivers.* Proc. ICRA IV, Ottawa 1984, Lecture Notes in Math., 1177, Springer-Verlag, Berlin–Heidelberg–New York (1986) 13–49.

[Bren2]   S. Brenner. *The almost split sequence starting with a simple module.* Arch. Math., 62 (1994) 203–206.

[BrenB1] S. Brenner and M. C. R. Butler. *The equivalence of certain functors occurring in the representation theory of artin algebras and species.* J. London Math. Soc., 14 (1976) 183–187.

[BrenB2] S. Brenner and M. C. R. Butler. *Generalization of the Bernstein–Gelfand–Ponomarev reflection functors.* Lecture Notes in Math., 832, Springer-Verlag, Berlin–Heidelberg–New York (1980) 103–169.

[BretT]   O. Bretscher and G. Todorov. *On a theorem of Nazarova and Roiter.* Proc. ICRA IV Ottawa 1984, Lecture Notes in Math., 1177, Springer-Verlag, Berlin–Heidelberg–New York (1986) 50–54.

[BurF]   W. D. Burgess and K. R. Fuller. *On quasihereditary rings.* Proc. Amer. Math. Soc., 106 (1989) 321–328.

[BurFGZ] W. D. Burgess, K. R. Fuller, E. Green and D. Zacharia. *Left monomial rings – a generalization of monomial algebras.* Osaka J. of Math., Vol 30, No 3, (1993) 543–558.

[But]   M. C. R. Butler. *Grothendieck groups and almost split sequences.* Proc. Oberwolfach, Lecture Notes in Math., 882, Springer-Verlag, Berlin–Heidelberg–New York (1980) 357–368.

[ButR]   M. C. Butler and C. M. Ringel. *Auslander–Reiten sequences with few middle terms and applications to string algebras.* Comm. in Algebra, 15, No. 1-2 (1987) 145–179.

[CR]   C. W. Curtis and I. Reiner. *Methods of Representation Theory, I & II.* John Wiley & Sons, New York, (1990).

[DlR1]   V. Dlab and C. M. Ringel. *On algebras of finite representation type.* J. Algebra, 33 (1975) 306–394.

[DlR2]   V. Dlab and C. M. Ringel. *Indecomposable representations of graphs and algebras.* Memoirs Amer. Math. Soc., 173 (1976).

[DoF]   P. Donovan and M. R. Freislich. *The representation theory of finite graphs and associated algebras.* Carleton Lecture Notes, 5, Ottawa, 1973.

[DoS]   P. Dowbor and A. Skowroński. *Galois coverings of representations-infinite algebras.* Comm. Math. Helv., 62 (1987) 311–337.

[EcH]   B. Eckmann and P. Hilton. *Homotopy groups of maps and exact sequences.* Comm. Math. Helv., 34 (1960) 271–304.

[Er]   K. Erdmann. *Blocks of Tame Representation Type and Related Algebras,* Lecture Notes in Math., 1428, Springer-Verlag, Berlin–Heidelberg–New York (1990).

[Fi]   U. Fischbacher. *Une nouvelle preuve d'un théorème de Nazarova et Roiter.* C. R. Acad. Sci. Paris Sér I, 9 (1985) 259–262.

[FoGR]   R. M. Fossum, P. A. Griffith, and I. Reiten. *Trivial Extensions of Abelian Categories,* Lecture Notes in Math., 456, Springer-Verlag, Berlin–Heidelberg–New York (1975).

[Fu]   K. R. Fuller. *Generalized uniserial rings and their Kupisch series.* Math. Z., 106 (1968) 248–260.

[FuZ]   K. R. Fuller and B. Zimmermann-Huisgen. *On the generalized Nakayama conjecture and the Cartan determinant problem.* Trans. Amer. Math. Soc., 294 (1986) 679–691.

[Ga1]   P. Gabriel. *Unzerlegbare Darstellungen I.* Manuscripta Math., 6 (1972) 71–103.

[Ga2]   P. Gabriel. *Auslander–Reiten sequences and representation-finite algebras.* Lecture Notes in Math., 831, Springer-Verlag, Berlin–Heidelberg–New York (1980) 1–71.

[GaRi]   P. Gabriel and C. Riedtmann. *Group representations without groups.* Comm. Math. Helv., 54 (1979) 240–287.

[GaRo]   P. Gabriel and A. Roiter. *Representation of Finite-Dimensional Algebras*, Algebra VIII, Encyclopedia of Math. Sci., 73, Springer-Verlag (1992).

[GrKK]   E. Green, E. Kirkman and J. Kuzmanowich. *Finitistic dimension of finite dimensional algebras*. J. Algebra, 136 (1991) 37–50.

[GrZ]   E. L. Green and B. Zimmermann-Huisgen. *Finitistic dimension of Artinian rings with vanishing radical cube*. Math. Z., 206 (1991) 505–526.

[Hap1]   D. Happel. *Composition factors for indecomposable modules*. Proc. Amer. Math. Soc., 86 (1982) 29–31.

[Hap2]   D. Happel. *Triangulated categories in the representations theory of finite dimensional algebras*, London Math. Soc. Lecture Note Series, 119 (1988).

[HapL]   D. Happel and S. Liu. *Module categories without short cycles are of finite type*. Proc. Amer. Math. Soc. 120 No. 2 (1994) 371–375.

[HapPR]   D. Happel, U. Preisel and C. M. Ringel. *Vinberg's characterization of Dynkin diagrams using subadditive functions with applications to D Tr-periodic modules*. Proc. ICRA II, Ottawa 1979, Lecture Notes in Math., 832, Springer-Verlag, Berlin–Heidelberg–New York (1980) 280–294.

[HapR]   D. Happel and C. M. Ringel. *Tilted algebras*, Trans. Amer. Math. Soc., 274 (1982) 399–443.

[HapRS]   D. Happel, I. Reiten, and S. O. Smalø. *Short cycles and sincere modules*. Proc. ICRA VI, Ottawa 1992, Canadian Math. Soc. Conf. Proc. 14 (1993) 233–237.

[HapV]   D. Happel and D. Vossieck. *Minimal algebras of infinite representation type with preprojective component*. Manuscr. Math., 42 (1983) 221–243.

[HarS]   M. Harada and Y. Sai. *On categories of indecomposable modules I*. Osaka J. Math., 8 (1971) 309–321.

[He]   A. Heller. *Indecomposable modules and the loop space operation*. Proc. Amer. Math. Soc., 12 (1961) 640–643.

[Hi]   D. Higman. *Indecomposable representations of characteristic p*. Duke J. Math., 21 (1954) 377–381.

[HiS]   P. J. Hilton and U. Stambach. *A Course in Homological Algebra*, Graduate Texts in Math., 4, Springer-Verlag, Berlin–Heidelberg–New York (1971).

[I]   K. Igusa. *Notes on the no loop conjecture*. J. Pure and Applied Algebra, 69 (1990) 161–176.

[IZ]   K. Igusa and D. Zacharia. *Syzygy pairs in a monomial algebra*, Proc. Amer. Math. Soc., 108 (1990) 601–604.

[J]   G. J. Janusz. *Indecomposable representations of groups with a cyclic Sylow subgroup*. Trans. Amer. Math. Soc., 125 (1966) 288–295.

[JL]   Chr. U. Jensen and H. Lenzing. *Model Theoretic Algebra*. Gordon and Breach, New York–London (1989).

418   Bibliography

[Ke]   O. Kerner. *Stable components of wild tilted algebras*. J. Algebra, 142
            (1991) 37–57.
[Kra1]   H. Krause. *The kernel of an irreducible map*. Proc. Amer. Math.
            Soc., 121 (1994) 57–66.
[Kra2]   H. Krause. *On the Four Terms in the Middle Theorem for almost
            split sequences*. Arch. Math., 62 (1994) 501–505.
[Kro]   L. Kronecker. *Algebraische Reduction der scharen bilinearen Formen*.
            Sitzungsber. Akad. Berlin (1890) 1225–1237.
[Kru]   S. A. Krugliak. *Representations of algebras with zero square radical*,
            Zap. Naučn. LOM I, 28 (1972) 32–41.
[Ku1]   H. Kupisch. *Beiträge zur Theorie nichthalbeinfacher Ringe mit Min-
            imalbedingung*. Crelles Journal, 201 (1959) 100–112.
[Ku2]   H. Kupisch. *Unzerlegbare Moduln endlicher Gruppen mit zyklischer
            p-Sylowgruppe*. Math. Z., 108 (1969) 77–104.
[Le]   H. Lenzing. *Nilpotente Elemente in Ringen von endlicher globaler
            Dimension*. Math. Z., 108 (1969) 313–324.
[Lin]   M. Linckelmann. *Modules in the sources of Green's exact sequences
            for cyclic blocks*. Invent. Math., 97 (1989) 129–140.
[Liu1]   S. Liu. *Degrees of irreducible maps and the shapes of Auslander–
            Reiten quivers*. J. London Math. Soc., 45 (1992) 32–54.
[Liu2]   S. Liu. *Tilted algebras and generalized standard Auslander–Reiten
            components*. Arch. Math. Vol 61 (1993) 12–19.
[Liu3]   S. Liu. *Almost split sequences for non-regular modules*. Fund. Math.
            143 (1993) 183–190.
[Lu]   J. Luo. *Uniserial modules over commutative artin local rings*. Ph. D.
            thesis, Brandeis Univ. (1993).
[M0]   R. Martinez-Villa. *Algebras stably equivalent to l-hereditary*, Repre-
            sentation Theory II, Proc. Ottawa, Carleton Univ. 1979, LNM 832,
            396–431.
[M1]   R. Martinez-Villa. *Almost projective modules over hereditary alge-
            bras*. An. Inst. Mat. Univ. Nac. Auton., Mex, 20 (1980), 1 (1982)
            1–89.
[M2]   R. Martinez-Villa. *Almost projective modules and almost split se-
            quences with indecomposable middle term*. Comm. in Algebra, 8
            (1980) 1123–1150.
[M3]   R. Martinez-Villa. *Properties that are left invariant under stable
            equivalence*. Comm. in Algebra, 18 (1990) 4141–4169.
[Nak]   T. Nakayama. *Note on uniserial and generalized uniserial rings*. Proc.
            Imp. Akad. Japan, 16 (1940) 285–289.
[Naz]   L. A. Nazarova. *Representations of quivers of infinite type*. Izv. Akad.
            Nauk SSSR, Ser. Mat., 37 (1973) 752–791.
[No]   D. G. Northcott. *A First Course of Homological Algebra*. Cambridge
            Univ. Press, London (1973).
[PeT]   J.A. de la Peña and M. Takane. *Spectral properties of Coxeter trans-
            formations and applications*. Arch. Math., 55 (1990) 120–134.
[Pr]   M. Prest. *Model Theory and Modules*. London Math. Soc. Lecture
            Note Series, 130, Cambridge Univ. Press, (1988).

[ReR]   I. Reiten and C. Riedtmann. *Skew group algebras in the representation theory of artin algebras.* J. Algebra, 92 (1985) 224–282.

[ReSS1] I. Reiten, A. Skowronski, and S. O. Smalø. *Short chains and short cycles of modules.* Proc. Amer. Math. Soc., 117 (1993) 343–354.

[ReSS2] I. Reiten, A. Skowronski, and S. O. Smalø. *Short chains and regular components.* Proc. Amer. Math. Soc., 117 (1993) 601–612.

[ReV]   I. Reiten and M. Van den Bergh. *Two-dimensional tame and maximal orders of finite representation type,* Memoirs Amer. Math. Soc., 408 (1989) 1–69.

[Rie]   C. Riedtmann. *Algebren, Darstellungsköcher, Überlagerungen und zurück.* Comm. Math. Helv., 55 (1980) 199–224.

[Rin1]   C. M. Ringel. *Representations of K-species and bimodules.* J. Algebra, 41 (1976) 269–302.

[Rin2]   C. M. Ringel. *Finite dimensional hereditary algebras of wild representation type.* Math. Z., 161 (1978) 235–255.

[Rin3]   C. M. Ringel. *Tame algebras and integral quadratic forms.* Lecture Notes in Math., 1099, Springer-Verlag, Berlin–Heidelberg–New York (1984).

[Rin4]   C. M. Ringel, *The canonical algebras.* Topics in algebra, Part 1, Banach Center Publ. 26 Part 1 (Warsaw 1990), 407–432.

[Roi]   A. V. Roiter. *The unboundedness of the dimension of the indecomposable representations of algebras that have an infinite number of indecomposable representations.* Izv. Akad. Nauk SSSR, Ser. Math., 32 (1968) 1275–1282, Transl. Math. USSR, Izv. 2 (1968) 1223–1230.

[Rot]   J. J. Rotman. *An Introduction to Homological Algebra.* Academic Press, New York–San Francisco–London (1979).

[Si]   D. Simson. *Linear Representations of Partially Ordered Sets and Vector Space Categories.* Gordon and Breach, London (1992).

[Sk1]   A. Skowronski. *Cycles in module categories.* Proc. CMS Annual Seminar/Nato Advanced Research Workshop (Ottawa 1992).

[Sk2]   A. Skowronski. *Cycle-finite algebras.* J. Pure and Applied Algebra, 103 (1995) 105–116.

[Sm]   S. O. Smalø. *The inductive step of the second Brauer–Thrall conjecture.* Can. J. Math., 32 (1980) 342–349.

[SmV]   S. O. Smalø and S. Venås, *Lengths of Endomorphism Rings of Finitely Generated Indecomposable Modules over Artin Algebras.* Preprint 10/1995, Trondheim.

[So]   Ø. Solberg. *Hypersurface singularities of finite Cohen–Macaulay type.* Proc. London Math. Soc., 58 (1989) 258–280.

[Ta]   H. Tachikawa. *Quasi-Frobenius rings and generalizations.* Lecture Notes in Math., 351, Springer-Verlag, Berlin (1973).

[To]   G. Todorov. *Almost split sequences for Tr D-periodic modules.* Proc. ICRA II, Ottawa 1979, Lecture Notes in Math., 832, Springer-Verlag, Berlin–Heidelberg–New York (1980) 579–599.

[V]   E. B. Vinberg. *Discrete linear groups generated by reflections.* Isv. Akad. Nauk SSSR, 35 (1971), Transl. Math. USSR Izv. 5 (1971) 1083–1119.

[W]   G.V. Wilson. *The Cartan map on categories of graded modules.* J.
               Algebra, 85 (1983) 390–398.
[Ya]   K. Yamagata. *On artinian rings of finite representation type.* J. Al-
               gebra, 50 (1978) 276–283.
[Yoshii]   T. Yoshii. *On algebras of bounded representation type.* Osaka Math.
               J., 8 (1956) 51–105.
[Yoshin]   Y. Yoshino. *Cohen–Macaulay Modules over Cohen–Macaulay Rings,*
               London Math. Soc. Lecture Notes Series., 146, Cambridge Univer-
               sity Press (1990).
[Z]   D. Zacharia. *On the Cartan matrix of an artin algebra of global
               dimension two.* J. Algebra, 82 (1983) 353–357.

Relevant conference proceedings

1. Springer Lecture Notes in Math., 488, Proceedings ICRA, Ottawa (1974).

2. Springer Lecture Notes in Math., 831, Proceedings Workshop ICRA II, Ottawa (1979).

3. Springer Lecture Notes in Math., 832, Proceedings ICRA II, Ottawa (1979).

4. Springer Lecture Notes in Math., 903, Proceedings ICRA III, Puebla, Mexico (1980).

5. Springer Lecture Notes in Math., 944, Proceedings Workshop ICRA III, Puebla, Mexico (1980).

6. Springer Lecture Notes in Math., 1177, Proceedings ICRA IV, Ottawa (1984).

7. Springer Lecture Notes in Math., 1178, Proceedings ICRA IV, Ottawa (1984).

8. London Math. Soc. Lecture Notes Series, 116, Proceedings Durham, England (1985).

9. London Math. Soc. Lecture Notes Series, 168, Proceedings Workshop ICRA V, Tsukuba, Japan (1990).

10. Canadian Math. Soc. Proceedings Series, 11, Proceeding ICRA V, Tsukuba, Japan (1990).

11. Topics in Algebra, Banach Center Publications, Warsaw 26 (1990).

12. Kluwer Academic Publishers, Proceedings 1992 CMS Annual Seminar–Nato Advanced Research Workshop, Ottawa.

13. Canadian Math. Soc. Proceedings Series 14, Proceedings ICRA VI, Ottawa, Canada 1992.

14. Canadian Math. Soc. Proceedings Series 18, Proceedings ICRA VII, Cocoyoc, Mexico 1994.
15. Canadian Math. Soc. Proceedings Series 19, Proceedings Workshop ICRA VII, Mexico 1994, Mexico City.

Index